SCIENTIFIC AND TECHNICAL TERMS IN BIOENGINEERING AND BIOLOGICAL ENGINEERING

Innovations in Agricultural and Biological Engineering

SCIENTIFIC AND TECHNICAL TERMS IN BIOENGINEERING AND BIOLOGICAL ENGINEERING

Megh R. Goyal, PhD, PE

APPLE
ACADEMIC
PRESS

Apple Academic Press Inc.   :   Apple Academic Press Inc.
3333 Mistwell Crescent   :   9 Spinnaker Way
Oakville, ON L6L 0A2 Canada   :   Waretown, NJ 08758 USA

Library and Archives Canada Cataloguing in Publication

Goyal, Megh Raj, author

Scientific and technical terms in bioengineering and biological engineering / Megh R. Goyal, PhD, PE.

(Innovations in agricultural and biological engineering)

Includes bibliographical references and index.
Issued in print and electronic formats.

ISBN 978-1-77188-659-8 (hardcover).--ISBN 978-0-203-71143-9 (PDF)

1. Bioengineering--Dictionaries. 2. Dictionaries. I. Title.

II. Series: Innovations in agricultural and biological engineering

| TA164.G69 2018 | 660.603 | C2017-906054-6 | C2017-906055-4 |

Library of Congress Cataloging-in-Publication Data

Names: Goyal, Megh Raj, author.
Title: Scientific and technical terms in bioengineering and biological engineering / Megh R. Goyal.
Description: Oakville, ON ; Waretown, NJ : Apple Academic Press, 2018. |
Includes bibliographical references and index.
Identifiers: LCCN 2017042649 (print) | LCCN 2017044614 (ebook) | ISBN 9781498709385 (ebook) |
ISBN 9781771886598 (hardcover : alk. paper) | ISBN 9780203711439 (eBook)
Subjects: | MESH: Bioengineering | Terminology
Classification: LCC R857.B54 (ebook) | LCC R857.B54 (print) | NLM QT 15 | DDC 610.28/4--dc23
LC record available at https://lccn.loc.gov/2017042649

CONTENTS

About the Senior Editor-in-Chief .... *vii*

List of Abbreviations .... *ix*

Preface .... *xi*

Foreword 1 by Jesús M. Román Vélez, MD .... *xv*

Foreword 2 by Taranjit Singh, MD, DHMS .... *xvii*

Book Reviews .... *xix*

Other Books from Apple Academic Press .... *xxi*

Editorial .... *xxiii*

1. **Bioengineering and Biological Engineering** **1**

2. **Timeline of Bioengineering and Biological Engineering** **55**

3. **Glossary of Scientific and Technical Terms in Bioengineering and Biological Engineering** **111**

   Index *563*

ABOUT THE SENIOR EDITOR-IN-CHIEF

Megh R. Goyal, PhD, PE, is, at present, a Retired Professor in Agricultural and Biomedical Engineering from the General Engineering Department in the College of Engineering at the University of Puerto Rico–Mayaguez Campus; and Senior Acquisitions Editor and Senior Technical Editor-in-Chief in Agricultural and Biomedical Engineering for Apple Academic Press, Inc.

He received his BSc degree in Engineering in 1971 from Punjab Agricultural University, Ludhiana, India; his MSc degree in 1977 and PhD degree in 1979 from the Ohio State University, Columbus; his Master of Divinity degree in 2001 from Puerto Rico Evangelical Seminary, Hato Rey, Puerto Rico, USA.

Since 1971, he has worked as Soil Conservation Inspector (1971); Research Assistant at Haryana Agricultural University (1972–1975) and the Ohio State University (1975–1979); Research Agricultural Engineer/ Professor at Department of Agricultural Engineering of UPRM (1979–1997); and Professor in Agricultural and Biomedical Engineering at General Engineering Department of UPRM (1997–2012). He spent one-year sabbatical leave in 2002–2003 at Biomedical Engineering Department, Florida International University, Miami, USA.

He was first agricultural engineer to receive the professional license in Agricultural Engineering in 1986 from College of Engineers and Surveyors of Puerto Rico. On September 16, 2005, he was proclaimed the "Father of Irrigation Engineering in Puerto Rico for the Twentieth Century" by the American Society of Agricultural and Biological Engineers, Puerto Rico Section, for his pioneering work on micro irrigation, evapotranspiration, agroclimatology, and soil and water engineering. During his professional career of 45 years, he has received awards such as: Scientist of the Year, Blue Ribbon Extension Award, Research Paper Award, Nolan Mitchell Young Extension Worker Award, Agricultural Engineer of the Year, Citations by Mayors of Juana Diaz and Ponce, Membership Grand Prize for ASAE Campaign, Felix Castro Rodriguez Academic Excellence, RashtryaRatan Award and Bharat Excellence Award and Gold Medal, Domingo Marrero Navarro Prize, Adopted son of Moca, Irrigation Protagonist of UPRM, Man of Drip Irrigation by Mayor of Municipalities of Mayaguez/Caguas/Ponce and Senate/Secretary of Agriculture of ELA, Puerto Rico.

The Water Technology Centre of Tamil Nadu Agricultural University in Coimbatore, India recognized Dr. Goyal as one of the experts "who rendered meritorious service for the development of the micro irrigation sector in India" by bestowing *"Award of Outstanding Contribution in Micro Irrigation."* This award was presented to Dr. Goyal during the inaugural session of the National Congress on "New Challenges and Advances in Sustainable Micro Irrigation on March 1, 2017, held at Tamil Nadu Agricultural University.

He has authored more than 200 journal articles and textbooks and edited over 50 books including: "Elements of Agroclimatology (Spanish) by UNISARC, Colombia"; two "Bibliographies on Drip Irrigation."

Apple Academic Press Inc. (AAP) has published his books, namely: "Management of Drip/Trickle or Micro Irrigation," and "Evapotranspiration: Principles and Applications for Water Management," 10-volume set on *"Research Advances in Sustainable Micro Irrigation."* Readers may contact him at: goyalmegh@gmail.com.

LIST OF ABBREVIATIONS

ACC	American College of Cardiology
AIMBE	American Institute for Medical and Biological Engineering
ANNA	American Nephrology Nurses Association
ASABE	American Society of Agricultural and Biological Engineers
ASAIO	American Society for Artificial Internal Organs
ASAO	Austrian Society for Artificial Organs
ASB	American Society of Biomechanics
BMES	Biomedical Engineering Society
BSN	British Society for Nanomedicine
CHF	congestive heart failure
CPB	cardiopulmonary bypass
EAMBES	The European Alliance for Medical and Biological Engineering and Science
EAPB	European Association of Pharma Biotechnology
EBSA	The European Biophysical Societies' Association
ECG	Electrocardiogram
EDTA	European Dialysis and Transplantation Association
EFB	The European Federation of Biotechnology
EMBS	Engineering in Medicine and Biology
ESAO	the European Society for Artificial Organs
ESB	European Society of Biomechanics
ESBRA	The European Society for Biomedical Research on Alcoholism
ESC	European Society of Cardiology
ESEM	European Society for Engineering and Medicine
FDA	Food and Drug Administration, USA
HSTAT	Health Services Technology Assessment Texts
IFAO	International Federation for Artificial Organs (formerly ISAO)
IMES	International Metabolic Engineering Society
ISABB	International Society for Artificial Cells, Blood Substitutes and Biotechnology
ISAO	International Society for Artificial Organs

ISFA	International Society for Apheresis
ISPD	International Society for Peritoneal Dialysis
ISRP	International Society for Rotary Blood Pumps
JSAO	The Japanese Society for Artificial Organs
NASA	National Aeronautics and Space Administration
NCBI	National Center for Biotechnology Information, USA
NIH	National Institute of Health, USA
NIHCBC	National Institute of Health Center for Biomedical Computation at Stanford University
NKF	National Kidney Foundation
NNI	National Nanotechnology Initiative
ORS	Orthopedic Research Society
RMTES	Regenerative Medicine and Tissue Engineering Society
SMB	Society for Mathematical Biology
SNN	Society of Nanoscience and Nanotechnology
SPBT	The Society of Pharmaceutical and Biotech Trainers (SPBT)
TERMIS	Tissue Engineering and Regenerative Medicine International

PREFACE

I have been teaching fluid mechanics to undergraduate and graduate students since 1971. I decided to apply principles of fluid mechanics to human body systems when my respiratory system collapsed in 1989 and I had three strokes in 2002 and my vagus nerve failed in 1999. In addidtion, my mother broke her hips, and the orthopedic surgeon did total hip replacement (THR). My mother survived 12 more years. My elder brother had a heart mitral valve replaced in 2003, and he feels perfectly fine. My mother-in-law passed away in 2010 with three heart valves damaged (she did not want to have a heart operation and she lived for next four years after diagnosis of heart disease). Our neighbor in Puerto Rico had total knee replacement (TKR) for both knees, and now she can do her daily activities. I thought why would this would happen to a human body, and I was able to understand that Almighty Supreme God grants serenity and knowledge to physicians so that we can live longer and happier. I also know a friend of mine with amputation of both legs, and he does his daily activities using a wheelchair. All this fascinates my soul. Let us all keep our spirits high so that our life is always dynamic to be happier today/tomorrow and forever. This is how the idea for this book volume was born.

Biotechnology, nanotechnology and bionanotechnology are sometimes used interchangeably with BME in general. Biotechnology is the use of biological processes, organisms, or systems to manufacture products intended to improve the quality of human life. The United Nations Convention on Biological Diversity defines biotechnology as "*any technological application that uses biological systems, living organisms, or derivatives thereof, to make or modify products or processes for specific use.*" Nanotechnology is the study and use of structures between 1 nanometer and 100 nanometers in size. For example a width of a human hair is equivalent to eight hundred 100-nanometer particles side-by-side. Nanotechnology is the engineering of functional systems at the molecular scale. Bionanotechnology or nanobiotechnology or nanobiology is the harnessing of biological processes on an ultra-small scale in the manufacture and alteration of materials and products.

The present era is already preparing the foundation for femtotechnology and biofemtotechnology for the 22nd century.

Bioengineering deals with living organisms: agricultural plants, animals, and humans. In this book, I present *scientific and technical terms in bioengineering and biological engineering* for all living organisms (agricultural plants, animals and humans). I have tried my best to include most of the terms. However, dictionary or glossary writing is a dynamic and continuous process that never finishes.

This book complements other similar books on the market that include terms in agricultural sciences. These books almost do not mention the technical terms in bioengineering and biomedical engineering. In 2001, the Food and Agriculture Organization of the United Nations published *FAO Glossary of Biotechnology for Food and Agriculture* (FAO Research and Technology Paper 9) by Zaid, H.G. Hughes, E. Porceddu, and F. Nicholas (http://www.fao.org/docrep/004/y2775e/ y2775e00.htm), which included 450 additional terms compared to the previous edition of 1999. This was the first book on glossary of biotechnology. In 2002, *Online Dictionary of Biotechnology Terms*, by Kimball R. Nill was published on the website www.biotechterms.org. Dr. Nill later published it as *Glossary of Biotechnology Terms* (2002 by CRC Press). In 2009, CRC Press published *Illustrated Dictionary of Biotechnology* (pages 450) by Himanshu Arora, which includes illustrations along with definitions thus enhancing understanding.

My book is titled *Scientific and Technical Terms in Bioengineering and Biological Engineering* which includes terms for plants, animals, and humans. Therefore, my book is unique, complete, and simple. The mission of my book is to serve as a reference manual for graduate and undergraduate students of biomedical engineering, biological engineering, biotechnology, nanotechnology, nursing, and medicine and health sciences. I hope that it will also be a valuable reference for professionals who work in agriculture, medicine and health sciences, for nursing institutes, and other agencies that work with human health. This volume is a must for physicians, scientists, educators, and students.

Chapter 1 introduces the topics in this book. It includes developments of bioengineering, biotechnology, and nanotechnology; and topics related to these areas. Chapter 2 presents the timeline that reveals historical discoveries/inventions and modifications, salient events, and predictions for these areas of science during 7000 B.C. to 2050 A.D. Chapter 3 includes the actual scientific and technical terms in bioengineering and biological engineering. At the end of the book, We include the book subject index.

This book would not have been written without the valuable cooperation of a group of engineers and scientists at the University of Puerto Rico – Mayagüez Campus (UPRM) and Florida International University (FIU) and physicians. I am grateful to my colleagues: Anthony McGoron, Ranu Jung, Paul Sundaram, Ricky Valentin, Alejandro Acevedo, Raj Bansal, Jesus M. Roman Velez, Antonio Padua, and Taranjit Singh. The reader would not have this book without the professional support of these colleagues.

I would like to thank editorial staff, Sandy Jones Sickels, Vice President, and Ashish Kumar, Publisher and President at Apple Academic Press, Inc., (http://appleacademicpress.com/contact.html) for making every effort to publish this book when human health is a major issue worldwide. Special thanks are due to the AAP production staff for typesetting the entire manuscript also.

I request that the readers offer me constructive suggestions that may help to improve the next edition.

Finally, a river of thanks flows from my heart and soul to my wife, Subhadra, for the understanding and collaboration of sharing the responsibility, time, and devotion necessary to prepare this book. With my whole heart and best affection, I dedicate this book to my *Godmother Late Sneh Jindal*, Clinical Psychologist. I adopted her as my spiritual guru (leader) an March 17, 1977 at the graduation ceremony for my master's degree at Ohio State University. She always motivated me to live longer and happier to serve the world community.

I also dedicate this book to *"those who want to live happily"* by making personal health a first priority. One should not hesitate to discuss any personal health issues with physicians who are dedicated and blessed by Almighty Supreme God to do the maximum for alleviating our bodily pain. Good health not only makes us happy but also makes happy everyone around us.

—*Megh R. Goyal, PhD, PE*
Senior Editor-in-Chief

IMPORTANT NOTE TO THE READER: If the definition or glossary of a technical term is not found in this book, the reader is recommended to search for such a term using internet search tools, namely *Google search, http://en.wikipedia.org/wiki/*, etc. Do not get frustrated, have patience and you will find it.

FOREWORD 1 BY JESÚS M. ROMÁN VÉLEZ, MD

In 1994, Dr. Megh R. Goyal taught me courses on soil and water management and farm machinery when I was an undergraduate student at the University of Puerto Rico – Mayaguez Campus. He was one of my favorite professors. After receiving my BSc degree in agriculture sciences, I decided to enter medical school to become a specialist in internal medicine, respiratory, and sleep mechanics. After reading and editing this manuscript, I feel honored to write a foreword for this book, thus paying tribute to my professor, who has excelled as an educator in bioengineering and biological engineering.

As a physician and a specialist in internal medicine, sleep, and pulmonary mechanics, I work closely with my colleagues, and I see them with a smiling face and joy every time each one of them comes out of an operating room. We all are here to extend the best treatment in the market to our patients so that they feel happy and do not feel the body pain. In our informal meetings, we generally also discuss medical applications of new technologies, such as bioengineering, biomedical engineering, biological engineering, biotechnology, and nanotechnology. During the last 20 years, medical and health sciences have benefitted significantly from these modern technologies.

Dr. Megh R. Goyal has requested me to write a foreword for this reference book. Glossaries are frequently found at the end of medical books. In medicine, we physicians prefer to use the term glossary instead of dictionary, because of its widespread coverage. Based on my professional expertise and my personal opinion, the glossary book can be a dictionary, but the dictionary cannot be a glossary book. A dictionary or glossary is used to find a definition or meaning of a word or a term. I strongly agree with Dr. Goyal to use "*scientific and technical terms*" instead of a dictionary or glossary.

A dictionary definition typically contains the meaning of a term (a word, phrase, or other set of symbols); an etymology and the language or languages of its origin, or obsolete meanings. In a medical dictionary, a definition should be simple and easy to understand, preferably even by the general public; useful clinically or in related areas where the definition will be used; specific, that is, by reading the definition only, it should ideally not be possible to refer to any other entity than the *definiendum*; measurable; and reflecting current scientific knowledge.

A glossary (Latin – *glōssarium*) is an alphabetical list of specialized terms peculiar to a field of knowledge with definitions or explanations. A glossary is commonly found at the back of a book or in software manual, to help readers and users understand terms, that may be unfamiliar. The glossary is used to provide a ready reference to a range of subjects. Internet texts make use of a form of glossary, using hyperlinks to explain particular subjects within an article or other text format.

Students and educators often use glossaries as study tools because they quickly cover a wide range of concepts with clear, concise definitions. Reading glossary at the back of a chapter of a textbook first can help to get a basic review of the concepts that will be covered, and then continuing to read the specific chapter with the definitions fresh in the mind. A glossary can also be used as a building block, as it highlights concepts that may turn up on a final.

The glossaries are very common in textbooks in medicine that encompass a variety of health care practices. The medical community will benefit from the glossary or a dictionary of technical terms in each of the focus areas of medicine. Apple Academic Press, Inc. can provide the leadership to fill this gap.

This book by Dr. Goyal is an attempt to present an up-to-date list of terms currently in use in bioengineering, biological engineering, biotechnology, nanotechnology, and closely allied fields. It is a convenient reference manual for educators. This book fulfills an expressed need. The author has done an exceptional job of including the terms that are otherwise found only in books, dictionaries, journals, and abstracts.

I hope that this concise book by Dr. Goyal and Apple Academic Press Inc. will be definitely appealing and valuable to the health sciences community. It is unique and user-friendly. I will like to see more glossary books on the focus areas in medicine.

Jesús Manuel Román Vélez, MD

Practicing Specialist in Internal Medicine,
Pulmonary and Sleep Mechanics
Fellow-American Society of Internal Medicine
Centro Neumologico Del Oeste
55-E Calle De Diego, Office 401, Mayaguez PR
00680; Tel: +(787) 832-0000

FOREWORD 2 BY
TARANJIT SINGH, MD, DHMS

Dr. Megh R. Goyal has visited me many times during visits to his home town Dhuri, Punjab, India, to discuss: various focus areas in medicine; borderlines between medicine and engineering; how the physicians have helped humanity to alleviate human pain; how engineers have advanced biotechnology to design and develop suitable artificial organs and prostheses to replace the diseased human body parts. During our informal meeting in January of 2014, he handed over to me the draft copy of his book titled, *Scientific and Technical Terms in Bioengineering and Biological Engineering,* and requested me to review and write a foreword for the book. After glancing through the book and thorough reading of this book, I attest that it is an invaluable contribution in this specialized discipline. To my knowledge, it is only book on this topic throughout the world. Dr. Goyal is a fluent communicator and educator in bioengineering and biological engineering. Homeopathic and medicine fraternities in the Asian continent will benefit from this compendium. The contribution is historic.

I have been practicing medicine since 1981, when I received my DHMS degree from the Council of Homeopathic System of Medicine (CHSM) in Punjab, India. Later I received my MD degree in 2006 from Calcutta. During the 1990s and 2000s, I have heard and read a lot on biotechnology and bioengineering that has contributed significantly to the homoeopathic medicine. During my school and college years, and throughout my professional career, I have used the English and medical dictionaries (print editions). It was a tedious and time-consuming task to find the meaning of a particular word in medicine. Many times, I and my colleagues were not successful in this task. Nowadays, one can find the meaning of a word using online dictionaries. The art of dictionary writing has revolutionized significantly.

The Oxford English Dictionary defines a dictionary as a "*book dealing with the individual words of a language (or certain specified class of them) so as to set forth their orthography, pronunciation, signification and use, their synonyms, derivation and history, or at least some of these facts, for convenience of reference the words are arranged in some stated order, now in most languages, alphabetical, and in larger dictionaries the information given in illustrated by quotations from literature.*" Therefore, a dictionary is collection of words in one or more specific languages, often listed alphabetically. It can be a specialized dictionary or general dictionary. There are other types of dictionaries such as: bilingual (translation) dictionary, dictionary of synonyms, or rhyming dictionary.

The age of the Internet brought online dictionaries to the desktop and, more recently, to the smart phone. An online dictionary is a dictionary that is accessible via the Internet through a web browser. They can be made available in a number of ways: free, a paid-only service, or free with a paid subscription for extended or more professional content. There exists a number of websites, that operate as online dictionaries, usually with a specialized focus. Some of them have exclusively user-driven content. Some of the more notable examples include: www.dictionary.com, double-tongued dictionary, free on-line dictionary of computing, LEO, logos dictionary, pseudodictionary, urban dictionary, wiktionary, WordNet, and WWWJDIC, in addition to other weblinks.

The glossary and dictionary are terms that are used interchangeably. Dr. Goyal neither uses dictionary or glossary in the title of this book. He diligently names his book as *Scientific and Technical Terms in Bioengineering and Biological Engineering*. The author has diligently described the technical terms in bioengineering and biological engineering. On behalf of all physicians in India, I commend the author for an extraordinary job of compiling the technical terms. I wish the author and Apple Academic Press, Inc. success on this project.

Taranjit Singh, MD, DHMS
Private Practicing Family Physician,
Near Telephone Exchange,
Dhuri – 148024,
Punjab, India
E-mail: dr_taranjit@yahoo.com

BOOK REVIEWS

....

"The use of a glossary or a dictionary has been valuable for a professor and researcher when reading or reviewing journal papers. Since research in biomedical engineering is invariably multidisciplinary and complex, it is common to find unfamiliar scientific and technical terms in papers in my own field. Dr. Megh Goyal has been a champion of biomedical engineering/biological engineering education and has been developing materials to promote its delivery since 1993. During his sabbatical visit to Florida International University in 2002–2003, he and I shared many ideas on pedagogy and the trends of the discipline. Relatively few textbooks are dedicated to the discipline, but Megh Goyal has already written three excellent volumes. The most recent, *Scientific and Technical Terms in Bioengineering and Biological Engineering* fills the urgent need of a concise but comprehensive source for critically important terminology in the field of biomedical engineering. I expect it to become an indispensable table reference for all biotechnology and bioengineering researchers, educators, and students."

—Anthony J. McGoron, PhD
Professor, Florida International University,
Department of Biomedical Engineering
E-mail: mcgorona@fiu.edu

"I have been educator in New Jersey Public Schools for more than twenty years. This reference book is a 'must have' for all users in bioengineering and biological engineering. Simple to complex definitions, glossaries, and meanings are included in this book. The definitions of the scientific terms are clear, concise, and understandable. Sometimes I felt I knew the definition of a certain word, but reading the term gave clearer and more definite understanding for me. The author has done a phenomenal job making the definitions (terms) understandable. I have read Dr. Goyal's books on bioengineering as I teach and inspire today's youth to go into the engineering arena. Engineers are the backbone of our society."

—Professor Raj Bansal
Piscataway School, New Jersey, USA
Private Consultant in English Language

"I have read this book that is informative, user-friendly and easy to read. The book introduces technical terms in bioengineering and biological engineering in nontechnical language."

—Miguel A. Muñoz, PhD
Ex-President of University of Puerto Rico, USA

OTHER BOOKS FROM
APPLE ACADEMIC PRESS

• •

Management of Drip/Trickle or Micro Irrigation
Evapotranspiration: Principles and Applications for Water Management

Book Series: Research Advances in Sustainable Micro Irrigation
Senior Editor-in-Chief: Megh R. Goyal, PhD, PE
  Volume 1: Sustainable Micro Irrigation: Principles and Practices
  Volume 2: Sustainable Practices in Surface and Subsurface Micro
          Irrigation
  Volume 3: Sustainable Micro Irrigation Management for Trees and Vines
  Volume 4: Management, Performance, and Applications of Micro
          Irrigation Systems
  Volume 5: Applications of Furrow and Micro Irrigation in Arid and Semi-
          Arid Regions
  Volume 6: Best Management Practices for Drip Irrigated Crops
  Volume 7: Closed Circuit Micro Irrigation Design: Theory and
          Applications
  Volume 8: Wastewater Management for Irrigation: Principles and Practices
  Volume 9: Water and Fertigation Management in Micro Irrigation
Volume 10: Innovation in Micro Irrigation Technology

Book Series: Innovations and Challenges in Micro Irrigation
Senior Editor-in-Chief: Megh R. Goyal, PhD, PE
Volume 1: Principles and Management of Clogging in Micro Irrigation
Volume 2: Sustainable Micro Irrigation Design Systems for Agricultural
          Crops: Methods and Practices
Volume 3: Performance Evaluation of Micro Irrigation Management:
          Principles and Practices
Volume 4: Potential Use of Solar Energy and Emerging Technologies in
          Micro Irrigation

Volume 5: Micro Irrigation Management: Technological Advances and
        Their Applications
Volume 6: Micro Irrigation Engineering for Horticultural Crops: Policy
        Options, Scheduling, and Design
Volume 7: Micro Irrigation Scheduling and Practices
Volume 8: Engineering Interventions in Sustainable Trickle Irrigation

Book Series: Innovations in Agricultural and Biological Engineering
Senior Editor-in-Chief: Megh R. Goyal, PhD, PE
Note: Alphabetical order

- Dairy Engineering: Advanced Technologies and their Applications
- Developing Technologies in Food Science: Status, Applications, and
  Challenges
- Emerging Technologies in Agricultural Engineering
- Engineering Interventions in Agricultural Processing
- Engineering Interventions in Foods and Plants
- Engineering Practices for Agricultural Production and Water
  Conservation: An Interdisciplinary Approach
- Flood Assessment: Modeling and Parameterization
- Food Engineering: Modeling, Emerging Issues, and Applications.
- Food Process Engineering: Emerging Trends in Research and Their
  Applications
- Food Technology: Applied Research and Production Techniques
- Modeling Methods and Practices in Soil and Water Engineering
- Novel Dairy Processing Technologies: Techniques, Management, and
  Energy Conservation
- Processing Technologies for Milk and Milk Products: Methods,
  Applications, and Energy Usage
- Soil and Water Engineering: Principles and Applications of Modeling
- Soil Salinity Management in Agriculture: Technological Advances and
  Applications
- Technological Interventions in Dairy Science: Innovative Approaches in
  Processing, Preservation, and Analysis of Milk Products
- Technological Interventions in Management of Irrigated Agriculture
- Technological Interventions in the Processing of Fruits and Vegetables

EDITORIAL

· ·

Apple Academic Press Inc., (AAP) is publishing various book volumes on the focus areas under book series titled *Innovations in Agricultural and Biological Engineering*. Over a span of 8 to 10 years, Apple Academic Press Inc., is publishing subsequent volumes in the specialty areas defined by American Society of Agricultural and Biological Engineers (www.asabe.org).

The mission of this series is to provide knowledge and techniques for agricultural and biological engineers (ABEs). The series aims to offer high-quality reference and academic content in *agricultural and biological engineering* (ABE) that is accessible to academicians, researchers, scientists, university faculty, and university-level students and professionals around the world. The following material has been edited/modified and reproduced below [From: *Goyal, Megh R., 2006. Agricultural and biomedical engineering: Scope and opportunities. Paper Edu_47 Presentation at the Fourth LACCEI International Latin American and Caribbean Conference for Engineering and Technology (LACCEI' 2006): Breaking Frontiers and Barriers in Engineering: Education and Research by LACCEI University of Puerto Rico – Mayaguez Campus, Mayaguez, Puerto Rico, June 21–23*].

WHAT IS AGRICULTURAL AND BIOLOGICAL ENGINEERING (ABE)?

"Agricultural Engineering (AE) involves application of engineering to production, processing, preservation and handling of food, fiber, and shelter. It also includes transfer of technology for the development and welfare of rural communities," according to www.isae.in. *"ABE is the discipline of engineering that applies engineering principles and the fundamental concepts of biology to agricultural and biological systems and tools, for the safe, efficient and environmentally sensitive production, processing, and management of agricultural, biological, food, and natural resources*

systems," according to www.asabe.org. "*AE is the branch of engineering involved with the design of farm machinery, with soil management, land development, and mechanization and automation of livestock farming, and with the efficient planting, harvesting, storage, and processing of farm commodities,*" definition by: http://dictionary.reference.com/browse/agricultural+engineering.

"*AE incorporates many science disciplines and technology practices to the efficient production and processing of food, feed, fiber and fuels. It involves disciplines like mechanical engineering (agricultural machinery and automated machine systems), soil science (crop nutrient and fertilization, etc.), environmental sciences (drainage and irrigation), plant biology (seeding and plant growth management), animal science (farm animals and housing) etc.,*" by: http://www.ABE.ncsu.edu/academic/agricultural-engineering.php.

"According to https://en.wikipedia.org/wiki/Biological_engineering: "*BE (Biological engineering) is a science-based discipline that applies concepts and methods of biology to solve real-world problems related to the life sciences or the application thereof. In this context, while traditional engineering applies physical and mathematical sciences to analyze, design and manufacture inanimate tools, structures and processes, biological engineering uses biology to study and advance applications of living systems.*"

SPECIALTY AREAS OF ABE

Agricultural and Biological Engineers (ABEs) ensure that the world has the necessities of life including safe and plentiful food, clean air and water, renewable fuel and energy, safe working conditions, and a healthy environment by employing knowledge and expertise of sciences, both pure and applied, and engineering principles. Biological engineering applies engineering practices to problems and opportunities presented by living things and the natural environment in agriculture. BA engineers understand the interrelationships between technology and living systems, have available a wide variety of employment options. The www.asabe.org indicates that "*ABE embraces a variety of following specialty areas.*" As new technology and information emerge, specialty areas are created, and many overlap with one or more other areas.

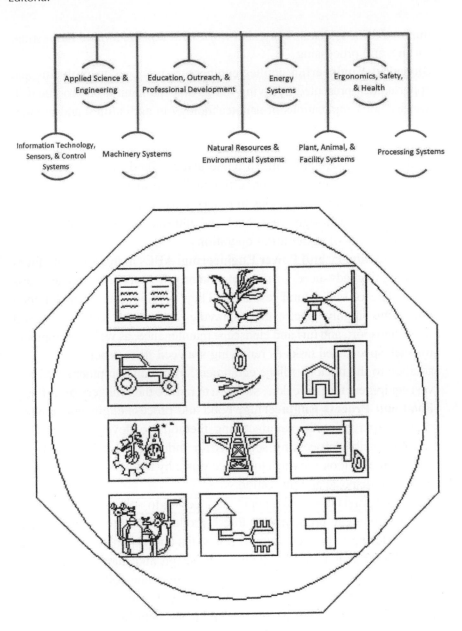

1. **Aquacultural Engineering**: ABEs help design farm systems for raising fish and shellfish, as well as ornamental and bait fish. They specialize in water quality, biotechnology, machinery, natural resources, feeding and ventilation systems, and sanitation. They seek ways to reduce pollution from aquacultural discharges, to reduce excess water use, and to

improve farm systems. They also work with aquatic animal harvesting, sorting, and processing.

2. **Biological Engineering** applies engineering practices to problems and opportunities presented by living things and the natural environment. It also includes applications of nanotechnology in agricultural and biological systems; beneficial effects of medicinal plants/fruits/vegetables/herbs.

3. **Energy:** ABEs identify and develop viable energy sources – biomass, methane, and vegetable oil, to name a few – and to make these and other systems cleaner and more efficient. These specialists also develop energy conservation strategies to reduce costs and protect the environment, and they design traditional and alternative energy systems to meet the needs of agricultural operations.

4. **Farm Machinery and Power Engineering**: ABEs in this specialty focus on designing advanced equipment, making it more efficient and less demanding of our natural resources. They develop equipment for food processing, highly precise crop spraying, agricultural commodity and waste transport, and turf and landscape maintenance, as well as equipment for such specialized tasks as removing seaweed from beaches. This is in addition to the tractors, tillage equipment, irrigation equipment, and harvest equipment that have done so much to reduce the drudgery of farming.

5. **Food and Process Engineering:** Food and process engineers combine design expertise with manufacturing methods to develop economical and responsible processing solutions for industry. Also food and process engineers look for ways to reduce waste by devising alternatives for treatment, disposal and utilization.

6. **Forest Engineering**: ABEs apply engineering to solve natural resource and environment problems in forest production systems and related manufacturing industries. Engineering skills and expertise are needed to address problems related to equipment design and manufacturing, forest access systems design and construction; machine-soil interaction and erosion control; forest operations analysis and improvement; decision modeling; and wood product design and manufacturing.

7. **Information and Electrical Technologies engineering** is one of the most versatile areas of the ABE specialty areas, because it is applied to virtually all the others, from machinery design to soil testing to food quality and safety control. Geographic information systems, global positioning systems, machine instrumentation and controls, electromagnetics, bioinformatics, biorobotics, machine vision, sensors, spectroscopy: These are

some of the exciting information and electrical technologies being used today and being developed for the future.

8. **Natural Resources:** ABEs with environmental expertise work to better understand the complex mechanics of these resources, so that they can be used efficiently and without degradation. ABEs determine crop water requirements and design irrigation systems. They are experts in agricultural hydrology principles, such as controlling drainage, and they implement ways to control soil erosion and study the environmental effects of sediment on stream quality. Natural resources engineers design, build, operate and maintain water control structures for reservoirs, floodways and channels. They also work on water treatment systems, wetlands protection, vertical farming, and other water issues.

9. **Nursery and Greenhouse Engineering**: In many ways, nursery and greenhouse operations are microcosms of large-scale production agriculture, with many similar needs – irrigation, mechanization, disease and pest control, and nutrient application. However, other engineering needs also present themselves in nursery and greenhouse operations: equipment for transplantation; control systems for temperature, humidity, and ventilation; and plant biology issues, such as hydroponics, tissue culture, and seedling propagation methods. And sometimes the challenges are extraterrestrial: ABEs at NASA are designing greenhouse systems to support a manned expedition to Mars!

10. **Safety and Health:** ABEs analyze health and injury data, the use and possible misuse of machines, and equipment compliance with standards and regulation. They constantly look for ways in which the safety of equipment, materials and agricultural practices can be improved and for ways in which safety and health issues can be communicated to the public.

11. **Structures and Environment:** ABEs with expertise in structures and environment design animal housing, storage structures, and greenhouses, with ventilation systems, temperature and humidity controls, and structural strength appropriate for their climate and purpose. They also devise better practices and systems for storing, recovering, reusing, and transporting waste products.

CAREERS IN AGRICULTURAL AND BIOLOGICAL ENGINEERING

One will find that university ABE programs have many names, such as biological systems engineering, bioresource engineering, environmental engineering, forest engineering, or food and process engineering. Whatever the title, the typical curriculum begins with courses in writing, social sciences, and economics, along with mathematics (calculus and statistics), chemistry, physics, and biology. Student gains a fundamental knowledge of the life sciences and how biological systems interact with their environment. One also takes engineering courses, such as thermodynamics, mechanics, instrumentation and controls, electronics and electrical circuits, and engineering design. Then student adds courses related to particular interests, perhaps including mechanization, soil and water resource management, food and process engineering, industrial microbiology, biological engineering or pest management. As seniors, engineering students work in a team to design, build, and test new processes or products.

For more information on this series, readers may contact:

Ashish Kumar,
Publisher and President
Sandy Sickels, Vice President
Apple Academic Press, Inc.,
Fax: 866–222–9549, E-mail:
ashish@appleacademicpress.com
http://www.appleacademicpress.com/
publishwithus.php

Megh R. Goyal, PhD, PE
Book Series Senior Editor-in-Chief
*Innovations in Agricultural and
Biological Engineering*
goyalmegh@gmail.com

CHAPTER 1

BIOENGINEERING AND BIOLOGICAL ENGINEERING

CONTENTS

1.1   Introduction..1

1.2   Biotechnology...2

1.3   Nanotechnology 2

1.4   Bionanotechnology, Nanobiotechnology, and Nanobiology..............16

1.5   Tissue Engineering...18

1.6   Biomedical Engineering/Bioengineering/Biological Engineering 19

Keywords 34

References..34

Appendix I 36

1.1   INTRODUCTION

This reference book, *Scientific and Technical Terms in Bioengineering and Biological Engineering*, is divided into three chapters: Chapter 1 defines biotechnology, nanotechnology, bionanotechnology, bioengineering/biomedical engineering, and molecular engineering; and topics related to these areas. Chapter 2 deals about the timeline of bioengineering, biotechnology, and nanotechnology and also includes the historical background and exceptional contributions by scientists and educationists. Finally, Chapter 3 deals with the glossary section of scientific and technical terms used in bioengineering and biological engineering.

This introduction chapter briefly describes the topics that are related to the bioengineering/biomedical engineering and biotechnology. It also includes the list of societies, journals, and software programs. At the end of this chapter, an appendix section is included that discusses biomedical engineering education at Florida International University.

1.2 BIOTECHNOLOGY

Biotechnology (BT) refers to the application of biological organisms, systems, or processes to learn about the life science and to improve value of pharmaceuticals, crops, animals, and humans. It is a relatively new and fast-developing field that integrates knowledge from the traditional sciences like biochemistry, chemistry, microbiology, engineering, and medicine.

Currently, there are more than 250 BT health care products and vaccines available to patients; most of them are for the previously untreatable diseases. More than 13.3 million farmers around the world use agricultural BT to increase yields, prevent damage from insects and pests, and reduce farming's impact on the environment. And more than 50 biorefineries are being built across North America to test and refine technologies for producing biofuels and chemicals from renewable biomass and for helping to reduce greenhouse gas emissions. Recent advances in BT are helping us to meet society's most pressing challenges.

1.3 NANOTECHNOLOGY

Nanotechnology (NT) is the science and technology that is precisely used for structuring and controlling the nanometer scale. According to bioengineering NT initiative (NIH), NT is the development and useful techniques for studying physical phenomena and construct structures in the nanoscale size range or smaller.

The NT is an emerging technology as a new field enabling the creation and application of materials, devices, and systems at atomic and molecular levels; and the exploitation of novel properties that emerge at the scale of 1 to 100 nanometers. For example, the potential uses of NT in biomedicine include the early detection and treatment of disease; and the development of smart and rejection–resistant implants that will respond appropriately to the physical health's requirements. The term NT was first coined by K. Eric Drexler in 1986 in his book *Engines of Creation*—although research in this field dates back to

Richard P. Feynman's classic talk in 1959. Many scientists are beginning to use the term molecular NT while discussing about NT at the molecular level.

Nanomedicine refers to the use of nanobots to address the medical problems, and to the use of molecular knowledge to maintain and improve health at a molecular scale. As a specialized field within NT, the nanomedicine would work towards body repair through the use of engineered, *in vivo* probes, and sensors that would operate, in a semi-permanent fashion, within the body.

The development of nanomedicine will have extraordinary implications for the veterinary profession, because it will change the definition of the disease and the way we do diagnosis and treatment for the medical conditions of animals. One of the areas of veterinary medicine that can benefit most from the NT research is the field of pharmacology. The most of the animal diseases in the near future will be addressed by the use of nanopharmaceuticals. The research in the area of nanopharmaceuticals can provide new advances in the area of *drug delivery systems*.

Biomedical engineering (BME) and NT are sometimes used interchangeably. However, NT more typically denotes the specific products which use "biological systems, living organisms, or their derivatives." Even some complex "medical devices" can be deemed from "biotechnology" depending on the degree to which such elements are central to the principle of operation. Biologics/biopharmaceuticals (e.g., vaccines, stored blood product), genetic engineering, and various agricultural applications are some of the major classes of biotechnology.

The pharmaceuticals are related to bioNT in two indirect ways: (i) certain major types (e.g., biologics) fall under both categories; and (ii) together they essentially comprise the "non-medical device" set of BME applications. *Nanotoxicology* is a branch of bionanoscience, which deals with the study and application of toxicity of nanomaterials. Nanomaterials become highly active at nanometer dimensions—even when made of inert elements like gold. Nanotoxicological studies can determine whether and to what extent these properties may pose a threat to the environment and to human beings. For instance, Diesel nanoparticles have been found to damage the cardiovascular system in a mouse model.

1.3.1 BRANCHES OF NANOTECHNOLOGY (NT)

1. **Bioengineering** (including biological systems engineering) is the application of concepts and methods of biology (and secondarily of physics,

chemistry, mathematics, and computer science) to solve real-world problems related to the life sciences or the application thereof, using engineering's own analytical and synthetic methodologies; and also its traditional sensitivity to the cost and practicality of the solution(s). In this context—while traditional engineering applies physical and mathematical sciences to analyze—design and manufacture inanimate tools, structures and processes, and biological engineering uses a primarily molecular biology to study and advance applications of living organisms.

2. **Ceramic engineering** is a science and technology of creating objects from inorganic and non-metallic materials.

3. **Contributing fields of NT** include: calculus, chemistry, computer science, engineering, miniaturization, physics, quantum mechanics, self-organization, science, and supramolecular chemistry.

4. **Green nanotechnology** enhances the environmental-sustainability of processes that are currently producing negative externalities.

5. **Materials science** is an interdisciplinary field applying the properties of matter to various areas of science and engineering.

6. **Nanoarchitectonics** deals with arranging nanoscale structural units in an intended configuration.

7. **Nanobiotechnology** is an intersection of nanotechnology and biology.

8. **Nanoengineering** is a practice of engineering on a nanoscale.

9. **Nanoelectronics** is the use of nanotechnology on electronic components, including transistors.

10. **Nanomechanics** is a branch of nanoscience studying fundamental mechanical (elastic, thermal, and kinetic) properties of physical systems at the nanometer scale.

11. **Nanophotonics** is the study of the behavior of light on the nanometer scale.

12. **Protein engineering** is the process of developing useful or valuable proteins.

13. **Tissue engineering** makes use of artificially stimulated cell proliferation by using suitable nanomaterial-based scaffolds and growth factors.

14. **Wet nanotechnology** involves working up to large masses from small ones.

1.3.2   TOPICS IN NANOTECHNOLOGY

The scope of NT is very wide that includes topics such as: advantages of NT, applications for carbon nanotube, atomic force microscope, carbon nanotube, disadvantages of NT, energy applications of NT, fullerenes, green NT, health implications of NT, mechanosynthesis, microelectromechanical systems, molecular assembler, molecular electronics, molecular NT (mechanosynthesis, molecular assembler, nanorobotics), molecular self-assembly (DNA NT, self-assembled monolayer, supramolecular assembly, nanoelectronics, nanoelectronics), molecular scale electronics/nanoelectronics (nanolithography, nanolithography), nanomaterials (carbon nanotubes, fullerenes, nanoparticles, nanomechanics, nanomedicine), nanometrology (atomic force microscopy, electron microscope, scanning tunneling microscope, super resolution microscopy), nanoparticles, nanophotonic, nanorobotics, nanoscience, nanosensors, NT education, NT in fiction, nanotoxicology, regulation of NT, scanning tunneling microscope, self-assembled monolayer, social implications of NT, supramolecular assembly, etc.

1.3.3 APPLICATIONS OF NANOTECHNOLOGY AND BIOTECHNOLOGY

The BT is currently being used in many areas including agriculture, bioremediation, food processing, energy production, and medicine. DNA fingerprinting is becoming a common practice in forensics. Similar techniques have been used recently to identify the bones of the last Czar of Russia and several members of his family. Production of insulin and other medicines has been accomplished through cloning of vectors that now carry the chosen gene. Immunoassays are used not only in medicine for drug level and pregnancy testing, but also by farmers to aid in detection of unsafe levels of pesticides, herbicides, and toxins on crops and in animal products. These assays also provide rapid field tests for industrial chemicals in ground water, sediment, and soil.

In agriculture, genetic engineering is being used to produce plants that are resistant to insects, weeds, and plant diseases. The new biotechnological techniques have permitted scientists to manipulate desired traits. Prior to the advancement of the methods of recombinant DNA, scientists were limited to the techniques of their time—cross-pollination, selective breeding, pesticides, and herbicides. The explosion in techniques has resulted in three major branches of BT:

- genetic engineering,
- diagnostic techniques, and
- tissue engineering.

The selected daily applications of NT/biotechnologies are listed in the Appendix. The various applications of NT are listed below:

1. Devices: micromachinery, nano-abacus, nanomotor, nanopore (nano-pore sequencing), quantum point contact, synthetic molecular motors, and carbon nanotube actuators.
2. Energy applications of NT.
3. Fullerene is a molecule composed entirely of carbon, in the form of a hollow sphere, ellipsoid, or tube. The fullerene spheres and tubes that have applications in NT are:
    *   allotropes of carbon;
    *   aggregated diamond nanorods;
    *   buckypaper;
    *   carbon nanofoam;
    *   carbon nanotube (nanoknot, nanotube membrane);
    *   fullerene chemistry (bingel reaction, endohedral hydrogen fullerene, prato reaction);
    *   fullerenes in popular culture;
    *   endohedral fullerenes;
    *   fullerite;
    *   graphene (graphene nanoribbon);
    *   potential applications of carbon nanotubes; and
    *   timeline of carbon nanotubes.
4. Microscopy: atomic force microscope, scanning tunneling microscope, scanning probe microscope, IBM millipede, and sarfus.
5. Molecular electronics.
6. Molecular NT: Grey goo, mechanosynthesis, molecular assembler, molecular modeling, nanorobotics (smartdust and utility fog), nanochondria, programmable matter, self reconfigurable, and self-replication.
7. Molecular self-assembly: DNA NT (DNA computing, DNA machine, and DNA origami), self-assembled monolayer, and supramolecular assembly.
8. Nanoelectronics: break junction, chemical vapor deposition, microelectromechanical systems (MEMS), nanocircuits, nanocomputer, nanoelectromechanical systems (NEMS), and surface micromachining.
9. Nanolithography: dip pen nanolithography, electron beam lithography, ion-beam sculpting, nanoimprint lithography, and photolithography.
10. Nanomaterials is the study of materials with morphological features on the nanoscale, and especially those that have special properties stemming from the nanoscale dimensions.

11. Nanomedicine: lab-on-a-chip, nanobiotechnology, nanosensor, and nanotoxicology.
12. Nanoparticles and colloids: nanoparticle, ceramics processing, colloid, colloidal crystal, diamondoids, nanocomposite, nanocrystal, nanostructure (nanocages, nanocomposite, nanofabrics, nanofiber, nanofoam, nanoknot, nanomesh, nanopillar, nanopin film, nanoring, nanorod, nanoshell, nanotube, quantum dot, quantum heterostructure, and sculptured thin film).
13. Quantum computing is a computation using quantum mechanical phenomena, such as superposition and entanglement, to perform data operations.

1.3.4 IMPLICATIONS OR RISKS OF NANOTECHNOLOGY

The increasing use of NT has benefited the humanity, but it has also raised some serious concerns (risks) to the humanity and that are discussed below:

a. **Grey goo** consists of devices that could potentially be able to make copies of themselves repeatedly without control from the creator. In other words, an uncontrollable, self-replicating nano-machine or robot. The concept of grey goo is largely similar to the vast amount of science-fiction movies. The term grey goo was devised by Eric Drexler in one of his books. He said that "dangerous replicators could easily be too tough, small, and rapidly spreading to stop." In his paper in a collaboration with Chris Phoenix, there is a reference to "safe exponential manufacturing." They indicate that grey goo cannot be produced through the software programs used for NT.

2. State agencies, criminals, and enterprises could use nanorobots to *eavesdrop on conversations held in private*.

3. **Health implications from nanoparticles (nanopollution or nanotoxicology)**: The extreme reactions and increased mobility reactions by nanoparticles are not suitable in every situation. Nanoparticles are very different in their natural environment as compared to their counterparts, and therefore it raises critical issues. Nanodevices for repairing surgical wounds can have side effects. Some nano liquids (nano aluminum, nano sulphur, or hydrochloric acid) can be harmful if they get in contact with human body.

4. **Environmental implications of NT:** The process of manufacturing nanodevices also generates some waste, which may be dangerous. It can be released to air and can easily penetrate in plants which make them

deoxygenated when gets reacted with carbon dioxide and animals. The cats and little animals can die because of inhaling these toxic gases and by drinking contaminated water or liquid released by the manufacturing plants.

5. **Societal impacts of NT** has potential benefits and pose challenges. Beyond the toxicity risks to human health and the environment, which are associated with first-generation nanomaterials, NT has broader societal implications and poses broader social challenges. Social scientists have suggested that the NT challenges should be factored into "upstream" research and decision making in order to ensure technology development that meets social objectives. Many social scientists and civil organizations suggest that technology assessment and governance should also involve public participation. Some observers suggest that NT will radically reshape our economies, our labor markets, international trade, international relations, social structures, civil liberties, and our relationship with the natural world. Those concerned with the *negative impact of NT* suggest that it will simply exacerbate problems stemming from existing socio-economic inequity and unequal distributions of power, creating greater inequities between rich and poor through an inevitable nano-divide and increased potential for bioweaponry, and the possibility of military applications of NT as well as enhanced surveillance capabilities through nano-sensors. The critics fear that chemical weapons developed from nanoparticles will be more dangerous than present chemical weapons.

6. **Link between commodities and poverty**: Many least developed countries are dependent on a few commodities for employment, government revenue, and export earnings. Many applications of NT are being developed that could impact global demand for specific commodities. For instance, certain nanoscale materials could enhance the strength and durability of rubber, which might eventually lead to a decrease in demand for natural rubber. Demand for titanium may increase as a result of new uses for nanoscale titanium oxides, such as titanium dioxide nanotubes. Various organizations have proposed that NT can only be effective in alleviating poverty and aid development "when adapted to social, cultural and local institutional contexts, and chosen and designed with active participation by citizens right."

7. **Risks to developing countries**: NT has also created risks for the developing countries, such as safe and clean drinking water, reliable energy,

clean air, and education. People should be properly aware of side effects of NT.

8. **Implications of molecular NT**: The molecular NT is one of the finest technology but its implications in certain scenarios are quite harmful to the living beings.

1.3.5 PUBLISHERS OF LITERATURE ON NANOTECHNOLOGY

*   American Chemical Society – Nano Letters;
*   American Scientific Publishers – Journal of Nanoscience and Nano-technology;
*   Apple Academic Press, Inc. – Innovations in Agricultural and Biological Engineering;
*   BCC Research – Nanotechnology Reports;
*   Encyclopedia of Nanoscience and Nanotechnology;
*   Erudite Scientific Publishers – Nano Communications International;
*   Erudite Scientific Publishers – Rapid Communications in Nano-technology;
*   Institute of Physics – Nanotechnology;
*   NanoTrends – A Journal of Nanotechnology and Its Applications; and
*   SPIE – International Society for Optics and Photonics – Journal of Nanophotonics.

1.3.6 NANOTECHNOLOGY RELATED SITES

*   Free Nano Science Magazine for School Children in Indian Regional Languages;
*   Nano Magazine – Nano Magazine;
*   Nanoacademia – NanoAcademia;
*   Nanodictionary – Dictionary of Nanotechnology;
*   NanoTechBank – NanoTechnology DataBank;
*   NanoTechMap – The online exhibition of nanotechnology featuring over 4000 companies;
*   Nanotechnology focusing on nano scale machines enthusiast/fan page;
*   Nanotechnology Now – Nanotechnology Now;
*   Nanowerk – Nanowerk;
*   NanoWiki – Tracking Nanotechnology; and
*   STATNANO – Nano Science, Technology, and Industry Scoreboard.

1.3.7 PROMOTERS OF NANOTECHNOLOGY

1. Government
   - National Center for Nanoscience and Technology (China);
   - National Institute for Nanotechnology (Canada);
   - Iranian Nanotechnology Laboratory Network;
   - EU Seventh Framework Programme and Action Plan for Nanosciences and Nanotechnologies 2005–2009;
   - Russian Nanotechnology Corporation;
   - National Nanotechnology Center (NANOTEC), Thailand;
   - National Nanotechnology Initiative (United States);
   - National Cancer Institute Alliance for Nanotechnology in Cancer;
   - National Institutes of Health Nanomedicine Roadmap Initiative;
   - American National Standards Institute Nanotechnology Panel (ANSI-NSP);
   - NanoNed (Netherland);
   - CCAN – Collaborative Centre for Applied Nanotechnology (Ireland); and
   - Instituto Zuliano de Investigaciones Tecnológicas – INZIT (Venezuela).
2. Advocacy and Information Groups
   - American Chemistry Council: Nanotechnology Panel;
   - American Nano Society;
   - American Scientific Publishers;
   - Bangladesh Nanotechnology Society, Dhaka, Bangladesh;
   - Biological Applications of Nanotechnology(BANTech), University of Idaho;
   - Center for Biological and Environmental Nanotechnology, Rice University;
   - Center for Responsible Nanotechnology;
   - Foresight Nanotech Institute;
   - Friends of the Earth Australia's Nanotechnology Project;
   - IFT International Food Nanoscience Conference, July 17, 2010, Chicago;
   - Institute of Environmental Sciences and Technology (IEST);
   - Institute of Nanotechnology, Stirling, Scotland, UK;
   - Institute of Occupational Medicine, Scotland, UK;
   - Intelligent Testing Strategies for Engineered Nanomaterials (ITS-NANO);
   - International Association of Nanotechnology;

- International Council on Nanotechnology (ICON);
- International Institute for Nanotechnology;
- Materials Research Society;
- Nano Science and Technology Consortium (NSTC);
- Nano Science and Technology Institute (NSTI);
- Nano Tech Training and Consultancy (NTTC);
- Nanotechnology Industries;
- Nanotechnology Now—Daily News and Information for the Field of Nanotechnology;
- Nanotechnology Research and Technical Data;
- Nanowerk—The Comprehensive Nanotechnology Portal;
- Project on Emerging Nanotechnologies;
- Red Venezolana de Nanotecnología, Venezuela;
- Research News in Nanotechnology;
- Russian Society of Scanning Probe Microscopy and Nanotechnology: NanoWorld;
- SAFENANO – Europe's Centre of Excellence on Nanotechnology Hazard and Risk;
- Schau-Platz NANO, Munich, Germany;
- STATNANO: Nano Science, Technology and Industry Scoreboard;
- The International Council on Nanotechnology (ICON) at Rice University;
- The Nanoethics Group;
- The National Nanomanufacturing Network (NNN);
- The National Nanotechnology Manufacturing Center (NNMC);
- The UK Nanosafety Partnership Group; and
- The Understanding Nanotechnology website.

1.3.8 NOTABLE ORGANIZATIONS IN NANOTECHNOLOGY

1. Government
   - National Cancer Institute (US);
   - National Institutes of Health (US);
   - National Institutes of Health and National Institute of Biomedical Imaging and Bioengineering (US);
   - National Nanotechnology Initiative (US);
   - Russian Nanotechnology Corporation (RU); and
   - Seventh Framework Program, FP7 (EU).

2. Advocacy and Information Groups
    *   American Chemistry Council (US);
    *   American Nano Society (US);
    *   Center for Responsible Nanotechnology (US);
    *   Foresight Institute (US); and
    *   Project on Emerging Nanotechnologies (global).

1.3.9   DISTINGUISHED SCIENTISTS IN NANOTECHNOLOGY

*   Adriano Cavalcanti – nanorobot expert working at CAN (http://www.canbiotechnems.com).
*   Akhlesh Lakhtakia – conceptualized sculptured thin films.
*   Alex Zettl – built the first molecular motor based on carbon nanotubes.
*   Andre Geim – discoverer of 2-D carbon film called graphene.
*   Carlo Montemagno – inventor ATP nanobiomechanical motor (UCLA).
*   Charles M. Lieber – pioneer on nanoscale materials (Harvard).
*   Chris Phoenix – co-founder of the Center for Responsible Nanotechnology.
*   Eric Drexler – was the first to theorize about nanotechnology in depth and popularized the subject.
*   Erwin Wilhelm Müller – invented the field ion microscope, and the atom probe.
*   Gerd Binnig – co-inventor of the scanning tunneling microscope.
*   Harry Kroto – co-discoverer of buckminsterfullerene.
*   Heinrich Rohrer – co-inventor of the scanning tunneling microscope.
*   Joseph Wang – pioneer in electrochemical sensors exploiting nanostructured materials; synthetic nanomotors.
*   Lajos P. Balogh – editor in chief of Nanomedicine: NBM journal.
*   Mike Treder – co-founder of the Center for Responsible Nanotechnology.
*   Norio Taniguchi – coined the term "nano-technology."
*   Phaedon Avouris – first electronic devices made out of carbon nanotubes.
*   Ralph Merkle – nanotechnology theorist.
*   Richard Feynman – gave the first mention of some of the distinguishing concepts in a 1959 talk.

- Richard Smalley – co-discoverer of buckminsterfullerene.
- Robert Freitas – nanomedicine theorist.
- Russell M. Taylor II – co-director of the UNC CISMM, *http://cismm. cs.unc.edu*.
- Sumio Iijima – discoverer of carbon nanotube.
- Vicki Colvin – Director for the Center for Biological and Environmental Nanotechnology, Rice University.

During the author's sabbatical leave in 2002–2003 at the Biomedical Engineering Department at Florida International University (*http:// www.bme.fiu.edu*), he had a chance to interview Dr. Venkatesan Renugopalakrishnan—whose research interests have been targeted towards understanding protein structure, dynamics, and engineering—for a long time. The author summarizes this interview here for the benefit of readers so that they are aware of typical research activities in NT of a university professor. Because of its interdisciplinary nature, his research encompasses chemistry, biology, physics, bioengineering, and NT. The 3-D structures of proteins are derived by circular dichroism, multi-dimensional multi-nuclear NMR and time-resolved spectroscopy. Protein expressions and modifications of proteins rely on state-of-the-art molecular biology. Computational methods are used in the derivation of 3-D structures from experimental data. Bio-NT applications of proteins in biomedical sciences, electronics are being pursued. According to him, his thrust areas of research in NT have been: construction, purification, expression of human rhodopsin; applications of human rhodopsin in technology like flat screen display; a spectroscopic studies of cytochrome-c and a double mutant [ALA93 and ASN96] bacteriorhodopsin; 3D-NMR studies of high tm bacteriorhodopsin mutants, FT-IR spectroscopic studies of cytochrome c–cytochrome c' molecular dynamics simulation and neutron scattering studies of cytochrome-c, plastocyanin and bacteriorhodopsins in lipid phase; 3D structure of human amelogenin; nanotemplates of immobilized proteins, AFM studies; nanobiosensors; flash memory using bacteriorhodopsin mutants, spin coating of Br; phospho and glycoproteins: structure, dynamics, and engineering neuropeptides, and proteins of the central nervous system and their interactions with receptor; receptor stereochemistry; time-resolved 2D-FT-IR and laser Raman spectroscopy at liquid N2 temperatures with isotope editing; new methods in structure determination of large proteins by multi-dimensional heteronuclear NMR; and nano bio-technology applications of proteins as biomaterials.

1.3.10 NANOTECHNOLOGY IN SCIENCE FICTION

NT is also useful in fiction science, which deals with the things that are beyond the boundaries of present technology. The NT in fiction explores fantasized things in the real world. Since the introduction of NT in 1987 by Eric Drexler, many publications have appeared that have supported the relationship between NT and fiction, such as: speculative fiction, insertion of bio-chips into talking robots, artificial dinosaurs in fiction movies, elf-replicating robots, total human-like robots, nanobots or nanties, gray groo, and nanolathes. Selected examples of NT in fiction include:

1. *The Tale of Cross-eyed Lefty from Tula and the Steel Flea* (1881) by Nikolai Leskov.
2. *The Next Tenants* (1956) by Arthur C. Clarke.
3. *Terminator 2: Judgement Day* (1991) by Randall Frakes.
4. *GURPS Robots* (1986) by David Pulver.
5. Novels: *The Invincible* (1964) and *Peace on Earth* (1984) by Stanislaw Lem; 1985 novel *Blood Music* by Greg Bear; *Diamond Age* novel (1995) by Neal Stephenson; *Decipher* (2001) by Stel Pavlou; *Prey* (2002) by Michael Crichton; *Lazarus Vendetta* (2005) by Robert Ludlum; *Book of Kur* (2011) by Lansing; The *Diamond Age* by Neal Stephenson; *Tomorrow* by Andrew Saltzman.
6. TV shows, episodes: *Outer Limits: The New Breed*; Sci-Fi Channel era of *Mystery Science Theater 3000*; *Stargate SG-1* and *Stargate Atlantis*; *Doctor Who* 2005; *Knight Rider*; NBC science fiction show *Revolution (TV series)*; manga series *Battle Angel Alita: Last Order*; *Dx13: Nano A Mano*, a manga series by Kirupagaren Kanni; *G.I. Joe Reinstated* series; Mystery Science Theater 3000; *Ghost in the Shell: Stand Alone Complex*; *Trinity Blood* series by Abel Nighroad; *Red Dwarf* VII to IX; *The Use of Nanoprobes* by Borg in *Star Trek: The Next Generation*, etc.
7. Movies: *Cowboy Bebop: The Movie* (2001); 2003 film *Agent Cody Banks*; 2004 film *I Robot*; 2008 film *The Day the Earth Stood Still*; 2009 film *G.I. Joe: The Rise of Cobra*.
8. Computer Games: *Total Annihilation, Mortal Kombat* series, *System Shock 2, Deus Ex*, MMORPG *Anarchy Online, Metal Gear Solid 2, Metal Gear Solid 4, Red Faction, Red Faction Guerilla, Ratchet & Clank* series, *Resident Evil 4, Crysis* 1-2-3, *Marvel: Ultimate Alliance 2*, and *SpaceChem*, etc.

9. E-books: *Nanomech* [Kindle Edition] by King, Amazon Digital Services, Inc. (2011); *Duty, Honor, Planet* [Kindle Edition], Partlow, Amazon Digital Services Inc., (2011); *New Mother Earth Series* by Olsen (2014 Kindle eBook) by Litzophreniacs3, LLC; *Nano Surveillance* by Mark Donovan (2014), Kindle eBook by Amazon Digital Services Inc.; *MEMergence: The HuMEM Series: Book* by Donnelly (2014, Kindle eBook) by Amazon Digital Services Inc.; The roleplaying game Splicers by Palladium Books.

10. *The Subatomic Fiber-Optic Deconstruction/Construction Transportation Chamber* (2012) by Dylan Otto Krider, http://dailysciencefiction.com/science-fiction/nanotech/dylan-otto-krider/the-subatomic-fiber-optic-deconstructionconstruction-transportation-chamber.

11. *Sabi, Wabi, Aware, Yugen* by Sam J Miller (2013). http://dailysciencefiction.com/science-fiction/nanotech/sam-j-miller/sabi-wabi-aware-yugen

11. *A Puddle of Dead by Grayson Bray Morris* (2011), http: // dailysciencefiction.com/science-fiction/nanotech/grayson-bray-morris/a-puddle-of-dead

12. *Mortal Coil* by Ian Nichols (2012), http://dailysciencefiction.com/science-fiction/nanotech/ian-nichols/mortal-coil

13. *Nanomite* by Patricia Duffy Novak (2012): *Oops, tactical error*. http://dailysciencefiction.com/science-fiction/nanotech/patricia-duffy-novak/nanomite

14. *Susumu Must Fold* by Tony Pi (2012), http://dailysciencefiction.com/science-fiction/nanotech/tony-pi/susumu-must-fold

15. *The Next Generation* by Michael Adam Robson (2014), http://dailysciencefiction.com/science-fiction/nanotech/michael-adam-robson/the-next-generation.

16. *The Ambiguity Clock* by Lavie Tidhar (2011), http://dailysciencefiction.com/science-fiction/nanotech/lavie-tidhar/the-ambiguity-clock.

17. *Everlasting* by Fran Wilde (2011), http://dailysciencefiction.com/science-fiction/nanotech/fran-wilde/everlasting.

18. *The Bittersweet Here and After* by Maggie Clark (2012), http://dailysciencefiction.com/science-fiction/nanotech/maggie-clark/the-bittersweet-here-and-after.

19. *MiracleMech* by Tim Deans (2012), http://dailysciencefiction.com/science-fiction/nanotech/tim-deans/miraclemech.

20. *Lets All Sing Like the Birdies Sing* by Di Filippo (2012), http://dailysciencefiction.com/science-fiction/nanotech/paul-g-di-filippo/lets-all-sing-like-the-birdies-sing.

21. *The Rules of the Regeneration Manual* by Findlay (2011), http://dailysciencefiction.com/science-fiction/nanotech/andrew-l-findlay/the-rules-of-the-regeneration-manual.
22. *Swarm, the Queen Commanded* by Gillett (2014), http://dailysciencefiction.com/science-fiction/nanotech/k-a-gillett/swarm-the-queen-commanded.
23. *The Professor's Boy* by Goranson (2012), http://dailysciencefiction.com/science-fiction/nanotech/erik-goranson/the-professors-boy.
24. *True Hollywood Story* by Gutierrez (2011), http://dailysciencefiction.com/science-fiction/nanotech/ryan-gutierrez/true-hollywood-story.
25. *Nathan and the Amazing TechnoPocket Nerd* Coat by Kabza (2012), http://dailysciencefiction.com/science-fiction/nanotech/kj-kabza/nathan-and-the-amazing-technopocket-nerdcoat.
26. *Bit Storm* by Lancer & Kind (2011), http://dailysciencefiction.com/science-fiction/nanotech/lancer-shelli-kind/bit-storm.

1.4   BIONANOTECHNOLOGY, NANOBIOTECHNOLOGY, AND NANOBIOLOGY

These terms refer to the intersection of nanotechnology and biology. Bionanotechnology (bioNT) and nanobiotechnology (nanoNT) serve as blanket terms for various related technologies.

This discipline helps to indicate the merger of biological research with various fields of NT (see Section 1.3). Nanobiology focuses on concepts such as nanodevices, nanoparticles, and nanoscale phenomena. BioNT does have many potential ethical issues associated with it.

The most important objectives involve applying nanotools to relevant medical/biological problems (plants, animals, and humans) and refining these applications. Developing new tools, such as peptoid nanosheets for medical and biological purposes, is another primary objective in NT. The imaging of native biomolecules, biological membranes, and tissues are also a major topic for researchers. Other topics concerning nanobiology include the use of cantilever array sensors and the application of nanophotonics for manipulating molecular processes in living cells. Recently, the use of microorganisms to synthesize functional nanoparticles has been of great interest.

BioNT refers to the ways that NT is used to create devices to study biological systems. BioNT is an essentially miniaturized biotechnology.

1.4.1 APPLICATIONS

Applications of bioNT are widely spread. It (sometimes referred to as nanobiology) helps modern medicine from treating symptoms to generating cures and regenerating biological tissues. Physicians have used nanobiology techniques to treat patients. Stem cell treatments have been used to fix heart diseases. Also, there is a research that allows patients to have new limbs without having to resort to prosthesis.

Another example of current nanobiotechnological research involves nanospheres coated with fluorescent polymers. And as an another example, from a different perspective, it would be the treatment of .anobacteria (25-200 nm) as is done by NanoBiotech Pharma.

The promising avenues of bioNT research are: utilization of inherent properties of nucleic acids like DNA to create useful materials; use of membrane properties to generate synthetic membranes; protein folding; and lipid NT to build nanodevices.

1.4.2 TOOLS

The related terms in bioNT are: nanobiopharmaceutics, nanobiomechanics, nanosub, nanomedicine, colloidal gold, gold nanoparticle, and gold nanobeacons.

The bioNT relies on a variety of research methods, including:

- Experimental tools (e.g., imaging, characterization via AFM/optical tweezers, etc.),
- X-ray diffraction-based tools, synthesis via self-assembly and characterization of self-assembly (e.g., using dual polarization interferometry, recombinant DNA methods, etc.),
- Theory (e.g., statistical mechanics, nanomechanics, etc.), and as well as
- Computational approaches (bottom-up multi-scale simulation, super-computing).

1.5 TISSUE ENGINEERING

Tissue engineering (cellular engineering, molecular engineering, and regenerative medicine) is the used for the combination of cells, engineering and materials methods, and suitable biochemical and physio-chemical factors to improve or replace biological functions. It is an interdisciplinary field integrating research in biology, chemical engineering, computer science, animal science, microbiology, and materials science. Understanding cellular function and cell manipulation are the basis for many ventures, such as: drug production from cell culture, methods for improved drug delivery, generation of artificial organs for replacement of diseased tissues, development of biofuels, and design of bioremediation processes for wastewater clean-up.

1.5.1 EXAMPLES OF TISSUE ENGINEERING

- Artificial bladders.
- Artificial bone marrow.
- Artificial bone.
- Artificial heart valve.
- Artificial pancreas.
- Artificial penis.
- Artificial skin.
- Bioartificial liver device.
- Bioartificial windpipe.
- Doris Taylor's heart in a jar.
- Foreskin.
- *In vitro* meat.
- Oral mucosa tissue engineering.
- Scaffold-free cartilage.
- Tissue-engineered airway.
- Tissue-engineered vessels.

1.6 BIOMEDICAL ENGINEERING/BIOENGINEERING/ BIOLOGICAL ENGINEERING

Biomedical engineering (BME) is the application of engineering principles and design concepts to the medicine and biology of healthcare purposes (e.g., diagnostic or therampeutic). This field seeks to close the gap between engineering and medicine: It combines the design and problem solving skills of engineering with medical and biological sciences to advance healthcare treatment, including diagnosis, monitoring, and therapy.

Much of the work in BME consists of research and development by spanning a broad array of subfields. Prominent BME applications include the development of biocompatible prostheses, various diagnostic and therapeutic medical devices ranging from clinical equipment to micro-implants, common imaging equipment such as MRIs and EEGs, regenerative tissue growth, pharmaceutical drugs, and therapeutic biologicals.

1.6.1 SUBDISCIPLINES OF BIOMEDICAL ENGINEERING

A biomedical engineer must have knowledge from the viewpoint of medical and engineering applications. As with many medical specialties (e.g., cardiology, neurology), some BME sub-disciplines are identified by their associations with particular systems of the human body, such as:

- Cardiovascular technology, which includes all drugs, biologics, and devices related with diagnostics and therapeutics of cardiovascular systems;
- Neural technology, which includes all drugs, biologics, and devices related with diagnostics and therapeutics of the brain and nervous systems;
- Orthopaedic technology, which includes all drugs, biologics, and devices related with diagnostics and therapeutics of skeletal systems;
- Cancer technology, which includes all drugs, biologics, and devices related with diagnostics and therapeutics of cancer.

But more often, subdisciplines within BME are classified by their association(s) with other more established engineering fields, which can include (at a broad level):

- Biochemical-BME, based on chemical engineering: often associated with biochemical, cellular, molecular and tissue engineering, biomaterials, and biotransport.
- Bioelectrical-BME, based on electrical engineering and computer science: often associated with bioelectrical and neural engineering, bioinstrumentation, biomedical imaging, medical devices, biomedical optics, bioinformatics, and bioimaging.
- Biomechanical-BME, based on mechanical engineering: often associated with biomechanics, biotransport, medical devices, and modeling of biological systems, and soft tissue mechanics.

One more way to sub-classify the discipline is on the basis of the products created:

- Biologics and biopharmaceuticals: often designed using the principles of synthetic biology.
- Pharmaceutical drugs (so-called "small-molecule" or non-biologic), which are commonly designed using the principles of synthetic chemistry and traditionally discovered using high-throughput screening methods at the beginning of the development process.
- Devices, which commonly employ mechanical and/or electrical aspects in conjunction with chemical and/or biological processing or analysis.
- Combination products, which involve *more than one of the above* categories in an integrated product (e.g., a microchip implant for targeted drug delivery).

Genetic engineering utilizes modern tools such as molecular cloning and transformation to directly alter the structure and characteristics of target genes.

Neural engineering (also known as neuroengineering) uses engineering techniques to understand, repair, replace, or enhance neural systems.

Pharmaceutical engineering is an interdisciplinary science that includes drug engineering, novel drug delivery and targeting, pharmaceutical technology, and the unit operations of Chemical Engineering, and Pharmaceutical Analysis. The ISPE is an international body, which certifies this, is now rapidly emerging in interdisciplinary science.

Medical device is intended for the use in the diagnosis of disease or other conditions, or in the cure, mitigation, treatment, or prevention of disease. Some examples of this kind include: pacemakers, infusion pumps, the heart-lung machine, dialysis machines, artificial organs, implants, artificial limbs, corrective lenses, cochlear implants, ocular prosthetics, facial prosthetics, somato prosthetics, and dental implants. Stereolithography is a practical example of medical modeling being used to create physical objects. Medical devices are regulated and classified (in USA) as follows:

1. Class I devices present minimal potential for harm to the user and are often simpler in design than Class II or Class III devices. Devices in this category include tongue depressors, bedpans, elastic bandages, examination gloves, hand-held surgical instruments, and other similar types of common equipment.
2. Class II devices are subject to special controls in addition to the general controls of Class I devices. Special controls may include special labeling requirements, mandatory performance standards, and postmarket surveillance. Devices in this class are typically non-invasive and include

x-ray machines, PACS, powered wheelchairs, infusion pumps, and surgical drapes.

3. Class III devices generally require premarket approval (PMA) or premarket notification (510k)—a scientific review to ensure the device's safety and effectiveness—in addition to the general controls of Class I. The examples include: replacement of heart valves, hip and knee joint implants, silicone gel-filled breast implants, implanted cerebellar stimulators, implantable pacemaker pulse generators, and endosseous (intrabone) implants.

Medical/biomedical imaging is a major segment of medical devices. This area deals with enabling clinicians to directly or indirectly view things that are not visible through normal sight (due to their size, and/or location). This can involve utilizing ultrasound, magnetism, UV, other radiology, and other means. Imaging technologies include: fluoroscopy, magnetic resonance imaging (MRI), nuclear medicine, positron emission tomography (PET), X-rays and CT scans, tomography, ultrasound, optical microscopy, and electron microscopy.

Implants: An implant is a kind of medical device designed to replace and act as a missing biological structure (as compared with a transplant, which indicates transplanted biomedical tissue).

The best contribution of bionics is in the field of BME. BME is the building of useful replacements for various parts of the human body. Biomedical engineers work hand-in-hand with doctors to build the artificial body parts.

Clinical engineering is the branch of BME dealing with the actual implementation of medical equipment and technologies in hospitals or other clinical settings. Major roles of clinical engineers include training and supervising biomedical equipment technicians (BMETs), selecting technological products/services and logistically managing their implementation, working with governmental regulators on inspections/audits, and serving as technological consultants for other hospital staff. Clinical engineers also advise and collaborate with medical device producers regarding prospective design improvements based on clinical experiences, as well as monitor the progression of the state-of-the-art, so as to redirect procurement patterns accordingly. Clinical engineering departments will sometimes hire not just biomedical engineers, but also industrial/systems engineers to help and address operations research/optimization, human factors, cost analysis, etc. The safety engineering includes procedures used to design safe systems in BME.

1.6.2   REGULATORY ISSUES

Practicing biomedical engineers must routinely consult and cooperate with regulatory law attorneys and other experts. Regulatory issues have been constantly increased in the last decades to respond to the many incidents caused by medical devices. During 2008–2011 in USA, there were 119 FDA recalls of medical devices classified as Class 1. According to U.S. Food and Drug Administration (FDA), Class 1 recall is associated to "a situation in which there is a reasonable probability that the use of, or exposure to, a product will cause serious adverse health consequences or death."

The FDA is the principal healthcare regulatory authority in the United States, having jurisdiction over *medical devices, drugs, biologics,* and *combination* products. The paramount objectives driving policy decisions by the FDA are *safety* and *effectiveness* of healthcare products that have to be assured through a *quality system* in place as specified under 21 CFR 829 regulation. In addition, because biomedical engineers often develop devices and technologies for "consumer" use, such as physical therapy devices (which are also "medical" devices), these may also be governed in some respects by the Consumer Product Safety Commission. The greatest hurdles tend to be 510K "clearance" (typically for Class 2 devices) or pre-market "approval" (typically for drugs and Class 3 devices).

Under the *European context,* safety effectiveness and quality are ensured through the *conformity assessment*; that is defined as "the method by which a manufacturer demonstrates that its device complies with the requirements of the European Medical Device Directive." The directive specifies different procedures according to the class of the device ranging from: the simple Declaration of Conformity for Class I devices to EC verification; production quality assurance; product quality assurance; and full quality assurance. The Medical Device Directive specifies detailed procedures for Certification.

In general terms, these procedures include tests and verifications that are to be contained in specific deliveries such as the risk management file, the technical file and the quality system deliveries. The risk management file is the first deliverable that conditions the following design and manufacturing steps. *Risk management* stage shall drive the product so that product risks are reduced at an acceptable level with respect to the benefits expected for the patients for the use of the device. The *technical file* contains all the documentation data and records supporting medical device certification. FDA technical file has similar content although organized in different structure.

The *Quality System* deliverables usually includes procedures that ensure quality throughout all product life cycle. The same standard (ISO EN 13485) is usually applied for quality management systems in US and worldwide.

In European Union, there are certifying entities named "Notified Bodies," accredited by European Member States. The Notified Bodies must ensure the effectiveness of the certification process for all medical devices apart from the Class 1 devices where a declaration of conformity produced by the manufacturer is sufficient for marketing. Once a product has passed all the steps, required by the Medical Device Directive, the device is entitled to bear a CE marking indicating that the device is believed to be safe and effective when used as intended, and therefore, it can be marketed within the European Union area.

1.6.2.1   RoHS 2

Directive 2011/65/EU, better known as RoHS 2, is a recast of legislation originally introduced in 2002. The original EU legislation *"Restrictions of Certain Hazardous Substances (RoHS) in Electrical and Electronics Devices (RoHS Directive 2002/95/EC)"* was replaced and superseded by 2011/65/EU published in July 2011 and commonly known as RoHS 2. RoHS seeks to limit the dangerous substances in circulation in electronics products, in particular toxins and heavy metals, which are subsequently released into the environment when such devices are recycled. The scope of RoHS 2 includes products previously excluded, such as medical devices and industrial equipment. In addition, manufacturers are now obliged to provide conformity risk assessments and test reports; or explain why they are lacking. For the first time, not only manufacturers, but also importers and distributors share a responsibility to ensure electrical and electronic equipment within the scope of RoHS comply with the hazardous substances limits and have a CE mark on their products.

1.6.2.2   IEC 60601

The new International Standard IEC 60601 for home healthcare electro-medical devices defining the requirements for devices used in the home healthcare environment. IEC 60601-1-11 (2010) must now be incorporated into the design and verification of a wide range of home use and point of care medical devices along with other applicable standards in the IEC 60601-3rd edition series. The North American agencies require these standards for new device submissions, while the EU will take the more severe approach of requiring all applicable devices being placed on the market to consider the home healthcare standard.

1.6.3 *TRAINING AND CERTIFICATION IN BIOMEDICAL ENGINEERING*

1.6.3.1 Education

Biomedical engineers require considerable knowledge of both engineering and biology, and typically have a Master's (MS, MTech, MSE, or MEng) or a Doctoral (PhD) degree in BME or another branch of engineering with considerable potential for BME overlap. As interest in BME increases, many engineering colleges now have a BME Department or Program, with offerings ranging from the undergraduate (BTech, BS, BEng or BSE) to doctoral levels. As mentioned in this chapter, BME has only recently been emerging as *its own discipline* rather than a cross-disciplinary hybrid specialization of other disciplines; and BME programs at all levels are becoming more widespread, including the Bachelor of Science in BME, which actually includes so much biological science content that many students use it as a "pre-med" major in preparation for medical school. The number of biomedical engineers is expected to rise as both a cause and effect of improvements in medical technology.

In the United States, an increasing number of undergraduate programs are also becoming recognized by ABET as accredited bioengineering/BME programs. Over 88 programs are currently accredited by ABET. In Canada and Australia, accredited graduate programs in BME are common, for example:
- McMaster University; the first Canadian undergraduate BME program at Ryerson University offering a four-year BME program; and
- Polytechnique in Montreal offering a bachelors' degree in BME.

Graduate education is a particularly important aspect in BME. While many engineering fields (such as mechanical or electrical engineering) do not need graduate-level training to obtain an entry-level job in their field, the majority of BME positions does prefer or even require graduate education. Since most BME-related professions involve scientific research, graduate education is almost a requirement. This can be either a Masters or Doctoral level degree; while in certain specialties a PhD is notably more common than in others. In fact, the perceived need for some kind of graduate credential is so strong that some undergraduate BME programs will actively discourage students from majoring in BME without an expressed intention and also to obtain a masters degree or apply to medical school afterwards. Graduate programs in BME, like in other scientific fields, are highly varied, and particular programs may emphasize certain aspects within the field. They may also feature extensive collaborative efforts with programs in other fields (such as the University's Medical School or other engineering divisions), owing again to the interdisciplinary nature of BME. The MS and PhD programs typically require applicants to have an undergraduate degree in BME, or another engineering discipline (plus certain life science coursework), or life science (plus certain engineering coursework).

The education in BME also varies greatly around the world. By virtue of its extensive biotechnology sector, numerous major universities, and relatively few internal barriers, the U.S. has progressed a great deal in the development of BME education and training opportunities. Europe, which also has a large biotechnology sector and an impressive education system, has encountered trouble in creating uniform standards as the European community attempts to supplant some of the national jurisdictional barriers that still exist. Recently, initiatives such as BIOMEDEA have sprung-up to develop BME-related education and professional standards. Other countries, such as Australia, are recognizing and moving to correct deficiencies in their BME education. Also, as high technology endeavors are usually the marks of developed nations, some areas of the world are prone to slower development in education, including in BME.

1.6.3.2   Licensure/certification

Engineering licensure in the U.S. is largely optional, and rarely specified by branch/discipline. As with other learned professions, each state has certain (fairly similar) requirements for becoming licensed as a registered Professional Engineer (PE), but in practice such a license is not required to practice in the majority of situations (due to an exception known as the private industry exemption, which effectively applies to the vast majority of American engineers).

This is notably not the case in many other countries, where a license is as legally necessary to practice engineering as it is for law or medicine. BME is regulated in some countries, such as Australia, but registration is typically only recommended and not required. In the U.K., mechanical engineers working in the areas of medical engineering, bioengineering, or BME can gain Chartered Engineer status through the Institution of Mechanical Engineers. The Institution also runs the Engineering in Medicine and Health Division.

The Fundamentals of Engineering exam now cover biology (although technically not BME). For the second exam, called the Principles and Practices (Part 2) or the Professional Engineering exam, candidates may select a particular engineering discipline's content to be tested on. Currently, there is no option for BME with it; that is, any biomedical engineers seeking a license must prepare to take this examination in another category. However, since 2009, the BME Society (BMES) is exploring the possibility of seeking to implement a BME-specific version of this exam to facilitate biomedical engineers pursuing licensure. Beyond governmental registration, certain private-sector professional/industrial organizations also offer certifications with varying degrees of prominence. One such example is the Certified Clinical Engineer (CCE) certification for Clinical engineers.

1.6.4   CAREER PROSPECTS IN BME

Biomedical engineers are expected to have excellent job prospects and earning potential in the near future, according to the US Bureau of Labor Statistics.

The job outlook for biomedical engineers is expected to grow much faster than average through 2018. Because the population is aging, the demand for better medical devices and equipment designed by biomedical engineers is rising. Along with the demand for more sophisticated medical equipment and procedures is an increased concern for cost efficiency and effectiveness that also will boost demand for biomedical engineers.

In 2008, the middle annual salary for biomedical engineers was $77,400. The middle 50% earned between $59,420 and $98,830. The lowest 10% earned less than $47,650 and the highest 10% earned more than $121,970.

1.6.5   IMPORTANT ADVANCES IN BME

Biomedical engineers have developed many important techniques and equipment such as:
- Angioplasty;
- Arthroscopic instrumentation for diagnostic and surgical purposes;

- Artificial articulated joint's;
- Bioengineered skin;
- Heart pacemakers;
- Heart-lung machines;
- Hip joint replacement;
- Kidney dialysis;
- Magnetic resonance imaging (MRI); and
- Time-release drug delivery.

1.6.6 FOUNDING FIGURES IN BME

- Alfred E. Mann: Physicist, entrepreneur, and philanthropist.
- Ascher Shapiro (deceased): Institute Professor at MIT, contributed to the development of the BME field, medical devices (e.g., intra-aortic balloons).
- Forrest Bird: Aviator and pioneer in the invention of mechanical ventilators.
- Frederick Thurstone (deceased): Professor Emeritus at Duke University, pioneer of diagnostic ultrasound.
- Herbert Lissner (deceased): Professor of Engineering Mechanics at Wayne State University. American Society of Mechanical Engineers' top award in BME is named as Herbert R. Lissner Medal.
- John G. Webster: Professor Emeritus at the University of Wisconsin– Madison, a pioneer in the field of instrumentation amplifiers for the recording of electrophysiological signals.
- John James Rickard Macleod (deceased): One of the co-discoverers of insulin at Case Western Reserve University.
- Kenneth R. Diller: Chaired and Endowed Professor in Engineering, University of Texas at Austin. Pioneer in bioheat transfer, mass transfer, and biotransport.
- Leslie Geddes (deceased): Professor Emeritus at Purdue University; Recipient of a National Medal of Technology in 2006 from President George Bush for his more than 50 years of contributions that have spawned innovations ranging from burn treatments to miniature defibrillators.
- Nicholas A. Peppas: Chaired Professor in Engineering, University of Texas at Austin, pioneer in drug delivery, biomaterials, hydrogels and nanobiotechnology.

- Otto Schmitt (deceased): Biophysicist with significant contributions in biomimetics.
- Robert Langer: Institute Professor at MIT; Runs the largest BME laboratory in the world, pioneer in drug delivery and tissue engineering.
- Robert Plonsey: Professor Emeritus at Duke University; pioneer of electrophysiology.
- U. A. Whitaker (deceased): Founder of Whitaker Foundation, which supported research and education in BME by providing over $700 million to various universities, helping to create 30 BME programs and helping finance the construction of 13 buildings.
- Willem Johan Kolff (deceased): Pioneer of hemodialysis as well as in the field of artificial organs.
- Y. C. Fung: Professor at the University of California, San Diego; founder of modern Biomechanics.

1.6.7   BIOMECHANICS

Biomechanics (Ancient Greek: βίος "life" and μηχανική "mechanics;" Modern Greek, εμβιομηχανική) is the study of the structure and function of biological systems such as humans, animals, plants, organs, and cells by means of the methods of mechanics. In this book, we shall discuss only biomechanics of human body systems. The word biomechanics was developed in the 1970s, while describing the application of engineering mechanics to biological and medical systems. Engineering biomechanics uses traditional engineering sciences to analyze biological systems. Applied mechanics, such as continuum mechanics, mechanism analysis, structural analysis, kinematics and dynamics play prominent roles in the study of biomechanics. Applied subfields of biomechanics include:

- Allometry;
- Animal locomotion and Gait analysis;
- Cardiovascular biomechanics;
- Ergonomy;
- Human factors engineering and occupational biomechanics;
- Implant (medicine)/Orthotics/Prosthesis;
- Kinesiology (kinetics + physiology);
- Musculoskeletal and orthopedic biomechanics;
- Rehabilitation;

- Soft body dynamics; and
- Sports biomechanics.

Sports biomechanics applies the laws of mechanics to gain a greater understanding of athletic performance and to reduce sport injuries as well.

Continuum biomechanics is used for the mechanical analysis of biomaterials and biofluids.

Biofluid mechanics is a study of body fluids that are at rest or in motion. Flow of body fluids can be modeled by the Navier–Stokes equations.

Biotribology is a study of friction, wear and lubrication of biological systems especially human joints such as hips and knees.

Comparative biomechanics is the application of biomechanics to nonhuman organisms.

Plant biomechanics is the application of biomechanical principles to crops and plant organs.

The study of biomechanics ranges from the inner workings of a cell to the movement and development of limbs, to the mechanical properties of soft tissue, and bones. Biomechanics is also applied to study human musculoskeletal systems. Biomechanics is widely used in orthopedic industry to design orthopedic implants for human joints, dental parts, external fixations, and other medical purposes. It also includes study of the performance and function of biomaterials used for orthopedic implants. It plays a vital role to improve the design and produce successful biomaterials for medical and clinical purposes. For in-depth study, the reader is advised to study following books (list may not be complete):

- Anthony C. Fischer-Cripps, *Introduction to Contact Mechanics*, ISBN 0-387-68187-6.
- Donald R. Peterson and Joseph D. Bronzino. *Biomechanics: Principles and Applications*, ISBN 0-8493-8534-2.
- Jagan N. Mazmudar. *Biofluid Mechanics*, ISBN 981-02-0927-4.
- Jay D. Humphrey. *Cardiovascular Solid Mechanics*, ISBN 0-387-95168-7.
- Lee Waite and Jerry Fine. *Applied Biofluid Mechanics*, ISBN 0-07-147217-7.
- Megh R. Goyal, 2013. *Biofluid Dynamics of Human Body Systems*. Apple Academic Press Inc., ISBN: 9781926895468; Out of print.
- Megh R. Goyal, 2014. *Biomechanics of Artificial Organs and Prostheses*. Apple Academic Press Inc., ISBN: 9781926895840; Out of print.

- Bruce R. Munson, Alric P. Rothmayer, Theodore H. Okiishi, and Wade W. Huebsch. *Fundamentals of Fluid Mechanics*. 7th Edition, 2013. John Wiley & Sons.
- Stephen C. Cowin. *Bone Mechanics Handbook*, ISBN 0-8493-9117-2.
- J. S. Temenoff and Antonios G. Mikos, 2008. *Biomaterials: The Intersection of Biology and Materials Science*, ISBN 978-0-13-009710-1.
- Y. C. Fung, *Biomechanics*, ISBN 0-387-94384-6.
- D. F. Young, B. R. Munson, T. H. Okiishi, and W. W. Huebsch. *A Brief Introduction to Fluid Mechanics*. 5th Edition, 2011. John Wiley.

Scientific Journals in Biomechanics (List may not be complete)

• Applied Bionics and Biomechanics	• Journal of Arthroplasty
• Biomechanics and Modeling in Mechanobiology	• Journal of Biomechanical Engineering
• Clinical Biomechanics	• Journal of Biomechanics
• Computer Methods in Biomechanics and Biomedical Engineering	• Journal of Bone and Joint Surgery
• Footwear Science	• Journal of Electromyography and Kinesiology
• Gait and Posture	• Journal of Experimental Biology
• Journal of Applied Biomechanics	• Journal of Experimental Zoology
• Journal of Applied Physiology	• Journal of Morphology
	• Sports Biomechanics
	• The Journal of Experimental Biology

Biomechanics and Bioengineering Societies, http://www.asbweb.org/html/links/links.html

American College of Sports Medicine	International Society of Biomechanics
American Physiological Society	International Society of Biomechanics in Sports
American Society of Biomechanics	International Society of Electrophysiology and Kinesiology
American Society of Mechanical Engineers	Japanese Society of Biomechanics
Australian and New Zealand Society of Biomechanics	Orthopaedic Research Society
Biophysical Society	Societe de Biomecanique
Canadian Society for Biomechanics	Society for Integrative and Comparative Biology

European Society of Biomechanics

Federation of American Societies for Experimental Biology

Gait and Clinical Movement Analysis Society

German Society of Biomechanics

Hellenic Society of Biomechanics

Human Factors and Ergonomics Society

Instituto de Biomecánica de Valencia

Society for Mathematical Biology

Taiwanese Society of Biomechanics

The American Society of Biomechanics

The American Society of Bone and Mineral Research

The Biomedical Engineering Society

The Orthopaedic Research Society

Biomechanics and Bone Websites

Biomechanics course materials, <uoregon.edu/~karduna>

Biomechanics World Wide <per.ualberta.ca>

Biomed experts http://www.biomedexperts.com/

Boise State University—*Center for Orthopaedic and Biomechanics Research*

Bone Net

Boston University—*Orthopaedic and Developmental Biomechanics*

Columbia University—*Bone Bioengineering*

Columbia University—*Musculoskeletal Biomechanics*

NIH National Resource Center for Osteoporosis and Bone Related Diseases

NIOSH Musculoskeletal Disorder Program, www.cdc.gov

Ohio State University—*Biodynamics*

Oregon State University—*Biomechanics*

Powerful Bones. Powerful Girls.

Rice University—*Computational Biomechanics*

The ASBMR bone curriculum

U.S. Center for Disease Control,, www.cdc.gov

Cornell University

Johns Hopkins

Mayo Clinic—*Biomechanics and Motion Analysis*

Michigan State University—*Orthopaedic Biomechanics*

Milk Matters

National Osteoporosis Foundation

UC San Francisco *Orthopaedic Bioengineering*

University of California Berkeley—*Orthopaedic Biomechanics Laboratory*

University of Michigan—*Orthopaedic Research Laboratory*

USC Engineering Neuroscience and Health Seminar series (www.bbdl.usc.edu/enh)

Whitaker Foundation, www.whitaker.org

Societies: Artificial Organs, http://www.esao.org/links.php

ACC American College of Cardiology	ASAO Austrian Society for Artificial Organs
AIMBE American Institute For Medical and Biological Engineering	EDTA European Dialysis and Transplantation Association
ANNA American Nephrology Nurses Association	EMBS Engineering in Medicine and Biology
ASAIO American Society for Artificial Internal Organs	ESB European Society for Biomaterials
ESEM European Society for Engineering and Medicine	ISPD International Society for Peritoneal Dialysis
IFAO International Federation for Artificial Organs – former ISAO	ISRP International Society for Rotary Blood Pumps
ISABI International Society for Artificial Cells, Blood Substitutes and Immobilization Biotechnology	JSAO Japanese Society for Artificial Organs (Japanese version)
ISFA International Society for Apheresis	NKF National Kidney Foundation
	TERMIS Tissue Engineering and Regenerative Medicine International

General Medical/Technology, http://www.esao.org/links.php

- HSTAT Health Services Technology Assessment Texts: Searchable NIH databases on health services and technologies.
- MedEc Interactive The Medical Economics site, with a searchable database of their publications, including the PDR.
- Medical Device Link – An industry-sponsored site full of info: News, publications, suppliers directory, job listings, WWW links, etc.
- National Inventors Hall of Fame
- NIH Technology Assessment Statements Full text search of consensus statements from NIH Technology Assessment Conferences and Workshops
- PubMed Free, fast, and efficient searching of the MEDLINE database. (A service of the National Center for Biotechnology Information (NCBI), which has its own site, well worth visiting.)
- The Visible Human Project "Complete, anatomically detailed, three-dimensional representations of the male and female human body. The current phase of the project is collecting transverse CT, MRI and cryosection images of representative male and female cadavers at one millimeter intervals." A project of the National Library of Medicine.

- WWW Home Pages about Artificial Organs Courtesy of the Japanese Society for Artificial Organs. Most listed sites are in Japan.
- Center for Apheresis Technology

Literature, http://www.esao.org/links.php
- International Journal for Artificial Organs, published at Wichtig-Verlag
- Journal of Artificial Organs, published at Springer
- Artificial Organs: The Journal of the IFAO
- MediConf®, The Database on Medical Conferences and Exhibitions Worldwide
- Journals of Biomaterials: link-list of several worldwide published journals
- Science-Directory.net: link-list of Science

Organizations to help and inform the general public about prosthetics
At the end of World War II, the National Academy of Sciences began to advocate better research and development of prosthetics. Through government funding, a research and development program was developed within the Army, Navy, Air Force, and the Veterans Administration. The following organizations were created to help and inform the general public about prosthetics:
- **American Orthotics and Prosthetic Association, American Board for Certification in Prosthetics and Orthotics, American Academy of Orthotics and Prosthetics**: These three educational groups work together and provide certification of individuals and facilities working with orthotics and prosthetics.
- **The International Society for Prosthetics and Orthotics**: Founded in 1970 and headquartered in Copenhagen, this association helps with the progression in research and clinical practice worldwide. They hold an international conference every three years and publish a technical journal.
- **Association of Children's Orthotic-Prosthetic Clinics**: The organization was started in 1950s to advocate research and development of children's prosthetics. They meet annually and publish technical articles.
- **Amputee Coalition of America:** The organization was created in 1990 to improve the lives of amputees. Advocate the improvement of amputee lifestyle through education and publish a magazine: "inMotion."

KEYWORDS

- **bioengineering**
- **biomechanics**
- **biomedical engineering**
- **Biomedical Engineering Education**
- **bionanotechnology**
- **biotechnology**
- **Canada**
- **clinical engineering**
- **Florida International University, FIU**
- **nanotechnology**
- **National Science Foundation, USA, NSF**
- **technology**
- **timeline**
- **tissue engineering**
- **United States of America, USA**

REFERENCES

1. Berger, E., & Roth, J., (1997). *The Golgi Apparatus*. Basel, Switzerland: Birkhäuser.
2. Blonder, R., (2010). The influence of a teaching model in nanotechnology on chemistry teachers' knowledge and their teaching attitudes. *Journal of Nano Education, 2*, 67–75.
3. Bly, R. W., (2005). *The Science In Science Fiction: 83 SF Predictions that Became Scientific Reality*. BenBella Books, Inc.
4. Brock, T. D., (1961). *Milestones in Microbiology*. Science Tech Publishers. Madison, Wisconsin. pp. 273.
5. Brock, T. D., (1990). *The Emergence of Bacterial Genetics*. Cold Spring Harbor Laboratory Press. Cold Spring Harbor, New York. pp. 346.
6. Bunch, B., & Hellemans, A., (1993). *The Timetables of Technology*. Simon & Schuster. New York, New York. pp. 490.
7. Colin, M., (2008). *Nanovision: Engineering the Future*. Duke University Press.
8. Feneque, J., (2000). Nanotechnology: a new challenge for veterinary medicine. *The Pet Tribune. 6*(5), 16.
9. Hellemans, A., & Bunch, B., (1988). *The Timetables of Science*. Simon & Schuster. New York, pp. 660.
10. http://bmes.org/cmbesig Cellular and Molecular Bioengineering Special Interest Group

11. http://bmes.org/faseb Federation of American Societies for Experimental Biology.
12. http://bse.unl.edu/history
13. http://dailysciencefiction.com/science-fiction/nanotech
14. http://svtc.org/our-work/nano/timeline/
15. http://timelines.ws/subjects/Technology.HTML
16. http://www.accessexcellence.org/RC/AB/BC/1977-Present.php
17. http://www.accessexcellence.org/RC/AB/BC/1977-Present.php
18. http://www.biotechinstitute.org/go.cfm?do=Page.View&pid=22
19. http://www.bme.fiu.edu/about/history/
20. http://www.dummies.com/how-to/content/nanotechnology-timeline-and-predictions. html
21. http://www.ehow.com/about_6364029_history-bioengineering.html
22. http://www.ehow.com/about_6364029_history-bioengineering.html#ixzz31PBIg8vV
23. http://www.greatachievements.org/?id=3824
24. http://www.greatachievements.org/?id=3837
25. http://www.historyworld.net/timesearch/default.asp?conid=static_timeline&timelineid =406&page=1&keywords=Engineering%20timeline
26. http://www.medscape.com/viewarticle/566133_2
27. http://www.nano.gov/timeline
28. http://www.nano.gov/timeline
29. http://www.nature.com/nmat/journal/v8/n6/fig_tab/nmat2441_F1.html
30. http://www.science-of-aging.com/timelines/cell-history-timeline-detail.php
31. http://www.timetoast.com/timelines/tissue-engineering-and-regenerative-medicine
32. http://www.wifinotes.com/nanotechnology/nanotechnology-in-fiction.html
33. https://engineering.purdue.edu/BME/AboutUs/History
34. Lopez, J., (2004). Bridging the Gaps: Science Fiction in Nanotechnology. *International Journal for Philosophy of Chemistry, 10*(2), 129–152.
35. Murphy, A., & Judy, P., (1993). *A Further Look at Biotechnology*. Princeton, NJ: The Woodrow Wilson National Fellowship Foundation, Woodrow Wilson Foundation Biology Institute.
36. Patrick, D., (2007). Tiny Tech, Transcendent Tech – Nanotechnology, Science Fiction, and the Limits of Modern Science Talk. *Science Communication, 29*(1), 65–95.
37. Peters, P., (1993). *From Biotechnology: A Guide To Genetic Engineering*. Wm. C. Brown Publishers, Inc., 1993.
38. Peterson, C., (1995). *Nanotechnology: From Concept to R&D Goal*. HotWired.
39. Peterson, C., (1995). Nanotechnology: evolution of the concept. In: *Prospects in Nanotechnology: Toward Molecular Manufacturing* by ed. Markus Krummenacker and James Lewis. Wiley.
40. Peterson, C., (2000). Molecular Nanotechnology: the Next Industrial Revolution. IEEE Computer, January issue.
41. Peterson, C., (2004). Nanotechnology: from Feynman to the grand challenge of molecular manufacturing. IEEE Technology and Society, Winter issue.
42. Regenerative medicine glossary. Regenerative Medicine *4* (4 Suppl): 81–88. July 2009.
43. Regenerative Medicine, 2008, *3*(1), 1–5.
44. Regis, E., (1995). *Nano: The Emerging Science of Nanotechnology*. Little Brown.
45. Schwarz, J. A., Contescu, C. I., & Putyera, K., (2004). *Dekker Encyclopedia of Nanoscience and Nanotechnology*. CRC Press.

46. Sweeney, A. E., (2008). Developing a viable knowledge base in nanoscale science and engineering. In: *Nanoscale Science and Engineering Education*, by A. E. Sweeney and S. Seal. (eds.). American Scientific: Stevenson Ranch, CA, USA, pp. 1–35.
47. Toumey, C., (2008). The Literature of Promises. *Nature Nanotechnology, 3*(4), 180–181.

APPENDIX I: BIOMEDICAL ENGINEERING EDUCATION AT FLORIDA INTERNATIONAL UNIVERSITY

1. INTRODUCTION

The Biomedical Engineering Department at Florida International University (BME-FIU) is a prime resource for biomedical engineering education, training, research, and technology development in Florida and is nationally recognized as a model for servicing the needs of the clinical medicine and the biomedical industries through workforce and technology development (Figure 1.1).

FIGURE 1.1 Overview of BME-FIU.

Reproduced with permission. http://www.bme.fiu.edu

Since top-quality health care is still out of reach for many, it is time to train the next generation of leaders and innovators to shape the technological and economic landscapes of tomorrow with a focused academic curriculum. The interdisciplinary nature of biomedical engineering (BME) affords the discipline a tremendous opportunity to bring forth advances in technological development and new discoveries in life sciences to foster improved quality of life for all individuals. The U.S. Bureau of Labor and Statistics projects a 72% growth in biomedical engineers from 2008 to 2018 with tremendous impact of biomedical engineering in all the sectors they track.

The mission of the BME-FIU is:

- to bridge engineering, science, and medicine;
- to educate and train the next diverse generation of biomedical engineers;
- to conduct research leading to significant discoveries in medical sciences;
- to conduct design and development of innovative medical technology;
- to translate scientific discovery and medical technology to industry or clinical practice for delivery of health care; and
- to engage with the local to global community for knowledge dissemination.

Wallace H. Coulter Foundation Seminar Series

Through the generous support of the Wallace H. Coulter Foundation, the BME-FIU facilitates weekly lecture series each year during academic terms. Experts in all areas of biomedical engineering are invited to campus to provide a research seminar and to meet with faculty and students and to tour our academic and research facilities. The interactions supported by this fund provide exposure to the department as well as foster collaborations among faculty at FIU and at other institutions. The Department has been very proactive in running a scientific WH Coulter Lecture Series supported by the Coulter endowment. There are dozens of lectures have been presented by international and national invited experts.

Academic Advisory Board

- Eduardo N. Warman, PhD, CRDM Monitoring and Diagnostics Medtronic, Inc.

- George Wittenberg, MD, PhD, http://medschool.umaryland.edu
- James Hickman, PhD, http://www.nanoscience.ucf.edu
- Jennifer L. West, PhD, http://www.bme.duke.edu
- Leonard Pinchuk, PhD, DSc, http://innovia-llc.theinnoport.com
- Norberto M. Grzywacz, PhD, http://bme.usc.edu
- Omowunmi A. Sadik, PhD, http://chemiris.chem.binghamton.edu
- Ravi Bellamkonda, PhD, Georgia Institute of Technology, Atlanta, GA, http://www.bme.gatech.edu
- Stephen Snowdy, PhD, http://www.vtivision.com
- Warren Grill, PhD, http://www.bme.duke.edu

2. FACILITIES

The BME department has multiple labs maintained by individual professors, with applications including medical imaging, tissue engineering, microfluidics, and mathematical modeling. Another enticing element of FIU's Engineering Center is its access to the Motorola Nanofabrication Research Facility, located onsite. This is a separate facility that is a resource to BME. The $15 million Motorola Nanofabrication Research Facility is an integral part of the Advanced Materials Engineering Research Institute (AMERI), FIU's broader materials research program. This facility boasts:
- specialized equipment required to develop new and novel fabrication techniques unique to the creation of functional materials and devices that are no greater than 100 nanometers (i.e., 1000 times smaller than the diameter of a human hair);
- a full complement of standard semiconductor processing equipment to leverage the capabilities of robust and proven techniques; and
- state-of-the-art analytical tools to study and characterize these nano-sized devices, as well as the materials and processes used to make them.

FIU's main campus (Modesto Maidique Facilities) hosts a IACUC-approved animal testing facility with modern sterilization and life-support systems as well as constant veterinarian care. This facility can hold numerous animal models and has a procedure-room on site for optical imaging.

BME-FIU Research Laboratories

Adaptive Neural Systems Laboratory (ANS): The ANS is focused on developing and utilizing new scientific knowledge and engineering technology to address the complex physiological, medical, and societal problems presented by neurological disability. Its research agenda is at the intersection between bioengineering, neuroscience, and rehabilitation.

Nanobioengineering/Bioelectronics Lab: The research of this group interfaces with biomedical engineering, nanobiotechnology, electrochemistry, BioMEMS, biochemistry, nanomedicine, surface science, and materials science. The work done here looks ahead to the next generation of nano-electrical components such as protein nanowires, DNA transistors as well as end use electronic devices such as Lab-on-Chip, biosensors, and enzymatic biofuel cells.

Laboratory of Vascular Physiology and Biotransport: The main focus of this laboratory is on the mechanisms that regulate blood flow and pressure in the human body. Scientists investigate the physiology of the microcirculation through the parallel development of theoretical and experimental models. Mathematical modeling guides experimentation and assist in data analysis while *in vitro* experimental studies provide important modeling parameters and promote further model development.

Optical Imaging Laboratory (OIL): Researchers focus on optical imaging instrumentation, tomography studies with various biomedical applications such as breast cancer imaging and functional brain mapping.

Tissue Engineered Mechanics, Imaging, and Materials (TEMIM): The primary research interests of this facility are in heart valve tissue engineering, cardiovascular mechanobiology, and evaluation of functionality and hemocompatibility of cardiovascular devices (e.g., stents and heart valve prosthetics).

3. FIU BIOMEDICAL ENGINEERING ACADEMIC PROGRAMS, BS/MS/PHD

3.1 Undergraduate Program: Bachelor of Science in Biomedical Engineering

The undergraduates at BME-FIU engage in the rich South Florida culture that advance study in engineering, medicine, or other sciences and develop the professional practice as is expected from a biomedical engineer in the industry. Bringing engineering and medicine together, biomedical engineering is one of the fastest-growing occupations in the U.S. for the decade ending in 2018. Biomedical engineers conduct research, design, develop, and evaluate systems; and deliver products to solve health-related problems. They apply science and technology to support improved healthcare. Graduates are trained in:

- clinical application of biomedical engineering tools and product development;
- manufacturing and commercialization in the biomedical industry;
- participation in diverse teams; and
- biomedical research.

The BME-FIU is endowed by the Wallace H. Coulter Foundation, the Ware Foundation and the State of Florida. BME-FIU has a growing alignment with the *Herbert Wertheim College of Medicine*, the College of Nursing and Health Sciences, and the *College of Arts and Sciences*. BME-FIU is the only department in the nation offering an accredited BS in Biomedical Engineering at a Minority- and Hispanic-serving institution.

Degree Requirements

The student will need 128 credit hours to graduate (Table 1.1). Lower-division requirements include at least 60 hours of pre-engineering credits. Up to 90 academic credits may be transferred from an ABET-accredited institution of higher education. The student must also meet the University Foreign Language Requirement, pass the CLAST or have it waived and meet all of state and university requirements for graduation.

Admission Requirements

The student will need a high school degree from an accredited institution, official SAT/ACT scores, transcripts from all other colleges. If the student is transferring from a public university or college in Florida, passing scores from the College Level Academic Test (CLAST) must be submitted. Application Materials consist of:

- high school degree from an accredited institution;
- official SAT/ACT scores; and
- transcripts from all other colleges you have attended;

TABLE 1.1 Bachelor of Science in Biomedical Engineering Degree with 128 Credits

First Semester – Fall: (17)	Second Semester – Spring: (17)
MAC 2311 Calculus I	BME 1008 Intro to Biomed Eng
CHM 1045 General Chemistry I CHM 1045 General Chemistry I	MAC 2312 Calculus II
	CHM 1046 General Chemistry II
CHM 1045L General Chemistry I Lab	CHM 1046L Gen Chemistry II Lab
ENC 1101 Writing and Rhetoric I	PHY 2048 Physics I w/ Calc
EGN 1100 Intro to Engineering	PHY 2048L General Physics I Lab
SLS 1501 Freshman Experience	ENC 1102 Writing and Rhetoric II
Humanities Elective 3	

Third Semester – Fall: (18)	Fourth Semester – Spring: (15)
MAC 2313 Multi-variable Calculus	MAP 2302 Differential Equations
CHM 2210 Organic Chemistry I	STA 3033 Intro Probability Statistics
CHM 2210L Organic Chemistry I Lab	BME 2740 BME Modeling & Simulation
BSC 1010 General Biology I	Engineering or Science Elective
BSC 1010L Gen Biology I Lab	Humanities
PHY 2049 Physics II w/ Calc	
PHY 2049L Physics II Lab	

Fifth Semester – Fall: (16)	Sixth Semester – Spring: (15)
BME 3721 BME Data Evaluation Principles	BME 3404 Eng Analysis Biological Systems II
BME 3403 Eng Analysis Biological Systems	
EEL 3110 Circuit Analysis	EGM 3503 Applied Mechanics
EEL 3110L Circuits Lab	EEE 4202C Medical Inst Design
Engineering or Science Elective	BME 4011 Clinical Rotations
Social Science Elective	Art Elective

Seventh Semester – Fall: (17)	Eighth Semester – Spring: (13)
BME 4050L BME Lab I	BME 4908 Senior Design Project
BME 3632 BME Transport	BME 4332 Cell & Tissue Engineering
BME 4100 Biomaterials Science	Engineering or Science Elective
BME 4090 Design Project Organization	Engineering or Science Elective
BME 4800 Design Biomedical Systems and Devices	BME 4930 Undergraduate Seminar
Engineering or Science Elective	
Humanities/Historical Elective 3	

TABLE 1.1 (Continued)

BME Suggested Electives:	Suggested Electives for Premedical:
BME students have the choice of several areas of specialization guiding them to their intended field of study. They must complete atleast 15 credit hours.	CHM 2211 Organic Chemistry II CHM 2211L Organic Chemistry II Lab BCH 3033 General Biochemistry I BCH 3033L General Biochemistry I Lab MCB 3010 General Microbiology OTH 4418 Impact of Neurological Dysfunction on Human Performance
Suggested Electives for Tissue Engineering and BioMaterials Area:	**Suggested Electives for BioSignals and Systems Area:**
BME 4311 Molecular Imaging EGN 3365 Materials in Engineering EMA 3066 Polymer Science and Engineering	BME 4401 Medical Imaging BME 4562 Biomedical Optics BME 4730 Analysis of Self Regulation EEL 3135 Signals and Systems EEL 3657 Control Systems I EEL 4510 Introduction to Digital Signal Processing
Suggested Electives for BioMechanics Area:	
BME 4311 Orthopedic Biocmechanics BME 4260 Engineering Hemodynamics EGM 3311 Analysis of Engineering Systems EML 3036 Sim Software for Mechanical Engineers EML 4807 Introduction to Mechatronics	

3.2 Graduate Programs: MS, PhD, BS+MS

The BME-FIU offers advanced study for the degrees of Master of Science in Biomedical Engineering, and Doctor of Philosophy in Biomedical Engineering. A graduate degree in Biomedical Engineering can enhance the career opportunities or prepare the student for further success in academia. The graduate programs in the BME-FIU consist of: Doctor of Philosophy (PhD) Biomedical Engineering; Master of Science (MS) Biomedical Engineering; or 5 year combined (BS/MS) Biomedical Engineering.

The graduates at BME-FIU are prepared for academic, clinical, or industrial research and development in: Basic Research in Engineered Tissue Model Systems, Diagnostic Bioimaging, and Sensor Systems, or Therapeutic and Reparative Neurotechnology.

A. Doctor of Philosophy in Biomedical Engineering, PhD

The PhD program provides students and practicing engineers with the theoretical and practical experience needed to succeed. The PhD program has three objectives:

1. to provide highly trained professionals to serve in academic institutions, government agencies, research labs and manufacturing and service industries;
2. to provide minority students a great opportunity for advanced graduate studies; and
3. to supply additional minority graduates to the biomedical engineering field, where they are highly underrepresented.

Clinical and industrial practice is integrated into the doctoral academic program through the Biomedical Engineering Partnership Program. The student will gain valuable exposure to clinical practice and research and get real experience in product development and commercialization. The PhD program prepares graduates for academic, clinical, or industrial research and development in one or more of three cluster areas of specialization: Basic Research in Engineered Tissue Model Systems; or Diagnostic Bioimaging and Sensor Systems; or Therapeutic and Reparative Neurotechnology. These three areas are served by technological advancements in: bio-imaging, bio-signal processing, and computational modeling; bio-instrumentation, devices, and sensors; biomaterials and bio-nano technology; and, cellular and tissue engineering.

Prospective students must satisfy all university admission requirements as well as the specific program requirements. Meeting the minimum requirements does not guarantee admission into the program. FIU and the College of Engineering and Computing offer a variety of fellowships, assistantships, and scholarships to qualified domestic and international students. Amounts vary depending on the type of award; however, they may provide full tuition and a monthly stipend.

The PhD program requires 90 credits beyond a BS degree: a minimum of twenty-seven (27) credit hours coursework and a minimum of 24 credits of dissertation. Applicants with a Master's Degree in Biomedical Engineering, or closely related field, from an accredited institution may be given a maximum of 30 transferred credit hours.

B.   Master of Science in Biomedical Engineering, MS

Many biomedical engineers, particularly those whom work in research labs, need a graduate degree. The Master of Science in Biomedical Engineering provides students and practicing engineers with the theoretical and practical experience needed to succeed. MS program has the following three objectives:

1. to provide highly trained professionals to serve in academic institutions, government agencies, research labs and manufacturing and service industries;
2. to provide minority students a great opportunity for advanced graduate studies; and
3. to supply additional minority graduates to the biomedical engineering field, where they are highly underrepresented.

The BME-FIU offers two tracks: The Professional Track is for engineers currently practicing in the biomedical industry and students interested in pursuing a management career in biomedical industry. It requires 27 credit hours of course work, a three-credit hour capstone project, and a final oral examination. The *Research Track* prepares graduates for further study or a career in biomedical research. On this track student will need a minimum of 24 credit hours of course work, a minimum of 6 credit hours of Master's Thesis or a 3 credit hours of capstone project, and one semester of the Biomedical Engineering Seminar.

The new students must satisfy all university admission requirements as well as the specific program requirements. Meeting the minimum requirements does not guarantee admission into the program.

C.   Combined Bachelor of Science and Master of Science in Biomedical Engineering, BS+MS

Always just a little bit ahead of the game? Eager to learn and ready for a challenge? If the student has the academic might, then the student can earn a bachelor's and master's degree in less time. One can count graduate courses toward *both* degrees, cutting down your time to graduate by a semester or more. Also the students pay less and are a more marketable professional. The student will also be able to take courses and do research which are not available normally to undergraduates. The student can complete a Bachelor of Science degree in biomedical, mechanical, or electrical engineering combined with the Master of Science Degree in biomedical engineering. Students who have completed a minimum of 90 hours towards their BS degree in bio-medical engineering and have earned at least a 3.3 grade point average with a combined GRE score of 1,000 (verbal ≥ 350, and quantitative ≥ 650) may apply for admission in the combined BSMS program.

Table 1.2 indicates student population in BME-FIU at the beginning of fall semester of an academic year. Table 1.3 shows number of students who have received BS, MS, or PhD degree during a specific academic year.

TABLE 1.2 Headcount of enrollment at the beginning of an academic year.

STUDENT LEVEL	Fall 2007	Fall 2008	Fall 2009	Fall 2010	Fall 2011	Fall 2012
GRAD II	14	22	25	34	32	30
LOWER	105	102	84	123	143	200
UPPER	139	165	175	189	213	223
GRAD I	48	27	22	18	15	20
Biomedical Engineering: Total	306	316	306	364	403	473
Biomedical Engineering/ CENGR Total	306	316	306	364	403	473

TABLE 1.3 Degree Count: Number of Graduates for the Academic Year

Department	Student Level	2007-2008	2008-2009	2009-2010	2010-2011	2011-2012	2012-2013
Biomedical Engineering	BACHELORS	32	29	38	39	31	46
	MASTERS	11	20	6	5	10	7
	DOCTORAL	1	2	3	5	3	9
	TOTAL DEGREES	44	51	47	49	44	62
Biomedical Engineering/CENGR BACHELORS		32	29	38	39	31	46
Biomedical Engineering/CENGR MASTERS		11	20	6	5	10	7
Biomedical Engineering/CENGR DOCTORAL		1	2	3	5	3	9
Biomedical Engineering/CENGR: TOTAL DEGREES		44	51	47	49	44	62

D.   *The FIU Biomedical Engineering Student Learning Outcomes*

1. Ability to apply knowledge of mathematics (including differential equations and statistics), physical and life sciences, and engineering to carry out analysis and design to solve problems at the interface of engineering and biology;
2. Ability to design and conduct experiments, as well as to measure, analyze and interpret data from living systems;
3. Ability to design a system, component, or process to meet desired needs, including systems that involve the interaction between living and non-living materials, within realistic constraints such as economic, environmental, social, political, ethical, health and safety, manufacturability, and sustainability;
4. Ability to identify, formulate, and adapt engineering solutions to unmet clinical/biological needs;
5. Ability to use modern engineering techniques, skills, and tools necessary for engineering practice, including the ability to model and analyze physiological/biological systems as engineering systems;
6. Ability to function on multi-disciplinary teams;
7. Ability to communicate effectively;
8. Awareness of the characteristics of responsible professional engineering practice, including ethical conduct, consideration of the impact of engineering solutions on society in a global and contemporary context, and the value of life-long learning.

E.   *Biomedical Engineering Accreditation and Assessment*

All academic programs of Florida International University are approved by the Florida Board of Education, the FIU Board of Trustees and the Florida Board of Governors. The university is accredited by the Commission on Colleges of the Southern Association of Colleges and Schools to award baccalaureate, master's, and doctoral degrees. The Bachelor of Science in Biomedical Engineering is accredited by the Engineering Accreditation Commission of ABET (www.abet.org). ABET accreditation provides assurance that a college or university program meets the quality standards established by the profession for which the program prepares its students.

F.   Contact Information (See Table 1.4)

The contact information for BME-FIU is given in Table 1.4.

TABLE 1.4   Contact information for BME-FIU.

Dr. Ranu Jung
Professor and Head
Biomedical Engineering
The Engineering Center
Florida International
University
10555 West Flagler St.,
EC 2600
Miami, Florida 33174
Phone: 305-348-6950
Fax: 305-348-6954

BME Information:
bmeinfo@fiu.edu
Undergraduate Contact:
Michael C. Christie, PhD
mchristi@fiu.edu
Graduate Contact:
bmeinfo@fiu.edu
Webmaster Contact:
Amado.Gonzalez@fiu.edu

**Graduate Program
Director
Shuliang Jiao, PhD**
Associate Professor
E-mail: shjiao@fiu.edu
Office: AHC4 Room 332
Phone: (305) 348-4984
Fax: (305) 348-6954

**Undergraduate
Program Director
Anuradha
Godavarty, PhD**
Associate Professor
E-mail: Godavart@
fiu.edu
Office: EC 2675
Phone: (305) 348-
7340; Fax: (305)
348-6954

4.   BME RESEARCH OVERVIEW

The research conducted by the members of the BME-FIU is clustered into three focus areas (Figures 1.2–1.14):

1. basic research in engineered tissue model systems,
2. diagnostic bioimaging and sensor systems; and
3. therapeutic and reparative neurotechnology. These three focus areas are served by technological advancements in:
   • *Bio-imaging, Bio-signal Processing and Computational Modeling:* Malek Adjouadi, PhD; Armando Barreto, PhD; Kenneth Horch, PhD; Ranu Jung, PhD; Wei-Chiang Lin, PhD; Anthony J. McGoron, PhD; and Jorge Riera, PhD.

- *Bio-instrumentation, Devices and Sensors:* Armando Barreto, PhD; Kenneth Horch, PhD; Ranu Jung, PhD; Chenzhong Li, PhD; and Wei-Chiang Lin, PhD.
- *Biomaterials and Bio-nano Technology:* Michael C. Christie, PhD; Yen-Chih Huang, PhD; Chenzhong Li, PhD; Anthony J. McGoron, PhD; and Sharan Ramaswamy, PhD.
- *Cellular and Tissue Engineering:* Yen-Chih Huang, PhD; Sharan Ramaswamy, PhD; and Nikolaos Tsoukias, PhD.

5. ILLUSTRATIONS: CURRENT STATUS OF BME-FIU, MIAMI

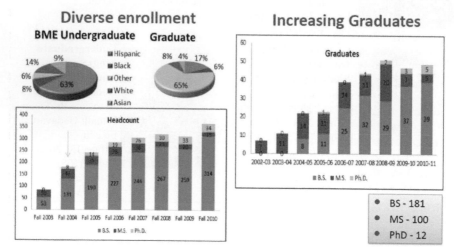

FIGURE 1.2   Student population and number of graduates at BME-FIU.

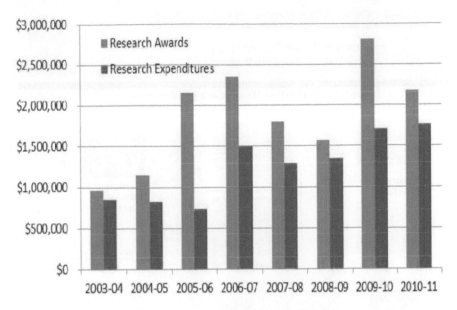

FIGURE 1.3 Research funding at BME-FIU.

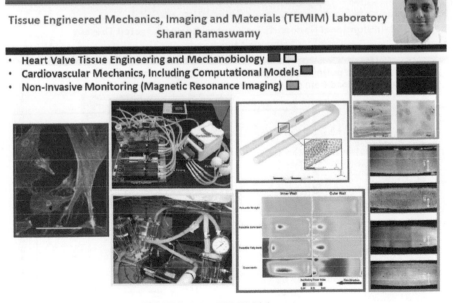

FIGURE 1.4 TEMIM laboratory.

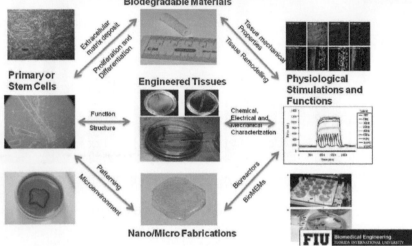

FIGURE 1.5   Manipulation at molecular, cellular and tissue levels.

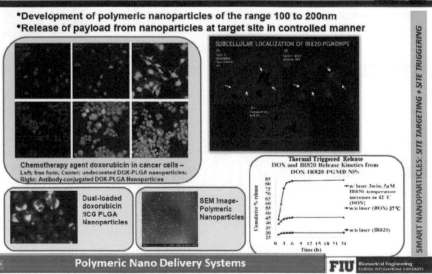

FIGURE 1.6   Polymeric nano delivery systems.

Basic Research in Engineered Tissue Model Systems

Laboratory of Vascular Physiology and Biotransport
Nikolaos M. Tsoukias

- Systems biology of microcirculation
- Multiscale models that reach down to the subcellular level

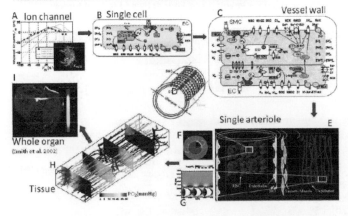

FIGURE 1.7   Systems of biology of microcirculation.

Diagnostic Imaging and Sensor Systems

Nanobioengineering and Bioelectronics Laboratory
Chenzhong Li

- **Biomarker Detection: Cancer (BRCA 1, VEGF, PSA, etc); ROS (8-OhDG , NO etc); Alzheimer's Disease (Tau, Abeta, Dopamine, etc).**
- **Whole Cell Analysis: Pathogen Bacteria Cells, Cancer Cells**

**Nanomedicine
Biosensors
Bioenergy**

Synthesis and
Characterization
of Nanomaterials
(Nps, Nws, CNTs, etc)

Bio-Nano-
Conjugation

Macro/Nano
Fabrication
MEMS/NEMS

DNA Sensors

Cell Sensors

From nano-scaled materials to prototype biomedical devices

Point-of-Care Testing: From Electronic Devices to Paper Based Strips

FIGURE 1.8   Biomarker detection and whole cell analysis.

Diagnostic Imaging and Sensor Systems

Targeted Image Guided Therapy: Reduced Cardiac Toxicity
Anthony McGoron

- Chemotherapy drug encapsulated into 100 nm nanoparticle (liposomes or polymers)
- NIR absorbing dye used for imaging and later activated to heat tumor
- Studies done in isolated perfused hearts and tumor bearing rats

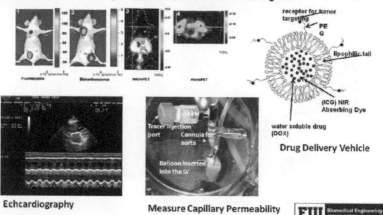

FIGURE 1.9   Targeted image guided therapy.

Diagnostic Imaging and Sensor Systems

Pre-clinical Rodent Model for Brain Mapping
Jorge Riera

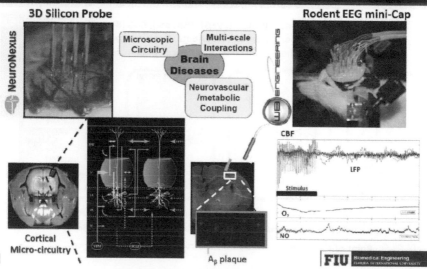

FIGURE 1.10   Preclinical rodent model.

Diagnostic Imaging and Sensor Systems

Biomedical Optics Laboratory
Wei-Chiang Lin

- Optical diagnostic technology for
  - Intraoperative brain tumor demarcation
  - Physiological characterization of epileptic cortex
  - Characterization of myocardial infarction
  - Detection of mitochondrial malfunction

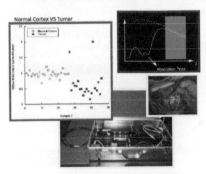

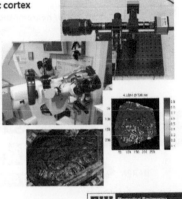

FIGURE 1.11   Optical diagnostic technology.

Diagnostic Imaging and Sensor Systems

Optical-Based Molecular Imaging and Tomography
Anuradha Godavarty

Hand-Held Based Optical Imager
- Towards Breast Cancer Diagnosis

Real-Time Coregistered Imaging   3D Tomography

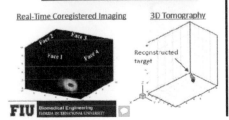

Diffuse Optical Imaging
- Towards Functional Brain Mapping

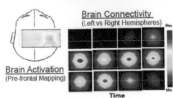

Brain Connectivity
(Left vs Right Hemispheres)

Brain Activation
(Pre-frontal Mapping)

Applications: Autism & Cerebral Palsy

FIGURE 1.12   Optical based molecular imaging.

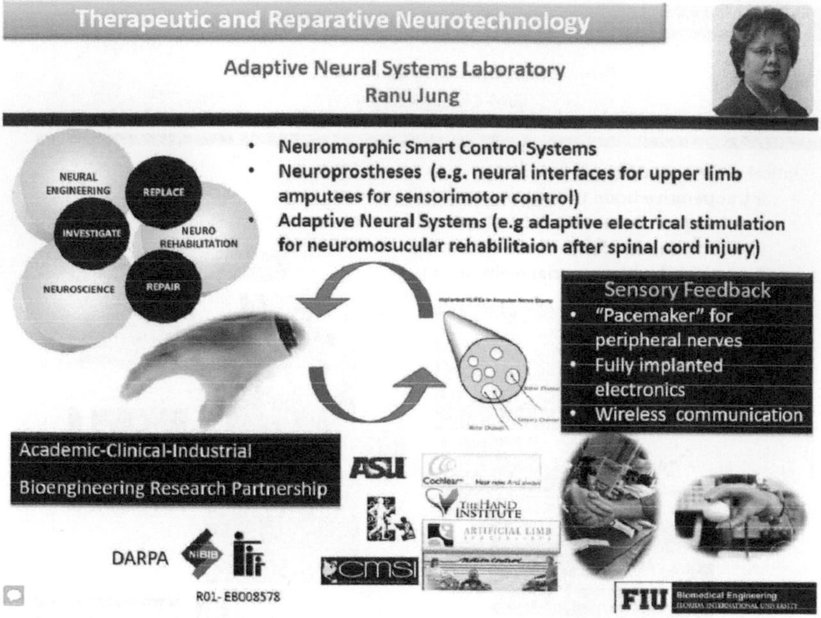

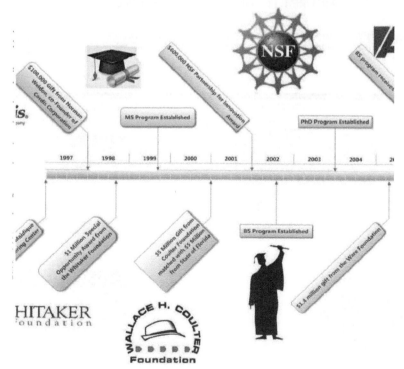

FIGURE 1.13   Adaptive neural systems.

FIGURE 1.14   Milestones in Biomedical Engineering Education during 1996–2007 at FIU (http://www.fiu.edu).

CHAPTER 2

TIMELINE OF BIOENGINEERING AND BIOLOGICAL ENGINEERING

CONTENTS

2.1   Introduction 55
2.2   Inventions, Modifications, and Significant Events
       in Chronological Order 56
2.3   Summary 107
Keywords 107
References 107

2.1   INTRODUCTION

In the past, numerous developments have been made in bioengineering, biological engineering, biotechnology, molecular engineering (tissue engineering), nanotechnology, and bionanotechnology. The timeline in this chapter will help the reader to understand the past, present, and future of these focus areas of science. I have tried my best to list the complete timeline in this chapter. The developments or events for the timeline in this chapter were compiled from the information in the books, journals, glossaries, dictionaries, manuals, websites, and personal files of my colleagues. The historical application of biotechnology/bioengineering is provided in the Section 2.2 in the chronological order. These discoveries, inventions, and modifications are evidence of the evolution of these disciplines of science before and after the Christian era.

2.2   INVENTIONS, MODIFICATIONS, AND SIGNIFICANT EVENTS IN CHRONOLOGICAL ORDER

Year	Event/Concepts, etc.	Who
B.C.		
7000	Making of beer through fermentation.	Chinese
6000	Use of lactic acid bacteria (LAB) in the preparation of cheese/yogurt.	–
6000	Sumerians and Babylonians used to make beer from yeast.	Sumerians & Babylonians
4000	Baking of leavened bread using yeast.	Egyptians
4000	Cheese was produced using molds.	China
4000	Fermentation process was used to produce vinegar and wine	China
3000	Possible origin of BME: A mummy was unearthed from ancient Thebes by archaeologists (2000 A.D.)	German
1000	Pollination of date palm trees was celebrated.	Babylonians
500	Use of moldy soybean curd to treat boils.	China
440	Introduction of concept of human wellbeing based on ethics.	Socrates
424	Birth of mechanics with the conceptualization of mathematics as the life force of science.	Plato
420	Why children do not always resemble their parents	Socrates
400	Women carry fluid similar to semen. Child's heredity is carried in the semen and receive traits equal proportion from each.	Hippocrates
384	May be first biomechanician. First book on "De Motu Animalium"	Aristotle
320	The semen determines the baby's form. Mother merely provides the material and media from which the baby is made.	Aristotle
250	Crop rotation for maximum soil fertility.	Greeks
100	Powdered chrysanthemum was used to prepare insecticide.	Chinese
A.D. or A.E.C.		
100	The nature of reproduction and inheritance.	The Hindu
100	Mares may be fertilized by the wind.	Romans
160	First treatise: On the Movements of the Heart and Lung.	Galen
192	Biomechanician – Major treatise on: *the Usefulness of the Parts of the Body* (in 17 books: meaning the parts of the human body); the Natural Faculties, etc.	Galen
400	4th Century – Lycurgus Cup: Colloidal gold and silver in the glass allow it to look opaque green when lit from outside but translucent red when light shines through the inside.	Rome

Year	Event/Concepts, etc.	Who
600	6th–15th Century – Gold nanoparticles as photocatalytic air purifiers. Use of gold and silver in glazes.	Europeans
900	9th–17th Century – Glowing, glittering "luster" ceramic glazes.	Muslims
1000	Laws of Manu: Child inherits all their parents' characteristics; Certain diseases may run in the family.	Hindus
1100	Maggots were supposed to arise from horse-hair.	–
1300	Algae was harvested from lakes as a food source by the Aztecs	Mexico
1300	Damascus saber blades contained carbon nanotubes and cementite nanowires.	–
1400	Spirits from fermented grain.	Africans
1452	Analysis of muscle forces and joint function. Development of a mechanical lion capable of walking.	Leonardo da Vinci
1514	Concept of a heliocentric solar system.	Copernicus
1539	An essay on bloodletting.	Andreas Vesalius
1543	*Basel De humani corporis fabrica libri septem* (Seven Books on the Construction of the Human Body) was published.	Andreas Vesalius
1590	Microscope was invented.	Jansen
1600	Theory of how living organisms form is explained by Spontaneous Generation.	
1604	The hydrostatic balance and universal law of acceleration were developed.	Galileo Galilei
1612	*Discourse on Bodies in Water* was published	Galileo Galilei
1632	Galileo father of mechanics and biomechanician.	Galileo Galilei
1640	Measurements of the orbits of satellites of Jupiter.	Giovanni Borelli
1630	plants and animals alike reproduce in a sexual manner: males contribute pollen or sperm; females contribute eggs.	William Harvey
1653	The cell was discovered by looking at a thin slice of cork through a microscope at 50x.	Robert C. Hooke
1660	He used a microscope to study blood circulation in capillaries; described the nervous system as bundles of fibers connected to the brain by the spinal cord; detailed the anatomy of the silkworm; described the development of the chick in its egg; and published his work on plant anatomy.	Marcello Malpighi
1665	He discovered called cells and wrote his book *Micrographia*.	Robert Hooke
1668	He observed that meat covered to exclude flies did not develop maggots, while similar uncovered meat did.	Francesco Redi
1673	He described protozoa and bacteria and recognized that such microorganisms might play a role in fermentation.	Anton van Leeuwenhoek

Year	Event/Concepts, etc.	Who
1680	First works on *De Motu Animalium*.	Borelli
1682	He observed fish blood cells with a microscope at 250x.	Leeuwenhoek
1700	They found that the human body is full of cells.	Luigi Galvani et al.
1701	He practiced "inoculation" – intentionally giving children smallpox to prevent a serious case later in life.	Giacomo Pylarini
1724	Cross-fertilization in corn was discovered.	–
1728	He developed an improved drill. He favored tin foil or lead cylinders for fillings. Father of modern dentistry,	Pierre Fauchard
1748	He concluded "there is a vegetative force in every microscopic Point of Matter...," in support of the idea of spontaneous generation.	Turbevill Needham
1750	Farmers increased the cultivation of leguminous crops and began rotation method to increase yield and land use.	Europe
1766	He observed that bacteria are replicated by dividing in a process known as binary fission. He developed methods to keep bacteria alive.	Abraham Trembley
1774	He designed the first hard-baked, rot-proof porcelain dentures.	Dubois de Chemant
1798	He published his book comparing vaccination to inoculation.	Edward Jenner
1799	He described experiments using "hermetically sealed" flasks heated in boiling water to test the possibility of using heat to kill all the microbes in an infusion.	Lazaro Spallanzani
1802	The first recorded use of the word biology.	–
1805	All life is based on individual cells	Oken
1809	He devised a technique using heat to can and sterilize food.	Nicolas Appert
1816	Amalgam was developed, with fears of leaky dental fillings and the effect of mercury on health surfacing quickly.	Auguste Taveau
1895	He standardized both cavity preparation and amalgam manufacturing.	G. V. Black
1816	Tariff Act excluded foreign plants and trees from U.S. import duties.	USA
1816	Rolled-up newspaper was used to listen to a patient's chest,	France
1820	He proposed that all cells form from other cells.	François-Vincent Raspail
1820	Prior to the eating of first tomato by him, the tomato was believed to be poisonous.	Robert Gibbon Johnson
1824	He discovered that tissues are composed of living cells.	Henri Dutrochet
1827	The worldwide search for the elusive mammalian egg ended with the first observation of canine eggs.	–

Year	Event/Concepts, etc.	Who
1831	He suggested importance of cell nucleus in fertilization by looking at cells of Asclepiads and orchids through a microscope.	Robert Brown
1837	He invented an improved porcelain tooth.	Claudius Ash
1838	Protein was discovered, named and recorded.	Gerardus Johannes Mulder
1838	They formulated the so-called "Cell Theory" based on their microscopic findings.	Schleiden and Schwann
1839	He proposed that all living things are entirely made of cells.	T. Schwann
1839	US Congress invested $1,000 in the Congressional Seed Distribution Program.	USA
1839	He discovered vulcanized rubber whose moldable base was available for false teeth.	Charles Goodyear
1844	Discovery of nitrous oxide for painless teeth extraction, dentures.	Horace Well
1850	He called the 19th century as the gait century.	Benno Nigg
1850	He correlated ground reaction forces with movement and pioneered modern motion analysis.	Etienne Marey
1850	They significantly advanced the science using recent advances in engineering mechanics.	Wilhelm Braune et al.
1850	Industrial revolution	France and Germany
1850	They compared the stress patterns in a human femur with those in a similarly shaped crane.	Karl Culmann et al.
1850	Wolff's Law of bone remodeling was described.	Julius Wolff
1850	He proposed the hypothesis that child-bed fever can be spread from mother to mother by physicians. He proposed that physicians wash their hands after examining each patient.	Ignaz Semmelweis
1850	Horse drawn harrows, seed drills, corn planters, horse hoes, 2-row cultivators, hay mowers, and rakes became popular.	Europe
1852	An international "Corn Show."	Paris
1855	He developed the cohesive gold foil method for dentistry.	Robert Arthur
1855	He developed a method to isolate the cell membrane.	Robert Remak
1855	The *Escherichia coli* bacterium was discovered.	–
1855	He began working with yeast, thus showed that these are living organisms.	Pasteur
1856	He discovered a technique for keeping animal organs alive outside the body,	Karl Ludwig
1856	He asserted that microbes are responsible for fermentation.	Louis Pasteur

Year	Event/Concepts, etc.	Who
1857	He discovered colloidal "ruby" gold, demonstrating that nanostructured gold under certain lighting conditions produces different-colored solutions.	Michael Faraday
1858	He described his ideas about cell formation using words, *"Omnis cellula e cellula..."*	Rudolf Virchow
1858	He endorsed cell division and its role in pathology.	Virchow
1859	He hypothesized that animal populations adapt their forms over time to best exploit the environment, a process he referred to as "natural selection." He theorized that only the creatures best suited to their environment survive to reproduce. "On the Origin of Species" was published by him.	Charles Darwin
1862	The Morrill Act established the "land-grant" colleges of agriculture and mechanical arts.	USA
1862	He discovered the bacterial origin of fermentation.	Louis Pasteur
1863	He proved that a fungus causes potato blight.	Anton de Bary
1863	He invented the process of pasteurization,	Louis Pasteur
1863	He discovered the laws of inheritance.	Gregor Mendel
1864	First centrifuge to separate cream from milk.	Antonin Prandtl
1864	Decayed organisms are found as small organized 'corpuscles' or 'germs' in the air.	Louis Pasteur
1865	Disinfectants were first used in wound care and surgery.	Joseph Lister
1865	He investigated silkworm disease.	Louis Pasteur
1868	He used heat treatment to cure a plant of bacterial infection.	Davaine
1868	He isolated nuclein.	Fredrich Miescher
1870	He discovered mitosis.	W. Flemming
1871	*"The Descent of Man and Selection Relation to Sex"* was published.	Darwin
1871	DNA was isolated from the sperm of trout found in the Rhine River.	–
1871	He discovered invertase.	Ernst Hoppe-Seyler
1873	He investigated anthrax and developed techniques to view, grow, and stain organisms.	Robert Koch
1874	He discovered the important influence of granulation tissue on wound healing.	Thiersch
1875	He proposed the idea of "gemmules."	Charles Darwin

Year	Event/Concepts, etc.	Who
1878	He described the "most probable number" technique to isolate pure cultures of bacteria.	Joseph Lister
1878	He suggested that weeds were actually plants "whose virtues have not yet been discovered."	Ralph Waldo Emerson
1878	He discovered chromatin.	Walther Flemming
1879	He began his studies of nuclein.	Albrecht Kossel
1879	He developed crosses of corn.	William James Beal
1881	He used attenuation to develop vaccines against the bacterial pathogens of fowl cholera and anthrax in chickens.	Pasteur
1881	He described bacterial colonies growing on potato slices, on gelatin medium, and on agar medium.	Robert Koch
1882	He observed phagocytes surrounding microorganisms in starfish larvae.	Ilya Metchnikoff
1882	He described the bacterium that causes tuberculosis in humans.	Robert Koch
1882	He recorded the first extensive study on the origins and history of cultivated plants.	Alphonse de Candolle
1882	He reported his discovery on chromosomes and mitosis.	Walther Flemming
1883	He coined the term "germ-plasm."	August Weismann
1883	He coined the term "eugenics."	Francis Galton
1884	He developed a rabies vaccine.	Pasteur
1884	He described Gram stain.	Christian Gram
1884	He stated his "postulates" for testing whether a microbe is the causal agent of a disease.	Robert Koch
1885	1885–1895: They identified a host of human disease-causing organisms.	Robert Koch et al.
1885	He developed the first antitoxin, for diphtheria.	Emil von Behring
1885	They built and used the first artificial heart-lung apparatus for organ perfusion studies.	M. von Frey and M. Gruber
1886	He demonstrated that pear fire blight is a bacterial disease.	J.C. Arthur
1887	He discovered that each species has a fixed number of chromosomes.	Edouard-Joseph-Louis-Marie van Beneden

Year	Event/Concepts, etc.	Who
1887	He described Petri plates.	R.J. Petri
1888	The chromosome was discovered.	Waldyer
1891	1891–1950s: There was little improvement in knee replacement surgery from the late 19th century through the 1950s. Gluck's ivory and plaster technique was updated several times using metal and plastic components, but they were still formed into a hinge-type device.	Gluck
1891	He performed first knee surgery.	Theophilus Gluck
1892	Allometry was first outlined.	Otto Snell
1892	He reported that the causal agent of the tobacco mosaic disease is transmissible and can pass through filters that trap the smallest bacteria.	Ivanovsky
1893	Scientists isolated the diphtheria antitoxin.	Lister Institute
1895	He described complex organ perfusion apparatus.	C. Jacob
1895	He standardized both dental cavity preparation and amalgam manufacturing.	G. V. Black
1895	He discovered a form of electromagnetic radiation that could pass through the body and leave an image of its bones or organs on a photographic plate.	Wilhelm C. Roentgen
1895	He demonstrated nitrogen fixation in the absence of oxygen by Clostridia bacteria.	Winogradski
1896	He elaborated on August Weismann's chromosome theory of heredity.	E. B. Wilson
1896	He developed cholera and typhoid vaccines.	Wilhelm Kolle
1897	He demonstrated that fermentation can occur with an extract of yeast in the absence of intact yeast cells.	Eduard Buchner
1897	They coined the term "filterable viruses."	Friedrich Loeffler and P. Frosch
1897	He first reported the idea of growing cells outside the human body.	Loeb
1897	He discovered Plasmodium.	Ronald Ross
1898	He discovered the Golgi apparatus.	Camillo Golgi
1899	He proposed that the virus becomes incorporated into the protoplasm of the host plant.	Beijerinck
1899	He indicated that application of an electrical impulse to the human heart in asystole caused a ventricular contraction.	J. A. McWilliam
1900	1900–1953: Mechanism of reproduction, the nature and structure of DNA were explored.	–

Year	Event/Concepts, etc.	Who
1900	Key industrial chemicals (glycerol, acetone, and butanol) could be generated using bacteria.	–
1900	Introduction of large-scale sewage purification systems based on microbial activity.	–
1900	The science of genetics was born.	Hugo DeVries, Erich Von Tschermak, and Carl Correns
1900	He established that yellow fever is transmitted by mosquitoes.	Walter Reed
1900	He observed homologous pairs of chromosomes in grasshopper cells.	William Sutton
1901	He identified free-living aerobic nitrogen fixers.	Beijerinck
1901	He discovered "a new substance indispensable for the development of yeast."	E. Wildiers
1902	He developed the connection between Mendelian heredity and the biochemical pathways of reproduction in the individual organism.	Archibald Garrod
1902	He stated that chromosomes are paired and may be the carriers of heredity; and developed the chromosomal theory of heredity.	Walter S. Sutton
1903	He developed first electrocardiograph machine.	Willem Einthoven
1903	They proposed that each egg or sperm cell contains only one of each chromosome pair.	Walter Sutton and Theodor Boveri
1904	He demonstrated that some characteristics are not independently inherited. This led to the concept of gene linkage and genetic maps.	William Bateson
1905	He demonstrated that some genes can modify the action of other genes.	William Bateson
1905	They proposed the idea that separate X and Y chromosomes determine sex.	Edmund Wilson and Nellie Stevens
1906	First hygienist course in 1913 in Bridgeport–Connecticut, USA.	USA
1906	He discovered the beneficial properties of Salvarsan	Paul Erlich
1907	He grow frog ectodermal cells *in vitro*.	Harrison
1907	He established mutation theory.	Thomas Morgan
1908	He described "inborn errors of metabolism."	A. E. Garrod
1908	He developed a vaccine against TB (BCG).	Calmette and Guerin

Year	Event/Concepts, etc.	Who
1909	He coined the terms gene, genotype, and phenotype.	Wilhelm Johannsen
1909	He discovered that the sugar ribose is found in some nucleic acids.	Phoebus Levene
1909	He first applied Mendel's laws to animals.	William Bateson
1910	He proved that genes are carried on chromosomes. He also demonstrated the existence of sex-linked genes,	Thomas Morgan
1911	He began to map the positions of genes on chromosomes of the fruit fly.	Thomas Morgan
1912	He grew pieces of chick embryo in various media,	Carrel
1912	He discovered that X-rays can be used to study the molecular structure of simple crystalline substances.	Lawrence Bragg
1913	He constructed the first gene map by analyzing mating results for fruit flies; and calculated the exact percentage of crossing-over between the genes.	Alfred H. Sturtevant
1913	They first described *in vivo* hemodialysis of rabbits, dogs.	J. J. Abel et al.
1914	Bacteria was used to treat sewage.	England
1915	He discovered a bacterial disease called "glassy transformation" and showed the disease agent is transmissible and filterable.	Frederick Twort
1915	They reported the use of a screen oxygenator for perfusion of isolated organs in which venous blood flows by gravity down a cloth in an oxygen-rich atmosphere.	N. Richards et al.
1915	Phages or bacterial viruses were discovered.	–
1916	He published inaugural issue of the journal Genetics.	George H. Shull
1916	They discovered that trypsin is capable of degrading matrix proteins.	Rous and Jones
1916	Both discovered anticoagulant heparin.	McLean and Howell
1917	He analyzed the inheritance of coat colors in guinea pigs, mice, rats, rabbits, horses, and other mammals. He suggested that each step in this biochemical process was mediated by a different, specific enzyme.	Sewall Wright
1917	He demonstrated the rearrangement of chromosomes known as 'crossing over.'	Plough
1918	World-wide epidemic of influenza killed 20 million people	
1918	He discovered that human cells contained 48 chromosomes (of course, not true).	Herbert M. Evans

Year	Event/Concepts, etc.	Who
1918	Yeast was grown in large quantities to produce glycerol; and activated sludge was produced in large quantities for sewage treatment.	–
1919	He coined The term "biotechnology." Ereky foresaw a biochemical age.	Károly Ereky
1920	Plant hybridization became widespread.	USA
1921	He published on the nature of the gene.	Hermann J. Muller
1925	Congress voted to cut the Seed Distribution Program.	USA
1925	He performed first clinical hemodialysis of 5 patients, using a modification of the Hopkins artificial kidney.	G. Haas
1926	He founded the Hi-Bred Company: a hybrid corn seed producer.	Henry A. Wallace
1926	He discovered that x-rays induced genetic mutations in fruit flies 1,500 times more quickly than under normal circumstances	Hermann Muller
1926	They devised a portable apparatus in which "One pole was applied to a skin pad soaked in strong salt solution" while the other pole was plunged into the appropriate cardiac chamber."	Edgar H Booth, Mark C. Lidwell
1926	He developed the first "sandwich" artificial kidney using a biological membrane consisting of peritoneal membrane of calf. Dogs were dialyzed.	H. Necheles
1926	He published 'the theory of the gene'	Thomas H. Morgan
1927	They devised first respirator using an iron box and two vacuum cleaners.	Philip Drinker et al.
1928	He described physical laws governing how atoms are arranged within molecules.	Linus Pauling
1928	He discovered penicillin. He noticed that all the bacteria in a radius surrounding a bit of mold in a petri dish had died.	Alexander Fleming
1928	He noticed that a rough type of bacterium changed to a smooth type when an unknown "transforming principle" from the smooth type was present.	Fredrick Griffiths
1928	They described a double perfusion pump (for pulmonary and systemic circulations) relying on compressible diaphragms to circulate de-fibrinated blood during organ perfusion experiments	H. H. Dale et al.
1928	He showed that ultraviolet radiation can also cause mutations.	Lewis Stadler
1928	He discovered a deoxyribose in nucleic acids that do not contain ribose.	Phoebus Levene

Year	Event/Concepts, etc.	Who
1929	They maintained temporary function of guillotined dogs' heads using donor lungs for gas exchange and a bellows-type pump for blood circulation.	S. Brukhonenko et al.
1930	1930s: He invented an artificial pacemaker to resuscitate patients.	Albert S. Hyman
1930	1930s: They constructed a hand-cranked apparatus with a spring motor that turns a magnet to supply an electrical impulse.	Hyman & his brother
1930	1930s: Physicians were looking into the possibility of extracorporeal circulation, or blood flow outside of the body.	–
1931	He made his mission to come up with an artificial heart-lung machine that would keep a patient alive during heart surgery.	John H. Gibbon
1931	Thirty states in the U.S. had adopted compulsory sterilization laws.	USA
1932	He described an electro-mechanical instrument, powered by a spring-wound hand-cranked motor; thus artificial pacemaker was introduced.	Albert Hyman
1933	He wrote that he had charted perceptible differences among chromosomes under the microscope – differences.	T.S. Painter
1933	They discovered that while a low-voltage shock can cause ventricular fibrillation, or arrhythmia, a second surge of electricity, or counter-shock, can restore the heart's normal rhythm and contraction.	William B. Kouwenhoven et al.
1934	He showed that giant molecules, such as proteins, can be studied using x-ray crystallography.	Desmond Bernal
1934	He built a prototype of his heart-lung machine and tested its function on cats.	John H. Gibbon
1934	He described dual-roller pump for transfusion of blood.	M. E. DeBakey
1934	He purified bacteriophage and found about equal amounts of protein and DNA.	Martin Schlesinger
1934	He isolated DNA in the pure state for the first time.	A. N. Belozersky
1935	They transplanted tissue between larvae of fruit flies bred with various mutations of eye color	George Beadle et al.
1935	They developed a machine that could replace the function of heart and lungs of a cat for 20 min.	Gibbon & his wife.
1935	He crystallized tobacco mosaic virus. He also isolated nucleic acids from the tobacco mosaic virus.	Wendell M. Stanley
1936	He invented the field emission microscope, allowing near-atomic-resolution images of materials.	Erwin Müller

Year	Event/Concepts, etc.	Who
1937	He built the first heart-lung machine that had sufficient capacity to perfuse organs in cats.	John H. Gibbon
1937	He discovered that tobacco mosaic virus contains RNA.	Fred C. Bawden
1937	He employed an extracorporeal assist device for 5.5 hours to substitute for the cardiac function of a dog.	V. Demikhov
1937	They isolated the antimicrobial agent penicillin.	Howard Florey et al.
1938	Proteins and DNA were studied with X-ray crystallography.	–
1938	The term "molecular biology" was coined.	–
1939	He showed that both DNA and RNA are always present in bacteria.	A. N. Belozersky
1939	He cultivated carrot callus cultures.	Gauteret
1939	He performed the ligation of a patent ductus arteriosus.	Robert E. Gross
1939	He performed the first hemodialysis of a dog using cellophane membrane and heparin anticoagulation.	W. Thalheimer
1940	Large-scale production of penicillin.	–
1940	He developed a screw type dental implan40	Formiggini
1940	Heparin is used for the first time in human patients.	Gordon Murray
1941	They developed the "one-gene-one-enzyme" hypothesis. They hypothesized that the X-rays had damaged the genes that synthesized the proteins.	George Beadle et al.
1941	The term "genetic engineering" was coined.	Denmark
1942	The electron microscope is used to identify and characterize a bacteriophage – a virus that infects bacteria.	–
1943	Cortisone was first manufactured in large amounts.	–
1943	They performed "the fluctuation test": the first quantitative study of mutation in bacteria.	Salvador Luria et al.
1943	He developed a rotating drum artificial kidney.	W. Kolff
1943	He developed dialyzer.	W. Kolff
1944	He discovered that genes can be transposed from one position to another on a chromosome.	Barbara McClintock
1944	He used chromatography to determine the amino acid sequences of the bovine insulin molecule.	Frederick Sanger
1944	They determined that DNA is the hereditary material involved in transformation in pneumococcus bacteria.	Oswald T. Avery et al.
1944	He isolated streptomycin, an effective antibiotic for TB.	Waksman

Year	Event/Concepts, etc.	Who
1945	Isolated animal cell cultures were grown in laboratories.	–
1945	He developed a stationary drum artificial kidney.	G. Murray
1945	They developed cardiopulmonary bypass machine without the filter.	Gibbon & Watson
1945	They added the 300 mesh filter to their cardiopulmonary bypass machine.	Gibbon & Watson
1945	They developed a model using phage for studying how genetic information is transferred to host bacterial cells.	Max Delbruck et al.
1945	He developed first stationary drum artificial kidney and the first artificial ultrafiltration kidney.	N. Alwall
1945	The U.N. Food and Agriculture Organization (FAO) was formed in Quebec.	Canada
1945	He successfully treated a dying patient with his artificial kidney.	Willem J. Kolff
1946	From Japan, he sent home Norin 10: source of the dwarfing gene.	D. C. Salmon
1946	They showed that bacteria sometimes exchange genetic material directly, in a process they called conjugation.	Edward Tatum et al.
1946	They discovered that the genetic material from different viruses can be combined to form a new type of virus.	Max Delbruck et al.
1946	They helped Gibbon to develop a heart-lung machine that minimized hemolysis and prevented air bubbles from entering the circulation.	Watson et al.
1947	She first reported on "transposable elements" – known today as "jumping genes."	Barbara McClintock
1947	They discovered the semiconductor transistor.	John Bardeen et al.
1948	He received a patent for a plastic contact lens designed to cover only the eye's cornea.	Kevin Touhy
1949	They developed the first practical flat-plate dialyzer.	Skeggs & Leonards
1950	He introduced a lens that was molded to fit the cornea's contours rather than lie flat atop it.	George Butterfield
1950	An external pacemaker was designed and built by him.	John Hopps
1950	Artificial insemination of livestock using frozen semen was successfully accomplished.	–
1950	They studied monkey, mouse, and chick cells in cell cultures.	Earle and Enders
1950	Chargaff's Rules: In DNA the amounts of adenine and thymine are about the same, as are the amounts of guanine and cytosine.	Erwin Chargaff

Year	Event/Concepts, etc.	Who
1950	They developed the theory and a process for growing mono-disperse colloidal materials.	Victor La Mer et al.
1950	They manufactured first commercial dialyzing machine: the Baxter/Travenol recirculating U-200 twin-coil dialyzer.	Baxter Corp.
1950	He applied engineering principles to orthopedics and developed the first artificial hip replacement procedure, or arthroplasty.	Charnley
1950	Organ transplantation with identical twins.	–
1950	The resistance of muscle and nervous tissue to electrical current was discovered by him, leading to electrophysiology.	Roentgen
1951	He developed an artificial heart valve and performed the first artificial valve implantation surgery in a human patient (1952).	Charles Hufnagel
1951	He replaced the abdominally aorta.	Dubost
1951	She discovered lambda phage, a virus of *E. coli*.	Esther M. Lederberg
1952	They performed the "blender experiments" using phages.	Alfred Hershey et al.
1952	Arterial graphs made of cloth were first described by them.	Voorhees et al.
1952	Discovery of Osseo-integration. He also tested a type of dental implant using pure titanium screws.	Brånemark
1952	Electron microscopy showed the inside of cells to be filled with minute but well-formed anatomical structures, including vast numbers of a complex molecular organ now termed the ribosome.	–
1952	He contributed greatly to the use of human embryonic cells.	Enders
1952	He introduced the term plasmid to describe the bacterial structures, which contained extrachromosomal genetic material.	J. Lederberg
1952	RNA, a nucleic acid, plays a part in the synthesis of proteins.	Jean Brachet
1952	They showed that bacteria sometimes exchange genes by an indirect method called transduction.	J. Lederberg et al.
1952	Kolff-Brigham artificial kidney was used by the U.S. Army in Korea.	Korea
1952	They published clear images of 50 nanometer diameter tubes made of carbon.	Radushkevich et al.
1952	They developed the first successful cardiac pacemaker.	Paul M. Zoll et al.
1952	A battery-powered external pacemaker was developed.	Earl Bakken et al.
1952	He discovered conjugation, the process whereby one bacterial cell pipes a copy of some of its genes into a second bacterial cell.	William Hayes

Year	Event/Concepts, etc.	Who
1953	1953–1976: The boundaries of DNA research were expanded. The 3-D structure of DNA was revealed.	–
1953	Gibbon's heart-lung machine was used in open-heart surgery by Gibbon Jr. The patient, Cecelia Bavolek, whose heart was connected to the machine for 45 min, recovered fully from the operation.	J. Gibbon, Jr.
1953	He developed the HeLa human cell line.	Gey
1953	They discovered the F factor in *E. coli*.	L. Cavalli et al.
1953	The double-stranded, helical, complementary, anti-parallel model for DNA was proposed by them.	Watson & Crick
1953	He discovered that plasmids can be used to transfer introduced genetic markers from one bacterium to another.	William Hayes
1954	The word bioengineering was coined by him.	Heinz Wolff
1954	His team performed the first human kidney transplant.	Joseph E. Murray
1954	Intra-cardiac surgery was performed by him using cross circulation from a healthy donor.	C. Laughton Lillehei
1954	His original papers on knee replacement surgery were published in the Journal of Bone and Joint Surgery.	Leslie G. Percival-Shiers
1955	American Society of Artificial Internal Organs was organized, "promote the increase of knowledge about artificial internal organs and of their utilization." First meeting of ASAIO was held at the Hotel Chelsea in Atlantic City, New Jersey with 67 founding members.	www.asaio. com
1955	Landmark publications: Clowes, G. H. A. Jr., Hopkins, A. L., & Kolobow, T. Oxygen diffusion through plastic films. *Tr Am Soc Artif Intern Org 1*, 23–24; Gibbon, J. H. Jr. Artificial heart-lung machines: Chairman's address. *Tr Am Soc Artif Intern Org 1*, 58–62; Kolff, W. J. The artificial kidney – past, present and future. *Tr Am Soc Artif Intern Org 1*, 1–7.	Trans. ASAIO
1955	He devised an experimental setup to map mutations within a short genetic region of a particular bacterial virus.	Seymour Benzer
1956	They developed a disposable twin-coil dialyzer.	W. Kolff et al.
1956	He introduced concepts of "molecular engineering" as applied to dielectrics, ferroelectrics, and piezoelectrics.	Arthur von Hippel
1956	He pioneered the field ion microscope.	Erwin Müller
1956	He developed first successful membrane oxygenator.	G. Clowes
1956	He took apart and reassembled the tobacco mosaic virus, demonstrating "self-assembly."	Heinz F. Conrat

Year	Event/Concepts, etc.	Who
1956	Clark, L. C. Jr. Monitor and control of blood and tissue oxygenation. *Tr Am Soc Artif Intern Org, 2*, 41–45.	Trans. ASAIO
1956	The silicon transistor was developed and available commercially.	–
1957	Helix reservoir disposable bubble oxygenator was developed.	De Wall-Lillehei
1957	First transistorized, battery-powered, wearable pacemaker was developed.	William L. Weirich
1957	They explained how DNA functions to make protein. They also suggested that genetic information flows only in one direction, from DNA to messenger RNA to protein.	Francis Crick et al.
1957	Edwards, W. S., Tapp, J. S. Two and a half years' experience with crimped nylon grafts. *Tr Am Soc Artif Intern Org, 3*, 70–77.	Trans. ASAIO
1957	They demonstrated the replication mechanism of DNA.	Meselson et al.
1957	They successfully implanted a hydraulic, polyvinyl chloride total artificial heart in a dog, keeping the animal alive for 90 minutes.	Kolff et al.
1958	He constructed an external pacemaker weighing 45 kg and powered by a 12-volt auto battery.	Jorge R. Pombo
1958	He introduced many concepts and coined the term "molecular engineering."	Arthur Von Hippel
1958	He discovered and isolated DNA polymerase.	Coenberg
1958	He produced first wearable external pacemaker.	Earl Bakken
1958	They designed fully implantable pacemaker in Sweden.	Rune Elmqvist et al.
1958	He was the world's first implantable pacemaker patient.	Arne Larsson
1958	Hufnagel, C., Villegas, A. Aortic valvular replacement. *Tr Am Soc Artif Intern Org., 4*, 235–239.	Trans. ASAIO
1958	The term bionics was coined by him.	Jack E. Steele
1959	He delivered his talk describing molecular machines building with atomic precision.	Feynman
1959	They established the existence of genetic regulation—mappable control functions located on the chromosome in the DNA sequence—which they named the repressor and operon.	Francois Jacob et al.
1959	He demonstrated first temporary transvenous pacing with a catheter electrode inserted via the patient's basilic vein.	Richard Furman
1959	He built the first integrated circuit, for which he received the Nobel Prize in 2000.	Jack Kilby
1959	He introduced hybrid corn to the Soviet Union.	Nikita Krushchev

Year	Event/Concepts, etc.	Who
1959	He gave first lecture on *There's Plenty of Room at the Bottom* at American Physical Society meeting.	Richard Furman
1959	Systemic fungicides were developed.	–
1959	The steps in protein biosynthesis were delineated.	–
1960	They discussed friction properties of carbon due to rolling sheets of graphene in nature.	Bollmann et al.
1960	The Rockefeller and Ford Foundation jointly established the International Rice Research Institute (IRRI).	Rockefeller
1960	Further laboratory research studying function and improvement of membrane lungs was undertaken.	Kolobow et al.
1960	Biocompatible inert material, ceramic, was developed and was used as dental implant soft tissue interface.	–
1960	He used cooling blankets and ice bags to cool the dogs to 20°C.	Bill Bigelow
1960	It was considered safe to use the CPB along with hypothermia to perform CABG surgery. Heat exchangers were used to cool and rewarm the blood.	–
1960	He developed the first totally internal pacemaker using two commercial silicon transistors.	Wilson Greatbatch
1960	He implanted Greatbatch's device into 10 fatally ill patients. The first lives for 18 months, another for 30 years.	William Chardack
1960	First successful intracardiac prosthetic valve operation was conducted by him.	Albert Starr
1960	Quinton, W., Dillard, D., Scribner, B. H. Cannulation of blood vessels for prolonged hemodialysis. *Tr Am Soc Artif Intern Org, 6*, 104–113; Scribner, B. H. et al. The treatment of chronic uremia by means of intermittent hemodialysis: A preliminary report. *Tr Am Soc Artif Intern Org, 6*, 114–122.	Trans. ASAIO
1960	He developed first hollow fiber dialyzer.	R. Stewart
1960	He developed a double-cage design of heart valve, in which the outer cage separates the valve struts from the aortic wall.	Dwight Harken
1960	They designed a silicone ball inside a cage (heart valve) made of stellite-21	Starr & Edwards
1960	Their shunt made chronic renal dialysis possible.	Quinton & Scribner
1960	His success with hip replacement.	John Charnley
1960	In 1960s, attempts were made to design knee replacements. Knee replacements with hinged implants were available. However, hinged implants did not allow bending or rotation of knees. These would often come apart soon after surgery and had a high infection rate.	USA

Year	Event/Concepts, etc.	Who
1960	His design of knee allowed for unicompartmental operations.	Leonard Marmor
1961	He built a strand of mRNA comprised only of the base uracil, called poly-u. He also discovered that UUU is the codon for phenylalanine.	Marshall Nirenberg
1961	International Union for the Protection of New Varieties of Plants	Paris
1962	He introduced screw type of dental implant made of chrome cobalt.	Chercheve
1962	They developed the subcutaneous arterio-venous shunt for chronic hemodialysis.	Cimino et al.
1962	He devised a low-friction, high-density polythene suitable for artificial hip joints and pioneered the use of methyl methacrylate cement for holding the metal prosthesis, or implant, to the shaft of the femur.	John Charnley
1962	Japanese Society for Artificial Organs	www.jsao.org
1962	The first successful kidney transplant in unrelated humans is performed by him.	Joseph Murray
1962	Green Revolution grains.	Mexico
1962	The discovery of the double helical structure of DNA.	Rosalind Franklin
1963	They fabricated a tubular left ventricular assist device (LVAD), which was implanted by S. Crawford and M.E. DeBakey.	D. Liotta et al.
1963	He began working with a ruby laser photo-coagulator to treat diabetic retinopathy.	Francis L'Esperance
1965	They designed an argon laser for eye surgery.	Esperance et al.
1963	Kolobow, T., Bowman, R.L. Construction and evaluation of an alveolar membrane heart lung. *Tr Am Soc Artif Intern Org, 9,* 238–245; Nosé, Y. et al. An experimental artificial liver using extracorporeal metabolism with sliced or granulated canine liver. *Tr Am Soc Artif Intern Org., 9,* 358–362.	Trans. ASAIO
1963	Pacemaker with capability to synchronize ventricular stimuli to atrial activation was introduced by him.	Parsonnet
1964	National Heart and Lung Institute established U.S. Artificial Heart Program.	USA
1964	The National Heart, Lung and Blood Institute sets a goal of designing a total artificial heart by 1970.	USA
1964	Gott, V. L. et al. Techniques of applying a graphite-benzalkonium-heparin coating to various plastics and metals. *Tr Am Soc Artif Intern Org., 10,* 213–217.	Trans. ASAIO

Year	Event/Concepts, etc.	Who
1964	Proportioning pump dialysis machine with safety monitors was developed in Seattle. First long term unattended overnight dialysis with this machine begun in Seattle.	USA
1964	Crop Research and Introduction Center at Izanir, Turkey was established.	FAO
1964	The first commercial myoelectric arm was developed by the Central Prosthetic Research Institute of the USSR.	USSR
1964	The genetic code was understood.	–
1965	They successfully fused mouse and human cells.	Harris and Watkins
1965	He foresaw several trends in the field of electronics. Moore's Law described the density of transistors on an integrated chip (IC) doubling every 12 months.	Gordon Moore
1965	Genes conveying antibiotic resistance in bacteria are often carried on small, supernumerary chromosomes called plasmids.	–
1966	He successfully implants a partial artificial heart at Houston.	Michael DeBakey
1966	He successfully implanted into a patient an aortic U-shaped auxiliary ventricle.	A. Kantrowitz
1966	Chang, T. M. Semi-permeable aqueous micro capsules (artificial cells) with emphasis on experience in extracorporeal systems. *Tr Am Soc Artif Intern Org., 12,* 13–19. Eschbach, J. W. et al. Unattended overnight home hemodialysis. *Tr Am Soc Artif Intern Org, 12,* 346–366.	Trans. ASAIO
1966	*Artificial Organ History: A Selective Timeline,* by http://www.echo.gmu.edu/bionics/exhibits.htm	George Mason University
1966	He developed a blade type dental implant of chromium nickel and vanadium.	Linkow
1966	He performed the first clinical implantation of a ventricular assist device (a pneumatically driven para-corporeal diaphragm pump).	M.E. DeBakey
1967	He demonstrated clinical effectiveness of the intraaortic balloon pump in cardiogenic shock patients, with its potential for treatment in acute heart failure.	A. Kantrowitz
1967	He conducted a study using one strand of natural viral DNA to assemble 5,300 nucleotide building blocks.	Arthur Kornberg
1967	He performed the world's first adult heart transplant on Louis Washkansky	Christiaan Barnard
1967	He performed the first pediatric heart transplant in the world.	Adrian Kantrowitz

Year	Event/Concepts, etc.	Who
1967	He used acrylic resin in the form of a tooth and tested these dental implants on monkeys.	Hodosh
1967	Henderson, L. W. et al. Blood purification by ultrafiltration and fluid replacement (diafiltration). *Tr Am Soc Artif Intern Org., 13,* 216–225; Menno, A. D. et al. An evaluation of the radial arterio-venous fistula as a substitute for the Quinton shunt in chronic hemodialysis. *Tr Am Soc Artif Intern Org., 13,* 62–76.	Trans. ASAIO
1967	They pioneered in human gene mapping with the publication of a technique for using human cells and mouse cells grown together in one culture. This was called somatic-cell hybridization.	Mary Weiss et al.
1967	The double-leaflet prosthetic cardiac valve was described by them.	Lillehei et al.
1967	The first automatic protein sequencer was perfected.	–
1967	The first biological engineering program in USA was created at Mississippi State University	USA
1967	The first clinical use of the capillary fiber kidney developed by Stewart.	R. Stewart
1968	First bone-marrow transplant.	–
1968	A nuclear powered energy cell for implantable cardiac devices in animals was described by him at ASAIO meeting.	J. Norman
1968	The first meeting of the Biomedical Engineering Society was held at the Ritz-Carlton Hotel in Atlantic City on April 17, 1968.	bmes.org
1968	He performed the first adult heart transplant in the United States on January 6, 1968	Norman Shumway
1968	Chardack, W. M. et al. Experimental observations and clinical experiences with the correction of complete heart block by an implantable self-contained pacemaker. *Tr Am Soc Artif Intern Org, 7,* 286–295. Kantrowitz, A. et al. Clinical experience with cardiac assistance by means of intra-aortic phase-shift balloon pumping. *Tr Am Soc Artif Intern Org, 14,* 344–348.	Trans. ASAIO
1968	The first successful human liver transplant was reported by him.	Thomas Starzl
1969	encasing The pacemaker generator was encased in a hermetically sealed titanium metal case by Telectronics of Australia (1969) and Cardiac Pacemakers Inc. (1972).	Australia
1969	Programmable pacemakers were introduced in the mid-1970s.	USA
1969	An enzyme was synthesized *in vitro* for the first time.	–
1969	Dr. Denton Cooley planted an artificial heart in him who lived 65 days.	Haskell Karp

Year	Event/Concepts, etc.	Who
1969	He implanted a pneumatically powered heart designed by D. Liotta.	D. A. Cooley
1969	Cooley, D. A. et al. First human implantation of cardiac prosthesis for staged total replacement of the heart. *Tr Am Soc Artif Intern Org., 15*, 252–263; Kwan-Gett, C. S. et al. Total replacement artificial heart and driving system with inherent regulation of cardiac output. *Tr Am Soc Artif Intern Org, 15*, 245–251.	Trans. ASAIO
1969	He developed the fluorescence-activated cell sorter, which could identify up to 5,000 closely related animal cells.	Leonard Herzenberg
1970	They first isolated "reverse transcriptase" a restriction enzyme that cuts DNA molecules at specific sites.	Howard Temin et al.
1970	Schuder, J. C. et al. Experimental ventricular defibrillation with an automatic and completely implanted system. *Tr Am Soc Artif Intern Org., 16*, 207–212.	Trans. ASAIO
1970	They discovered the first oncogene in a virus.	Peter Duesberg et al.
1970	They published the first method for staining human or other mammalian chromosomes.	T. Caspersson et al.
1970	1970s to early 1980s: The "Geometric" design and Condylar Knee design by John Insall found favor. Total condylar knee was developed at Massachusetts General Hospital. By 1974, replacing the patella femoral joint and either preserving or sacrificing the cruciate ligaments became a standard practice.	John Insall & others
1974	The mobile-bearing knee replacement was designed and developed by his team.	Fred Buechel
1970	1970s: Advances in fiber-optics technology give surgeons a view into joints and other surgical sites through an arthroscope.	
1970	1970s: artificial skin and biohybrid pancreas were developed from cells combined with biomaterials.	
1971	The dynamic aortic patch (CardioVad) was implanted.	A. Kantrowitz
1971	Softlens, the first soft contact lens, license was granted.	Bausch & Lomb
1971	They invented "hydrophilic" gel: suitable for eye implants.	Otto Wichterle et al.
1972	They requested the National Institutes of Health to enact guidelines for DNA splicing.	Paul Berg & others
1972	The first successful DNA cloning experiments were performed.	California
1972	CAT or CT scan was introduced for medical filming.	USA
1972	End-Stage Renal Disease Act was passed ensuring federal support for chronic kidney disease management.	USA

Year	Event/Concepts, etc.	Who
1972	He reported observing genes that control immune responses to foreign substances.	Hugh McDevitt
1972	They reported first successful clinical case using extracorporeal membrane oxygenation (ECMO) for respiratory failure.	Hill et al.
1972	He isolated and employed a restriction enzyme to cut DNA. He used ligase to paste two DNA strands together to form a hybrid circular molecule	Paul Berg
1972	He donated $500,000 to establish biomedical engineering research at Purdue	William Hillenbrand
1972	Grace Showalter endowed the Showalter Distinguished Professorship of Biomedical Engineering.	Showalter
1972	The DNA composition of chimpanzees and gorillas was discovered to be 99% similar to that of humans.	–
1972	First Lithium iodide cell powered pacemaker was developed by Cardiac Pacemakers Inc., Minneapolis, USA.	USA
1972	They performed first recombinant DNA experiment, using bacterial genes.	Cohen and Boyer
1973	Ames's test to identify chemicals that damage DNA.	Bruce Ames
1973	They "spliced" sections of viral DNA and bacterial DNA with the same restriction enzyme, creating a plasmid with dual antibiotic resistance.	Stanley Cohen et al.
1973	Scientists for the first time successfully transferred DNA from one life form into another.	–
1973	The first human-gene mapping conference took place.	–
1974	European Society for Artificial Organs, Geneva	www.esao.org
1974	He defined "Biomechanics is the study of the structure and function of biological systems by means of the methods of mechanics."	Herbert Hatze
1974	Dobell, W. H. et al. The directions for future research on sensory prostheses. *Tr Am Soc Artif Intern Org., 20,* 425–429.	Trans. ASAIO
1974	Lung Division of the National Heart and Lung Institute proposed a multicenter prospective randomized study of ECMO	USA
1974	His team founded the Biomedical Engineering Center. Leslie Geddes was the first Director.	Charles Babbs et al.
1974	The first bio-cement for industrial applications was invented.	
1974	They demonstrated the expression of a foreign gene implanted in bacteria by recombinant DNA methods	Cohen & Boyer
1974	He coined the term nanotechnology to describe precision machining of materials to within atomic-scale dimensional tolerances.	Norio Taniguchi

Year	Event/Concepts, etc.	Who
1975	His team developed dental implants made of a biocompatible material, vitreous carbon.	Hodosh et al.
1975	They fused cells together to produce monoclonal antibodies.	Kohler and Milstein
1975	Five different membrane oxygenators for ECMO were manufactured.	–
1975	A moratorium on recombinant DNA experiments was called for at an international meeting at Asilomar, California,	USA
1975	Bartlett, R. H. et al. Extracorporeal membrane oxygenations (ECMO) cardiopulmonary support in infancy. *Tr Am Soc Artif Intern Org, 22,* 80–93.	Trans. ASAIO
1975	Method for producing monoclonal antibodies was developed by them.	Köhler and Milstein
1975	First biomedical engineering course (EE522) was developed	Purdue Univ.
1975	Norman Weldon received first grant of $25,000, to create the best automatic implanted cardioverter defibrillator.	Purdue Univ.
1975	They reported first successful clinical case of neonatal ECMO.	Bartlett et al.
1975	The first NIH-sponsored multicenter clinical trial for temporary support in acute ventricular dysfunction (primarily postcardiotomy), using the model 7 and model 10 Axio-symmetric LVADs (Thermo Electron Corp) and Pierce-Donachy LVAD (Thoratec).	NIH – USA
1976	They founded Genentech, Inc.,	Boyer & Swanson
1976	International Society for Artificial Cells, Blood Substitutes, and Biotechnology	ISABB
1976	They showed that oncogenes appear on animal chromosomes, and alterations in their structure or expression can result in cancerous growth.	Bishop & Varmus
1976	They report CVD (Chemical Vapor Deposition) growth of nanometer-scale carbon fibers.	Oberlin et al.
1976	The Biomedical Engineering Center moved into the newly completed Potter Engineering Building.	Purdue
1976	The NIH released the first guidelines for recombinant DNA experimentation.	USA
1976	Medical Device Amendments to the Food, Drug, and Cosmetic Act were passed.	FDA
1976	FDA began regulating medical devices.	FDA
1976	Toxic Substances Control Act was passed.	U.S. Congress
1977	1977–1999: The Dawn of Biotech	
1977	He originates molecular nanotechnology concepts at MIT.	Drexler

Year	Event/Concepts, etc.	Who
1977	American Society of Biomechanics	ASB
1977	They isolated the gene for rat insulin.	Rutter and Goodman
1977	Genentech, Inc., reported the production of the first human protein manufactured in a bacteria.	Genentech
1977	A synthetic, recombinant gene was used to clone a protein.	–
1977	International Society for Artificial Organs that was later renamed as International Federation for Artificial Organs in 2004	www.ifao.org
1977	Rohde, T. D. et al. Protracted parenteral drug infusion in ambulatory subjects using an implantable infusion pump. *Tr Am Soc Artif Intern Org, 23,* 13–16.	Trans. ASAIO
1977	Sixteen bills were introduced in to regulate recombinant DNA research; and these were not passed.	US Congress
1977	The Heart and Vascular Division of the National Heart and Lung Institute issued a request for proposals for components of a totally implantable LVAD, and subsequently awarded multiple contracts, establishing the current generation of pulsatile systems.	NHLI
1977	They devised a method for sequencing DNA using chemicals rather than enzymes.	Gilbert & Maxam
1977	The similarity of human DNA composition to that of chimps and gorillas was noted.	–
1978	BioMedicus Biopump disposable centrifugal pump became available as an alternative to the roller pump for CPB.	H. Kletschka
1978	Continuous ambulatory parenteral dialysis was reported by him.	Oreopoulous
1978	Genentech, Inc. announced the successful laboratory production of human insulin using recombinant DNA technology.	Genentech
1978	He carried out first cochlear implant surgery.	Graeme Clarke
1978	Researchers used genetic engineering techniques to produce rat insulin.	Harvard
1978	A synthetic version of the human insulin gene was constructed and inserted into the bacterium *Escherichia coli.*	Herbert Boyer Lab.
1978	Portner, P. M. et al. An implantable permanent left ventricular assist system for man. *Tr Am Soc Artif Intern Org., 24,* 98–103.	Trans. ASAIO
1978	He showed that it was possible to introduce specific mutations at specific sites in a DNA molecule.	Hutchison et al.
1978	Scientists successfully transplanted a mammalian gene.	Stanford
1978	They found that when a restrictive enzyme is applied to DNA from different individuals, the resulting sets of fragments sometimes differ markedly from one person to the next.	Botstein & his team

Year	Event/Concepts, etc.	Who
1978	The first Dental implant Consensus Conference.	Harvard
1978	He performed heart transplant on Tony Huesman, who lived 11 more years.	Norman Shumway
1978	They cloned a coat protein of the virus that causes hepatitis B.	Rutter's lab.
1979	His fiction novel, Fountains of Paradise, popularized the idea of a space elevator	Arthur C. Clarke's
1979	Continuous arterial venous hemofiltration was described.	Kramer
1979	Cloning the gene for human growth hormone.	John Baxter
1980	Comprehensive Environmental Response, Compensation and Liability Act (CERCLA or Superfund Law) enacted into U.S. federal law.	USA
1980	Continuous AV hemofiltration was used for acute renal failure.	Paganini
1980	They invented a technique for multiplying DNA sequences *in vitro*, by the polymerase chain reaction (PCR).	Mullis & his team
1980	Researchers successfully introduce a human gene – one that codes for the protein interferon – into a bacterium.	USA
1980	The U.S. Supreme Court ruled that engineered life forms could be patented.	USA
1980	CMBE-SIG was formed which later became a special interest group under BMES in 2011.	BMES
1980	The U.S. patent for gene cloning is awarded.	Cohen and Boyer
1980	The U.S. Supreme Court ruled that genetically altered life forms can be patented.	USA
1980	Scientists found that by cooling the heart to below 28°C and treating it with the right cocktail of chemicals, the heart could be stopped for many hours.	St Thomas
1980	In the early 1980s: Leads for the pacemaker were made available that would deliver a steroid drug from the tip of the electrode to minimize inflammation of the heart wall. Pacemaker with a "rate responsive" feature also became available	USA
1980	He invented the "Contoured Adducted Trochanteric-Controlled Alignment Method (CATCAM) socket."	John Sabolich
1980	In the late 1980s: The term tissue engineering appears in literature.	USA
1980	He develops the controlled drug delivery technology.	Robert Langer
1981	Both published a report on a yeast expression system to produce the hepatitis B surface antigen.	Rutter & Valenzuela

Year	Event/Concepts, etc.	Who
1981	First technical paper on molecular engineering to build with atomic precision.	–
1981	Genentech, Inc. cloned interferon gamma.	Genentech
1981	They invented the scanning tunneling microscope (STM).	Binnig and Rohrer
1981	Willebrordus Meuffels received an externally powered artificial heart.	US
1981	Malchesky, P. S. et al. Macromolecule removal from blood. *Tr Am Soc Artif Intern Org, 27,* 439–444.	Trans. ASAIO
1981	They mapped the gene for insulin.	Mary Harper et al.
1981	He discovered nanocrystalline.	Alexei Ekimov
1981	Scientists produced the first transgenic animals by transferring genes from other animals into mice.	Ohio State
1981	The first commercial MRI (magnetic resonance imaging) scanner arrives on the medical market.	–
1981	The introduction of the Seattle Foot (Seattle Limb Systems) revolutionized the field, bringing the concept of an Energy Storing Prosthetic Foot (ESPF).	USA
1981	The Biotechnology Center was created by the NC General Assembly as the nation's first state-sponsored initiative to develop biotechnology. Thirty-five other states followed also. The first gene-synthesizing machines are developed. The first genetically engineered plant is reported. Mice are successfully cloned.	North Carolina
1981	He held a series of hearings on the relationship between academia and commercialization in the arena of biomedical research.	Al Gore
1981	He reminded the biotech industry that "the most important long-term goal of biomedical research is to discover the causes of disease in order to prevent disease."	Jonathan King
1981	Massachusetts General Hospital received $70 million to build a new Department of Molecular Biology under the Harvard Medical School.	Harvard
1982	He carried out a series of five implants of the Jarvik total artificial heart.	William DeVries
1982	They introduced the first commercial gas phase protein sequencer.	Applied Biosystems, Inc.
1982	December 2: He received a Jarvik-7 artificial heart in Salt Lake City and lived 112 days.	Barney Clark
1982	Extracorporeal CO_2 removal was described by him.	Kolobow

Year	Event/Concepts, etc.	Who
1982	Humulin (human insulin drug) produced by genetically engineered bacteria for the treatment of diabetes received approval from FDA.	Genentech
1982	In Toronto: He presented his 15 year research on the application of Osseo integration.	Brånemark
1982	He requested government permission to test genetically engineered bacteria to control frost damage to potatoes and strawberries.	Lindow
1982	He developed a procedure for making precise amino acid changes anywhere in a protein.	Michael Smith
1982	His team called for the prohibition of the use of RNA technologies in the development of biological weapons.	Richard Goldstein
1983	The study of an extended family with Huntington's chorea demonstrated that family members with the disease show a distinct and characteristic pattern of restriction fragment lengths, leading to a new screening test.	Venezuela
1983	Company received a license to make insulin.	Eli Lilly
1983	His team isolated the AIDS virus at almost the same moment it was isolated at the Pasteur Institute in Paris and at the NIH.	Jay Levy
1983	Joyce, L. D. et al. Response of the human body to the first permanent implant of the Jarvik-7 total artificial heart. *Tr Am Soc Artif Intern Org, 29*, 81–87.	Trans. ASAIO
1983	He devised a method to construct fragments of DNA of predetermined sequence from five to about 75 base pairs long.	Marvin Carruthers
1983	Both invented instrument that could make such fragments automatically.	Carruthers & Hood
1983	Stanford International Research Institute filed for a patent for an *E. coli* expression vector.	Stanford
1983	The Corporation received FDA approval for a monoclonal antibody-based diagnostic test for Chlamydia trachomatis.	Syntex
1983	The Polymerase Chain Reaction (PCR) technique was conceived. The first artificial chromosome was synthesized.	–
1983	U.S. patents were granted to companies with genetically engineering plants.	US
1984	He introduced technique for DNA fingerprinting to identify individuals.	Alec Jeffreys
1984	Cal-Bio scientists described the isolation of a gene for anaritide acetate.	California
1984	They developed pulsed-field gel electrophoresis.	Charles Cantor et al.
1984	The company announced the first cloning and sequencing of the entire human immunodeficiency virus (HIV) genome.	Chiron Corporation

Year	Event/Concepts, etc.	Who
1984	First successful bridge-to-transplantation of Novacor LVAS by him.	P. Portner
1984	University received a product patent for prokaryote DNA.	Stanford
1984	The Plant Gene Expression Center was established.	USA
1984	The role of starvation and nutrition in mortality from acute renal failure was described by him at ASAIO meeting.	Mault
1985	April 7: He received an artificial heart in Sweden, and survived 229 days until November 21.	Leif Stenberg
1985	April 14: He received an artificial heart at Humana. He died 10 days later of bleeding complications.	Jack Burcham
1985	He reported the sequencing of the human insulin receptor in Nature.	Axel Ullrich
1985	His team described the sequencing in cell.	Bill Rutter
1985	He discovered colloidal semiconductor nanocrystals (quantum dots) at Bell Lab.	Louis Brus
1985	Cal Bio cloned the gene that encodes human lung surfactant protein.	California
1985	The company developed GeneAmp PCR technology, which could generate billions of copies of a targeted gene sequence in only hours.	Cetus Corporation
1985	One-half of all CPB procedures were performed with disposable membrane oxygenators, one-half with bubble oxygenators.	USA
1985	February 17: He became the third recipient of the Jarvik-7; and died after 488 days.	Murray Haydon
1985	Genetic fingerprinting enters the US court.	USA
1985	Fullerenes discovered.	USA
1985	Genetic Sciences Inc., injected genetically engineered microbes into trees.	USA
1985	Genetically engineered plants resistant to insects, viruses, and bacteria were field tested for the first time.	USA
1985	November 25: He was the second Jarvik-7 recipient and died after 620 days.	William Schroeder
1985	The Biomedical Engineering Center was formally renamed the William A. Hillenbrand Biomedical Engineering Center by the University Trustees.	Purdue
1985	The team discovered the Buckminsterfullerene (C60), more commonly known as the buckyball.	Harold Kroto et al.
1985	Guidelines for performing experiments in gene therapy on humans were approved.	NIH

Year	Event/Concepts, etc.	Who
1985	The US FDA approved his implantable cardioverter defibrillator (ICD),	Michel Mirowski
1985	He coined the term nanotechnology.	Norio Taniguchi
1986	First book on NT was published. AFM was invented. First NT organization was formed.	USA
1986	Scientists and technicians at Caltech and Applied Biosystems, Inc., invented the automated DNA fluorescence sequencer.	CalTech
1986	An artificial liver using porcine hepatitides was described by him.	DeMetriou
1986	His team at IBM's Zurich lab invented the scanning tunneling microscope.	Gerd Binnig et al.
1986	They invented the atomic force microscope to view upto fractions of a nanometer in size.	Gerd Binnig et al.
1986	Orthoclone OKT3 (Muromonab-CD3) was approved for reversal of acute kidney transplant rejection.	Orthoclone
1986	It approved the release of the first genetically engineered crop, gene-altered tobacco plants.	EPA
1986	The FDA granted a license for the first recombinant vaccine (for hepatitis) to Chiron Corp.	Chiron
1986	He described how to combine antibodies and enzymes to create pharmaceuticals.	Peter Schultz
1987	Advanced Genetic Sciences, Inc. conducted a field trial of a recombinant organism, a frost inhibitor, on a Contra Costa County strawberry patch.	Genetic Sciences
1987	Bard cardiopulmonary support system (CPS) was introduced for rapid deployment, portable, percutaneous access for cardiac emergencies.	Bard
1987	The company received a patent for the tomato polygalacturonase DNA sequence, that was used to produce an antisense RNA sequence.	Calgene Inc.
1987	Centocor's CA 125, diagnostic serum tumor marker test for ovarian cancer, was approved by the FDA.	Centocor
1987	Shulman, L. B., Driskell, T. D. *Dental Implants: A Historical Perspective.* Implants in Dentistry; Philadelphia – PA; USA: W.B. Saunders.	L. B. Shulman
1987	First protein was engineered. First university symposium on NT was held.	USA
1987	He implanted a deep-brain electrical stimulation system into a patient with advanced Parkinson's disease.	Alim-Louis Benabid
1987	The company received FDA approval to market rt-PA (genetically engineered tissue plasminogen activator) to treat heart attacks.	Genentech

Year	Event/Concepts, etc.	Who
1987	The U.S. Patent and Trademark Office issued to him the patent: for a carbon nanotube; and for graphitic, hollow core "fibrils."	Howard G. Tennent
1987	His team invented "yeast artificial chromosomes (YACs)," expression vectors for large proteins.	Maynard Olson
1987	He performed first laser surgery on a human cornea, after perfecting his technique on a cow's eye.	Steven Trokel
1987	Leslie Geddes creates the Geddes-Laufman-Greatbatch Graduate Student Award for productivity and service.	Purdue
1987	Recombivax-HB (recombinant hepatitis B vaccine) approved by FDA.	FDA
1988	First university on NT course.	USA
1988	Genencor International, Inc. received a patent for a process to make bleach-resistant protease enzymes to use in detergents.	Genencor
1988	They received awarded the first patent for a genetically altered animal, a mouse that was highly susceptible to breast cancer.	Philip Leder et al.
1988	Hoffmann-La Roche, Inc. and Cetus Corp. negotiated a licensing agreement for two anti-cancer drugs, Interleukin-2 and Polyethylene Glycol Modified IL-2.	Hoffmann
1988	September 23: He had a heart transplant at Shands Hospital, Gainesville.	Kenneth Claus
1988	SyStemix Inc. received a patent for the SCIDHU Mouse, an immune-deficient mouse with a reconstituted human immune system.	SyStemix
1988	The axial-flow blood pump was described by him at ASAIO.	Wampler
1988	National Center for Human Genome Research was created under him, to oversee the $3 billion U.S. effort to map and sequence all human DNA by 2005.	James Watson
1989	Discovery of the gene responsible for developing cystic fibrosis.	USA
1989	At IBM's Almaden Research Center, they manipulated 35 individual xenon atoms to spell out the IBM logo.	Don Eigler et al.
1989	Epogen (Epoetin alfa) a genetically engineered protein was introduced.	USA
1989	The Brånemark Osseointegration Center (BOC) was founded in Gothenburg, Sweden. He is called father of modern dental implantology	Brånemark
1989	Brånemark, P. I. *Introduction to osseointegration.* In: *Tissue-Integrated Prostheses – Osseointegration in Clinical Dentistry*, edited by P. I. Brånemark, G. Zarb, and T. Albrektsson; Quintessence Publishing, Co. Inc., Chicago – IL, USA.	Brånemark
1989	Scientists developed a recombinant vaccine against the deadly rinderpest virus.	USA

Year	Event/Concepts, etc.	Who
1990	Actimmune (Interferon gamma-1b) was approved by FDA for treatment of chronic granulomatous disease.	FDA
1990	Adagen (adenosine deaminase) was approved for treatment of severe combined immunodeficiency disease.	FDA
1990	Amputee Coalition of America was created to improve the lives of amputees. ACA publishes a magazine: "inMotion."	USA
1990	At http://bse.unl.edu/history: The department changed its name to Biological Systems Engineering – the first accredited program of its kind in the country. Degrees in Biological Systems Engineering and Water Science were initiated.	University of Nebraska–Lincoln
1990	He received a heart transplant in 1990, then in 1996, a living donor kidney transplant from his son. He died May 10, 2012 at the age of 89.	Carroll Shelby
1990	First nanotechnology journal. Japan's STA began funding nanotech projects.	Japan
1990	GenPharm International, Inc. created the first transgenic dairy cow. The cow was used to produce human milk proteins for infant formula.	GenPharm
1990	She reported the discovery of the gene linked to breast cancer in families.	Mary Claire King
1990	He reported the stable transformation of corn using a high-speed gene gun.	Michael Fromm
1990	Publication of his novel *Jurassic Park*, in which bioengineered dinosaurs roam a paleontological theme park.	Michael Crichton
1990	Researchers began the Human Genome Project, coordinated by the U.S. Department of Energy and the National Institutes of Health.	USA
1990	His team discovered the Buckminsterfullerene (C_{60}).	Harold Kroto et al.
1990	The FDA licensed Chiron's hepatitis C antibody test to help ensure the purity of blood bank products.	Chiron Corporation
1990	The first gene therapy took place, on a four-year-old girl with an immune-system disorder called ADA deficiency.	USA
1990	The first field trial of genetically engineered cotton plants was conducted by Calgene Inc. The plants had been engineered to withstand use of the herbicide Bromoxynil.	Calgene (company)
1990	UCSF and Stanford University were issued their 100th recombinant DNA patent license.	Stanford University
1990	US \$3.5 billion were invested worldwide in tissue engineering	–

Year	Event/Concepts, etc.	Who
1990	1990s: Early nanotechnology companies began to operate – Nanophase Technologies in 1989, Helix Energy Solutions Group in 1990, Zyvex in 1997, Nano-Tex in 1998, etc.	Nanophase
1990	In the late 1990s: Many surgeons began to use a minimally invasive surgery that required only a 3-to-5-inch incision. This technique allowed faster recovery and fewer complications.	–
1990	1990s: Pacemakers like microcomputers, were smaller than earlier devices (1/2 the size). "mode switching" was introduced. The significant contribution in the following years is note-worthy by the pioneers: Bob Anderson of Medtronic Minneapolis; J.G. Davies of St George's Hospital-London; B. Berkovits and Sheldon Thaler of American Optical; Geoffrey Wickham of Telectronics – Australia; Walter Keller of Cordis Corporation – Miami; Hans Thornander and Elmquist of in Sweden; Janwillem van den Berg of Holland and Anthony Adducci of Cardiac Pacemakers Inc.	USA
1991	While analyzing chromosomes from women in cancer-prone families, she finds evidence that a gene on chromosome 17 causes the inherited form of breast cancer and also increases the risk of ovarian cancer.	Mary-Claire King
1991	Japan's MITI announced bottom-up "atom factory." IBM endorsed bottom-up path in NT. Carbon nanotube was discovered.	Japan
1991	Research on nanotubes: Discovery in the soot of arc discharge by Sumio Iijima. Discovery in CVD by Al Harrington et al. leading to development of a method to synthesize monomolecular thin film nanotube coatings.	Sumio Iijima et al.
1991	Leslie Geddes "retires" as Center Director and Showalter Professor to become the Showalter Distinguished Professor Emeritus of Biomedical Engineering.	Purdue
1991	September 3: He received an experimental HeartMate (LVAD) and lived for 16 months.	Mike Templeton
1991	Iijima shared the Kavli Prize in Nanoscience in 2008 for discovery on carbon nanotube (CNT).	Sumio Iijima
1991	"Mendelian Inheritance in Man" work was made available online computer network.	–
1992	American and British scientists unveil a technique for testing embryos *in vitro* for genetic abnormalities such as cystic fibrosis and hemophilia.	USA
1992	Biomolecular engineering was defined by the National Institutes of Health as "research at the interface of chemical engineering and biology with an emphasis at the molecular level."	NIH

Year	Event/Concepts, etc.	Who
1992	Herceptin became the first drug designed by a biomolecular engineering approach and was approved by the FDA.	FDA
1992	Biomolecular Engineering was a former name of the Journal "New Biotechnology."	USA
1992	His team discovered the nanostructured catalytic materials MCM-41 and MCM-48, now used for drug delivery.	C.T. Kresge et al.
1992	First textbook on NT was published. First US Congressional testimony was given on NT.	USA
1992	First theoretical predictions of the electronic properties of single-walled carbon nanotubes by groups at Naval Research Laboratory, USA, etc.	USA
1992	Alumnus Neal Fearnot created the Fearnot Award to recognize the student giving the most outstanding summer seminar presentation on his research.	Purdue
1992	The U.S. Army begins collecting blood and tissue samples from all new recruits as part of a "genetic dog tag" program.	USA
1992	Founder of Human Genome Sciences: Coined the term "Regenerative Medicine"	William Haseltine
1993	First Feynman Prize in Nanotechnology was awarded for modeling a hydrogen abstraction tool useful in nanotechnology. First coverage of nanotech from White House.	Feynman
1993	His team produced a rough map of all 23 pairs of human chromosomes.	Daniel Cohen
1993	Chiron's Betaseron was approved by FDA as the first treatment for multiple sclerosis in 20 years.	Chiron
1993	The Biotechnology Industry Organization (BIO) was created.	USA
1993	The company launched a $10 million *nationwide communications network program* to enable high school biology teachers across the country to access their peers as well as experts.	Genentech
1993	Scientists cloned human embryos and nurture them in a Petri dish for several days.	www.gwu.edu
1993	His team discovered single-wall carbon nanotubes and methods to produce them using transition-metal catalysts.	Donald S. Bethune
1993	Nobel Prize in Chemistry for inventing the technology of PCR.	Kary Mullis
1993	He invented a method for controlled synthesis of nanocrystals (quantum dots).	Moungi Bawendi
1993	The first microprocessor-controlled prosthetic knees became available in the early 1990s	–
1993	The Intelligent Prosthesis was released in Great Britain.	Chas. A. Blatchford

Year	Event/Concepts, etc.	Who
1994	A multitude of genes, human and otherwise, were identified and their functions described: Ob, a gene predisposing to obesity; BCR, a breast cancer susceptibility gene; BCL-2, a gene associated with apoptosis (programmed cell death); hedgehog genes; and Vpr, a gene governing reproduction of the HIV virus.	–
1994	Fight over who owns what parts of the genome.	USA
1994	The scientists and research corporations have worked out a way to share access to a computerized database detailing 35,000 human genes.	USA
1994	Centocor's ReoPro was cleared for marketing in United States by the FDA and by the European Union's regulatory body; CPMP for patients undergoing high-risk balloon angioplasty.	Centocor
1994	Approval of the pneumatically driven HeartMate LVAD (by Thermo Cardio systems, Inc.) for bridge to transplantation	FDA
1994	First successful systemic selective inhibition of gene expression using antisense oligonucleotides.	USA
1994	Nutropin was approved for the treatment of growth hormone deficiency.	Genentech
1994	The first breast cancer gene was discovered.	USA
1994	Researchers successfully transferred the CFTR (cystic fibrosis transmembrane conductance regulator) gene into the intestines of mice.	–
1994	Genzyme's Ceredase/Cerezyme (alglucerase/recombinant alglucerase) approved for type 1 Gaucher's disease.	FDA
1994	Linkage studies identified genes for a variety of ailments including: bipolar disorder, cerulean cataracts, melanoma, hearing loss, dyslexia, thyroid cancer, sudden infant death syndrome, prostate cancer and dwarfism.	USA
1994	Nanosystems textbook was used in first university course. US Science Advisor advocated nano-technology.	USA
1994	Recombinant GM-CSF was approved for chemotherapy-induced neutropenia.	FDA
1994	Researchers reported that the enzyme telomerase appears to be responsible for the unchecked growth of cells seen in human cancers.	Texas
1994	The approval of genetically engineered version of human DNAase.	FDA

Year	Event/Concepts, etc.	Who
1994	The BRCA-1 gene also appeared to play a role in much more common types of non-inherited breast cancers.	
1994	The first crude but thorough linkage map of the human genome appeared in *Science*, volume 265, September 30.	*Science*
1994	The first genetically engineered food product, the Flavr Savr tomato, gained FDA approval.	FDA
1995	Research by his team led to the first isolation of human embryonic stem and human embryonic germ cells.	Michael D. West
1995	Researchers identified a genetic defect, which appeared to underlie the most common cause of deafness.	Europe
1995	His group became the controversial and outspoken critic of the biotechnology industry.	Jeremy Rifkin
1995	Religious groups in the US sought to overturn laws allowing the patenting of genes used for medical and research applications.	USA
1995	International Human Genome Project: STS gene mapping could greatly speed the work of geneticists.	–
1995	A new transgenic mouse carrying a gene for human Alzheimer's disease was developed.	–
1995	A single gene was identified that appeared to control the growth and development of eyes.	–
1995	Intelligent Prosthesis Plus was released.	
1998	Adaptive Prosthesis was released.	Blatchford (company)
1995	First US think tank report includes NT.	USA
1995	First industry analysis of military applications of NT.	USA
1995	He was found not guilty in a high-profile double-murder trial in which PCR and DNA fingerprinting played a prominent.	O.J. Simpson
1995	Gene therapy, immune system modulation and genetically engineered antibodies enter the clinic in the war against cancer.	USA
1995	The agency (Center of Disease Control [CDC]) confirmed that the Ebola virus was behind outbreak of hemorrhagic fever in Zaire.	CDC, USA
1995	Leptin appeared to cause weight loss in experimental animals.	
1995	Researchers transplanted hearts from genetically altered pigs into baboons, proving that cross-species operations are possible.	Duke Univ.
1995	Researchers demonstrated the electron emission properties of carbon nanotubes.	Sweden
1995	His team predicted emission property of carbon nanotubes.	Till Keesmann et al.

Year	Event/Concepts, etc.	Who
1995	The first full gene sequence of a living organism other than a virus was completed for the bacterium Hemophilus influenzae.	USA
1995	Researchers reported that unrecognized properties of RNA added further support to the idea that RNA was the central molecule in the origin of life.	USA
1996	First European conference on NT. NASA begins work in computational nanotech. First nanobio conference.	
1996	Scientists reported the sequencing of the complete genome of a complex organism, Saccharomyces cerevisiae (baker's yeast).	USA
1996	The complete sequencing of the largest genome to date – more than 12 million base pairs of DNA.	USA
1996	Inexpensive diagnostic biosensor test for the first time allowed instantaneous detection of the toxic strain of *E. coli* strain 0157:H7.	USA
1996	Avonex drug was approved for the treatment of multiple sclerosis.	Biogen
1996	Biomedical Engineering Graduate Program approved by the Board Trustees and sanctioned by the Indiana Commission for Higher Education. Alumni David and Stephen Grubbs funded an annual research stipend for students.	Purdue
1996	REMATCH Trial with HeartMate VE was initiated; the 2002 report showed mortality reduction of 50% at one year as compared to patients receiving optimal medical therapy.	E. Rose
1996	Surveys indicated the fear and mistrust of the public on the research into the workings of the human genome and gene therapy.	1996
1996	T-cell researchers determined the three-dimensional structure of these critical components of the immune system.	USA
1996	The discovery of a gene associated with Parkinson's disease provided potential treatment of the debilitating neurological ailment.	USA
1996	Tissue Engineering Society was founded (now Tissue Engineering Regenerative Medicine International Society, TERMIS).	https://termis.org/
1996	UK government announces that 10 people may have become infected with the BSE agent through exposure to beef.	UK
1996	10 people may have become infected with the BSE agent through exposure to beef.	
1997	Cloning of two Rhesus monkeys.	Oregon
1997	A new DNA technique combined PCR, DNA chips, and computer programming.	USA

Year	Event/Concepts, etc.	Who
1997	*E. coli* vaccine for prevention of urinary tract infections developed.	USA
1997	Artificial human chromosomes created for the first time.	USA
1997	Carticel autologous cartilage implant approved by US FDA.	FDA
1997	Clock, the first gene providing the circadian rhythm of mammalian life identified.	USA
1997	The first suggestion of using carbon nanotubes as optical antennas is made in the patent application.	Robert Crowley
1997	First carbon nanotube single-electron transistors.	Delft Univ
1997	First company Zyvex was founded in NT. First design of nanorobotic system.	Zyvex
1997	$250,000/year for three years from a Quality Improvement Program was awarded by President Maidique to establish Cardiovascular Engineering Center at FIU.	FIU, Miami
1997	Follistim was approved for treatment of infertility.	FDA
1997	Orasure, a bloodless HIV-antibody test using cells from the patient's gums, was approved.	FDA
1997	His team reported that they have cloned a sheep (named Dolly) using DNA from two cells of an adult ewe.	Ian Wilmut
1997	The complete genome of the Lyme disease pathogen, *Borrelia burgdorferi*, was sequenced, along with the genomes for *E. coli* and H pylori.	–
1997	The FDA approved Rituxan, the first antibody-based therapy for cancer (for patients with non-Hodgkin's lymphoma).	FDA
1997	The Otto Bock orthopedic industry introduced the C-Leg during the world congress on orthopedics in Nuremberg in 1997; and in USA in 1999.	Nuremberg
1997	TransCyte becomes first US FDA approved product.	FDA
1997	Using a bit of DNA and some common biological laboratory techniques, researchers engineered the first DNA computer "hardware" ever: logic made of DNA.	USA
1998	First DNA-based nanomechanical device.	–
1998	The Biotechnology Institute is founded by BIO.	BIO
1998	Cloned vain RT with fully active polymerase and minimized Rnase H activity was engineered.	–
1998	Favorable results with a new antibody therapy against breast cancer, HER2neu (Herceptin).	USA
1998	First clinical application of next-generation continuous-flow assist devices. DeBakey (Micromed Inc.) axial-flow pump was implanted by his team.	R. Hetzer et al.
1998	First NSF forum, held in conjunction with Foresight Conference.	NSF

Year	Event/Concepts, etc.	Who
1998	$100,000 gift from Norman Weldon, co-founder of Cordis Corporation, to support biomedical engineering undergraduate research assistants.	–
1998	Fomivirsen became the first approved therapeutic agent developed with antisense medical technology.	–
1998	Human embryonic stem cells were isolated.	–
1998	The Department of Biomedical Engineering was formed with George Wodicka as head of the new department with funds from Whitaker Foundation.	Purdue
1998	Research with tumor starving biologicals including angiostatin and endostatin begins to show promise in the clinic.	–
1998	Scientists at Japan's Kinki University cloned eight identical calves using cells taken from a single adult cow.	Japan
1998	FDA approval of HeartMate VE (Thermo Cardiosystems) and Novacor LVAS (Baxter Healthcare Corp.).	FDA
1998	The development of stem cell lines formed the basis of modern tissue engineering.	–
1998	The FDA grants marketing clearance to Remicade™ (infliximab).	FDA
1998	The first complete animal genome the *C. elegans* worm was sequenced.	–
1998	The Interagency Working Group on Nanotechnology (IWGN) was formed under the National Science and Technology Council.	USA
1998	Two research teams succeeded in growing embryonic stem cells.	USA
1998	Scientists cloned three generations of mice from nuclei of adult ovarian cumulus cells.	Hawaii
1998	US FFDA approved Apigraf: first allogenic TE product.	FDA
1998	Three generations of mice were cloned.	USA
1999	The complete genetic code of human chromosomes was decoded.	USA
1999	They probed secrets of chemical bonding by assembling a molecule [iron carbonyl $Fe(CO)_2$] from constituent components [iron (Fe) and carbon monoxide (CO)] with a scanning tunneling microscope.	Wilson Ho et al.
1999	October 23: Dr. R. Bruce Martin, ASB Founding Member and President (1998–1999) described the history of biomechanics at the annual conference of the American Society of Biomechanics at University of Pittsburgh, Pittsburgh, PA. He said, "Interest in biomechanics is ancient. From its earliest manifestations, science has looked inward as well as outward. This involved questions of epistemology but also human physiology, including biomechanics. The interest in biomechanics springs from the same source as Aristotle's – curiosity about the human body." USA: http://www.asbweb.org/html/biomechanics/genealogy/genealogy.htm.	www.asbweb.org

Year	Event/Concepts, etc.	Who
1999	A new medical diagnostic test allowed quick identification of BSE/CJD a rare but devastating form of neurologic disease transmitted from cattle to humans.	–
1999	A new technique based on unique individual antibody profiles offered an alternative to current DNA fingerprinting methods.	USA
1999	He invented dip-pen nanolithography (DPN).	Chad Mirkin
1999	LionHeart LVAS was implanted in 67 year-old male recipient by them.	R. Koerfer et al.
1999	First Nanomedicine book was published. First safety guidelines on NT. US Congressional hearings on proposed National Nanotechnology Initiative (NNI).	USA
1999	$1 million Special Opportunity Award from the Whitaker Foundation to establish the Biomedical Engineering Institute as the administrative unit for the biomedical engineering programs and MS program established: 75+ graduates, currently enrolls 35 students.	FIU, Miami
1999	National Science Foundation IGERT grant in Therapeutic and Diagnostic Devices was awarded.	Purdue
1999	He was credited with discovering the carbon nanotube (CNT).	Sumio Iijima
1999	There are 1,274 biotechnology companies in the United States alone, with at least 300 biotechnology drug products and vaccines currently in human clinical trials and hundreds more in early development.	USA
1999	1999 to early 2000s: Consumer products making use of nanotechnology began appearing in the marketplace, including lightweight nanotechnology-enabled automobile bumpers, golf balls, tennis rackets, baseball bats with better flex and "kick," nano-silver antibacterial socks, clear sunscreens, wrinkle- and stain-resistant clothing, deep-penetrating therapeutic cosmetics, scratch-resistant glass coatings, faster-recharging batteries for cordless electric tools, and improved displays for televisions, cell phones, and digital cameras, etc.	
1999	At the 1999 annual meeting of ASB, R. Bruce Martin indicated that "biomechanics should represent the broad interplay between the two, including the sub-disciplines defined in our bylaws, and more."	R. Bruce Martin
2000	The U.S. National Nanotechnology Initiative was formed.	USA
2000	A man in Israel became the first recipient of the Jarvik 2000, the first total artificial heart that could maintain blood flow in addition to generating a pulse.	Israel
2000	They discovered the nanostructured catalytic materials MCM-41 and MCM-48.	Kresge et al.

Year	Event/Concepts, etc.	Who
2000	Completion of a "rough draft" of the human genome in the Human Genome Project.	–
2000	First demonstration proving that bending carbon nanotubes changes their resistance.	
2000	He invented a method for controlled synthesis of nanocrystals (quantum dots).	Moungi Bawendi
2000	President Clinton announces U.S. National Nanotechnology Initiative. The NSET Subcommittee of the NSTC was designated as the interagency group responsible for coordinating the NNI.	USA
2000	Department BME received the prestigious Tony and Mary Hulman Health Achievement Award from the Indiana Public Health Foundation.	Purdue
2000	Time magazine names Tissue Engineer (TE) as the hottest job for the future: 3000 persons pursue TE careers.	Time magazine
2000	US$580 million spent annually on tissue engineering R&D; public TE companies valued at US$2.5 billion.	USA
2001	First report on a technique for separating semiconducting and metallic nanotubes.	USA
2001	Celera Genomics and the Human Genome Project create a draft of the human genome sequence. It is published by Science and Nature Magazine.	Celera
2001	Doctors implanted first self-contained, mechanical heart replacement (AbioCor)	Louisville
2001	U.S. announces first center for military applications of NT.	USA
2001	$10 million gift from the Wallace H. Coulter Foundation (including 100% match from the state of Florida) to support endowed faculty positions, scholarships and fellowships, research seed funding, lecture series, and infrastructure for biomedical engineering.	FIU, Miami
2001	An undergraduate biomedical engineering program was approved by the Purdue Board of Trustees and the Indiana Commission for Higher Education. Joint MD/PhD program was formed with the Indiana University School of Medicine.	Purdue
2001	The AbioCor (totally implantable, electrically powered) TAH was implanted into patient Robert Tools by them.	L. Gray et al.
2001	The sequence of the human genome is published in Science and Nature.	Nature
2001	US President Bush restricts federal funding for embryonic stem-cell research.	USA

Year	Event/Concepts, etc.	Who
2002	He did an implant with interfacing directly into Warwick's nervous system. The signals produced were detailed enough that a robot arm was able to mimic the actions of Warwick's own arm	Kevin Warwick
2002	Circe bio-artificial liver completed Phase III clinical trial.	Circe
2002	Ranawat indicated, "*Over 19 companies in the United States distribute total knee implants of three different types: cruciate-preserving, cruciate-substituting, and TC-III. Six major companies are actively involved in designing mobile-bearing knees. Future developments, such as navigation-guided surgery, enhanced kinematics, and wear-resistant bearing surfaces with better fixation, promise a consistent evolution for the total knee replacement.*" [Ranawat, C. S. History of total knee replacements. Journal South Orthopaedic Association, Winter Issue, 11(4), 218–226].	C. S. Ranawat
2002	FDA approval of Medtronic's INFUSE Bone Graft.	FDA
2002	Regional nanotech efforts multiplied in US.	USA
2002	$600,000 NSF Partnership for Innovation Award to establish the Biomedical Engineering Partnership Program: the membership in 2002 includes 35+ clinical organizations and biomedical companies. Also established the Institute for Technology Innovation, a unit of the Pino Global Entrepreneurship CenterBS program established – 50+ graduates, currently enrolls 150 students.	FIU, Miami
2002	Integra Dermal Regeneration Template by Integra Life Sciences approved for treatment of severe burns.	USA
2002	Multi-walled nanotubes demonstrated to be fastest known oscillators (>50 GHz).	USA
2002	Organogenesis and ATS, both previously valued at US$1 billion file for bankruptcy.	USA
2002	State of Indiana funding for new BME building and program growth secured.	USA
2002	Rice became the first crop to have its genome decoded.	USA
2002	TE activity halved since 2000, loss of 800 fulltime employees, capital value of public – traded TE corporations dropped from US$2.5 billion to US$300 million.	USA
2002	There was 42% increase in stem-cells firms, coining of term "Regenerative Medicine."	–
2003	Her team developed gold nanoshells that could be used to destroy cancer cells without harming adjacent healthy tissue.	Naomi Halas et al.

Year	Event/Concepts, etc.	Who
2003	At http://bse.unl.edu/history: The Biological Systems Engineering Department is being administered essentially in the same format as set forth in 1908. The Biological Systems and Agricultural engineering undergraduate programs are administered by the College of Engineering. www.bse.unl.edu	University of Nebraska–Lincoln
2003	The Department of Biomedical Engineering (Florida International University [CIU]) was established to replace the Institute as the administrative unit for the biomedical engineering programs.	FIU, Miami
2003	Organogenesis emerges from bankruptcy.	USA
2003	Whitaker Foundation funding for building was granted; Ronald W. Dollens Graduate Scholarship was established by the Guidant Foundation for graduate students in biomedical engineering and pharmacy. Groundbreaking took place on a new biomedical engineering building.	Purdue University
2003	NEC announced stable fabrication technology of carbon nanotube transistors.	NEC (company)
2003	The Human Genome Project was completed, providing information on the locations and sequence of human genes on all 46 chromosomes.	USA
2003	US Congress enacted the 21st Century Nanotechnology Research and Development Act (P.L. 108–153).	USA
2003	US Congressional hearings on societal implications of NT.	USA
2003	The sheep, cloned in 1997, was euthanized due to lung disease.	–
2003	National Institute of Health Center for Biomedical Computation at University was established.	Stanford Univ.
2003	OpenSim, a free software, was designed to propel biomechanics research by providing a common framework for investigation and a vehicle for exchanging complex musculoskeletal models. Since then, this software (OpenSim) is being continuously developed by Simbios. OpenSim 1.0 was released on August 20, 2007; OpenSim 2.2.1 was released on April 11, 2011. More information can be found at weblinks.	Stanford Univ.
2004	At http://bse.unl.edu/history: A state-of-the-art Biomedical Imaging and Biosignal Analysis Laboratory was started. Dr. Greg Bashford's lab features equipment used for medical imaging studies and biosignal analysis, such as ultrasound mammography for breast cancer screening, echodentography, cardiovascular flow quantification, ECG/EEG instrumentation, and evoked potentials for neurological experiments. (www.bse.unl.edu)	University of Nebraska–Lincoln
2004	BMES published its "Historical Perspective" in 2004, celebrating its 35-year history and growth.	www.bmes.org

Year	Event/Concepts, etc.	Who
2004	Britain's Royal Society and the Royal Academy of Engineering published Nanoscience and Nanotechnologies: Opportunities and Uncertainties advocating the need to address potential health, environmental, social, ethical, and regulatory issues associated with nanotechnology.	Britain
2004	First policy conference on advanced nanotech. First center for nanomechanical systems.	USA
2004	PhD program established – enrolls 26 students; Miami Children's Hospital funds the MCH Professorship in NeuroEngineering, a joint faculty position between MCH and FIU Biomedical Engineering.	FIU, Miami
2004	March: Nature published a photo of an individual 4 cm long single-wall nanotube (SWNT).	USA
2004	In recognition of the generosity of the Weldon family, the Purdue Board of Trustees elevated the Department of Biomedical Engineering to the Weldon School of Biomedical Engineering.	Purdue
2004	SUNY Albany launched the first college-level education program in nanotechnology in the United States, the College of Nanoscale Science and Engineering.	SUNY
2004	The European Commission adopted the Communication "Towards a European Strategy for Nanotechnology," COM(2004)338, which proposed institutionalizing European nanoscience and nanotechnology R&D efforts within an integrated and responsible strategy.	EC
2004	The National Cancer Institute establishes the Alliance for Nanotechnology in Cancer.	USA
2005	At Nanoethics meeting, Roco announced nanomachine/nanosystem project count has reached 300.	Roco
2005	August: University of California finds Y-shaped nanotubes to be ready-made transistors.	Univ. CA
2005	General Electric announced the development of an ideal carbon nanotube diode that operates at the "theoretical limit"	GE
2005	Nanotube sheet synthesized with dimensions 5 × 100 cm.	USA
2005	His team developed theories for DNA-based computation and "algorithmic self-assembly" in which computations are embedded in the process of nanocrystal growth.	Erik Winfree
2005	May: A prototype high-definition 10-centimetre flat screen made using nanotubes was exhibited.	USA

Year	Event/Concepts, etc.	Who
2005	The Biomedical Entrepreneurship graduate certificate program was started in collaboration with the Krannert School of Management through the generous support of the Guidant Foundation and the C.R. Bard Foundation. The Havel and Decker Families endowed the first undergraduate scholarship at the School. Biomedical Science graduate program with Veterinary Medicine was launched. Biomedical Engineering undergraduate industrial internship program was created.	Purdue
2005	Researchers at IBM developed vertical nanowire transistors.	USA
2005	Microprocessor-controlled knee prostheses are: Ossur's Rheo Knee, released in 2005; the Power Knee by Ossur, introduced in 2006; the Plié Knee from Freedom Innovations; and DAW Industries' Self Learning Knee (SLK).	Ossur
2005	The Federal Environmental Protection Agency (U.S. EPA) holds its first public meeting to explore the feasibility of a Nanoscale Materials Stewardship Program (NMSP), which was not implemented till 2008.	EPA
2005	Shaped crystal (diamondoid) manufacturing.	USA
2006	By 2006, there were an estimated 32,000 bioengineers addressing a variety of areas related to health technology, according to The Whitaker Foundation.	Whitaker
2006	$1.4 million gift from the Ware Foundation (including $400k match from the state of Florida) was established to permanently support a post-doctoral fellow, in the Laboratory for Brain Research and Neuro-Engineering Applications; BS program received ABET accreditation in fastest time possible, with no concerns or weaknesses. Annual research awards total over $1.8 million to $200,000 per FTE tenure-track faculty line.	FIU, Miami
2006	In absence of federal or state regulation, the city of Berkeley, passed a city ordinance requiring *"all facilities that manufacture or use manufactured nanoparticles ... submit a separate written disclosure of the current toxicology, to the extent known, and how the facility will safely handle, monitor, contain, dispose, track inventory, prevent release and mitigate such materials."*	City of Berkeley
2006	His team built a nanoscale car made of oligo (phenylene-ethynylene) with alkynyl axles and four spherical C60 fullerene (buckyball) wheels.	James Tour et al.
2006	July: Nanotubes were alloyed into the carbon fiber bike that won the 2006 Tour de France.	France
2006	June: Gadget could sort nanotubes by size and electrical properties.	Rice Univ.

Year	Event/Concepts, etc.	Who
2006	Company built an electronic circuit around a CNT.	IBM
2006	Method of placing nanotube accurately was developed.	IBM
2006	National Academies Nanotechnology (NANT) report called for experimentation toward molecular manufacturing	USA
2006	The new home of the Weldon School of Biomedical Engineering was dedicated. Cook Group, Inc. endowed Leslie A. Geddes Professorship.	Purdue
2006	Low-cost method of using nanoparticles to remove arsenic in drinking water was developed.	Rice Univ.
2006	TE bladder appeared in the Lancet.	Lancet
2006	The National Institute of Health began a 10-year, 10,000-patient study using a genetic test that predicts breast-cancer recurrence.	NIH
2006	The company developed the genetic test, Oncotype DXTM	Genomic Health
2007	The Univ. released free software http://opensim.stanford.edu/ for biomechanical modeling: OpenSim 1.0 on August 20, 2007; OpenSim 1.1 on December 11, 2007; OpenSim 2.2.1 on April 11, 2011; and OpenSim on October 10 of 2011.	www.stanford.edu
2007	Her team built a lithium-ion battery with a common type of virus that is non-harmful to humans.	Angela Belcher
2007	Creation of induced pluripotent stem cells from adult human skin cells.	USA
2007	$1 million was awarded from the Florida 21st Century World Class Scholar Program to establish joint BME/College of Medicine faculty position for the new Wallace H. Coulter Professor, Dr. Joe Leigh Simpson.	FIU, Miami
2007	Nanotechnology Products Began to be seen on the market, e.g., golf balls that fly straighter, nano-silver antibacterial socks, clear sunscreens, wrinkle- and stain-resistant clothing, deep-penetrating therapeutic cosmetics and improved displays for televisions, cell phones, and digital cameras.	USA
2007	Osiris named Biotech Company of the year.	Osiris
2007	Leslie Geddes receives the National Medal of Technology in a White House ceremony.	Purdue
2007	Korean Institute of Science and Technology Nanomedicine research partnership was created.	Korea
2007	University of Puerto Rico-Mayaguez graduate education partnership was founded.	www.uprm.edu
2007	Purdue Imaging Center created in partnership with GE Healthcare and Innervision. Institute for Biomedical Development launched. Inaugural biomedical engineering undergraduate class graduates.	Purdue

Year	Event/Concepts, etc.	Who
2007	The 170 companies offering TE products or services, sales in excess of US$1.3 billion; >1 million patients treated;	USA
2007	The first Nanoradio was described by him.	Alex Zettl
2008	At http://bse.unl.edu/history: A new, state-of-the art lab for Nonviral Gene Delivery and Cell Culture was officially dedicated in February. The lab, directed by Dr. Angela Pannier, is located in the lower lever of Chase Hall.	www.bse.unl.edu
2008	HP announced the development of a switching memristor for use in computer memory and logic applications.	Hewlett Packard
2008	Implantation of tracheal segment was engineered from decellularized tissue.	USA
2008	Japanese astronomers launched the first Medical Experiment Module called "Kibo," to be used on the International Space Station.	Japan
2008	His team performed the first tissue engineered trachea (wind pipe) transplantation.	Paolo Macchiarini
2008	He was briefly ruled ineligible to compete in the 2008 Summer Olympics. Later he qualified for the 2011 World Championship in South Korea.	Oscar Pistorius
2008	His company DEKA developed the "Luke arm," an advanced prosthesis.	Dean Kamen
2008	Undergraduate biomedical engineering program received ABET accreditation. National Science Foundation funded scholarship program in quantitative physiology. National Institutes of Health Clinical and Translational Science Institute (CTSI) grant with Indiana University School of Medicine was awarded. MD/PhD Program received National Institutes of Health Medical Scientist Training Program (MSTP) funding. The new building was formally renamed the Martin C. Jischke Hall of Biomedical Engineering.	Purdue
2008	Researchers discovered negative refraction to bend light in materials made from nanowires, for use in developing optical lenses with much higher resolution than conventional lenses.	Berkeley
2008	September 20: He received heart transplant.	Glen Gondrezick
2008	Technology Roadmap for Productive Nanosystems released. Protein catalysts designed for non-natural chemical reactions.	USA
2008	Feynman Prize in Nanotechnology awarded for work in molecular electronics and the synthesis of molecular motors and nanocars, and for theoretical contributions to nanofabrication and sensing.	Feynman

Year	Event/Concepts, etc.	Who
2008	The first official NNI Strategy for Nanotechnology-Related Environmental, Health, and Safety (EHS) Research was published, based on a two-year process of NNI-sponsored investigations and public dialogs. This strategy document was updated in 2011, following a series of workshops and public review.	USA
2008	M5 Fiber with 9.8 GPa tensile strength was developed.	USA
2008	Gecko mimicing wallcrawling suits for military and enthusiasts.	USA
2008	November 20: He had a heart–lung transplant.	Roberto Julio Sánchez
2008	U.S. Army World Class Athlete, won a gold medal in the 100 m with a time of 12.15 sec at the Endeavor Games in Oklahoma – USA.	Sgt. Jerrord Fields
2009	A single-walled carbon nanotube was grown by chemical vapor deposition across a 10-micron gap in a silicon chip, then used in cold atom experiments, creating a blackhole like effect on single atoms.	USA
2009	An improved walking DNA nanorobot. Structural DNA nanotechnology arrays devices to capture molecular building blocks.	USA
2009	Organizing functional components on addressable DNA scaffolds.	USA
2009	Design 'from scratch' of a small protein that performed the function performed by natural globin proteins.	USA
2009	April: Nanotubes incorporated in virus battery.	USA
2009	Heart Institute used modified SAN heart genes to create the first viral pacemaker in guinea pigs, known as iSAN's.	Cedars-Sinai
2009	His team created several DNA-like robotic nanoscale assembly devices. Later he was the Kavli Prize in Nanoscience in 2010.	Nadrian Seeman
2009	DNA assembly line was developed by him.	Nadrian Seeman
2009	Online Professional Master's degree program was launched.	Purdue
2009	His team aimed to improve in vivo-like conditions for 3D tissue via "stacking and de-stacking layers of paper impregnated with suspensions of cells in extracellular matrix hydrogel, making it possible to control oxygen and nutrient gradients in 3D, and to analyze molecular and genetic responses	Ratmir et al.
2009	All nanomaterials, other than CNTs, remained completely unregulated.	California
2009	SENS Foundation was launched, with its aim as "the application of regenerative medicine – defined to include the repair of living cells and extracellular material in situ – to the diseases and disabilities of ageing."	SENS

Year	Event/Concepts, etc.	Who
2009	According to the report by Utah Hip and Knee Center, the research was being done to develop a knee replacement technology in which the bones actually grow into the device and hold it together.	Utah
2009	Unicondylar, or partial, knee replacements were introduced.	USA
2009	US President Obama lifts ban on Federal Funding of embryonic stem-cell research.	USA
2009	At http://bse.unl.edu/history: The Translational and Regenerative Medicine Imaging laboratory (TREM), using MRI, was dedicated in April. The lab is under the direction of Dr. Shadi Othman and is located in the lower level of Chase Hall.	www.bse.unl.edu
2010	DNA-based 'robotic' assembly began.	–
2010	Researchers demonstrated a nanoscale transistor to measure electrical activity in a human heart cell.	Harvard Univ.
2010	IBM used a silicon tip measuring only a few nanometers at its apex to chisel away material from a substrate to create a complete nanoscale 3D relief map of the world one-one-thousandth the size of a grain of salt—in 2 minutes and 23 seconds.	IBM
2010	Her team developed gold nanoshells.	Naomi Halas
2010	Johns Hopkins University Applied Physics Laboratory developed Neural prosthetic Proto 1 and Proto 2.	Johns Hopkins
2010	Center for Implantable Devices was formed. Purdue chapter of Alpha Eta Mu Beta, the Biomedical Engineering Honor Society were created. Internship program at National University of Ireland-Galway was initiated.	Purdue
2010	Researchers at New York University and China's Nanjing University demonstrated an assembly-line method using nanorobots built from DNA strands.	NY Univ.
2010	Researchers created the first synthetic cell.	J. Craig Venter Institute
2010	Complex products were manufactured through nanotech.	–
2011	Published an article: A First: Organs Tailor-Made With Body's Own Cells.	NY Times
2011	A windpipe was created with body's own cells, that was suggested by a doctor at the Karolinska Institute.	Sweden
2011	Britain's Royal Society and the Royal Academy of Engineering published Nanoscience and Nanotechnologies: Opportunities and Uncertainties advocating the need to address potential health, environmental, social, ethical, and regulatory issues associated with nanotechnology.	Britain

Year	Event/Concepts, etc.	Who
2011	He developed a process for scraping cells from the lining of a pig's bladder: Decellularizing the tissue and then drying it to become a sheet or a powder.	Stephen Badylak
2011	The cellular matrix powder was used to regrow the finger of Lee Spievak.	Stephen Badylak
2011	Nicknamed "pixie-dust," the powdered extracellular matrix is being used success to regenerate tissue lost and damaged due to traumatic injuries.	US military
2011	His team investigated "*3D-plotting technique to produce poly-L-Lactide macroporous scaffolds with two different pore sizes*" via solid free-form fabrication (SSF) with CAD.	El-Ayoubi et al.
2011	First programmable nanowire circuits for nanoprocessors.	USA
2011	DNA molecular robots learned to walk in any direction along a branched track.	USA
2011	Mechanical manipulation of silicon dimers on a silicon surface.	USA
2011	Surgeons performed the first implantation of a synthetic trachea.	Sweden
2011	His team improved the 2008 implant by transplanting a laboratory-made trachea seeded with the patient's own cells.	Paolo Macchiarini
2011	Active U.S. patents exceeded 100, with 50 active corporate licenses. Biomedical Engineering cumulative licensing revenues exceeded $20 million. 12th start-up company based upon biomedical engineering discovery was launched.	Purdue
2011	The NSET Subcommittee updated both the NNI Strategic Plan and the NNI Environmental, Health, and Safety Research Strategy, drawing on extensive input from public workshops and online dialog with stakeholders from Government, academia, NGOs, and the public, and others.	NSET
2012	She delivered the lecture at the Annual Meeting of the Regenerative Medicine and Tissue Engineering Society and said: "*From a humanistic standpoint, the major aim of creating artificial organs is to give an end to the human organ trafficking, a transnational organized crime, that is rising in third world countries and has become a lucrative facet of economic development by annihilating the need for real organs.*"	Eleni V. Antoniadou
2012	They developed theories for DNA-based computation and "algorithmic self-assembly" in which computations are embedded in the process of nanocrystal growth.	Erik Winfree et al.
2012	January: IBM creates 9 nm carbon nanotube transistor that outperforms silicon.	IBM
2012	They focused on whether Laser-assisted BioPrinting can be used to build multicellular 3D patterns in natural matrix, and whether the generated constructs are functioning and forming tissue.	Koch et al.

Year	Event/Concepts, etc.	Who
2012	LaBP arranged small volumes of living cell suspensions in set high-resolution patterns.	LaBP
2012	Dick Chiney, Vice President under George W. Bush, received his heart transplant on March 24, 2012	Dick Chiney
2012	They created several DNA-like robotic nanoscale assembly devices.	Nadrian Seeman
2012	Nanotech Conference & Expo June 18–21 in Santa Clara.	California
2012	A faculty member was elected to AIMBE College of Fellows. Institute for Accessible Science becomes newest biomedical research center in Discovery Park. 7th biomedical engineering faculty member named a Purdue Faculty Scholar.	Purdue
2012	The NNI launched two more Nanotechnology Signature Initiatives (NSIs): Nanosensors and the Nanotechnology Knowledge Infrastructure (NKI) – bringing the total to five NSIs.	NKI
2012	The NSET Subcommittee updated both the NNI Strategic Plan and the NNI Environmental, Health, and Safety Research Strategy, drawing on extensive input from public workshops and online dialog with stakeholders from Government, academia, NGOs, and the public, and others.	NSET
2012	Thirty one year-old person successfully used a nervous system controlled bionic leg to climb the Chicago Willis Tower.	Zac Vawter
2012	Apple Academic Press Inc., published, "Goyal, Megh R., 2013. *Biofluid Dynamics of Human Body Systems*. ISBN 978192689546-8."	Megh R. Goyal
2013	Researchers developed cell laden fibers up to a meter in length and on the order of 100 μm in size. These fibers were created using a microfluidic device that forms a double coaxial laminar flow.	Tokyo
2013	In recent times, models have been created that can stand-alone and provide a permanent replacement for a heart with a functional impairment. This new type of artificial heart is currently in the process of being evaluated and it is thought that it will be ready for widespread live use beginning in the future.	USA
2013	Researchers successfully reprogrammed adult cells in a living animal for the first time, creating stem cells that have the ability to grow into any tissue found in the body. Until now these induced pluripotent stem cells have only ever been created in Petri dishes in the laboratory after being removed from the animal. However, researchers at the Spanish National Cancer Research Centre in Madrid, Spain, were able to create these cells in the bodies of living mice.	Spain
2013	September: Researchers build a Carbon nanotube computer.	Stanford

Year	Event/Concepts, etc.	Who
2013	The heart lung-machine, which was invented and perfected through a series of animal experiments, is commonly used to do the job of the heart and lungs for many hours, enabling complex and time-consuming cardiac surgeries to take place.	USA
2013	Using 3-D scaffolding in various configurations, substantial pancreatic organoids were produced *in vitro* (Matrigel)	
2013	Reprogrammable nanofactories.	–
2014	Resource Magazine by ASABE 21(3), 17–19 reports: "The first successful in-orbit plant experiments were flown in Biosatellite – 2, that was in orbit for three days in 1967. The NASA Plant Growth Unit first flew on the STS-3 space shuttle mission in 1982. The Biomass Production System with four independently plant growth chambers was flown abroad the International Space Station from April 8 to June 19, 2002. Plant habitat is currently in development for a 2016 space flight."	NASA
2014	The NNI release the updated 2014 Strategic Plan.	NNI
2014	Eric Brynjolfsson and Andrew McAfee authored *The Second Machine Age: Work, Progress, and Prosperity in a Time of Brilliant Technologies*.	Purdue University
2014	Apple Academic Press Inc. published, Goyal, Megh R., 2014. *Biomechanics of Artificial Organs and Prostheses*. ISBN 978192689584-0.	Megh R. Goyal
2014	Carbon nanotube fiber inexpensive and with over 50 GPa tensile strength.	USA
2014	Biosensors capable of detecting a single molecule.	USA
2020	An estimated 12 million Americans will be limited in daily activities because of arthritis.	USA
2014	"Orthopaedic surgeons started to use a robotic knee replacement process using a series of CT scans to create a customized knee replacement plan based on the needs of an individual patient. It consists of a computer-assisted planning program and robotic instruments used to make incisions and position implants. The robotic instruments are able to manipulate tools and implants much more accurately than the human eye. Because of the precise nature of this technology, many common knee replacement problems are avoided." http://echo.gmu.edu/bionics/exhibits.htm.	www.echo. gmu.edu
2020	Direct construction of artificial human organs.	USA
2020	Possible development of self-replicating nanotech.	USA
2020	Self-replicating nanofactories or nanorobots (nanoassemblers).	USA
2025	Use of nanomachines inside the body for diagnosis and therapy.	USA
2030	NT uses for diagnosis and therapy of Parkinson's & Alzheimer's conditions.	USA

2.3 SUMMARY

This chapter includes the chronological events/developments in bioengineering and biotechnology during 7000 B.C. to 2030 A.D.

KEYWORDS

- **2030 A.D.**
- **7000 B.C.**
- **bioengineering**
- **biological engineering**
- **biomedical engineering**
- **Biomedical Engineering Society, BMES**
- **bionanotechnology**
- **biotechnology**
- **DNA**
- **Florida International University**
- **molecular engineering**
- **nanomedicine**
- **nanotechnology**
- **regenerative medicine**
- **technology**
- **timeline**
- **tissue engineering**
- **Whitaker Foundation**

REFERENCES

1. Berger, E., & Roth, J. (1997). *The Golgi Apparatus*. Basel, Switzerland: Birkhäuser.
2. Bly, R. W. (2005). *The Science in Science Fiction: 83 SF Predictions that Became Scientific Reality*. BenBella Books, Inc.
3. Brock, T. D. (1961). *Milestones in Microbiology*. Madison, WI, USA: Science Tech Publishers. 273 pages.
4. Brock, T. D., (1990). *The Emergence of Bacterial Genetics*. Cold Spring Harbor, New York: Cold Spring Harbor Laboratory Press. 346 pp.

5. Bunch, B., & Hellemans, A., (1993). *The Timetables of Technology*. New York, NY: Simon & Schuster. 490 pp.
6. Colin, M., (2008). *Nanovision: Engineering the Future*. Duke University Press.
7. Feneque, J., (2000). Nanotechnology: a new challenge for veterinary medicine. *The Pet Tribune*. 6(5), 16.
8. Hellemans, A., & Bunch, B., (1988). *The Timetables of Science*. New York: Simon & Schuster. 660 pp.
9. http://bmes.org/cmbesig Cellular and Molecular Bioengineering Special Interest Group.
10. http://bmes.org/faseb Federation of American Societies for Experimental Biology.
11. http://bse.unl.edu/history
12. http://dailysciencefiction.com/science-fiction/nanotech
13. http://svtc.org/our-work/nano/timeline/
14. http://timelines.ws/subjects/Technology.HTML
15. http://www.accessexcellence.org/RC/AB/BC/1977-Present.php
16. http://www.accessexcellence.org/RC/AB/BC/1977-Present.php
17. http://www.biotechinstitute.org/go.cfm?do=Page.View&pid=22
18. http://www.bme.fiu.edu/about/history/
19. http://www.dummies.com/how-to/content/nanotechnology-timeline-and-predictions.html
20. http://www.ehow.com/about_6364029_history-bioengineering.html
21. http://www.ehow.com/about_6364029_history-bioengineering.html#ixzz31PBIg8vV
22. http://www.greatachievements.org/?id=3824
23. http://www.greatachievements.org/?id=3837
24. http://www.historyworld.net/timesearch/default.asp?conid=static_timeline&timelineid=406&page=1&keywords=Engineering%20timeline
25. http://www.medscape.com/viewarticle/566133_2
26. http://www.nano.gov/timeline
27. http://www.nano.gov/timeline
28. http://www.nature.com/nmat/journal/v8/n6/fig_tab/nmat2441_F1.html
29. http://www.science-of-aging.com/timelines/cell-history-timeline-detail.php
30. http://www.timetoast.com/timelines/tissue-engineering-and-regenerative-medicine
31. http://www.wifinotes.com/nanotechnology/nanotechnology-in-fiction.html
32. https://engineering.purdue.edu/BME/AboutUs/History
33. Lopez, J., (2004). Bridging the Gaps: Science Fiction in Nanotechnology. *International Journal for Philosophy of Chemistry*, 10(2), 129–152.
34. Murphy, A., & Judy, P., (1993). *A Further Look at Biotechnology*. Princeton, NJ: The Woodrow Wilson National Fellowship Foundation Biology Institute.
35. Patrick, D., (2007). Tiny Tech, Transcendent Tech – Nanotechnology, Science Fiction, and the Limits of Modern Science Talk. *Science Communication*, 29(1), 65–95.
36. Peters, P., (1993). *From Biotechnology: A Guide to Genetic Engineering*. William C. Brown Publishers, Inc., 1993.
37. Peterson, C., (1995). *Nanotechnology: From Concept to R&D Goal*. HotWired.
38. Peterson, C., (1995). Nanotechnology: evolution of the concept. In: *Prospects in Nanotechnology: Toward Molecular Manufacturing* by Markus Krummenacker and James Lewis (eds.). Wiley, USA.
39. Peterson, C., (2000). Molecular nanotechnology: the next industrial revolution. *IEEE Computer*, January Issue.

40. Peterson, C., (2004). Nanotechnology: from Feynman to the grand challenge of molecular manufacturing. *IEEE Technology and Society*, Winter Issue.
41. Regenerative medicine glossary. *Regenerative Medicine 4* (4 Suppl): 81–88. July 2009.
42. Regenerative Medicine, 2008, *3*(1), 1–5.
43. Regis, E., (1995). *Nano: The Emerging Science of Nanotechnology*. Little Brown.
44. Schwarz, J. A., Contescu, C. I., & Putyera, K., (2004). *Dekker Encyclopedia of Nanoscience and Nanotechnology*. CRC Press.
45. Toumey, C., (2008). The Literature of promises. *Nature Nanotechnology*, *3*(4), 180–181.

CHAPTER 3

GLOSSARY OF SCIENTIFIC AND TECHNICAL TERMS IN BIOENGINEERING AND BIOLOGICAL ENGINEERING

3.1 INTRODUCTION

Bioengineering and biological engineering are the dynamic areas of science and have evolved especially in the last 25 years, and both have now become an integral part of our daily life. The speciality areas in these disciplines are mentioned and described in the introductory pages and in Chapters 1 and 2 of this book. Each one of these disciplines includes a series of scientific and technical terms or words. This chapter defines a series of commonly used terms in these disciplines of science. This will help students, scientists, engineers, physicians, technicians, and other people to broaden their knowledge in bioengineering and biotechnology.

The technical words are often used daily, and therefore *Scientific and Technical Terms in Bioengineering and Biological Engineering* brings these words together and describes them using common English (US) language. The reader can use this as a reference tool. It should be useful to researchers, managers, government officials, students, and the general public, and all those who are involved in these specialized areas. A common understanding of these terms will make communication easier among the readers and scientists. The ability to communicate effectively and efficiently is extremely important and will become increasingly so. Use of consistent terminology is critical in presenting clear and meaningful information. Because the purpose of this glossary is to aid communication and understanding within and among the communities, *the definitions contained within are not intended for legal use.* This glossary is provided for *general information only.*

The glossaries, dictionaries, atlases, technical papers, websites, and several other sources used in this chapter are listed in the bibliography. For the most part, the definitions and glossary used by various sources were accepted without modifications. Finally, in the interest of readability, many of the technical terms in this chapter have been edited for brevity and clarity while retaining the original intent as closely as possible. When more than one definition or a glossary for a particular term was encountered, then the one that describes it most adequately was selected.

Many times, it is difficult to define a term in bioengineering and biological engineering. Therefore, it is necessary to describe it in a paragraph so that the reader will understand its use and meaning. Throughout this book, I prefer to use the word *glossary* instead of *dictionary*.

Scientific and Technical Terms are described in the following pages.
µTAS is an abbreviation for microTAS.

A

2 µm Plasmid is a naturally occurring, double-stranded, circular DNA plasmid (6318 bp) found in the nuclei of yeast.

22q deletion syndrome refers to the missing section of DNA on chromosome 22.

2D PAGE is an abbreviation for two-dimensional polyacrylamide gel electrophoresis (PAGE) that is a common technique for protein separation.

3'-end refers to the end of a polynucleotide with a free (or phosphorylated) 3'-hydroxyl group.

3'-extension refers to a short single-stranded nucleotide sequence on the 3'-hydroxyl end of a double-stranded DNA molecule.

3'-Hydroxyl end is the hydroxyl group that is attached to the 3' carbon atom of ribose or deoxyribose of the terminal nucleotide of a nucleic acid molecule.

5'-end is the end of a polynucleotide with a free (or phosphorylated or capped) 5'-hydroxyl group.

5' end refers to the phosphate group that is attached to the 5' carbon atom of ribose or deoxyribose of the terminal nucleotide of a nucleic acid molecule.

5'-extension is a short single-stranded nucleotide sequence on the 5'-hydroxyl end of a double-stranded DNA molecule.

A is an abbreviation for adenine.

a.k.a. is an abbreviation for *also known as*.

Ab is an abbreviation for antibody. Ab is also known as immuno-globulin (Ig).

ab initio **calculations** are quantum chemical calculations using exact equations with no approximations that involve the whole electronic population of the molecule [IUPAC Computational].

ab initio **gene prediction** are gene prediction programs that rely only on the statistical qualities of exons.

ab initio **quantum chemistry** involves the calculation of chemical properties directly from the molecular Schrodinger equation.

ab initio **quantum mechanical methods** are methods of quantum mechanical calculations independent of any experiment other than the determination of fundamental observables; and are based on the use of the full Schrödinger equation.

ab initio **quantum mechanical modeling** is the application of *ab initio* modeling cross diverse fields such as condensed matter physics, materials science and chemistry.

ab initio refers to models devised without experimental data.

ABC is an abbreviation for Approximate Bayesian Computation. It is a statistical framework using simulation modeling to approximate the Bayesian posterior distribution of parameters of interest often by using multiple summary statistics.

ABC model incorporates the *Arabidopsis* genes required for flower organ identity.

Abdominal aorta is the portion of the aorta in the abdomen.

Abdominal ultrasound is a diagnostic imaging technique which creates images from the rebound of high frequency sound waves in the internal organs.

Abdominal x-ray is a simple study that will give the physician an idea of how the internal organs look.

Abiotic refers to an absence of living organisms.

Abiotic stress refer to outside (non-living) factors, which can cause harmful effects to plants.

Ablation is a technique to treat abnormal heart rhythms, or arrhythmias, and can be done surgically or nonsurgically. It is the removal of material from the surface of an object by vaporization.

Ablution is an act of washing or bathing.

Abruptio placenta is a retro placental blood clot formation.

Absces is a bacterial infection that may be introduced from the bloodstream in cases of generalized or distant infection or form contiguous infection following a skull fracture.

Abscisic acid is a plant growth regulator involved in abscission, dormancy, stomatal opening/closure, and inhibition of seed germination.

Abscissa refers to the horizontal axis of a graph.

Absorb refers to take in. In the cell, materials are taken in (absorbed) from a solution.

Absorption is the movement of nutrients and other substances through the wall of the digestive tract and into the blood. It is the process of absorbing; taking up of water and nutrients by assimilation or imbibition. In animals: solubilized food material is absorbed into the circulatory system through cells lining the alimentary canal.

Abzyme is related to catalytic antibody. It is a monoclonal antibody with catalytic activity. Abzymes are potential tools in biotechnology to perform specific actions on DNA. They are also useful in hydrolysis of esters.

Acaricide is a pesticide used to kill or control mites or ticks.

ACC synthase is an abbreviation for 1-aminocyclopropane-1-carboxylase.

Acceptor control refers to the regulation of the rate of respiration by the availability of ADP as a phosphate acceptor.

Acceptor junction site refers to the junction between the 3' end of an intron and the 5' end of an exon.

Accession number is a unique identifier given to a sequence when it is submitted to one of the DNA repositories (GenBank, EMBL, DDBJ).

Accessory bud is a lateral bud that is at the base of a terminal bud or at the side of an axillary bud.

Acclimatization refers to the adaptation of a living organism to a changed environment under physiological stress.

Accommodation is an adaptation by the sensory receptors to various stimuli over an extended period of time.

Acellular describes tissues or organisms that are not made up of separate cells but often have more than one nucleus.

Acentric chromosome refers to chromosome fragment lacking a centromere.

Acetyl CoA is an abbreviation for acetyl co-enzyme A.

Acetyl co-enzyme A (acetyl CoA) is a compound formed in the mitochondria when an acetyl group combines with the thiol group of co-enzyme A.

aCGH is a technique involving the competitive hybridization of "test" and "reference" DNA probes to target genomic (or cDNA clones) immobilized on a microarray. Applications include detection of copy number variation (CNV), gene annotation and diagnostics.

Achondroplasia is a form of short limbed dwarfism characterized by a normal trunk size with disproportionally short arms and legs, and a disproportionally large head.

Acoustic-spectrum is the range of frequencies and wavelengths of sound waves.

ACP is an abbreviation for acyl carrier protein.

Acquired refers to one that is developed in response to the environment, not inherited.

Acridine dyes is a class of positively charged polycyclic molecules that intercalate into DNA and induce frameshift mutations.

Acrocentric is a chromosome and chromatid with a centromere near one end.

Acropetal refers to: (1) Developing or blooming in succession towards the apex; (2) Transport or movement of substances towards the apex.

Acrosome is an organelle that develops over the anterior half of the head in the spermatozoa of many animals.

Acrylamide gel is also called polyacrylamide gel that is used in biotechnology laboratories for a process called electrophoresis, which is used to separate molecules of different sizes.

Act is a law made by Parliament or a provincial legislature.

Actin is one of the two contractile proteins in muscle (the other being myosin).

Action potential is a recorded change in electrical potential between the inside and outside of a nerve cell.

Activated carbon is also called as activated charcoal.

Activated charcoal refers to the charcoal that has been treated to remove hydrocarbons and to increase its adsorptive properties.

Active collection is a collection, which complements a base collection and is a collection from which seed samples are drawn for distribution, exchange, multiplication and evaluation.

Active electrode is an electrode at which greatest current density occurs.

Active pharmaceutical ingredient (API) or pharmacon or pharmakon (from Greek: φάρμακον, adapted from pharmacos) is the substance in a pharmaceutical drug or a pesticide that is biologically active. The similar terms API and bulk active are also used in medicine.

Active transport is the movement of a molecule or groups of molecules across a cell membrane, which requires the expenditure of cellular energy, because the direction of movement is against the prevailing concentration gradient.

Acute implies severe; sharp; begins quickly.

Acute transfection refers to short-term transfection.

Acyl carrier protein (ACP) is a class of molecules that bind acyl intermediates during the formation of long-chain fatty acids.

Adaptation refers to an adjustment of a population to changes in environment over generations, associated (at least in part) with genetic changes resulting from selection imposed by the changed environment.

Adaptation traits are related to reproduction and survival of the individual in a particular production environment.

Adaptive radiation refers to the evolution of new forms, sub-species or species from one species of plant or animal in order to exploit new habitats or food sources.

Adaptor refers to: (1) A synthetic double-stranded oligonucleotide that has a blunt end, while the other end has a nucleotide extension that can base pair with a cohesive end created by cleavage of a DNA molecule with a specific type II restriction endonuclease. (2) A synthetic single-stranded oligonucleotide that produces a molecule with cohesive ends and an internal restriction endonuclease site.

Addendum (plural: addenda) is an item or a constituent substance to be added.

Addition rule establishes that "*when two events, A and B, are mutually exclusive, the probability that A or B will occur is the sum of the probability of each event*: $P(A \text{ or } B) = P(A) + P(B)$." **Example:** A single 6-sided die is rolled. What is the probability of rolling a 2 or a 5? Applying addition rule in the example, we get: $P(2) = 1/6$, $P(5) = 1/6$, $P(2 \text{ or } 5) = P(2) + P(5) = 1/6 + 1/6 = 2/6 = 1/3$ or 0.333.

Additive allelic effects refer to the effects of alleles at a locus, where the heterozygote is exactly intermediate between the two homozygotes.

Additive gene effects are additive allelic effects summed across all the loci that contribute to genetic variation in a quantitative trait.

Additive genes are genes whose net effect is the sum of their individual allelic effects. They show neither dominance nor epistasis.

Additive genetic variance refers to the net effect of the expression of additive genes, and thus the chief cause of the resemblance between relatives.

Additive genetic variation is the proportion of genetic variation that responds to natural selection.

Adenilate cyclase is the enzyme that catalyzes the formation of cyclic AMP.

Adenine ($C_5H_5N_5$, symbol: A) is a white crystalline purine base. It refers to constituent of DNA and RNA and nucleotides such as ADP and ATP.

Adenosine refers to the (ribo) nucleoside resulting from the combination of the base adenine and the sugar D-ribose.

Adenosine diphosphate (adenosine 5'-diphosphate) is abbreviated as ADP.

Adenosine monophosphate (adenosine 5'-monophosphate) is abbreviated as AMP.

Adenosine triphosphate (adenosine 5'-triphosphate, ATP) consists of adenosine with three phosphate groups, linked together linearly. It is the major carrier of chemical energy in all living organisms and is also required for RNA synthesis. ATP is regenerated by the phosphorylation of AMP and ADP.

Adenovirus is a group of DNA viruses, which cause diseases in animals. They are used in gene cloning, as vectors for expressing large amounts of recombinant proteins in animal cells.

Adenylic acid (Synonym for adenosine monophosphate) is a (ribo)nucleotide containing the nucleoside adenosine.

ADEPT is an abbreviation for antibody-directed enzyme pro-drug therapy. It is a way to target a drug to a specific tissue.

Adhesion is the attraction of dissimilar molecules for each other.

Adjuvant (Latin, *adiuvare*, to aid) is a pharmacological and/or immunological agent that modifies the effect of other agents. The most commonly used adjuvants include aluminum hydroxide and paraffin oil.

ADME is an abbreviation for "absorption, distribution, metabolism, and excretion." ADME describes the disposition of a pharmaceutical compound within an organism.

Admixture is the formation of novel genetic combinations through hybridization of genetically distinct groups.

A-DNA is a right-handed DNA double helix that has 11 base pairs per turn. DNA exists in this form when partially dehydrated.

Adoptive immunization is the transfer of an immune state from one animal to another by means of lymphocyte transfusions.

ADP is an abbreviation for adenosine diphosphate.

Adrenal cortex is the outer region of each adrenal gland; and it secretes steroid hormones.

Adrenal glands are paired endocrine glands, one located just superior to each kidney.

Adrenergic blocking agent inhibits response to sympathetic impulses by blocking the alpha (alpha-adrenergic blocking a.) or beta (beta-adrenergic blocking a.) receptor sites of effector organs.

Adrenergic neuron blocking agent inhibits the release of norepinephrine from postganglionic adrenergic nerve endings.

Adsorbent is a substance to which compounds adhere. A common

adsorbent in tissue culture is activated charcoal.

Adsorption is the formation of a layer of gas, liquid or solid on the surface of a solid.

Adult cloning refers to the creation of identical copies of an adult animal by nuclear transfer from differentiated adult tissue.

Adult stem cells are found in different tissues of the developed, adult organism that remain in an unspecialized state. Their natural role is to replenish dying cells and regenerate damaged tissue.

Advanced maternal age refers to a woman over age 34 (age 35 at delivery) at increased risk for nondisjunction trisomy in fetus.

Advanced refers to an organism or a part thereof, implying considerable development from the ancestral stage.

Adventitious (*L. adventitius*) is a structure arising at sites other than the usual ones, e.g., embryos from any cell other than a zygote.

Aerate refers to supply with or mix with air or gas. The process is aeration.

Aerobe is a micro-organism that grows in the presence of oxygen. Opposite is anaerobe.

Aerobic is one that is active in the presence of free oxygen.

Aerobic bacteria is a bacteria that can live in the presence of oxygen.

Aerobic respiration is a type of respiration in which foodstuffs are completely oxidized to carbon dioxide and water, with the release of chemical energy.

Afferent is a conduction of a nerve impulse toward an organ.

Affinity chromatography is a method for separating molecules by exploiting their ability to bind specifically to other molecules. There are several types of biological affinity chromatography: immuno-affinity chromatography; pseudo-affinity chromatography; and metal affinity chromatography.

Affinity tag (purification tag) refers to an amino acid sequence that has been engineered into a protein to make its purification easier.

AFI is an abbreviation for Amniotic Fluid Index.

Aflatoxin are toxic compounds that bind to DNA and prevent replication and transcription. Aflatoxins can cause acute liver damage and cancer.

AFLP is an abbreviation for amplified fragment length polymorphism. The AFLP is a PCR-based tool used in genetics research, DNA fingerprinting, and in the practice of genetic engineering.

Ag is an abbreviation for antigen.

Agamospermy is the asexual formation of seeds without fertilization in which mitotic division is sometimes stimulated by male gametes.

Agar is a polysaccharide solidifying agent used in nutrient media preparations and obtained from certain types of red algae (*Rhodophyta*).

Agarose is the main functional constituent of agar.

Agarose gel electrophoresis is a method to separate DNA and RNA molecules on the basis of their size, in which samples are subjected to an electric field applied to a gel made with agarose.

Agent Orange is a herbicide containing 2,4,5-T and 2,4-D and the contaminant dioxin.

Aggregate refers to: (1) A clump or mass formed by gathering or collecting units; (2) A body of loosely associated cells, such as a friable callus or cell suspension; (3) Coarse inert material, such as gravel; and (4) A serological reaction (aggregation) in which the antibody and antigen react and precipitate out of solution.

Agonist is a drug, hormone or transmitter substance that forms a complex with a receptor site that is capable of triggering an active response from a cell.

AGP is a file that describes how primary sequences can be assembled to make a non-redundant, contiguous sequence. The sequence being assembled may be a contig or a chromosome.

Agrobacterium is a genus of bacteria that includes several plant pathogenic species, causing tumor-like symptoms.

Agrobacterium rhizogenes causes hairy root disease in some plants.

Agrobacterium tumefaciens is a bacterium that causes crown gall disease in some plants. The bacterium infects a wound, and injects a short stretch of DNA into some of the cells around the wound.

Agrobacterium tumefaciens mediated transformation is the process of DNA transfer from *Agrobacterium tumefaciens* to plants.

Agrobiodiversity refers to that component of biodiversity that is relevant to food and agriculture production. The term agrobiodiversity encompasses within-species, species and ecosystem diversity.

AHG is an abbreviation for antihaemophilic globulin.

AI is an abbreviation for artificial insemination.

AIDS (acquired immunodeficiency syndrome) refers to fatal human disease in which the immune system is destroyed by a retrovirus (Human Immunodeficiency Virus, HIV). The virus infects and destroys helper T-cells, which are essential for combating infections.

Airlift fermenter is a cylindrical fermentation vessel in which the cells are mixed by air introduced at the base of the vessel and that rises through the column of culture medium.

Albinism (Latin *albus*, "white"; also called achromia, achromasia, or achromatosis) is a congenital disorder characterized by the complete or partial absence of pigment in the skin, hair and eyes due to absence or defect of tyrosinase, a copper-containing enzyme involved in the production of melanin.

Albino refers to: (1) An organism lacking pigmentation, due to genetic factors; (2) A conspicuous plastome (plastid) mutant involving loss of chlorophyll.

Alcoholism is a chronic and progressive condition characterized by the inability to control the consumption of alcohol.

Aleurone is the outermost layer of the endosperm in a seed.

Algal biomass refers to single-celled plants, such as *Chlorella* spp. and *Spirulina* spp., that are grown commercially in ponds to make feed materials.

Alginate (Alginic acid or algin) refers to the polysaccharide gelling agent that is distributed widely in the cell walls of brown algae.

Alien species refers to a non-native or nonindigenous species.

Alkylating agent refers to the chemicals that transfer alkyl (methyl, ethyl, etc.) groups to the bases in DNA; and this inhibits cell division by reacting with DNA.

Allantois is an extraembryonic membrane, endoderm in origin extension from the early hindgut, then cloaca into the connecting stalk of placental animals, connected to the superior end of developing bladder.

Allele (Gr. *Allelon*; adj.: allelic, allelomorphic) refers to one of a pair, or series, of variant forms of a gene that occur at a given locus in a chromosome. Alleles are symbolized with the same basic symbol (e.g., *B* for dominant and *b* for recessive).

Alleles refer to alternate forms of a gene or DNA sequence, which occur on either of two homologous chromosomes in a diploid organism.

Allele frequency is the relative number of copies of an allele in a population, expressed as a proportion of the total number of copies of all alleles at a given locus.

Allele-specific amplification (ASA) refers to the use of polymerase chain reaction (PCR) at a sufficiently high stringency that only a primer with exactly the same sequence as the target DNA will be amplified.

Allele-specific oligonucleotide hybridization is the use of an oligonucleotide probe to determine which of the two alternative nucleotide sequences is contained in a DNA molecule.

Allelic diversity is a measure of genetic diversity based on the average number of alleles per locus present in a population.

Allelic exclusion refers to a phenomenon whereby only one functional allele of an antibody gene can be assembled in a given B lymphocyte.

Allelic richness is a measure of the number of alleles per locus; and it allows comparison between samples of different sizes by using various statistical techniques (e.g., rarefaction).

Allelic series is a collection of distinct mutations that affect a single locus.

Allelopathy refers to the phenomenon by which the secretion of chemicals, such as phenolic and terpenoid compounds, by a plant inhibits the growth or reproduction of other plant species with which it is competing.

Allergen is an antigen that provokes an immune response.

Allergy is a condition in which the body is not able to tolerate eating certain foods, or exposure to certain animals, plants, or other substances.

Allogamy is a cross fertilization in plants.

Allogeneic transplantation is a process when the cell, tissue or organ transplants from one member of a species to a genetically different member of the same species.

Allogenic refers to the differing at one or more loci, although belonging to the same species. Thus an organ transplant from one human donor to another is allogeneic, whereas a transplant from a baboon to a human would be xenogeneic.

Allometric refers to when the growth rate of one part of an organism differs from that of another part or of the rest of the body.

Allopatric refers to the inhabiting distinct and separate areas, in the context of natural populations of animals or plants.

Allopatric species or populations occur in geographically separate areas.

Allopatric speciation refers to speciation (Orator F. Cook) occurring at least in part because of geographic isolation. There are four geographic modes of speciation in nature: allopatric, peripatric, parapatric, and sympatric.

Allopolyploid (Gr. *allos*) is a polyploid organism (usually a plant) having multiple sets of chromosomes derived from different species.

Allosome is synonym for sex chromosome.

Allosteric control refers to allosteric regulation.

Allosteric enzyme has two structurally distinct forms, one of which is active and the other inactive. Active forms tend to catalyse the initial step in a pathway leading to the synthesis of molecules. The end product of this synthesis can act as a feedback inhibitor, converting the enzyme to the inactive form. Synonym is allozyme.

Allosteric regulation is a catalysis-regulating process in which the binding of a small effector molecule to one site on an enzyme affects the activity at another site.

Allosteric site refers to that part of an enzyme molecule where the non-covalent binding of an effector molecule can affect the enzyme's catalytic activity.

Allosteric transition is a reversible interaction of a small molecule with a protein molecule, resulting in a change in the shape of the protein and a consequent alteration of the interaction of that protein with a third molecule.

Allotetraploid is an organism with four genomes derived from hybridization of different species: Two of the four genomes are from one species and two are from another species.

Allotype is a classification of antibody molecules according to the antigenicity of the constant regions.

Allozygote is a diploid individual that is homozygous at a locus in which the two genes are not identical by descent from a common ancestor.

Allozygous is an individual whose alleles at a locus are descended from different ancestral alleles in the base population. Allozygotes may be either homozygous or heterozygous in state at this locus.

Allozyme is an allelic enzyme detected through protein electrophoresis.

Alpha fetoprotein (AFP) is a protein excreted by the fetus into the amniotic fluid and from there into the mother's bloodstream through the placenta.

Alpha lactalbumin is a protein component of milk.

Alpha particles (α or α^{++}) consist of two protons and two neutrons bound together into a particle identical to a helium nucleus.

Alternative mRNA splicing refers to the inclusion or exclusion of different exons to form different mRNA transcripts.

Alu repetitive sequence is the most common dispersed repeated DNA sequence in the human genome accounting for 5% of human DNA. The name is derived from the fact that these sequences are cleaved by the restriction endonuclease Alu.

Alu sequence is a highly repeated family of 300-bp long sequences dispersed throughout the human genome. They are released by the digestion of genomic DNA with the restriction of endonuclease AluI.

Alveoli (plural for alveolus) are air sacs in the lungs where oxygen and carbon dioxide are exchanged.

Alveolus is an air sac of the lung through which the gas exchange with the blood takes place.

Ambient temperature is an air temperature at a given time and place.

Amino acid (Gr. *Ammon,* from the Egyptian sun god) is an acid containing the group NH_2. According to the side group R, they are subdivided into: polar or hydrophilic (serine, threonine, tyrosine, asparagine and glutamine); non-polar or hydrophobic (glycine, alanine, valine, leucine, isoleucine, proline, phenylalanine, tryptophan and cysteine); acidic (aspartic acid and glutamic acid) and basics (lysine, arginine, hystidine). The sequence of amino acids determines the shape, properties and the biological role of a protein.

Amino acid sequence is the linear order of the amino acids in a protein or peptide.

Amino acids are the building blocks of proteins. There are 20 known amino acids found in living organisms.

Aminoacyl site (A-site) is one of two sites on ribosomes to which the incoming aminoacyl tRNA binds.

Aminoacyl tRNA synthetase is an enzyme that catalyzes the attachment of an amino acid to its specific tRNA molecule.

Amitosis refers to cell division (cytokinesis), including nuclear division through constriction of the nucleus, without chromosome differentiation as in mitosis.

Amniocentesis is a procedure used in prenatal diagnosis to look at the chromosomes of the developing fetus.

Amniocyte are cells obtained by amniocentesis.

Amnion is a thin membrane that lines the fluid-filled sac in which the embryo develops in higher vertebrates, reptiles and birds.

Amniotic fluid is the fluid which surrounds the fetus and provides a shock absorber and a secondary vehicle for the exchange of body chemicals with the mother.

Amniotic sac is a thin-walled sac that surrounds the fetus during pregnancy. The sac is filled with amniotic fluid and the amnion.

Amniotic stem cells are cells found in the amniotic fluid that surrounds a fetus. Research has shown that they can differentiate into more cell types than adult stem cells.

Amorph is a mutation that abolishes gene function. Synonym is null mutation.

AMOVA is an analysis of molecular variation.

AMP is an abbreviation for adenosine monophosphate.

Amphidiploid is a plant derived from doubling the chromosome number of an interspecific F_1 hybrid.

Amphimixis is a true sexual reproduction involving the fusion of male and female gametes and the formation of a zygote.

Amphiphilic (Greek αμφις, amphis: both and φιλία, philia: love, friendship) describes a chemical compound that possesses both hydrophilic and lipophilic properties.

Ampicillin (beta-lactamase, $C_{16}H_{19}N_3O_4S$) is a penicillin-derived antibiotic that prevents bacterial growth by interfering with synthesis of the cell wall.

Amplicon is the product of a DNA amplification reaction.

Amplification refers to: (1) Treatment designed to increase the proportion of plasmid DNA relative to that of bacterial (host) DNA; (2) Replication of a gene library in bulk; (3) Duplication of gene(s) within a chromosomal segment; and (4) Creation of many copies of a segment of DNA by PCR.

Amplified fragment length polymorphism (AFLP) refers to a type of DNA marker, generated by digestion of genomic DNA with two restriction enzymes to create many DNA fragments, ligation of specific sequences of DNA (called adaptors) to the ends of these fragments, amplification of the fragments via PCR, and visualization of fragments via gel electrophoresis.

Amplify refers to increase the number of copies of a DNA sequence, *in vivo* by inserting into a cloning vector that replicates within a host cell, or *in vitro* by PCR.

Amplitude is the intensity of current flow as indicated by the height of the waveform from baseline.

Amylase is a group of enzymes that degrade starch, glycogen and other polysaccharides, producing a mixture of glucose and maltose. Amylase is present in the saliva of humans and some other mammals, where it begins the chemical process of digestion. Specific amylase proteins are designated by different Greek letters. All amylases are glycoside hydrolases and act on α-1,4-glycosidic bonds: α-amylase and β-amylase.

Amylolytic is the capability of enzymatically degrading starch into sugars.

Amylopectin is a polysaccharide comprising highly branched chains of glucose molecules.

Amylose is a polysaccharide consisting of linear chains of 100 to 1000 glucose molecules.

Anabolic pathway is a pathway by which a metabolite is synthesized.

Anabolism is one of the two subcategories of metabolism, referring to the building up of complex organic molecules from simpler precursors.

Anaemia is a condition caused by a reduced number of red blood cells or haemoglobin.

Anaerobe is an organism that can grow in the absence of oxygen.

Anaerobic digestion refers to the digestion of materials in the absence of oxygen.

Anaerobic is an environment in which molecular oxygen is not available for chemical, physical or biological processes.

Anaerobic respiration is the respiration in which foodstuffs are partially oxidized, with the release of chemical energy, in a process not involving atmospheric oxygen.

Anaerobic refers to an environment or condition in which molecular oxygen is not available for chemical, physical or biological processes.

Anagenesis are evolutionary changes that occur within a single lineage through time.

Anal fissure is a small tear in the anus that can cause bleeding, itching, or pain.

Analgesia is a loss of sensibility to pain.

Analgesic is any drug intended to alleviate pain.

Analogous refers to the features of organisms or molecules that are superficially or functionally similar but have evolved in a different way or contain different compounds.

Analysis of molecular variation (AMOVA) is a statistical approach to partition the total genetic variation in a species into components within and among populations or groups at different levels of hierarchical subdivision. Analogous to ANOVA in statistics.

Anaphase (Gr. *ana*, up + *phais*, appearance) refers to the stage of mitosis or meiosis during which the daughter chromosomes migrate to opposite poles of the cell (toward the ends of the spindle). Anaphase follows metaphase and precedes telophase.

Anastomosing is a connecting the end of an artery with a similar vessel.

Anastomosis refers to the connection between peripheral blood vessels without an intervening capillary bed.

Anchor gene is a gene that has been positioned on both the physical map and the linkage map of a chromosome, and thereby allows their mutual alignment.

Androgen is any hormone that stimulates the development of male secondary sexual characteristics.

Androgenesis refers to male parthenogenesis: the development of a haploid embryo from a male nucleus. Androgenesis is detected by cytological staining.

Anemia refers to not having enough red blood cells (RBC) in the human body.

Anesthesia refers to the loss of sensation.

Aneuploid is a cell where the total number of chromosomes is not an exact multiple of 23. The most common aneuploid numbers are 45 (one chromosome is missing) and 47 (one chromosome is added).

Aneuploidy (Gr. *aneu*, without + ploid) is an organism or cell having a chromosome number that is not an exact multiple of the monoploid *(x)* with one chromosome being present in greater (e.g., trisomic $2n + 1$)

or lesser (e.g., monosomic $2n - 1$) number than the normal diploid number.

Aneurysm refers to the bulge in the blood vessel. It affects the large arteries throughout the body.

Angelman syndrome is a condition characterized by severe mental deficiency, developmental delay and growth deficiency, puppet-like gait and frequent laughter unconnected to emotions of happiness.

Anger camera is also called a gamma camera or a scintillation to image gamma radiation emitting radioisotopes, a technique known as scintigraphy.

Angina refers to a chest pain.

Angiogenesis describes the development of new vessels from already existing vessels.

Angiogenin is one of the human angiogenic growth factors.

Angiography is an x-ray technique that makes use of a dye injected into the coronary arteries to study blood circulation through the vessels and to measure the degrees of obstruction to blood flow.

Angioplasty is a nonsurgical technique for treating diseased arteries by temporarily inflating a tiny balloon inside an artery.

Angiosperm is a division of the plant kingdom that includes all flowering plants.

Angiotensins are a group of hormones that are powerful vaconstrictors.

Animal cell immobilization refers to an entrapment of animal cells in some solid material in order to produce some natural product or genetically engineered protein.

Animal genetic resources databank is a databank that contains inventories of farm animal genetic resources and their immediate wild relatives, including any information that helps to characterize these resources.

Animal genome (gene) bank refers to a planned and managed repository containing animal genetic resources.

Animal model is a laboratory animal with a specific disease that researchers experiment with to find out more about the causes of a disease, its diagnosis in humans, and to investigate or trial new treatments or preventative actions.

Anion is a negatively charged ion.

Anneal is the pairing of complementary DNA or RNA sequences, via hydrogen bonding, to form a double-stranded polynucleotide.

Annealing refers to the process of heating (de-naturing step) and slowly cooling (re-naturing step) double-stranded DNA to allow the formation of hybrid DNA or complementary strands of DNA or of DNA and RNA.

Annotation is a process of adding biological information to genome sequence. Features that are added to the genome often include gene models, SNPs, and STSs.

Annual refers to: (1) (adj:) Taking one year, or occurring at intervals of one year; (2) Noun: In botany, a plant that completes its life cycle within one year.

Anode is positively charged electrode in a direct current system.

Anonymous DNA marker is a DNA marker detectable by virtue of variation in its sequence.

Anorectal atresia refers to the absence of a normal opening between the anus and the rectum.

ANOVA stands for Analysis of variance.

Antagonism is an interaction between two organisms in which the growth of one is inhibited by the other.

Antagonist is a compound that inhibits the effect of an agonist in such a way that the combined biological effect of the two becomes smaller than the sum of their individual effects.

Anterior choroidal artery (ACA) usually arises from the ICA just beyond the origin of the PoCA.

Anther culture refers to the aseptic culture of immature anthers to generate haploid plants from microspores via androgenesis.

Anther is the upper part of a stamen, containing pollen sacs within which the pollen develops and matures.

Anthesis refers to the flowering period or efflorescence.

Anthocyanin refers to water-soluble blue, purple and red flavonoid pigments found in vacuoles of cells.

Anthropocentrism is a view that regards humans as the central element of the universe.

Antiacids are medicines that neutralize stomach acid.

Antiauxin is a chemical that interferes with the auxin response.

Antibiosis refers to the prevention of growth or development of an organism by a substance or another organism.

Antibiotic is a class of natural and synthetic compounds that inhibit the growth of, or kill some micro-organisms.

Antibiotic resistance is the ability of a microorganism to produce a protein that disables an antibiotic or prevents transport of the antibiotic into the cell.

Antibiotic resistance marker gene (ARMG) refer to genes (usually of bacterial origin) used as selection markers in transgenesis, because their presence allows cell survival in the presence of normally toxic antibiotic agents.

Antibody (Gr. *anti*, against + body) refers to an immunological protein (called an immunoglobulin, Ig) produced by certain white blood cells (lymphocytes) of the immune system of an organism in response to a contact with a foreign substance (antigen). The antibody can be cut by proteases into several fragments, known as Fab, Fab,' and Fc.

Antibody binding site is the part of an antibody that binds to the antigenic determinant.

Antibody class refers to the class to which an antibody belongs, depending on the type of heavy chain present. In mammals, there are five classes of antibodies: IgA, IgD, IgE, IgG, and IgM.

Antibody structure are antibodies with a well-defined structure. Each antibody has two identical "light" chains and two identical "heavy" chains. Each chain comprises a constant region, and a variable region.

Antibody-mediated immune response is the synthesis of antibodies by B-cells in response to an encounter of cells of immune system with a foreign antigen. Synonym is humoral immune response.

Anticholinergics are medicines that help calm spasms in the intestine.

Anticlinal is the orientation of cell wall or plane of cell division perpendicular to the surface. Opposite is periclinal.

Anticoagulant is any drug that keeps blood from clotting.

Anticoding strand is the DNA strand used as template for transcription. Synonym is template strand.

Anticodon is a sequence of three bases in a molecule of transfer RNA (tRNA) that binds to a complementary codon in messenger RNA (mRNA).

Antidiarrheals are medicines that help control diarrhea.

Antiemetics are medicines that help prevent and control nausea and vomiting.

Antigen (Ag) is any substance that stimulates the production of antibodies in the body. Synonym is immunogen.

Antigenic determinant is a surface feature of a microorganism or macromolecule, such as a glycoprotein.

Antigenic switching refers to the altering of a micro-organism's surface antigens through genetic re-arrangement, to elude detection by the host's immune system.

Antihaemophilic globulin (AHG) is one of the blood clotting factors, a soluble protein that causes the fibrin matrix of a blood clot to form. Synonym is antihaemophilic factor VIII.

Anti-idiotype antibodies are antibodies which recognize the binding sites of other antibodies.

Antimicrobial agent is any chemical or biological agent that inhibits the growth and/or survival of micro-organisms.

Antinutrient refers to compounds that inhibit the normal uptake or utilization of nutrients.

Anti-oncogene is a gene whose product prevents the normal growth of tissue.

Antioxidant refers to compounds that slow the rate of oxidation reactions.

Antioxidant solution refers to the pre-treatment solution (e.g., Vitamin C; citric acid) that retards senescence and browning of tissue.

Antiparallel orientation refers to the normal arrangement of the two strands of a double-stranded DNA molecule, and of other nucleic-acid duplexes (DNA-RNA, RNA-RNA), in which the two strands are oriented in opposite directions so that the 5'-phosphate end of one strand is aligned with the 3'-hydroxyl end of the complementary strand.

Antisense DNA is one of the two strands of double-stranded DNA, usually that which is complementary (hence "anti") to the mRNA, i.e. the non-transcribed strand.

Antisense gene is a gene that produces an mRNA complementary to the transcript of a normal gene.

Antisense RNA is a RNA sequence that is complementary to all or part of a functional mRNA molecule, to which it binds, blocking its translation.

Antisense therapy refers to the *in vivo* treatment of a genetic disease by blocking translation of a protein with a DNA or an RNA sequence that is complementary to a specific mRNA.

Antiseptic is any substance that kills or inhibits the growth of disease-causing micro-organism (a micro-organism capable of causing sepsis), but is essentially non-toxic to cells of the body.

Antiserum is the fluid portion of the blood of an immunized animal, which retains any antibodies.

Antispasmodics are medicines that help reduce or stop muscle spasms in the intestines.

Anti-terminator is a type of protein which enables RNA polymerase to ignore certain transcriptional stop or termination signals and read through them to produce longer mRNA transcripts.

Antitranspirant is a compound applied to the leaves of plants to reduce transpiration. Examples include phenylmercury acetate, abscisic acid (ABA), and aspirin.

Antixenosis is the modification of the behavior of an organism by a substance or another organism. Particularly used in the context of a plant's apparent resistance against insect feeding, when the insects are presented with a choice of plant genotypes.

Anus is an opening at the end of the digestive tract where bowel contents leave the body.

Aorta is the largest and main systemic artery of the body; arises from the left ventricle and branches to distribute blood to all parts of the body except the lungs.

Aortic valve regulates blood flow from the heart into the aorta. Outlet valve from left ventricle to aorta.

Apert syndrome is a condition caused by the premature closure of the sutures of the skull bones, resulting in an altered head shape, with webbed fingers and toes.

Apex (plural: apices) refers to the tip, point or angular summit.

Apical cell is a meristematic initial in the apical meristem of shoots or roots of plants.

Apical dominance refers to the phenomenon of inhibition of growth of lateral (axillary) buds in a plant by the presence of the terminal (apical) bud on the branch, due to auxins produced by the apical bud.

Apical meristem is a region of the tip of each shoot and root of a plant in which cell division is continually occurring to produce new stem and root tissue, respectively.

Apoenzyme is an inactive enzyme that has to be associated with a specific organic molecule called a co-enzyme in order to function. The apoenzyme/co-enzyme complex is called a holoenzyme.

Apomixes (adj.: apomictic) refer to the asexual production of diploid offspring without the fusion of gametes.

Apomixis is the production of an embryo in the absence of meiosis.

Apoptosis (Ancient Greek ἀπό *apo*, away from and πτῶσις *ptōsis*, falling) is the process of programmed cell death (PCD) that may occur in multicellular organisms.

AP-PCR is an abbreviation for arbitrarily primed polymerase chain reaction.

Appendectomy is an operation to remove the appendix.

Appendicitis refers to the irritation, inflammation, and pain in the appendix, caused by infection, scarring, or obstruction (blockage).

Appendix is a small pouch attached to the first portion of the large intestine (the cecum).

Applicator is the electrode used to transfer energy in microwave diathermy.

Approximate Bayesian computation (ABC) is a statistical framework using simulation modeling to approximate the Bayesian posterior distribution of parameters of interest often by using multiple summary statistics.

Aptamers (Latin *aptus*, fit, and Greek *meros*, part) are oligonucleic acid or peptide molecules that bind to a specific target molecule.

Aquaculture (also known as fish or shellfish farming) refers to the breeding, rearing, and harvesting of plants and animals in all types of water environments including ponds, rivers, lakes, and the ocean.

Arabidopsis refers to the genus of flowering plants in the Cruciferae. *A. thaliana* is used in research as a model plant because it has a small fully sequenced genome.

Arabidopsis thaliana **refers to** the 'lab rat' of plants. This small, flowering plant is ideal for studying plant biology and genetics as it has a short life cycle; it is easy to transform and its genome is small and has been fully sequenced.

Arbitrarily primed polymerase chain reaction (AP-PCR) is an application of the PCR to generate DNA fingerprints. The technique uses arbitrary primers to amplify anonymous stretches of DNA.

Arbitrary primer is an oligonucleotide primer whose sequence is chosen at random, rather than one whose sequence matches that of a known locus.

Archaea refers to the single-celled life forms adapted for existence in high pressure, anaerobic, environments such as at extreme ocean depths.

ARMG is an abbreviation for antibiotic resistance marker gene.

Arrhenius equation relates collision frequency, steric factor. It defines rate constant, k, by the equation: $k = A[e^{Ea/RT}]$, where: A = product of the collision frequency and a steric factor, and $[e^{Ea/RT}]$ = fraction of collisions with sufficient energy to produce a reaction.

Arrhythmia is an alteration in rhythm of the heartbeat either in time or force.

ARS is an abbreviation for autonomous(ly) replicating segment (or sequence).

ARS (autonomous replicating sequence) is any eukaryotic DNA sequence that initiates and supports chromosomal replication; they have been isolated in yeast cells. Also called autonomous(ly) replicating segment.

Arterioles is a small, muscular branches of arteries. When they contract, they increase resistance to blood flow, and blood pressure in the arteries increases.

Artery is a vessel that carries oxygen-rich blood to the body.

Artery, Anterior Cerebral (ACA) passes anteromedially via the horizontal plane to enter the interheimispheric fissure, anastomoses with the contralateral ACA via the anterior communicating artery forming the anterior portion of the circle of Willis.

Artery, Anterior choroidal originates occasionally from the PoCA or the middle cerebral artery (MCA).

Artery, External Carotid is the artery that supplies blood to the jaw, face, neck and meninges.

Artery, Internal Carotid starts at the carotid sinus at bifurcation of CCA (Common Carotid Artery) at the level of the upper border of the thyroid cartilage al the level of the fourth cervical vertebra.

Artery, Middle Cerebral is the largest branch of ICA and appears almost as its direct continuation.

Artery, Ophthalmic passes through the optic canal to supply the eye and other structures of the orbit.

Artery, Posterior Cerebral (PCA) refers to the basilar artery ends by dividing into the two posterior cerebral arteries.

Artery, Posterior Communication arises just before the termination of the ICA and passes backward to the join the first part of the posterior cerebral artery PCA.

Arthritis is one of the most pervasive diseases in the United States and is the leading cause of disability.

Artificial inembryonation refers to non-surgical transfer of embryo(s) to a recipient female. As *in vitro* embryo technology develops, artificial inembryonation may replace artificial insemination.

Artificial insemination (AI) is the deposition of semen, using a syringe, at the mouth of the uterus to make conception possible.

Artificial seed refers to the encapsulated or coated somatic embryos that are planted and treated like seed.

Artificial selection is the practice of choosing individuals from a population for reproduction, usually because these individuals possess one or more desirable traits.

ASA is an abbreviation for allele-specific amplification. ASA utilizes an allele-specific oligonucleotide, at least partially complementary to more than one variant of the target sequence, but having a 3'-terminal nucleotide complementary to only one variant of the target sequence and having at least one nucleotide with a base covalently modified at the exocyclic amino group.

Ascending colon refers to the portion of the large intestine that is on the right side of the abdomen.

Ascertainment bias is the selection of loci for marker development (e.g., SNPs or microsatellites) from an unrepresentative sample of individuals, or using a particular method, which yields loci that are not representative of the spectrum of allele frequencies in a population.

Ascites refers to an abnormal accumulation of fluid in the peritoneal cavity, occurring naturally as a complication of cirrhosis of the liver, among other conditions.

Ascorbic acid (vitamin C: $C_6H_8O_6$) is a water-soluble vitamin that is present naturally in some plants, and also synthetically produced. Aside from its role as a vitamin, it is used as an anti-oxidant in plant tissue culture.

Ascospore is one of the spores contained in the ascus of certain fungi.

Ascus (plural: asci) refers to the reproductive sac in the sexual stage of a type of fungi (*Ascomycetes*) in which ascospores are produced.

Aseptic refers to sterile, free of contaminating organisms.

Asexual is any type of reproduction not involving meiosis or the union of gametes.

Asexual embryogenesis refers to the sequence of events whereby embryos develop from somatic cells.

Asexual propagation refers to vegetative, somatic, non-sexual

reproduction of a plant without fertilization.

Asexual reproduction refers to reproduction that does not involve the formation and union of gametes from the different sexes or mating types.

Asian Technology Information Program (ATIP) is carried out by Asia Pacific Nanotechnology Initiatives (APNI).

A-site is an abbreviation for amino-acyl site (or correctly: amino-acyl-tRNA site). It refers to the site on a ribosome to which the incoming aminoacyl-tRNA is bound during protein synthesis.

ASN is an abbreviation for asparagine ($C_4H_8N_2O_3$); and is also known as **asparamide**. **Asparagine** is an α-amino acid that is found in many proteins, particularly in plant proteins, such as in asparagus. Asparagine is closely related to the amino acid aspartic acid, into which it is easily hydrolized.

Aspartic acid (ASP, $C_4H_7NO_4$) is an amino acid necessary for nucleotide synthesis and occasionally included in plant tissue culture media.

Assay refers to: (1) To test or evaluate; (2) The procedure for measuring the quantity of a given substance in a sample (chemically or by other means);

and (3) The substance to be analysed.

Assignment test is a statistical method using multilocus genotypes to assign individuals to the population from which they most likely originated.

Assisted human reproduction (AHR) is any activity undertaken for the purpose of facilitating human reproduction, e.g., *in vitro* fertilization, donor insemination and intra-cytoplasmic sperm injection (ICSI).

Assisted reproductive technologies (ART) refer to advanced fertility techniques, such as *in vitro* fertilization (IVF), used to bring eggs and sperm together to help achieve pregnancy.

Associative overdominance is also known as pseudo-overdominance. It refers to an increase in fitness of heterozygotes at a neutral locus.

Assortative mating is a preferential mating between individuals with a similar (or a different) phenotype.

Astrocyt is a type of supporting (glial) cell found in the nervous system.

Asymmetric hybrid is a hybrid formed, usually via protoplast fusion, between two donors, where the chromosome complement of one of the donors is incomplete.

Asynapsis refers to the failure or partial failure in the pairing of homologous chromosomes during the first meiotic prophase.

Atherosclerosis is a form of arteriosclerosis. It is the reduction in blood flow through the arteries caused by plaques that form on the insides of arteries and partially restrict the flow of blood.

Atomic mass unit (a.m.u.) is a unit of mass to express relative atomic masses. It is equal to 1/12 of the mass of an atom of the isotope carbon-12 and is equal to 1.66033×10^{-27} kg.

Atomic number (also proton number Z) is a number of protons within the atomic nucleus of a chemical element.

Atomic weight is the weighted average mass of the atoms in a naturally occurring element.

Atonic colon is a lack of normal muscle strength in the large intestine; and is caused by overuse of laxatives or by a disease called Hirschsprung's disease. **Atonic** refers to without tone.

ATP (adenosine triphosphate) refers to a nucleotide of fundamental importance as a carrier of chemical energy in all living organisms. It consists of adenosine with three phosphate groups, linked together linearly.

ATPase is an enzyme that brings about the hydrolysis of ATP, by the cleavage of either one phosphate group with the formation of ADP and inorganic phosphate, or of two phosphate groups, with the formation of AMP and pyrophosphate.

Atresia is a lack of a normal opening, from the esophagus, the intestines, or the anus.

Atria refer to the two upper or holding chambers of the heart.

Atrioventicular valve refers to a valve between each atrium and its ventricle that prevents back flow of blood. The right trioventricular valve is the tricuspid valve; the left atrioventricular valve is the mitral valve.

Atrioventricular node refers to mass of specialized cardiac tissues that receive and impulse from sinoatrial node (peacemaker) and conducts it to the ventricles.

Atrium is an anatomical cavity or passage; especially a main chamber of the heart into which blood returns from circulation.

Attenuated vaccine refers to a virulent organism that has been modified to produce a less virulent form, but nevertheless retains the ability to elicit antibodies against the virulent form.

Attenuation (or extinction) is the gradual loss in intensity of any kind of flux through a medium. It is a mechanism for controlling gene expression in prokaryotes that involves premature termination of transcription.

Attenuator is a nucleotide sequence in the 5' region of a prokaryotic gene (or in its RNA) that causes premature termination of transcription.

Aureofacin is an antifungal antibiotic produced by a strain of *Streptomyces aureofaciens*.

Authentic protein is a recombinant protein that has all the properties of its naturally occurring counterpart.

Autocatalysis refers to catalysis in which one of the products of the reaction is a catalyst for the reaction.

Autoclave refers to: (1) An enclosed chamber in which substances are heated under pressure to sterilize utensils, liquids, glassware, etc., using steam; and (2) A pressure cooker used to sterilize growth medium and instruments for tissue culture work.

Autogamy refers to self-fertilization in a hermaphroditic species where the two gametes fused in fertilization come from the same individual.

Autogenous control is the action of a gene product to inhibit (negative autogenous control) or enhance (positive autogenous control) the expression of the gene that codes for it.

Auto-immune disease is a disorder in which the immune systems of affected individuals produce antibodies against molecules that are normally produced by those individuals (called self antigens).

Auto-immunity is a disorder in the body's defence mechanism in which an immune response is elicited against its own (self) tissues.

Autologous cells are cells taken from an individual, cultured (or stored), and, possibly, genetically manipulated before being transferred back into the original donor.

Autologous transplantation is a process when the cell, tissue or organ transplants from one individual back to the same individual. Such transplants do not induce an immune response and are not rejected, because the transplanted tissue is genetically identical to the recipient.

Autolysis is the process of self destruction of a cell, cell organelle, or tissue.

Autonomous refers to any biological unit that can function on its own.

Autonomous(ly) replicating segment (or sequence) (ARS) refers to any eukaryotic **DNA** sequence that initiates and supports chromosomal replication.

Autopolyploid is a polyploid whose constituent genomes are derived from the same or nearly the same progenitor. In an autotetraploid, each chromosome is present in four copies.

Autoradiograph is a technique for visualizing the presence, location and intensity of radioactivity in histological preparations, paper chromatograms or electrophoretic gel separations, obtained by overlaying the surface with X-ray film and allowing the radiation to form an image on the film.

Autoradiography is a technique that captures the image formed in a photographic emulsion as a result of the emission of either light or radioactivity from a labeled component that is placed next to unexposed film.

Autosomal is a locus that is located on an autosome (i.e., not on a sex chromosome).

Autosomal dominant (Autosomal) refers to a non-sex chromosome.

Autosomal recessive describes a type of inheritance where an individual must inherit a mutation in both copies of a gene in order to develop the associated trait or disorder.

Autosome is a chromosome that is not involved in sex determination.

Autotroph is an organism capable of self-nourishment utilizing carbon dioxide or carbonates as the sole source of carbon and obtaining energy from radiant energy or from the oxidation of inorganic elements, or compounds such as iron, sulphur, hydrogen, ammonium and nitrites. Opposite is heterotroph.

Autotrophic refers to self-nourishing organisms capable of utilizing carbon dioxide or carbonates as the sole source of carbon and obtaining energy for life processes from radiant energy or from the oxidation of inorganic elements, or compounds such as iron, sulphur, hydrogen, ammonium and nitrites.

Autotrophy is the capacity of an organism to use light as the sole energy source in the synthesis of organic material from inorganic elements or compounds.

Autozygosity is a measure of the expected homozygosity where alleles are identical by descent.

Autozygous are individuals whose alleles at a locus are identical

by descent from the same ancestral allele.

Auxin is a group of plant growth regulators (natural or synthetic) which stimulate cell division, enlargement, apical dominance, root initiation, and flowering. One naturally produced auxin is indole-acetic acid (IAA).

Auxin-cytokinin ratio (ACR) is the relative proportion of auxin to cytokinin present in plant-tissue-culture media. Varying the relative amounts of these two hormone groups in tissue culture formulae affects the proportional growth of shoots and roots *in vitro*. This relationship was first recognized by C.O. Miller and F. Skoog in the 1950s.

Auxotroph is a mutant cell or micro-organism lacking one metabolic pathway present in the parental strain, and that consequently will not multiply on a minimal medium, but requires for growth the addition of a specific compound, such as an amino acid or a vitamin.

Availability is a reflection of the form and location of nutritional elements and their suitability for absorption.

Avian influenza refers to a "bird flu," this is a highly contagious influenza virus that can infect any bird.

Avidin is a glycoprotein present in egg white, which has a strong affinity to biotin.

Avidity is a measure of the binding strength of an antibody to its antigen.

Avirulence gene (avr gene) refers to when many plants contain *R* genes, which confer simply-inherited resistance to a specific pathogen race.

Avogadro's number is number of atoms in exactly 12 g of pure C-12, and is equal to 6.022×10^{23}.

Avr gene is an abbreviation for avirulence gene.

Axenic culture refers to one free of external contaminants and internal symbionts; generally not possible with surface sterilization alone, sometimes used incorrectly to indicate aseptic culture.

Axial pump is a rotary pump with axial flow

Axilla is the cavity beneath the junction of the arm and the body, better known as the armpit.

Axillary bud is a bud found at the axil of a leaf. Synonym is lateral bud.

Axillary bud proliferation refers to propagation in culture using protocols and media which promote axillary (lateral shoot) growth.

B

. . . .

B cell is an important class of white blood cells that mature in bone marrow and produce antibodies.

B chromosome is a supernumerary (exceeding the usual number) chromosome present in some individuals (both plant and animal). They are smaller than the normal chromosomes.

B lymphocyte is an important class of lymphocytes that mature in bone marrow (in mammals) and the Bursa of Fabricius (in birds) and produce antibodies. Synonym is B cell.

BABS is an abbreviation for biosynthetic antibody binding sites. BABS, which incorporated the variable domains of anti-digoxin monoclonal antibody 26-10 in a single polypeptide chain, can be produced in *Escherichia coli* by protein engineering.

BAC (bacterial artificial chromosome) is a cloning vector constructed from bacterial fertility (F) factors; like YAC vectors, they accept large inserts of size 200 to 500 kb.

BAC and PAC (Bacterial Artificial Chromosome and P1 Artificial Chromosom) refer commonly to cloning vectors for the human genome project.

BAC end sequence refers to the ends of BACs that are sequenced and the clone association information is retained.

Bacillus is a rod-shaped bacterium.

Bacillus thuringensis (bt) is a naturally occurring soil bacterium, which makes an endotoxin that is toxic to larvae. The gene for this endotoxin has been incorporated into corn to produce a genetically modified corn plant that can defend itself against the corn borer.

Back mutation is a second mutation at the same site in a gene as the original mutation.

Backcross refers to the crossing an organism with one of its parent organisms.

Backflow is the flow of blood backward through the heart

Bacteria refer to large group of single-celled organisms that do not have organelles enclosed in membranes. Bacteria is plural of bacterium.

Bacterial artificial chromosome is a plasmid vector that can be used to clone large inserts of DNA (up to 500 kb).

Bacterial toxin is a toxin produced by a bacterium, such as Bt toxin of *Bacillus thuringiensis*.

Bacteriocide (bactericide) is a chemical or drug that kills bacterial cells.

Bacteriocin is a protein produced by bacteria of one strain and active against those of a closely related strain.

Bacteriophage is a virus that infects bacteria. Also called simply phage. Altered forms are used in DNA cloning work, where they are convenient vectors. The bacteriophages most used are derived from two "wild" phages, called M13 and lambda.

Bacteriostat is a substance that inhibits or slows down growth and reproduction of bacteria.

Bacterium (plural: bacteria) is a common name for the class Schizomycetes: minute, unicellular organisms, without a distinct nucleus. They are classified on the basis of their oxygen requirement (aerobic vs anaerobic) and shape (spherical = coccus; rodlike = bacillus; spiral = spirillum; comma-shaped = vibrio; corkscrew-shaped = spirochaete; filamentous).

Baculovirus is a class of insect virus which have been used to make DNA cloning vectors for gene expression in eukaryotic cells.

Baculovirus expression vector (BEV) is a method for the *in vitro* production of complex recombinant eukaryotic proteins.

Balanced lethal system is a system for maintaining a recessive lethal allele at each of two loci on the same pair of chromosomes.

Balanced polymorphism refers to two or more types of individuals maintained in the same breeding population.

Balancing selection refers to diversifying selection that maintains polymorphism resulting from such mechanisms as frequency-dependent selection, spatially heterogeneous selection, or heterozygous advantage.

Balloon urethroplasty is a thin tube with a balloon is inserted into the opening of the penis and guided to the narrowed portion of the urethra, where the balloon is inflated to widen the urethra and ease the flow of urine.

Band gap energy (Eg) is the energy that lie between the valence and conduction bands, for semiconductors and insulators; for intrinsic materials, electrons are forbidden to have energies within this range.

Bar is a unit used for pressure of fluid. 1 bar = 10^5 Pa = 1 atmosphere = 100 cbar.

Barium enema is a procedure done to evaluate the large intestine for abnormalities.

Barium is a liquid used to coat the inside of organs so they will show up on an x-ray.

Barnase is a bacterial ribonuclease, which, when transformed into plants and expressed in the anthers, generates a male sterile phenotype.

Barotropic is fluid is one whose pressure and density are related by an equation of state that does not contain the temperature as a dependent variable.

Barr bodies are inactivated X chromosomes in female mammals that condense to form a darkly colored structure in the nuclei of somatic cells.

Barr body is the condensed single X-chromosome seen in the nuclei of somatic cells of female mammals.

Barrett's esophagus is a condition in which normal cells that line the esophagus, called squamous cells, turn into abnormal cells, called specialized columnar cells.

Barstar protein is a polypeptide inhibitor of barnase.

Basal refers to: (1) Located at the base of a plant or a plant organ;

and (2) A fundamental formulation of a tissue culture medium containing nutrients but no growth promoting agents.

Basal body is a small granule to which a cilium or flagellum is attached.

Basal cells are more or less rounded cells located to the basal lamina. They partly sheathe the first portion of the axon of the olfactory cell that extends from the epithelium to the underlying connective tissue.

Base refers to a cyclic, nitrogen-containing compound that is one of the essential components of nucleic acids. Exists in five main forms: adenine, A; guanine, G; thymine, T; cytosine, C; uracil, U.

Base analogue is a non-natural purine or pyrimidine base that differs slightly in structure from the normal bases, but can be incorporated into nucleic acids.

Base collection is a collection of seed stock or vegetative propagating material (ranging from tissue cultures to whole plants) held for long-term security in order to preserve the genetic variation for scientific purposes.

Base pair (bp) refers to two strands that constitute DNA are held together by specific hydrogen bonding between purines and pyrimidines (A pairs with T; and G pairs with C). The size of a nucleic acid molecule is often described as number of base pairs or thousand base pairs (kilobase pairs, kb) it contains.

Base pairs refer to pairs of complementary bases that form each ring of the DNA double helix.

Base sequence is a partnership of organic bases found in DNA and RNA.

Base substitution is a replacement of one base by another in a DNA molecule.

Basic fibroblast growth factor is abbreviated as BFGF.

Basipetal refers to developing, in sequence, from the apex towards the base.

Basophil is a type of white blood cell (leucocyte), produced by stem cells in the red bone marrow.

Batch culture is a suspension culture in which cells grow in a finite volume of liquid nutrient medium and follow a sigmoid pattern of growth. Synonym is batch fermentation.

Batch fermentation refers to a process in which cells or microorganisms are grown for a limited time.

Bayesian inference is a procedure of statistical inference in which observed data are interpreted not as frequencies or proportions, but rather are used to compute the probability that a hypothesis is true, given what was observed. Bayes' theorem is named after the Reverend Thomas Bayes.

Baysian analysis is a mathematical method to further refine recurrence risk taking into account other known factors.

Becker muscular dystrophy refers to X-linked condition characterized by progressive muscle weakness and wasting; manifests later in life with progression less severe than Duchenne muscular dystrophy.

Bench scale process is a small- or laboratory-scale process.

Benign prostatic hyperplasia (BPH or Benign prostatic hypertrophy) is an enlargement of the prostate caused by disease or inflammation.

Benign tumors are abnormalities on the neuroglia cells.

Bernoulli Equation refers to statement of the conservation of energy in a form useful for solving problems involving fluids. For a non-viscous, incompressible fluid in steady flow, the sum of pressure, potential and kinetic energies per unit volume is constant at any point.

Beta-DNA is the form of DNA generally found in nature.

Beta-emission is a form of nuclear decay where a neutron splits into a proton plus electron plus neutrino set. The proton stays in the nucleus but the electron (beta ray) is ejected.

Beta-endorphin is a peptide consisting of amino acid sequence 61-91 of the endogenous pituitary hormone beta-lipotropin.

Beta-error is a statistical error (type II) made in testing when it is concluded that something is negative when it really is positive. Beta error is often referred to as a false negative.

Beta-galactosidase (beta-gal or β-gal) is a hydrolase enzyme that catalyzes the hydrolysis of β-galactosides into monosaccharides.

Beta-glucuronidase (GUS) is an enzyme produced by certain bacteria, which catalyzes the cleavage of a whole range of beta-glucuronides.

Beta-lactamase is a enzyme produced by some bacteria that provide resistance to beta-lactam antibiotics like penicillins, cephamycins, and carbapenems (ertapenem), although carbapenems are relatively resistant to beta-lactamase.

Beta-lipotropin is a pituitary hormone containing β-endorphin and enkephalin and having opiate activity.

Beta particles are high-energy, high-speed electrons or positrons emitted by certain types of radioactive nuclei such as potassium-40.

BEV is an abbreviation for baculovirus expression vector.

BFGF is an abbreviation for basic fibroblast growth factor.

Bicuspid aortic valve is the aortic valve normally has three leaflets or cusps. Occasionally an individual is born with a valve having only two cusps – called a bicuspid valve.

Biennial (*L. biennium*, a period of two years) is a plant which completes its life cycle within two years and then dies.

Bifunctional vector is a deoxyribonucleic acid vector able to replicate in two different organisms, and therefore able to shuttle foreign nucleic acids between two different hosts. Also known as shuttle vector.

Bilayer lipid membrane (BLM) is a structure found in most biological membranes, in which two layers of lipid molecules are arranged in such a way so that their hydrophobic parts interpenetrate, whereas their hydrophilic parts form the two surfaces of the bilayer.

Bile is a digestive fluid made by the liver and stored in the gallbladder which helps digest fats.

Bile ducts are tubes that take bile from the liver to the gallbladder and small intestine to aid in digestion.

Biliary atresia are bile ducts that do not have normal openings, preventing bile from leaving the liver. This causes jaundice (a yellow skin color) and liver damage known as cirrhosis.

Bilirubin is a normal substance produced when red blood cells break down and are excreted by the liver. Bilirubin gives bile its yellow-green color. Too much bilirubin in the blood causes jaundice.

Binary vector system is a two-plasmid system in *Agrobacterium tumefaciens* for transferring into plant cells a segment of T-DNA that carries cloned genes.

Binding is the ability of molecules to stick to each other because of the exact shape and chemical nature of parts of their surfaces. Binding can be characterized by a *binding constant* or *association constant* (K_a), or its inverse, the *dissociation constant* (K_d).

Binding affinity characterizes the interaction of most ligands with their **binding** sites.

Binomial expansion refers to the probability that an event will occur 0, 1, 2, ..., n times out of n is given by the successive terms of the expression $(p + q)^n$, where p is the probability of the event occurring, and $q = 1 - p$.

Binomial nomenclature refers to each species is generally identified by two terms: the first is the genus to which it belongs, and the second is the specific epithet that distinguishes it from others in that genus (e.g., *Quercus suber*, cork oak).

Binomial proportion refers to a population that will be in binomial proportions when it conforms to the binomial distribution so that the occurrence of a given event X, r i times with a probability (pi) of success, in a population of n total events, is not significantly different than that which would be expected based on random chance alone.

Bio- is a prefix derived from *bios* and used in scientific words to associate the concept of "living organisms." Usually written with a hyphen before vowels, for emphasis or in neologisms; otherwise usually without a hyphen.

Bioaccumulation refers to the concentration of materials which are not components critical for the survival of an organism. Usually it refers to the accumulation of metals or other compounds (e.g., DDT).

Bioassay is a procedure for the assessment of a substance by measuring its effect in living cells or on organisms.

Bioaugmentation refers to increasing the activity of bacteria that decompose pollutants.

Bioavailability is a subcategory of absorption and is the fraction of an administered dose of unchanged drug that reaches the systemic circulation, one of the principal pharmacokinetic properties of drugs.

Biocatalysis is the use of enzymes to improve the efficiency of chemical reactions.

Biochemistry is also called biological chemistry. It is the study of chemical processes within and relating to living organisms.

Biocomplexity is the investigation of complex adaptive systems in all disciplines of biology.

Biocontainment is a process aimed at keeping biological organisms within a limited space or area.

Biocontrol refers to the control of living organisms (especially pests) by biological means.

Bioconversion refers to the conversion of one chemical into another by living organisms, as opposed to their conversion by enzymes (which is biotransformation) or by chemical processes.

Biodegradation refers to the breakdown by living organisms of a compound to its chemical constituents. Materials that can be easily biodegraded are termed as *biodegradable.*

Biodegrade is the breakdown by micro-organisms of a compound to simpler chemicals.

Biodesulphurization is the removal of organic and inorganic sulphur from coal by bacterial and soil micro-organisms.

Biodiversity refers to: (1) The variety of species (species diversity) or other taxa of animals, micro-organisms and plants in a natural community or habitat, or of communities in a particular environment (ecological diversity), or of genetic variation in a species (genetic diversity); and (2) The variety of life in all its forms, levels and combinations, encompassing genetic diversity, species diversity and ecosystem diversity.

Bioenergetics is the study of the flow and the transformation of energy that occur in living organisms.

Bioenergy is an energy choices using a wide range of biomass sources (for example, agriculture, forestry, industry and municipal waste) and conversion technologies such as fermentation (alcohol production) and co-firing (co-combustion of biomass and coal).

Bioengineering is a branch of engineering applied to biological and medical systems, such as biomechanics, biomaterials and biosensors. Bioengineering also includes biomedical engineering.

Bioengineering or Biomedical Engineering (BME) is the application of engineering principles and design concepts to medicine and biology. This field seeks to close the gap between engineering and medicine: It combines the design and problem solving skills of engineering with medical and biological sciences to improve healthcare diagnosis, monitoring and therapy. BME has only recently emerged as its own discipline, compared to many other engineering fields. Such an evolution is common as a new field transitions from being an interdisciplinary specialization among already-established fields, to being considered a field in itself. Much of the work in BME consists of research and development, spanning a broad array of subfields. Prominent BME applications include the development of biocompatible prostheses, various diagnostic and therapeutic medical devices ranging from clinical equipment to micro-implants, common imaging equipment such as MRIs and EEGs, regenerative tissue growth, pharmaceutical and therapeutic drugs. In BME there is continual change and creation of new areas due to rapid advancement in technology. However, some of the well-established specialty areas of BME are: bioelectricity; bioheat transfer; bioinformatics; bioinstrumentation; biomass transport; biomaterials; biomechanics; biotechnology/bionanotechnology; biosensors; cellular, tissue, genetic engineering, and transport phenomena; clinical engineering; medical imaging; orthopaedic surgery; physiologic modeling; prosthetic devices and artificial organs; rehabilitation engineering; and systems physiology. These specialty areas frequently depend on each other. Often, the biomedical engineer who works in an applied field will use knowledge gathered by engineers working in other areas. For example, the design of an artificial hip is greatly aided by studies on anatomy, bone biomechanics, gait analysis, and biomaterial compatibility. The forces that are applied to the hip can be considered in the design and material selection for the prosthesis. Similarly, the design of systems to electrically stimulate paralyzed muscle to move in a controlled way uses knowledge of the behavior of the human musculoskeletal system. The selection of appropriate materials used in these devices falls within the realm of the biomaterials engineer. Now, let us define some of these areas.

Bioenrichment refers to adding nutrients or oxygen to increase microbial breakdown of pollutants.

Bioethics deals with the life sciences and their potential impact on society.

Biofilms is a layer of micro-organisms growing on a surface, in a bed of polymeric material which they themselves have made.

Biofouling refers to living organisms that attach to and start living on any object that is submerged in the sea.

Biofuel is a gaseous, liquid or solid fuel that contains energy derived from a biological source.

Biogas is a mixture of methane and carbon dioxide resulting from the anaerobic decomposition of waste such as domestic, industrial and agricultural sewage.

Biogenesis refers to the principle that a living organism can only arise from other living organisms similar to itself and can never originate from non-living material.

Biogeography is the study of the geographic distribution of species and the principles and factors influencing these distributions.

Biohazard is a biological agent, such as an infectious micro-organism, or a condition that constitutes a threat to humans, especially in biological research or experimentation.

Bioheat transfer refer to the transfer of fluids from the surface of living bodies (plants, animals or humans). The human body is composed of 70% of fluids which help in the thermoregulatory processes for maintaining a constant temperature of 37.4°C.

Bioinformatics is the generation/creation, collection, storage (in databases), and efficient use of data/information from genomics from biological research to accomplish an objective.

Bioinstrumentation is the application of electronics and measurement techniques to develop devices used in diagnosis and treatment of disease.

Bioleaching is the recovery of metals from the ores, using the action of micro-organisms, rather than chemical or physical treatment.

Biolistics is a technique to insert DNA into cells.

Biological ageing (Senescence, Latin: senescere, meaning "to grow old" or biological aging) is the gradual deterioration of function characteristic of most complex lifeforms, arguably found in all biological kingdoms, that on the level of the organism increases mortality after maturation.

Biological containment refers to restricting the movement of (genetically engineered) organisms by arranging barriers to prevent them from growing outside the laboratory.

Biological control is the control of a population of one organism by another organism.

Biological diversity refers to the biodiversity.

Biological half-life or elimination half-life is the time it takes for a substance (drug, radioactive nuclide, or other) to lose one-half of its pharmacologic, physiologic, or radiological activity.

Biological oxygen demand (BOD) is the dissolved oxygen required for the respiration of a population of **aerobic** organisms present in water. BOD is expressed in terms of the oxygen consumed in water at a temperature of 20°C per unit time.

Biological products refers to any virus, therapeutic serum, toxin, antitoxin, or analogous product used in the prevention, treatment or cure of diseases or injuries in humans.

Biological species concept (BSC) refers to groups of naturally occurring interbreeding populations that are reproductively isolated from other such groups or species.

Biological valves are artificial valves made from humans or animals, rather than from metal. Examples of biological valves are porcinexenografts, human homografts or allografts, and pulmonary autografts.

Biologics are agents, such as vaccines, that give immunity to diseases or harmful biotic stresses.

Bioluminescence refers to the enzyme-catalyzed production of light by a number of diverse organisms.

Biomagnification is related to bioaccumulation.

Biomass concentration is the amount of biological material in a specific volume.

Biomass refers to: (1) The cell mass produced by a population of living organisms; (2) The organic matter that can be used either as a source of energy or for its chemical components; (3) All the organic matter that derives from the photosynthetic conversion of solar energy.

Biomaterial is a material which is in whole or in part composed of living matter, such as a polymer scaffolding perfused with cells. Biomaterials may be used as a medical device which augments or replaces natural tissue for therapeutic effect.

Biomaterials include both living tissue and artificial materials used for implantation.

Biome is a major ecological community or complex of communities, extending over a large geographical area and characterized by a dominant type of vegetation.

Biomechanics applies classical mechanics (statics, dynamics, fluids, solids, thermodynamics, and continuum mechanics) to biological or medical problems. It includes the study of motion, material deformation, flow within the body and in devices, and transport of chemical constituents across biological and synthetic media and membranes.

Biomedical Applications Group (GAB) at the PSC pursues leading edge research in high performance computing in the biomedical sciences and fosters exchange between PSC expertise in computational science and biomedical researchers nationwide. The Biomedical Applications Group also engages in outreach activities at both the local and national level. Scientists in the Biomedical Applications Group develops computational methods and tools and conduct research on biomedical systems at the cell and tissue level with a focus on neural systems such as the brain and the central nervous system.

Biomedical Engineering is an application of principles and practices of engineering science to biomedical research and health care.

Biomedical Engineering or Bioengineering (BME) is also termed as clinical (or medical) engineering. According to Pacela's *Bioengineering Education Directory* (Quest Publishing Co., 1990), the scope of bioengineering is broader. Bioengineering is usually a basic research-oriented closely related to biotechnology/bionanotechnology and genetic engineering with a mission to improve agriculture plants or animals cells or human body cells. **BME** is the application of engineering principles and design concepts to medicine and biology. This field seeks to close the gap between engineering and medicine: It combines the design and problem-solving skills of engineering with medical and biological sciences to improve healthcare diagnosis, monitoring, and therapy. BME is, therefore, an interdisciplinary branch of engineering that ranges from theoretical, non-theoretical knowledge to state of the art application in medicine and human health. BME has only recently emerged as its own discipline, compared to many other engineering fields. Much of the work in BME consists of academics, research, and development. Prominent BME applications include the development of biocompatible prostheses, various diagnostic and therapeutic medical devices ranging from clinical equipment to micro-implants, common imaging equipment such as MRIs and EEGs, regenerative tissue growth, and pharmaceutical and therapeutic drugs. In BME there is continual change and creation of new areas due to rapid advancement in technology. However, some of the well-established specialty areas of BME are: bioelectricity; bioheat transfer; bioinformatics; bioinstrumentation; biomass transport; biomaterials; biomechanics; biotechnology/bionanotechnology; biosensors; cellular, tissue, genetic engineering, and transport phenomena; clinical engineering; medical imaging; orthopedic surgery; physiologic modeling; prosthetic devices and artificial organs; rehabilitation engineering; and systems physiology. These specialty areas frequently depend on each other. BMES has been the lead society to define some of these areas as below:

Biomedical engineers develop devices and procedures that solve medical and health-related problems. Many do research, along with life scientists, chemists, and medical scientists, to develop and evaluate systems and products for use in the fields of biology and health, such as artificial organs, prostheses (artificial devices that replace missing body parts), instrumentation, medical information systems, and health management and care delivery systems. Biomedical engineers design devices used in various medical procedures, such as the computers used to analyze blood or the laser systems used in corrective eye surgery. They develop artificial organs, imaging systems such as magnetic resonance, ultrasound, and x-ray, and devices for automating insulin injections or controlling body functions. Most engineers in this specialty require a sound background in one of the basic engineering specialties, such as mechanical or electronics engineering, in addition to specialized biomedical training. Some specialties within biomedical engineering include biomaterials, biomechanics, medical imaging, rehabilitation engineering, and orthopedic engineering.

BioMEMS (Biological MicroElectro Mechanical Systems) highlight the technical advances in the field that are leading a revolution in medicine, and creating a new generation of analytical devices for medical diagnostics.

Biometrics refers to the quantifiable data (or metrics) related to human characteristics and traits. Biometrics identification (or biometric authentication) is used in computer science as a form of identification and access control.

Biometry is the application of statistical methods to the analysis of continuous variation in biological systems. Synonym is biometrics.

Biomimetic materials is employed to describe synthetic analogues of natural materials with advantageous properties.

Biomimetics is an interdisciplinary field in Materials Science, Engineering, and Biology, studying the use of biological principles for synthesis or fabrication of biomimetic materials.

Bionanotechnology includes molecular motors, biomaterials, single molecule manipulation technologies, biochip technologies, etc.

Biopesticide is a compound that kills organisms by virtue of specific biological effects rather than as a broader chemical poison.

Biopharmaceuticals refer to biologic drugs produced through rDNA technology, but essentially they also fall under the regulatory definition of a biologic.

Biopharming is the use of genetically transformed crop plants and livestock animals to produce valuable compounds, especially pharmaceuticals.

Synonym is molecular pharming.

Biopiracy refers to the patenting of genetic stocks, and the subsequent privatization of genetic resources collections. The term implies a lack of consent on the part of the originator.

Biopolymer is any large polymer (protein, nucleic acid, polysaccharide) produced by a living organism. Synonym is biological polymer.

Bioprocess refers to any process that uses complete living cells or their components (e.g., enzymes, chloroplasts) to effect desired physical or chemical changes.

Biopsy is a procedure in which tissue samples are removed (with a needle or during surgery) from the body for examination under a microscope; to determine if cancer or other abnormal cells are present.

Bioreactor is a tank in which cells, cell extracts or enzymes carry out a biological reaction. Often refers to a fermentation vessel for cells or micro-organisms.

Biorecovery is the use of micro-organisms for the recovery of valuable materials (metals or particular organic compounds) from complex mixtures.

Bioremediation refers to the use of plants and micro-organisms to consume or otherwise help remove materials (such as toxic chemical wastes and metals) from contaminated sites (especially from soil and water).

Biosafety protocol is an internationally agreed protocol set up to protect biological diversity from the potential risks posed by the release of genetically modified organisms. Synonym is Cartagena protocol. It is related to *Convention on biological diversity*.

Biosafety refers to the avoidance of risk to human health and safety, and to the conservation of the environment, as a result of the use for research and commerce of infectious or genetically modified organisms.

Biosensing is a technology for the detection of a wide range of chemical and biological agents, including bacteria, viruses and toxins, in the environment and humans.

Biosensor is a device that uses an immobilized biologically-related agent (such as an enzyme, antibiotic, organelle or whole cell) to detect or measure a chemical compound.

Biosilk is a biomimetic fiber produced by the expression of the relevant orb-weaving spider genes in yeast or bacteria, followed by the spinning of the expressed protein into a fiber.

Biosorbents refer to micro-organisms which, either by themselves or in conjunction with a substrate are able to extract and/or concentrate a desired molecule by means of its selective retention.

Biosphere is the part of the earth and its atmosphere that is inhabited by living organisms.

Biosynthesis refers to the synthesis of compounds by living cells, which is the essential feature of anabolism.

Biosynthetic antibody binding sites are abbreviated as BABS.

Biot number (Bi) is a dimensionless number used in unsteady-state and heat transfer calculations. It relates the heat transfer resistance inside and at the surface of a body.

Biotechnologists are scientists who use biological processes to develop novel products.

Biotechnology refers to: (1) A broad term generally used to describe the use of biology in industrial processes such as agriculture, brewing and drug development. The term also refers to the production of genetically modified organisms or the manufacture of products from genetically modified organisms; and (2) The use of plants, animals and micro-organisms to create products or processes. Modern biotechnology also includes the use of gene technology, which allows us to move genetic material from one species to another.

Biotechnology, nanotechnology and bionanotechnology are sometimes used interchangeably with BME in general. Biotechnology is the use of biological processes, organisms, or systems to manufacture products intended to improve the quality of human life. The *United Nations Convention on Biological Diversity* defines biotechnology as any technological application that uses biological systems, living organisms, or derivatives thereof, to make or modify products or processes for specific use. Nanotechnology is a study and use of structures between 1 nanometer and 100 nanometers in size. For example a width of a human hair is equivalent to eight hundred 100–nanometer particles side by side. Nanotechnology is the engineering of functional systems at the molecular scale. Bionanotechnology or nano-biotechnology or nanobiology is the harnessing of biological processes on an ultra-small

scale in the manufacture and alteration of materials and products.

Bioterrorism is the use of bacteria, viruses or toxins with the intent of causing harm to people, animals or food to achieve certain political, religious or ideological goals through intimidation.

Biotherapeutic strategy is the plan or program to contribute to the cure of disease or to general, especially mental, well-being.

Biotic factor refers to other living organisms that are a factor of an organism's environment, and form the biotic environment, affecting the organism in many ways.

Biotic stress is a stress resulting from living organisms which can harm plants, such as viruses, fungi, bacteria, parasitic weeds and harmful insects.

Biotin is also known as vitamin H or coenzyme R. It is a water-soluble B-vitamin (vitamin B_7). It is composed of a ureido (tetrahydroimidizalone) ring fused with a tetrahydrothiophene ring. A valeric acid substituent is attached to one of the carbon atoms of the tetrahydrothiophene ring. Biotin is a coenzyme for carboxylase enzymes, essential in the synthesis of fatty acids, isoleucine, and valine, and in gluconeogenesis.

The only human health condition for which there is strong evidence of biotin's potential benefit as a treatment is biotin deficiency. Biotin is attached to pyruvate carboxylase by a long, flexible chain like that of lipoamide in the pyruvate dehydrogenase complex. Adequate amounts are normally produced by the intestinal bacteria in animals.

Biotin labeling refers to the attachment of biotin to another molecule, especially DNA.

Biotinylated – DNA refers to a DNA molecule labeled with biotin by incorporation of a biotinylated nucleotide (usually uracil) into a DNA molecule. The detection of the labeled DNA is achieved by complexing it with streptavidin to which is attached a color-generating agent such as horseradish peroxidase that gives a fluorescent green color upon reaction with various organic reagents.

Biotope is a small habitat in a large community.

Biotoxin is a naturally produced toxic compound which shows pronounced biological activity and presumably has some adaptive significance to the organism which produces it.

Biotransformation refers to the conversion of one chemical or material into another using a biological catalyst: a near synonym is biocatalysis, and hence the catalyst used is called a biocatalyst. Usually the catalyst is an enzyme, or a whole, dead micro-organism that contains an enzyme or several enzymes.

Biotreatment is the treatment of a waste or hazardous substance using organisms such as bacteria, fungi and protozoa (see bioremediation).

Bivalent is a pair of synapsed or associated homologous chromosomes (one of maternal origin; the other of paternal origin) that have each undergone duplication. Each duplicated chromosome comprises two chromatids. Thus a bivalent comprises four chromatids.

Bladder is a triangle-shaped, hollow organ located in the lower abdomen that holds urine. It is held in place by ligaments that are attached to other organs and the pelvic bones. The bladder's walls relax and expand to store urine, and contract and flatten to empty urine through the urethra.

Bladder instillation (or bladder wash or bladder bath) is a process of filling the bladder with a solution that is held for varying periods of time, from a few seconds to 15 minutes, before being drained through a catheter.

BLAST (Basic Local Alignment Search Tool) is a method for performing sequence comparisons. Either protein sequences or nucleotide sequences can be used. This algorithm has been extended and now includes a suite of programs including megaBLAST and discontiguous megaBLAST.

Blast cell is a large, rapidly dividing cell that develops from a B cell in response to an antigenic stimulus. The blast cell then becomes an antibody-producing plasma cell.

Blastocoel or blastcele is the fluid-filled cavity inside the blastocyst, an early, preimplantation stage of the developing embryo.

Blastocyst is a spherical cell mass produced by cleavage of the zygote (fertilized egg) after approximately 5-7 days of cell divisions.

Blastocyst stage refers to a time period of four to five days after the union of the sperm and the egg, before the embryo implants in the uterus.

Blastomere is any of the cells in the early embryo produced as the result of cell division in the fertilized egg. A blastocyst is made up of many blastomeres.

Blastula refers to an early embryo form in animals that follows the morula stage; typically, a single-layered sheet (blastoderm) or ball of cells (blastocyst).

BLAT is a hashing algorithm developed by Jim Kent to allow rapid searching of large amounts of genome sequence.

Bleach is a fluid, powder or other whitening (bleaching) or cleaning agent, usually with free chlorine ions. Commercial bleach contains calcium hypochlorite or sodium hypochlorite.

Bleeding refers to: (1) Collection of blood from immunized animals; and (2) Used to describe the occasional purplish-black coloration of media due to phenolic products given off by (usually fresh) transfers.

Blocking agent is an agent that inhibits a biological action, such as movement of an ion across the cell membrane, passage of a neural impulse, or interaction with a specific receptor.

Blood Brain Barrier (BBB) prevents materials from the blood from entering the brain.

Blood clot is a jelly-like mass of blood tissue formed by clotting factors in the blood. Clots stop the flow of blood from an injury. They can also form inside an artery whose walls are damaged by atherosclerotic build-up and can cause a heart attack or stroke.

Blood components include: blood cells, platelets, plasma

Blood filter is a filter that is inserted into arterial line to remove gas and particulate emboli.

Blood is a fluid, circulating connective tissue that transport nutrients and others materials through the body. It circulates in the heart, arteries, capillaries and veins of a vertebrate animal carrying nourishment and oxygen to and taking away waste products from all parts of the body.

Blood pressure is the force exerted by blood against the inner walls of the blood vessel.

Blood products are derived from blood. These products are made from plasma like coagulation factors, plasma proteins and albumin.

Blot refers to: (1) As a verb, this means to transfer DNA, RNA or protein to an immobilizing matrix; and (2) As a noun, it usually refers to the autoradiograph produced during the Southern or northern blotting procedures. The variations on this theme depend on the molecules: Southern blot, Northern blot, Western blot, Southwestern blot, dot blot, and colony blot.

Blunt end refers to the end of a double-stranded DNA molecule in which neither strand extends beyond the other. Synonym is flush end.

Blunt-end cut refers to cleave phospho-diester bonds in the backbone of duplex DNA between the corresponding nucleotide pairs on opposite strands.

Blunt-end ligation refers to joining (ligation) of the nucleotides that are at the ends of two blunt-ended DNA duplex molecules.

B-lymphocytes (B cells) is an important class of lymphocytes that mature in bone marrow (in mammals) and the Bursa of Fabricius (in birds), that are largely responsible for the antibody-mediated or humoral immune response; they give rise to the antibody-producing plasma cells and some other cells of the immune system.

BMES *stands for Biomedical Engineering Society.*

BOD is an abbreviation for biological oxygen demand.

Body fluid, bodily fluid or biofluid is a fluid originating from inside the living bodies. Body fluid is a term most often used in medical and health contexts.

Body temperature refers to the degree of body heat in cold- and warm-blooded animals.

Body water is water content of the human body. They include fluids that are excreted or secreted from the body as well as body water that normally is not.

Bone marrow stromal cell is also known as mesenchymal stem cells. Bone marrow stromal cells are a mixed population of cells derived from the non-blood forming fraction of bone marrow.

Bonferroni correction is a correction used when several statistical tests are being performed simultaneously (since while a given Ã¡-value may be appropriate for each individual comparison, it is not for the set of all comparisons).

Bootstrap analysis is a nonparametric statistical analysis for computing confidence intervals for a phylogeny or a point estimate (e.g., of FST).

Boring platform refers to sterile bottom half of a Petri dish used for preparing explants with a cork borer.

Bottleneck is a special case of strong genetic drift where a population experiences a loss of genetic variation by temporarily going through a marked reduction in effective population size.

Bottom-up nanotechnology refers to when chemists are attempting to create structure by connecting molecules.

Bound water is a cellular water not released into the intercellular space upon freezing and thawing. Opposite is free water.

Boundary layer is the layer of fluid in the immediate vicinity of a bounding surface.

Bovine growth hormone is related to bovine somatotrophin.

Bovine somatotrophin (BST) is a bovine growth hormone. This protein is found naturally in cattle, and is the bovine counterpart of human growth hormone, one of the earliest biopharmaceutical products.

Bovine spongiform enecelophalopathy (BSE) is a cattle disease (or mad cow disease) caused by proteinaceous infectious particles.

Bowel is a small and large intestine.

Bowel movement is a passage of stool (body wastes) from the large intestine through the rectum and anus.

Bowman's capsule is a doubled wall sac of cells that surrounds the glomerulus of each nephron.

bp is an abbreviation for base pair.

Brachial relates to the arm or a comparable process.

Bract is a modified leaf that sub-
tends flowers or inflorescences
and may appear to be a petal.

Brain refers to the concentration
of nerve tissue in the front or
upper end of an animal's body.

Brain stem is the part of the brain
that includes the medulla,
pons, midbrain, thalamus and
hypothalamus.

Branch length is a length of
branches on a phylogenetic
tree. It is often proportional to
the amount of genetic divergence between species or
groups.

BRCA1 and BRCA2 are two
genes that are associated with
an increased risk of breast cancer when they have mutations.
A mutation in the BRCA1/2
genes can be inherited or it can
be acquired during our lifetime.

Breed at risk is any breed that may
become extinct if the factors
causing its decline in numbers
are not eliminated or mitigated.
FAO has established categories
of risk status: critical, endangered, critical-maintained,
endangered-maintained, and
not-at-risk.

Breed not-at-risk refers to the
breed where the total number
of breeding females and males
is greater than 1000 and 20
respectively; or the population size approaches 1000 and

the percentage of pure-bred
females is close to 100%, and
the overall population size is
increasing.

Breed refers to: As noun in AnGR,
either (1) a sub-specific group
of domestic livestock with
definable and identifiable
external characteristics that
enable it to be separated by
visual appraisal from other
similarly defined groups within
the same species; or (2) a
group of domestic livestock
for which geographical and/
or cultural separation from
phenotypically similar groups
has led to acceptance of its
separate identity.

Breeding is the process of sexual
reproduction and production of
offspring.

Breeding value is a quantitative
genetics term, describing that
part of the deviation of an
individual **phenotype** from
the population mean that is
due to the additive effects of
alleles.

Brewer's yeast refers to strains
of the yeast *Saccharomyces
cerevisiae* that are used for the
production of beer.

Brewing refers to the process by
which beer is made.

Bridge refers to a filter paper or
other substrate used as a wick
and support structure for a plant

tissue in culture when a liquid medium is used.

Brinell (indicated by HB), Knoop (HK) and Vickers (HV) are referred to as microhardness testing methods on the basis of load and indenter size.

Broad-host-range plasmid is a plasmid that can replicate in a number of different bacterial species.

Broad sense heritability (*H*B) is the proportion of phenotypic variation within a population that is due to genetic differences among individuals.

Bronchiole is an air duct in the lung that branches from a bronchus; divides to form air sacs Alveoli).

Bronchiolitis is an inflammation that involves the bronchioles (small airways).

Bronchoscopy is a fiberoptic, flexible tube is passed through the mouth into the bronchi to locate tumors or blockages, and to gather samples of tissue and/or fluid.

Bronchus are large airways that divide the lung into right and left bronchi. It is one of the branches of the trachea and its immediate branches within the lung.

Brood stock is the group of males and females from which fish are bred.

Browning is discoloration of freshly cut surfaces of plant tissue due to phenolic oxidation.

Brucellosis is a disease caused by infection with organisms of the genus *Brucella*.

BSA is an abbreviation for bovine serum albumin.

BSE is an abbreviation for bovine spongiform encephalopathy. The BSF is a progressive neurological disorder of cattle that results from infection by an unusual transmissible agent called a prion.

BST is an abbreviation for bovine somatotrophin. BST is a peptide hormone produced by cows' pituitary gland.

Bt is an abbreviation for *Bacillus thuringiensis*. Bt is a Gram-positive, soil-dwelling bacterium, commonly used as a biological pesticide.

Bt crops are crop plants that contain genes for *Bt* toxins.

Bt toxins are insecticidal proteins produced by the soil microorganism called *Bacillus thuringiensis*. *Bt* is an abbreviation of *Bacillus thuringiensis*.

BTR (Bridge to Recover) refers to a medical indication aiming not at a heart transplant, but at the recovery of the heart.

BTT (Bridge to Transplantation) refers to a medical indication referring to the bridging of the waiting period before a heart transplant.

Bubble column fermenter is a bioreactor in which the cells or micro-organisms are kept suspended in a tall cylinder by rising air, which is introduced at the base of the vessel.

Bubble oxygenator injects oxygen directly through a column of blood; direct blood-gas interface exists; and gas bubbles are formed that must be removed in defoaming section; traumatic to blood elements; contributes to protein denaturation and formation of microemboli.

Bud is a region of meristematic tissue with the potential for developing into leaves, shoots, flowers or combinations; generally protected by modified scale leaves.

Bud scar is a scar left on a shoot when the bud or bud scales drop.

Bud sport is a somatic mutation arising in a bud and producing a genetically different shoot.

Budding refers to: (1) A method of asexual reproduction in which a new individual is derived from an outgrowth (bud); (2) Among fungi, budding is characteristic of the brewers yeast *Saccharomyces cerevisiae*; and (3) A form of graft in which a single vegetative bud is taken from one plant and inserted into stem tissue of another plant so that the two will grow together.

Buffer is a compound that stabilizes the pH of a solution by removing or releasing hydrogen ions.

Bulk degradation occurs throughout the polymer structure in a rather random fashion.

Bulk micromachining is a process used to produce micromachinery or microelectromechanical systems (MEMS).

Bulked segregant analysis is a method to obtain markers linked to a target trait, in which DNA samples, prepared from a number of individuals of each of two contrasting phenotypes, are separately pooled and used to generate contrasting DNA fingerprints.

Bundle of his is a small band of cardiac muscle fibers transmitting the waves of depolarization from the atria to the ventricles during cardiac contraction.

Buoyant density is the intrinsic density which a molecule, virus or sub-cellular particle has when suspended in an aqueous solution of a salt, such as CsCl, or a sugar, such as sucrose.

Byte is a group of eight bits that can represent any of $2^8 = 256$ different entities.

C

C is an abbreviation for cytosine.

CAAT box (or CAT box) refers to a conserved sequence found within the promoter region of the protein-encoding genes of many eukaryotic organisms. It has the consensus sequence GGCCAATCT.

Cable electrode refers to an inductance type electrode in which the electrodes are coiled around a body part creating an electromagnetic field.

CADD stands for Computer Assisted Drug Design.

CAGE (Cap Analysis Gene Expression) is a technique for identifying transcription start sites and quantifying promoter usage in eukaryotic genomes.

Calcium channel blocking agent is any of a class of drugs that inhibit the influx of calcium ions across the cell membrane or inhibit the mobilization of calcium from intracellular stores; used in the treatment of angina, cardiac arrhythmias, and hypertension.

Calf scours refers to a watery diarrhoea in calves.

Calibration is a process of adapting a sensor output to a known physical quantity to improve sensor output accuracy.

Calicivirus refers to the virus that causes rabbit calicivirus disease (RCD) in rabbits. It is spread by mosquitoes and fleas.

Callipyge is an inherited trait in livestock (e.g., sheep) that results in thicker, meatier hindquarters, and hence a higher meat yield per animal.

Callus (*L. callum*, thick skin; plural: calluses or calli) refers to: (1) A protective tissue, consisting of parenchyma cells, that develops over a cut or damaged plant surface; (2) Mass of unorganized, thin-walled parenchyma cells induced by hormone treatment; and (3) Actively dividing non-organized masses of undifferentiated and differentiated cells often developing from injury (wounding).

Callus culture is a technique of tissue culture; it is usually on solidified medium and initiated by inoculation of small explants or sections from established organ or other cultures (the inocula).

Calorie is the amount of heat required to raise the

temperature of 1 gram of water from 14.5° to 15.5°C.

Calyx (Gr. *kalyx*, a husk, cup) refers to all the sepals of a flower considered collectively.

Cambial zone is a region in stems and roots consisting of the cambium and its recent derivatives.

Cambium (plural cambia) refers to one or two cells thick layer of plant meristematic tissue, between the xylem and phloem tissues, which gives rise to secondary tissues, thus resulting in an increase in the diameter of the stem or root.

CAMD stands for Computer Aided Molecular Design.

CAMM stands for Computer Assisted Molecular Modeling.

cAMP is an abbreviation for cyclic adenosine monophosphate.

CaMV is an abbreviation for cauliflower mosaic virus.

CaMV 35S is an abbreviation for cauliflower mosaic virus 35S, ribosomal DNA promoter.

Cancer is an abnormal, uncontrolled and rapid growth of cells that invade and destroy surrounding tissues. Cells from the tumor can break away (metastasise) and spread through the bloodstream or lymph system to other parts of the body, creating new tumors.

Cancer stem cells have been found to be the source of some, and possibly most cancers. The cancer stem cell hypothesis states that certain stem cells remain in tissues to replenish them after injury or disease, yet because they are self-renewing and can survive for a longer period of time, the adult stem cells can also accumulate mutations which would cause them to spin off cells that divide uncontrollably, forming a tumor.

Candida is a yeast that causes irritation and infection, especially of the mucous membranes of the body such as the mouth, vagina, and anus.

Candidate gene is a gene whose deduced function (on the basis of DNA sequence) suggests that it may be involved in the genetic control of an aspect of phenotype.

Candidate gene strategy refers to an experimental approach in which knowledge of the biochemistry and/or physiology of a trait is used to identify candidate genes. Synonym is functional gene cloning.

Canine refers to pertaining to domestic dogs, wolves, foxes, jackals, coyotes, and other dog-like mammals.

Cannula is a tube designed for insertion into a duct, body cavity, or blood vessel.

Canola is a specific subgroup of oilseed rape cultivars.

CAP is an abbreviation for catabolite activator protein. It refers to the structure found on the 5′-end of eukaryotic mRNA, and consisting of an inverted, methylated guanosine residue. Related terms are G cap, cap site.

Cap site is the site on a DNA template where transcription begins. It corresponds to the nucleotide at the 5′ end of the RNA transcript which accepts the G cap.

Capacitance is a charge-storing ability of a capacitor. It is a magnitude of charge stored on either plate divided by the applied voltage. A one-F capacitor charged to one V contains C of charge. One C is an amount of charge = 6.24×10^{18} electrons.

Capacitation is the final stage, inside the female genital tract, in the maturation process of a spermatozoon, as it penetrates the ovum.

Capacitor is a energy storage circuit element having two conductors separated by an insulator.

Capillaries refer to microscope blood vessels in the tissues the permit exchange of materials between cells and blood.

Capillary electrophoresis is a form of electrophoresis used widely in current large-scale DNA sequencing facilities, where the sample is passed through a long, very-narrow-bore tube containing a re-usable matrix.

CAPS is an abbreviation for cleaved amplified polymorphic sequence.

Capsid is the protein coat of a virus.

Capsule refers to carbohydrate coverings that have antigenic specificity, present on some types of bacteria and other micro-organisms. The capsule is usually composed of polysaccharides, polypeptides, or polysaccharide-protein complexes. These materials are arranged in a compact manner around the cell surface.

Carbohydrate is an organic compound based on the general formula $C_x(H_2O)_y$, performing many vital roles in living organisms. The simplest carbohydrates are the sugars (saccharides), including glucose and sucrose.

Carbon nanofoam is a new form of carbon: a spongy solid that is extremely lightweight and, unusually, attracted to magnets.

Carbon nanotubes are tiny tubes about 10,000 times thinner than a human hair – consist of rolled up sheets of carbon hexagons.

Carbowax (polyethylene glycol, PEG) is a polyether compound with many applications from industrial manufacturing to medicine. The structure of PEG is $H-(O-CH_2-CH_2)_n-OH$. PEG is also known as polyethylene oxide (PEO) or polyoxyethylene (POE), depending on its molecular weight.

Carboxypeptidase is a class of enzymes, which catalyze the cleavage of peptide bonds, requiring a free carboxyl group in the substrate.

Carboxypeptidases are two enzymes (A and B) that are found in pancreatic juice. Their role is to remove the C-terminal amino-acid from a peptide; the A form removes any amino acid; the B form removes only lysine or arginine.

Carcinogen is a substance capable of inducing cancer in an organism.

Carcinoma is a malignant tumor derived from epithelial tissue, which forms the skin and the outer cell layers of internal organs.

Cardiac arrest refers to the standstill of the heart, its action and diseases.

Cardiac cycle is a one complete heart beat.

Cardiac output is the amount of blood the heart pumps through the circulatory system in one minute.

Cardiac pertain to the heart.

Cardiology is the study of the heart, its action and diseases.

Cardioplegia infusion is used to cool and deliver cardioplegia solution; uses a separate roller head; when blood cardioplegia is used, system is connected to arterial blood supply.

Cardioplegia is a paralysis of the heart.

Cardiotomy suction is a process to aspirate blood from operative site and return to cardiotomy reservoir.

Cardiovascular (CV) pertains to the heart and blood vessels. The circulatory system of the heart and blood vessels is the cardiovascular system.

Cardiovascular disease refers to the diseases of the heart or blood vessels.

Carotene (*L. carota*, carrot) is a reddish-orange plastid pigment involved in light reactions in photosynthesis.

Carotenoid is a group of chemically similar red to yellow pigments responsible for the characteristic color of many plant organs or fruits, such as tomatoes, carrots, etc.

Carpal tunnel syndrome is a painful and disabling disorder characterized by inflammation and swelling in the tendons that run through the narrow carpal tunnel in the wrist, and one of the most common of repetitive motion injuries.

Carpel is a female reproductive organ of flowering plants, consisting of stigma, style and ovary.

Carrier is an individual who has a gene mutation for a recessive disease on one allele while the other allele is normally functioning.

Carrier DNA refers to DNA of undefined sequence, which is added to the transforming (plasmid) DNA used in physical DNA-transfer procedures.

Carrier gas refers to the gas that carries the sample in gas chromatography.

Carrier mediated transport refers to the transport of a molecule from point A, usually on one side of a cell (e.g., external to the cell), to point B via a trans-membrane carrier.

Carrier molecule refers to: (1) A molecule that plays a role in transporting electrons through the electron transport chain. Carrier molecules are usually proteins bound to non-protein groups and able to undergoing oxidation and reduction relatively easily, thus allowing electrons to flow; and (2) A lipid-soluble molecule that can bind to lipid-insoluble molecules and transport them across membranes. Carrier molecules have specific sites that interact with the molecules they transport.

Cartagena protocol is related to biosafety protocol.

Casein hydrolysate is the mixture of amino acids and peptides produced by enzymatic or acid hydrolysis of casein.

Casein refers to a group of milk proteins.

Catabolic pathway is a pathway by which an organic molecule is degraded in order to release energy for growth and other cellular processes; degradative pathway.

Catabolism refers to the metabolic breakdown of large molecules in living organism, with accompanying release of energy.

Catabolite activator protein (CAP) is a protein, which combines with cyclic AMP. The cAMP-CAP complex binds to the promoter regions of *E. coli* and stimulates transcription of the relevant operon. Synonyms are catabolite regulator protein (CRP), cyclic AMP receptor protein.

Catabolite repression refers to the glucose-mediated reduction in the rates of transcription of operons that encode enzymes involved in catabolic pathways.

Catalase is a metalloenzyme, present in both plants and animals, that catalyzes the decomposition of hydrogen peroxide to water and oxygen.

Catalysis is the process of increasing the rate of a chemical reaction by the addition of a substance that is not itself changed by the reaction (the catalyst).

Catalyst (Gr. *katalyein*, to dissolve) is a substance that promotes a chemical reaction by lowering the activation energy of a chemical reaction, without itself undergoing any permanent chemical change.

Catalytic antibody (abzyme) refers to an antibody selected for its ability to catalyze a chemical reaction by binding to and stabilizing the transition state intermediate.

Catalytic RNA (ribozyme) refers to a natural or synthetic RNA molecule that cuts an RNA substrate.

Catalytic site is the part of the surface of an enzyme molecule (usually only a small portion of the total) necessary for the catalytic process.

Catheter is a narrow tube, which can be passed inside blood vessels to the heart for diagnostic and treatment purposes.

Cathode is an electrode in an electrochemical cell or galvanic couple at which a reduction reaction occurs; in other words the electrode receiving electrons from an external circuit.

Cation is a positively charged ion.

Cauliflower mosaic virus (CaMV) refers to a DNA virus affecting cauliflower and many other dicot species.

CBD is an abbreviation for Convention on Biological Diversity.

cccDNA is an abbreviation for covalently-closed circle DNA.

CD44 is expressed in a large number of mammalian cell types. The standard isoform, designated CD44s, comprising exons 1–5 and 16–20 is expressed in most cell types. CD44 splice variants containing variable exons are designated CD44v.

CD molecules is any group of antigens that is associated with a specific sub-population of T cells. There are designations for surface molecules on various cells of the immune system, e.g., CD4 is present on the surface of helper T cells.

cDNA (complementary DNA) refers to the double-stranded

DNA complement of an mRNA sequence; synthesized *in vitro* from a mature RNA template using reverse transcriptase and DNA polymerase (to create the double-stranded DNA).

cDNA cloning is a method of cloning the coding sequence of a gene, starting with its mRNA transcript. It is normally used to clone a DNA copy of a eukaryotic mRNA.

cDNA library is a collection of cDNA clones that were generated *in vitro* from the mRNA sequences isolated from an organism or a specific tissue or cell type or population of an organism.

CDR (complementarity-determining regions) are regions of the variable (V) regions of light and heavy antibody chains that make contact with the antigen.

CDS (Coding sequence) is the portion of an mRNA or genomic sequence that encodes for a protein sequence.

Cecum is the beginning of the large intestine.

Cell (*L. cella*, small room) refers to the smallest structural unit of living matter capable of functioning independently; a microscopic mass of protoplasm surrounded by a semipermeable membrane, usually including one or more nuclei

and various non-living products, capable – either alone or by interacting with other cells – of performing all the fundamental functions of life.

Cell based therapies involve transplanting stem cells into damaged tissues to regenerate the various cell types of that tissue.

Cell culture is a growth of cells *in vitro* in an artificial medium for research or medical treatment.

Cell cycle refers to the sequence of stages that a cell passes through between one division and the next. The cell cycle oscillates between mitosis and the interphase, which is divided into G, S, and G_2.

Cell differentiation is the transition of cells (by the programmed activation and de-activation of the necessary genes) from an tissue-unspecific type, in which daughter cells are similarly undifferentiated, to a committed type in which the cell line specializes to become a recognizable tissue or organ.

Cell division is the formation of two or more daughter cells from a single parent cell. The nucleus divides first, followed by the formation of a cell membrane between the daughter nuclei. Division of somatic cells is termed mitosis.

Cell free is related to *in vitro*.

Cell free protein synthesis (in-vitro protein synthesis, CFPS) is the production of protein using biological machinery without the use of living cells.

Cell free transcription accurately transcribe genes has provided a means to investigate the mechanisms of eukaryotic transcription.

Cell free translation is a process of synthesizing recombinant proteins in cell-free extracts to verify the identity of cloned genes, to study protein-protein, protein-nucleic acid, and protein-drug interactions, and to carry out mutagenesis studies.

Cell fusion is the formation *in vitro* of a single hybrid cell from the coalescence of two cells of different species origin. Synonym is cell hybridization.

Cell generation time refers to the interval between the beginning of consecutive divisions of a cell. It refers to the time that it takes for a population of single-celled organisms to double its cell number.

Cell hybridization is the fusion of two or more dissimilar cells leading to the formation of a somatic hybrid.

Cell line refers to: (1) A cell lineage that can be maintained *in vitro*; and (2) A cell lineage that can be recognized *in vivo*.

Cell mediated immune response refers to the activation of T cells of the immune system in response to the presence of a foreign antigen.

Cell membrane separates the cell wall and the cytoplasm, and regulates the flow of material into and out of the cell.

Cell number is the number of cells per unit volume of a culture.

Cell plate is the precursor of the cell wall, formed at the beginning of cell division. The cell plate develops in the region of the equatorial plate and arises from membranes in the cytoplasm.

Cell sap refers to water and dissolved substances, sugar, amino acids, waste substances, etc., in the plant cell vacuole.

Cell selection refers to the process of selecting cells within a group of genetically different cells.

Cell strain is an *in vitro* culture initiated by asexual reproduction from a single cell. *Synonym*: single-cell line.

Cell suspension refers to cells in culture in moving or shaking liquid medium, often used to describe suspension cultures of single cells and cell aggregates.

Cell type is s specific subset of cells within the body, defined by their appearance, location and function. **Adipocyte:** the functional cell type of fat, or adipose tissue, that is found throughout the body, particularly under the skin. **Cardiomyocytes**: the functional muscle cell type of the heart that allows it to beat continuously and rhythmically. **Chondrocyte**: the functional cell type that makes cartilage for joints, ear canals, trachea, epiglottis, larynx, the discs between vertebrae and the ends of ribs. **Fibroblast**: a connective or support cell found within most tissues of the body. **Hepatocyte**: the functional cell type of the liver that makes enzymes for detoxifying metabolic waste, destroying red blood cells and reclaiming their constituents, and the synthesis of proteins for the blood plasma. **Hematopoietic cell**: the functional cell type that makes blood. Hematopoietic cells are found within the bone marrow of adults. **Neuron**: the functional cell type of the brain that is specialized in conducting impulses. **Osteoblast**: the functional cell type responsible for making bone. **Islet cell**: the functional cell of the pancreas that is responsible for secreting insulin, glucogon, gastrin and

somatostatin. Together, these molecules regulate a number of processes including carbohydrate and fat metabolism, blood glucose levels and acid secretions into the stomach.

Cell wall is a rigid external coat which surrounds plant cells.

Cellomics is a study that combines information from genomics and proteomics with the complex chemical and molecular relationships of cell components.

Cellular immune response involves the adaptive immune response in which antigen-specific T cells play the main role in controlling and eliminating the infection.

Cellular oncogene (proto-oncogene) is a normal gene that when mutated or improperly expressed contributes to the development of cancer.

Cellular or tissue or genetic engineering involves more recent attempts to solve biological problems/issues at the microscopic level. These areas utilize the anatomy, biochemistry, and mechanics of cellular and sub-cellular structures in order to understand disease processes and to be able to intervene at very specific sites. With these capabilities, miniature devices deliver compounds that can stimulate or inhibit cellular processes at precise target locations to promote healing or inhibit disease formation and progression.

Cellulose is a complex carbohydrate composed of long, unbranched chains of beta-glucose molecules, which contribute to the structural framework of plant cell walls.

Cellulose nitrate is related to nitrocellulose.

Cellulosome is a multi-protein aggregate present in some micro-organisms, which degrade cellulose.

Census population size is the number of individuals in a population.

Center of origin is the geographic locations, where a particular domesticated plant species originated.

CentiMorgan (cM) or **map unit** (m.u.) is a unit for measuring genetic linkage. It is defined as the distance between chromosome positions (also termed as loci or markers) for which the expected average number of intervening chromosomal crossovers in a single generation is 0.01.

Central dogma is the basic concept that, in nature, genetic information generally flows from DNA to RNA to protein.

Central mother cell is a subsurface cell located in a plant apical meristem and characterized by a large vacuole.

Central Nervous System (CNS) refers to the brain and spinal cord.

Centrifugal pump transforms potential energy generated by electromagnetic forces into kinetic energy; vortex principle.

Centrifugation is a process of separating molecules by size or density using centrifugal forces generated by a spinning rotor.

Centrifuge is a device in which solid or liquid particles of different densities are separated by rotating them in a tube in a horizontal circle.

Centriole is an organelle in many animal cells that appears to be involved in the formation of the spindle during mitosis.

Centromere is the portion of the chromosome to which the spindle fibers attach during mitotic and meiotic division.

Centrosome is a specialized region of a living cell, situated next to the nucleus, where microtubules are assembled and broken down during cell division.

Cephalic relates to the head.

Cephem-type antibiotic is an antibiotic that shares the basic chemical structure of cephalosporin.

Ceramic is a nonmetallic material made from clay and hardened by firing at high temperature; it contains minute silicate crystals suspended in a glassy cement.

Cerebellum is a convoluted subdivision of the brain concerned with the coordination of muscular movements, muscle tone, and balance.

Cerebral cortex is a layer of gray matter that constitutes the outer layer of the cerebrum and is responsible for integrating sensory impulses and for higher intellectual functions.

Cerebral Perfusion Pressure (CPP) is defined as the difference between mean arterial and intracranial pressure (ICP).

Cerebrospinal fluid (CSF) is a clear, colorless liquid that surrounds the brain and spinal cord and fills the spaces; and that bathes the central nervous system.

Cerebrum is a large, convoluted subdivision of the brain; it functions as the center for learning, voluntary movement, and interpretation of sensation.

Cervix is a lower part of the uterus that projects into the vagina.

Chain terminator refers to: (1) Codons, which do not code for an amino acid. They signal ribosomes to terminate protein synthesis. The codons are UAA, UAG and UGA. Also known as stop codons or termination codons.

Chakrabarty decision is a landmark legal case in the USA, in which it was held that the inventor of a new **microorganism** whose invention otherwise met the legal requirements for obtaining a patent, could not be denied a patent solely because the invention was alive.

Chaperone is a family of proteins that ensure the correct assembly and conformation of other polypeptides *in vivo* as they emerge from the ribosome, but are not themselves components of the functional assembled structures. The prokaryotic equivalents are known as chaperonins.

Character is a distinctive feature of an organism.

Characterization refers to: (1) **Of AnGR**: All activities associated with the description of AnGR aimed at better knowledge of these resources and their state; (2) **Of PGR**: Systematic recording of descriptors that are independent of environmental factors. AnGR is an abbreviation for animal genetic resources.

Charcoal is the black porous residue of partly burnt wood, bones, etc.

Charcot-Marie tooth disease is a condition characterized by degeneration of the motor and sensory nerves that control movement and feeling in the arm below the elbow and the leg below the knee.

Chelate is a cation bound to an organic molecule through electron pair donation from nitrogen and/or oxygen atoms in its structure.

Chelating agent refers to: **(1)** A compound that combines with metal ions to form stable ring structures; **(2)** A substance used to reduce the concentration of free metal ion in solution by complexing it.

Chelator describes a particular way that ions and molecules bind metal ions.

Chemical mutagen is a chemical or product capable of causing genetic mutation in living organisms exposed to it.

Chemical vapor deposition is abbreviated as CVD.

Chemically defined medium refers to when all of the chemical components of a plant tissue culture medium are fully known and defined.

Chemiluminescence refers to the emission of light during the course of a chemical reaction.

Chemoembolization is a combination of local delivery of chemotherapy and a procedure called embolization to treat cancer, most often of the liver.

Chemoluminescence is the emission of light (luminescence), as the result of a chemical reaction.

Chemostat is a continuous and open culture in which growth rate and cell density are maintained constant by a fixed rate of input of a growth-limiting nutrient.

Chemotaxis is a motion of a motile cell, organism or part towards or away from an increasing concentration of a particular substance.

Chemotherapy is a treatment for cancer that involves administering chemicals toxic to malignant cells.

Chemotherapy refers to the treatment of disease, especially infections or cancer, by means of chemicals.

Chiari malformation is an extra cerebellum crowding's at the outlet of brainstem/spinal cord from the skull on its way to the spinal canal.

Chiasma (Gr. *chiasma*, two lines placed crosswise; plural:

chiasmata) refers to a visible point of junction between two non-sister chromatids of homologous chromosomes during the first meiotic prophase.

Chimera (or chimaera) refers to an organism whose cells are not all derived from the same zygote.

Chimeraplasty is a method designed to create defined alterations in DNA sequence at a target locus, with potential both for gene therapy and for investigating gene function.

Chimeric DNA is a recombinant DNA molecule containing unrelated genes.

Chimeric gene is a semi-synthetic gene, consisting of the coding sequence from one organism, fused to promoter and other sequences derived from a different gene.

Chimeric proteins (Fusion proteins: literally made of parts from different sources) are proteins created through the joining of two or more genes that originally coded for separate proteins. **Chimeric selectable marker gene** is a gene that is constructed from parts of two or more different genes and allows the host cell to survive under conditions where it would otherwise die.

ChIP (Chromatin immunoprecipitation) is a method for identifying protein-DNA interactions. Genomic DNA and associated proteins are cross-linked, sheared, and immunoprecipitated with antibodies that recognize specific DNA proteins.

Chip is a die that is cut from a silicon wafer, incorporating semiconductor circuit elements such as: a sensor, actuator, resistor, diode, transistor, and/or capacitor.

ChIP/SAGE is the preparation of small tags from ChIP purified DNA and their subsequent SAGE analysis to achieve genome-wide identification of protein-DNA interactions.

ChIP/SEQ is a technique involving size selection, high throughput sequencing and mapping of ChIP purified DNA onto a reference genome to achieve genome-wide identification of protein-DNA interactions.

Chi-square test is a test of statistical significance based on the chi-squared statistic, which determines how closely experimental observed values fit theoretical expected values.

Chitin is a nitrogenous polysaccharide that gives structural strength to the exoskeleton of insects and the cell walls of fungi.

Chitinase is an enzyme which breaks down chitin.

Chitosan is a linear polysaccharide composed of randomly distributed β-(1-4)-linked D-glucosamine (deacetylated unit) and N-acetyl-D-glucosamine (acetylated unit). It is made by treating shrimp and other crustacean shells with the alkali sodium hydroxide.

Chitosanase is an enzyme with system name *chitosan N-acetylglucosaminohydrolase.*

Chloramphenicol is an antibiotic that interferes with protein synthesis.

Chlorenchyma (Gr. *chloros*, green + *enchyma*) refers to tissue containing chloroplasts, including leaf mesophyll and other parenchyma cells.

Chlorophyll (Gr. *chloros*, green + *phyllon*, leaf) refers to one of the two pigments responsible for the green color of most plants.

Chloroplast (Gr. *chloros*, green + *plastos*, formed) refers to specialized cytoplasmic organelle that contains chlorophyll. Lens-shaped and bounded by a double membrane, chloroplasts contain membranous structures (thylakoids) piled up into stacks, surrounded by a gel-like matrix (stroma).

Chloroplast DNA (cpDNA) refers to the DNA present in the chloroplast.

Chloroplast transit peptide (CTP) is a transit peptide that, when fused to a protein, acts to transport that protein into plant chloroplasts.

Chlorosis (Gr. *chloros*, green + *osis*) refers to failure of chlorophyll development, and appearance of yellow color in plants.

Cholangiography refers to x-rays series of the bile ducts.

Cholecystectomy is an operation to remove the gallbladder.

Cholesterol is a substance normally made by the body, but also found in foods from animal sources, like beef, eggs, and butter. Ideally, blood cholesterol levels should be less than 200 mg/dL.

Cholinergic blocking agent blocks or inactivates acetylcholine.

Chorioamnionitis (CA) is an intraamniotic puerperal infection described as having 3 forms: histologic, clinical (clinical chorioamnionitis, IAI), and subclinical.

Chorion is an extraembryonic membrane generated from trophoblast and extraembryonic mesoderm that forms placenta. Chorion and amnion are made by the somatopleure.

Chorion frondosum (frondosum = leafy) is the chorion found on conceptus oriented towards maternal blood supply where the majority of villi form and proliferate, will contribute the fetal component of the future placenta.

Chorion laeve (laeve = smooth) is the smooth chorion found on conceptus away from maternal blood supply (towards uterine epithelium and cavity) with very few villi present.

Chorionic cavity refers to the fluid-filled extraembryonic coelom (cavity) formed initially from trophoblast and extraembryonic mesoderm that forms placenta, chorion and amnion are made by the somatopleure.

Chorionic somatomammotropin (CSH, human lactogen) is a hormone synthesized within the placenta by syncytiotrophoblast cells. This protein hormone (190 amino acids) has a structure similar to pituitary growth hormone.

Chorionic villus sampling (CVS) is a procedure used in prenatal diagnosis to look at the chromosomes of the developing fetus.

Chromatid (chromosome + *id*, "L. suffix meaning "daughters of") refers to each of the two daughter strands comprising a duplicated chromosome.

Chromatin (Gr. *chroma*) refers to substance of which eukaryotic chromosomes are composed.

Chromatin fibers refer to basic organizational unit of eukaryotic chromosomes, consisting of DNA and associated proteins assembled into strands of 30 nm average diameter.

Chromatography (Greek χρῶμα, *chroma*, color and γράφειν, *graphein*, to write) is the collective term for a set of laboratory techniques for the separation of mixtures. The mixture is dissolved in a fluid called the *mobile phase,* which carries it through a structure holding another material called the *stationary phase.*

Chromocenter refers to a body produced by fusion of the heterochromatic regions of the chromosomes in the polytene tissues (e.g., the salivary glands) of certain *Diptera.*

Chromogenic substrate is a compound or substance that contains a color-forming group.

Chromomeres are small dense bodies identified by their characteristic size and linear arrangement along a chromosome.

Chromonema (plural: chromonemata) is an optically single thread forming an axial structure within each **chromosome**.

Chromoplast plastid containing pigments, such a chloroplast, or one in which carotenoids predominate.

Chromosomal aberration is an abnormal change in chromosome structure or number, including deficiency, duplication, inversion, translocation, aneuploidy, polyploidy, or any other change from the normal pattern.

Chromosomal integration site is a chromosomal location where foreign DNA can be integrated, often without impairing any essential function in the host organism.

Chromosomal polymorphism refers to the occurrence of one to several chromosomes in two or more alternative structural forms within a population.

Chromosomal rearrangement refers to events that are mediated by double-strand breaks and subsequent repair occurring in the genome. There are many different types of rearrangements: Deletion – the removal of a DNA sequence; Insertion – the addition of a DNA sequence; Translocation – fusing one part of a chromosome to another; and Inversion.

Chromosome (Gr. *chroma*, color + *soma*, body) refers to the nuclear bodies containing most of the genes largely responsible for the differentiation and activity of the cell, in eukaryotic cells. Each eukaryotic species has a characteristic number of chromosomes. Humans have 46 chromosomes in every cell of their body except the sperm and egg cells. We inherit 23 chromosomes from our mother and 23 from our father.

Chromosome aberration refers to an abnormal structure or number of chromosomes; includes deficiency, duplication, inversion, translocation, aneuploidy, polyploidy, or any other change from the normal pattern.

Chromosome banding refers to staining of chromosomes in such a way that light and dark areas occur along the length of the chromosomes in repeatable patterns.

Chromosome jumping is a technique that allows two segments of duplex DNA that are separated by thousands of base pairs (about 200 kb) to be cloned together.

Chromosome landing is an alternative to chromosome walking for positional cloning.

Chromosome mutation is a change in the gross structure of a chromosome, usually causing severely deleterious effects in the organism, but can be maintained in a population. The main types of chromosome mutation are: translocation, duplication, and inversion.

Chromosome theory of inheritance defines that chromosomes carry the genetic information and that their behavior during meiosis provides the physical basis for segregation and independent assortment.

Chromosome walking is a strategy for mapping or sequencing a chromosome segment and for positional cloning.

Chronic refers to a disease or disorder that usually develops slowly and lasts for a long period of time.

Chyme refers to the semifluid mass of partly digested food expelled by the stomach into the duodenum.

Chymosin is an enzyme that clots the milk, used in the manufacture of cheese.

Cilium (plural: cilia, adj.: ciliate) refers to hairlike locomotor structure on certain cells; a locomotor structure on a ciliate protozoan.

Circadian (Latin *circa*, about + *di*, day + an) refers to pertaining to rhythmic biological cycles recurring at approximately 24-hour intervals.

Circuit is a path of current from a generating source through the various components back to the generating source.

Circularization is the self-ligation of a linear DNA fragment having complementary ends, generally generated by digestion with a restriction endonuclease. Successful ligation produces a molecule in the form of a covalently-closed circle.

Circulatory system refers to the body system that functions in internal transport and protects the body from disease.

Cirrhosis is a chronic problem that makes it hard for the liver to remove toxins from the body.

cis **acting protein** is a protein with the particular property of acting only on the molecule of DNA from which it was expressed.

cis **acting sequence** is a nucleotide sequence that only affects the expression of genes located on the same chromosome.

cis **configuration** is related to coupling.

cis **heterozygote** is a double heterozygote that contains two mutations arranged in a *cis* configuration (e.g., $a + b + / a b$).

Cistron is a DNA sequence that codes for a specific polypeptide; a gene.

CITES stands for "Convention on International Trade in Endangered Species" of wild fauna and flora.

Clade is a species, or group of species that has originated and includes all the descendants from a common ancestor.

Cladistics is the classification of organisms based on phylogeny.

Cladogenesis is the splitting of a single evolutionary lineage into multiple lineages.

Cladogram is a diagram illustrating the relationship between taxa that is built using synapomorphies. This is also called a phylogeny.

Class switching is the process during which a plasma cell stops producing antibodies of one class and begins producing antibodies of another class.

Cleave refers to breaking phosphodiester bonds of double-stranded DNA, usually with a type II restriction endonuclease. Synonyms is cut or digest.

Cleaved amplified polymorphic sequence refers to a segment of DNA that can be amplified by polymerase chain reaction (PCR) and which contains a DNA sequence polymorphism.

Cleft lip/palate is a congenital condition with cleft lip alone, or with cleft palate; cause is thought to be multifactorial.

Cline is a variation in one or more phenotypic characters or allele frequencies across a geographical gradient.

Clinical Engineering is the application of technology to health care in hospitals. The clinical engineer is a member of the health care team along with physicians, nurses and other hospital staff. Clinical engineers are responsible for developing and maintaining computer databases of medical instrumentation and equipment records and for the purchase and use of sophisticated medical instruments. They may also work with physicians to adapt instrumentation to the specific needs of the physician and the hospital. This often involves the interface of instruments with computer systems and customized software for instrument control and data acquisition and analysis. Clinical engineers are involved with the application of the latest technology to health care.

Clinical trial refers to a medical research undertaken with informed and consenting human subjects in a controlled environment. The intent of a clinical trial is for the sponsoring company or research institution to gather information on the safety and effectiveness of new drugs or therapies before seeking approval of a procedure or product for use.

Clitoris is a small, erectile structure at the anterior part of the vulva in females; homologous to the male penis.

Clonal propagation is an asexual propagation of many new plants (ramets) from an individual (ortet); all have the same genotype.

Clonal selection is the production of a population of plasma cells all producing the same antibody in response to the interaction between a B lymphocyte producing that specific antibody and the antigen bound by that antibody.

Clone bank is related to gene bank.

Clone (Gr. *klon*, a twig or slip) refers to: (1) A group of cells or organisms that are genetically identical as a result of asexual reproduction, breeding of completely inbred organisms, or forming genetically identical organisms by nuclear transplantation; (2) Group of plants genetically identical in which all are derived from one selected individual by vegetative propagation, without the sexual process; (3) A population of cells that all carry a cloning vehicle with the same insert DNA molecule; (4) Verb: To clone. To insert a DNA segment into a vector or host chromosome; **(5) In AnGR:** A genetic replica of another organism obtained through a non-sexual (no fertilization) reproduction process.

Cloned strain or line is a strain or line descended directly from a clone.

Cloning is the process in which an organism produces one or more genetically identical copies of itself by asexual means. The term cloning can be applied to a group of cells undergoing replication by repetitive mitoses (cell divisions).

Cloning vector is a small, self-replicating DNA molecule – usually a plasmid or viral DNA chromosome – into which foreign DNA is inserted in the process of cloning genes or other DNA sequences of interest.

Closed continuous culture is a culture system, in which the inflow of fresh medium is balanced by the outflow of corresponding volumes of spent medium.

Clostridium difficile (C. diff or C. difficile) is a bacteria normally found in the large intestine, which can cause a serious intestinal infection and diarrhea.

Cluster of differentiation (cluster of designation or Classification Determinant, CD) is a protocol used for the identification and investigation of cell surface molecules providing targets for immunophenotyping of cells.

cM is an abbreviation for centiMorgan (map distance).

CMP is an abbreviation for cytidine monophosphate, that is also known as also known as 5'-cytidylic acid or simply cytidylate ($C_9H_{14}N_3O_8P$). CMP consists of the phosphate group, the pentose sugar ribose, and the nucleobase cytosine.

Coalescent is the point at which the ancestry of two alleles converge at a common ancestral sequence.

Coanda effect is the tendency of a stream of fluid to stay attached to a convex surface, rather than follow a straight line in its original direction.

Coat protein (capsid) is the coating of a protein that enclosed the nucleic acid core of a virus.

Coaxial cable is a heavy, well-insulated electrical wire.

Coccus is a spherical bacterium.

Cochlea is the structure of the inner ear that contains the auditory receptors (organ of Corti).

Co-cloning refers to the unintentional cloning of DNA fragments, along with the desired one, that can occur when the source of DNA being cloned is not sufficiently purified.

Coconut milk is a liquid endosperm of the coconut, often used to supply organic nutrients to *in vitro* cultures of plant cells and tissues.

Cocoon refers to a protective coverage for eggs and/or larvae produced by many invertebrates, such as the silkworm moth.

Co-culture refers to the joint culture of two or more types of cells, such as a plant cell and a micro-organism, or two types of plant cells.

Codex Alimentarius Commission is an international regulatory body (part of FAO) responsible for the definition of a set of international food standards.

Coding is the specification of a peptide sequence, by the code contained in DNA molecules.

Coding sequence refers to that portion of a gene which directly specifies the amino acid sequence of its protein product.

Coding strand refers to the strand of duplex DNA which contains the same base sequence (after substituting U for T) found in the mRNA molecule resulting from transcription of that segment of DNA. a.k.a. sense strand. The mRNA molecule is transcribed from the other strand, known as the template or antisense strand. Coding strand 5' ATGAAAGCTTTA-GTGGGCGCCCGTAT 3'; Template strand 3'

TACTTTCGAAATCA-
CCCGCGGGCATA
5'; mRNA 5'
AUGAAAGCUUUAGUGGG-
CGCCCGUAU 3'

Co-dominance refers to the situation in which both alleles in a heterozygous individual are expressed, so that the phenotype of heterozygotes incorporates the phenotypic effect of each allele.

Co-dominant alleles refers to alleles that produce independent effects when in the heterozygous condition.

Codon is one of the groups of three consecutive nucleotides in mRNA, which represent the unit of genetic coding by specifying a particular amino acid during the synthesis of polypeptides in a cell. Each codon is recognized by a tRNA carrying a specific amino acid. Related terms are genetic code, start codon, stop codon. Synonym is triplet.

Codon optimization is an experimental strategy in which codons within a cloned gene are changed by *in vitro* mutagenesis to the preferred codons, without changing the amino acids of the synthesized protein.

Codons are organic bases in sets of three that form the genetic code.

Coefficient refers to a number expressing the amount of some change or effect under certain conditions (e.g., the coefficient of inbreeding).

Coelocentesis is a sampling of extracoelomic fluid usually for an early prenatal diagnostic technique.

Coenzyme (cofactor) is an organic molecule, such as a vitamin, that binds to an enzyme and is required for its catalytic activity.

Co-evolution refers to the evolution of complementary adaptations in two species caused by the selection pressure that each exerts on the other.

Co-factor is an organic molecule or inorganic ion necessary for the normal catalytic activity of an enzyme. Synonym is co-enzyme.

Co-fermentation refers to the simultaneous growth of two micro-organisms in one bioreaction.

Cohesion refers to a force holding a solid to a solid or a solid to a liquid, owing to attraction between like molecules.

Cohesive ends refer to double-stranded DNA molecules with single-stranded ends which are complementary to each other, enabling the different molecules to join each other.

Coincidence is the ratio of the observed to the expected frequency of double cross-overs, where the expected frequency is calculated by assuming that the two cross-over events occur independently of one another.

Co-integrate refers to a chimeric DNA molecule formed by the incorporation at a single site of two different DNA molecules.

Co-integrate vector system refers to a two-plasmid system for transferring cloned genes to plant cells.

Colchicine (*L. colchicum*, meadow saffron) is an alkaloid obtained from *Colchicum autumnale*, autumn crocus or meadow saffron, which inhibits spindle formation in cells during mitosis, so that chromosomes cannot separate during anaphase, thus inducing multiple sets of chromosomes.

Cold-induced vasodilatation is a vasodilatation following cold application.

Coleoptile is a protective sheath covering the shoot apex of the embryo in the grasses.

Coleorhiza (Gr. *koleos*, sheath + *rhiza*, root) is a protective sheath surrounding the radicle of monocotyledenous plants.

Colic is a condition in an otherwise healthy baby characterized by excessive crying.

Co-linearity refers to the phenomenon whereby gene order is preserved between distinct species.

Colitis is an irritation of the colon (large intestine).

Collagen tissue is a fibrous insoluble protein found in connective tissue, bone, ligaments, and cartilage.

Collecting duct refers to the duct in the kidney that receives filtrate from several nephrones and conducts it to the renal pelvis.

Collenchyma (Gr. *kolla*, glue + *enchyma*, a suffix) is a tissue of living cells, the walls being unevenly thickened with cellulose and hemicellulose, but never lignified.

Colloid refers to the fluid suspension of the body's intercellular fluid.

Colon is a portion of the large intestine between the cecum and the rectum.

Colony hybridization is a technique that uses a nucleic acid probe to identify a bacterial colony with a vector carrying a specific cloned gene or genes.

Colony refers to: (1) An aggregate of identical cells (clones) derived from a single progenitor cell. (2) A group of interdependent cells or organisms.

Combinatorial library refers to many novel combinations that are generated when a heavy-chain library is combined by random pairing with a light-chain library.

Commensalism is the interaction of two or more dissimilar organisms, where the association is advantageous to one without affecting the other(s).

Common bile duct refers to a tube that moves bile from the liver to the small intestine.

Companion cell refers to a living cell associated with the sieve cell of phloem tissue in vascular plants.

Comparative gene mapping refers to the comparison of map locations of genes between species.

Comparative mapping is the comparison of map locations of genes and markers between species.

Comparative positional candidate gene relates to an indirect means of assigning function to a QTL (Quantitative trait loci).

Compartmental model is a type of mathematical model that describes the way materials or energies are transmitted among the *compartments* of a system.

Competence is the ability of a bacterial cell to take up DNA molecules and become genetically transformed.

Competency refers to an ephemeral state, induced by treatment with cold cations, during which bacterial cells are capable of taking up foreign DNA.

Competent refers to: (1) Bacterial cells able to take up foreign DNA molecules and thereby become genetically transformed; (2) A competent cell is capable of developing into a fully functional embryo.

Competent cell refers to the cell that is capable of developing into a fully functional embryo. The opposite is non-competent.

Complement proteins are proteins that bind to antibody-antigen complexes and help degrade the complexes by proteolysis.

Complementarity refers to the relationship between the two strands of a double helix of DNA.

Complementarity determining regions (CDRs) are part of the variable chains in immunoglobulins (antibodies) and T cell receptors, generated by B-cells and T-cells respectively, where these molecules bind to their specific antigen.

Complementary DNA (cDNA) is a DNA synthesized from a messenger RNA (mRNA) template in a reaction catalysed by the enzymes reverse transcriptase and DNA polymerase. cDNA is often used to clone eukaryotic genes in prokaryotes.

Complementary DNA or RNA refers to the matching strand of a DNA or RNA molecule to which its bases pair.

Complementary entity refers to: (1) One of a pair of nucleotide bases that form hydrogen bonds with each other. Adenine (A) pairs with thymine (T) [or with Uracil (U) in RNA], and guanine (G) pairs with cytosine (C); (2) One of a pair of segments or strands of nucleic acid that will hydridize (join by hydrogen bonding) with each other.

Complementary genes refer to two or more interdependent genes, such that the dominant allele from either gene can only produce an effect on the phenotype of an organism if the dominant allele from the other gene is also present; or (in the case of recessive complementarity) only double homozygous recessive show the effect.

Complementary homopolymeric tailing refers to the process of adding complementary nucleotide extensions to different DNA molecules, e.g., dG (deoxyguanosine) to the 3'-hydroxyl ends of one DNA molecule and dC (deoxycytidine) to the 3'-hydroxyl ends of another DNA molecule to facilitate, after mixing, the joining of the two DNA molecules by base pairing between the complementary extensions. This is also called dG–dC tailing, dA–dT tailing.

Complementary nucleotides are members of the pairs adenine-thymine, adenine-uracil, and guaninecytosine that have the ability to hydrogen bond to one another.

Complementation occurs when two strains of an organism with different homozygous recessive mutations that produce the same phenotype.

Complementation test (trans test, by Edward B. Lewis) refers to introduction of two mutant chromosomes into the same cell to determine whether the mutants are alleles of the same gene. If the mutations are non-allelic, the genotype will be m_1 $+/+ M_2$. In contrast, if they are allelic, the mutant phenotype will result.

Complete digest is the treatment of a DNA preparation with a restriction endonuclease for sufficient time for all of the potential target sites within that DNA to have been cleaved. Opposite is partial digest.

Component is a sequence used to construct a larger sequence (a sequence contig or a scaffold).

Composite transposon is a transposable element formed when two identical or nearly identical transposons insert on either side of a non-transposable segment of DNA, such as the bacterial transposon Tn5.

Compound chromosome is a chromosome formed by the union of two separate chromosomes, as in attached-X chromosomes or attached-X-Y chromosomes.

Compressible fluid flow is compressible if its density changes appreciably (typically by a few percent) within the domain of interest.

Computational gene recognition is a process of interpreting nucleotide sequences by computer, in order to provide tentative annotation on the location, structure and functional class of protein- coding genes.

Computed tomography (CT) is a technology that uses computer-processed x-rays to produce tomographic images (virtual slices) of specific areas of the scanned object, allowing the user to see what is inside it without cutting it open.

Computed tomography scan (CT or CAT scan) is a diagnostic imaging procedure that uses a combination of x-rays and computer technology to produce cross-sectional images

(often called slices), both horizontally and vertically, of the body.

Computer Aided Molecular Design (CAMD) involves all computer-assisted techniques used to discover, design and optimize compounds with desired structure and properties. Also known as molecular modeling or computational chemistry.

Computer-Assisted Drug Design involves all computer- assisted techniques used to discover, design and optimize biologically active compounds with a putative use as drugs. Broader term is drug design.

Computer-Assisted Molecular Design involves all computer-assisted techniques used to discover, design and optimize compounds with desired structure and properties.

Computer-Assisted molecular modeling is the investigation of molecular structures and properties using computational chemistry and graphical visualization techniques.

Concatemer is a DNA segment composed of repeated sequences linked end to end.

Concordance refers to the identity of matched pairs or groups for a given trait, such as sibs expressing the same trait.

Condenser electrode is an electrical current is conducted back and forth between the two electrodes.

Confocal laser scanning microscopy (CLSM or LSCM) is a technique for obtaining high-resolution optical images with depth selectivity. The key feature of confocal microscopy is its ability to acquire in-focus images from selected depths, a process known as optical sectioning.

Conditional lethal mutation refers to the mutation that is lethal under one set of environmental conditions (the restrictive conditions, commonly associated with high temperature) but is viable under another set of environmental conditions (the permissive conditions).

Conditional lethal mutation is a mutation that is lethal under one set of environmental conditions .

Conditioning refers to: (1) The effects on phenotypic characters of external agents during critical developmental stages; (2) The undefined interaction between tissues and culture medium resulting in the growth of single cells or small aggregates.

Conduction is a method of heat transfer in opaque solids. If the temperature at one end of a metal rod is raised by heating, then heat is conducted to the colder end.

Conductor is a metal that conducts electricity via the motion of electrons.

Conformation refers to the various three dimensional shapes that can be adopted by a given molecule.

Conformational analysis consists of the exploration of energetically favorable spatial arrangements (shapes) of a molecule (conformations) using molecular mechanics, molecular dynamics, quantum chemical calculations or analysis of experimentally-determined structural data, e.g., NMR or crystal structures.

Congenital hypothyroidism is an inherited trait that results in reduced activity of the thyroid gland, generally due to reduced production of thyroid stimulating hormone.

Congestion refers to the presence of an abnormal amount of blood in the vessels as a result of an increase in blood flow or obstructed venous return.

Conidium (plural: conidia) is an asexual spore produced by a specialized hypha in certain fungi.

Conjugate (graph theory) is an alternative term for a line graph, i.e. a graph representing the edge adjacencies of another graph. Conjugation in biochemistry refers to the turning of substances into a hydrophilic state in the body.

Conjugation refers to: (1) Union of gametes or unicellular organisms during fertilization; (2) The unidirectional transfer of plasmid DNA from one bacterium cell to another, involving cell-to-cell contact; (3) Attachment of sugar and other polar molecules to less polar compounds, thus making them more water soluble.

Conjugative functions refers to plasmid-based genes and their products that facilitate the transfer of a plasmid from one bacterium to another via conjugation.

Consanguinity is related by descent from a common ancestor.

Consensual heat vasodilatation is an increased blood flow that spreads to a remote area of the body as a result of localized heating.

Consensus sequence is the part of a gene or signal sequence that is shared over a wide range of members of a gene family, both within a given species, or in comparisons between species.

Conservation breeding refers to the efforts to manage plant and animal species breeding that do not necessarily involve captivity.

Conservation collection is a living collection of rare or endangered organisms established for the purpose of contributing to the survival and recovery of a species.

Conservation laws states that particular measurable properties of an isolated physical system do not change as the system evolves.

Conservation of farm animal genetic resources refers to all human activities, including strategies, plans, policies and actions, undertaken to ensure that the diversity of farm animal genetic resources is being maintained to contribute to food and agricultural production and productivity; now and in the future.

Conservation of Mass states that mass can neither be created or destroyed.

Conserved sequence is an identical or highly similar sequence of nucleotides or amino acids which occurs as part, or all of a number of different genes or proteins, in either the same or different species.

Conspecific refers to a member of the same species.

Constant domains are regions of antibody chains that have the same amino acid sequence in different members of a particular class of antibody molecules.

Constipation refers to hard, dry stools that are difficult to pass in a bowel movement, or having fewer than three bowel movements per week.

Constitutive is an organism is said to be constitutive for the production of an enzyme or other protein if that protein is always produced by the cells under all physiological conditions.

Constitutive enzyme is an enzyme that is synthesized continually regardless of growth conditions.

Constitutive gene is a gene that is continually expressed in all cells of an organism.

Constitutive promoter is an unregulated promoter that allows for continual transcription of its associated gene.

Constitutive synthesis is a continual production of RNA or protein by an organism.

Construct refers to an engineered chimeric DNA designed to be transferred into a cell or tissue.

Contaminant refers to bacterial, fungal or algal micro-organism accidentally introduced into a culture or culture medium.

Contig is a set of overlapping cloned DNA fragments that can be assembled to represent a defined region of the chromosome or genome from which they were obtained.

Contiguous genes are genes physically close on a chromosome that when acting together express a phenotype.

Contiguous map (contig map) is the alignment of sequence data from large, adjacent regions of the genome to produce a continuous nucleotide sequence across a chromosomal region.

Continuous characters are phenotypic traits that are distributed continuously throughout the population (e.g., height or weight).

Continuous culture is a suspension culture continuously supplied with nutrients by the inflow of fresh medium. The culture volume is normally constant.

Continuous distribution model of migration is a model when individuals are continuously distributed across the landscape; neighborhoods of individuals exist that are areas within which panmixia occurs, and across which genetic differentiation occurs due to isolation by distance.

Continuous fermentation is a process in which cells or

micro-organisms are maintained in culture in the exponential growth phase by the continuous addition of fresh medium that is exactly balanced by the removal of cell suspension from the bioreactor.

Continuous variation is a variation where individuals cannot be classified as belonging to one of a set of discrete classes. Characters showing continuous variation are referred to as quantitative. Opposite is discontinuous variation.

Contrast agent is a substance used to enhance the contrast of structures or fluids within the body in medical imaging. It is commonly used to enhance the visibility of blood vessels and the gastrointestinal tract.

Contrast bath refers to the hot (106°F) and cold (50°F) treatments in a combined sequence to stimulate superficial capillary vasodilatation or vasoconstriction.

Control refers to: (1) **Noun**: Unchanged (standard) protocol or treatment for comparison with the experimental treatment;. (2) **Verb**: To direct or regulate cultures with addition of plant growth regulators.

Controlled environment is a closed environment in which parameters, such as light,

temperature, relative humidity and sometimes the partial gas pressure (and possibly its composition), are fully controlled.

Controlling element refers to transposable elements which control the activity of standard genes. A controlling element may, in the simplest case, inhibit the activity of a gene through becoming integrated in, or close to, that gene.

Convention is a set of agreed, stipulated, or generally accepted standards, norms, social norms, or criteria, often taking the form of a custom.

Convention on Biological Diversity (CBD) refers to the international treaty governing the conservation and use of biological resources around the world.

Convention on International Trade in Endangered Species of Wild Fauna and Flora (CITES) refers to an agreement among 145 countries that bans commercial international trade in an agreed-upon list of endangered species, and that regulates and monitors trade in others that might become endangered.

Converged refers to the point where a MCMC simulation has become independent of starting parameter biases, or has been "burnt in."

Conversion is the development of a somatic embryo into a plant.

Coordinate repression is a correlated regulation of the structural genes in an operon by a molecule that interacts with the operator sequence.

Co-polymers are mixtures consisting of more than one monomer.

Copy DNA is the process of producing two identical replicas from one original DNA molecule. This biological process occurs in all living organisms and is the basis for biological inheritance.

Copy number is the number of a particular plasmid per bacterium cell, or gene per genome.

Copy Number Variation (CNV) refers to large-scale structural changes in DNA that vary from individual to individual. These include insertions, deletions, duplications and complex multi-site variants that range from kilobases to megabases in size.

Co-repressor is an effector molecule that forms a complex with a repressor and turns off the expression of a gene or set of genes.

Cornelia de Lange syndrome is a condition involving growth deficiency, significant developmental delay, anomalies of the extremities and a characteristic facial appearance.

Coronary arteries are two arteries arising from the aorta that arch down over the top of the heart and divide into branches. They provide blood to the heart muscle.

Coronavirus is a single-stranded RNA virus that resembles a crown when viewed under an electron microscope because of its petal-shaped projections.

Corpus collosum is a large bundle of nerve fiber interconnecting the two cerebral hemispheres.

Corpus is found below the tunica and is a part of the apical meristem. In the corpus, cells divide in all directions, giving them an increase in volume.

Corpus luteum is the temporary endocrine tissue in the ovary that develops from the ruptured follicle after ovulation.

Correlation is a statistical association between variables.

Corrosion is a deteriorative loss of a metal as a result of dissolution environmental reactions.

Cortex is the primary tissue of a stem or root, bounded externally by the epidermis and internally in the stem by the phloem and in the root by the pericycle.

Cortical relates to or consists of the cortex.

Corticosteroids are medications that reduce irritation and inflammation.

Cos ends refers to the 12-base, single-strand, complementary extensions of bacteriophage lambda DNA.

Cos site is the sequence that is cut to produce the cohesive, single-stranded extensions located at the ends of the linear DNA molecules of certain phages (e.g., lambda).

Co-segregation is the joint inheritance of two characters, usually the result of genetic linkage.

Cosmid is a plasmid vector which contains the two cos (cohesive) ends of phage lambda and one or more selectable markers such as an antibiotic resistance gene.

Co-suppression is a natural gene silencing phenomenon, which probably evolved as part of plants' defence against viral attack, but which has become important in the context of plant transgenesis.

Cot curve is a method to estimate the heterogeneity of sequence of a DNA preparation, based on the observation that the more homogenous the DNA, the more easily (and therefore faster) the annealing of single-stranded DNA will occur. The Cot curve plots the extent of annealing from a fully single-stranded preparation over time.

Co-transfection is the procedure by which the baculovirus and the transfer vector are simultaneously introduced into insect cells in culture, in baculovirus expression systems,.

Co-transformation is a protocol for producing transgenesis, in which host (plant or animal) cells are transformed simultaneously with two different plasmids, one of which carries a selectable marker, and the other the gene to be transferred.

Cotyledon refers to leaf-like structures at the first node of the seedling stem.

Coulomb is a measure of electrical charge, and one C is an amount of charge $= 6.24 \times 1018$ electrons.

Countergradient variation occurs when genetic effects on a trait oppose or compensate for environmental effects so that phenotypic differences across an environmental gradient among populations are minimized.

Coupling agent is a substance used as a medium for the transfer of sound waves.

Coupling is the phase state in which either two dominant or two recessive alleles of two different genes occur on the same chromosome. Synonym is cis configuration. Opposite is repulsion; trans configuration.

Covalent bond is a chemical bond that involves the sharing of electron pairs between atoms. The stable balance of attractive and repulsive forces between atoms when they share electrons is known as covalent bonding.

Covalently closed circle (CCC) is a double-stranded DNA molecule with no free ends. The two strands are interlinked and will remain together even after denaturation. In its native form, a CCC will adopt a supercoiled configuration.

Covalently-closed circular DNA (ccc DNA) refers to a DNA molecule in which the free ends have ligated to form a circle.

Covariance is a measure of the statistical association between variables; the extent to which two variables vary together.

CP4 EPSPS is an abbreviation for CP4 5-enolpyruvyl-shikimate-3-phosphate synthase, which is an enzyme that catalyzes the chemical reaction consisting of the two substrates of this enzyme as phosphoenolpyruvate and 3-phospho-shikimate, with its two products as phosphate and 5-enolpyruvylshikimate-3-phosphate.

CPB refers to cardiopulmonary bypass.

cpDNA is an abbreviation for chloroplast DNA. It refers to the DNA of plant plastids, including chloroplasts.

CpG islands are areas of multiple CG repeats in DNA.

Cranial nerves are ten to twelve pairs of nerves that emerge directly from the brain.

Craniocerebral trauma can cause bleeding into the brain.

Creep is a time-dependent permanent deformation that occurs under stress; for most materials it is important only at elevated temperatures.

Cri-du-chat syndrome refers to a chromosomal condition (monosomy 5p). Name comes from the distinctive mewing cry of affected infants; characterized by significant mental deficiency, low birth weight, failure to thrive and short stature; deletion of a small section of the short arm of chromosome 5.

Critical breed refers to the breed where the total number of breeding females is less than 100 or the total number of breeding males is less than or equal to five; or the overall population size is close to, but slightly above 100 and decreasing, and the percentage of purebred females is below 80%.

Critical-maintained breed (endangered-maintained breed) refers to the categories where critical or endangered breeds are being maintained by an active public conservation program or within a commercial or research facility.

Crohn's disease is a chronic illness that causes irritation in the digestive tract. It occurs most commonly in the ileum (lower small intestine) or in the colon (large intestine). It is a form of inflammatory bowel disease.

Cross hybridization is the annealing of a single-stranded DNA sequence to a single-stranded target DNA to which it is only partially complementary. Often, this refers to the use of a DNA probe to detect homologous sequences in species other than the origin of the probe.

Cross is the mating of two individuals or populations.

Cross pollination efficiency is the ease with which cross pollination can be achieved. Generally measured by the number of hybrid progeny generated per flower pollinated.

Cross pollination is the application of pollen from one plant to another to effect the latter's fertilization.

Cross-breeding is the mating between members of different populations (lines, breeds, races or species).

Crossing over is a process in which homologous chromosomes exchange material through the breakage and reunion of two chromatids.

Crossing-over unit is a measure of distance between two loci on genetic maps that is based on the average number of crossing-over events that take place in the interval between those two loci during meiosis.

Cross-linked polymer is a polymer with adjacent linear molecular chains that are joined at various positions by covalent bonds.

Cross-sensitivity is the influence of one measurand on the sensitivity of a sensor, another measurand.

Crown is the region at the base of the stem of cereals and forage species from which tillers or branches arise.

Crown gall refers to a bulbous growth that occurs at the base of certain plants as result of infection, especially by *Agrobacterium tumefaciens*.

CRP is an abbreviation for catabolite regulator protein.

Cry proteins are a class of crystalline proteins produced by strains of Bacillus thuringiensis, and engineered into crop plants to give resistance against insect pests. Synonym is delta endotoxins.

Cryoablation is a process that uses extreme cold (cryo) to destroy or damage tissue (ablation).

Cryobiological preservation (cryopreservation; freeze preservation) refers to the preservation of germplasm resources in a dormant state by cryogenic techniques, as currently applied to storage of plant seeds and pollen, micro-organisms, animal sperm, and tissue culture cell lines.

Cryogenic refers to one that is at very low temperature.

Cryokinetics (Cryotherapy) is the use of cold and exercise in the treatment of pathology or disease.

Cryopreservation or cryoconservation is a process where cells, whole tissues, or any other substances susceptible to damage caused by chemical reactivity or time are preserved by cooling to sub-zero temperatures.

Cryoprotectant refers to the compound preventing cell damage during successive freezing and thawing processes. Cryoprotectants are agents with high water solubility and low toxicity.

Cryptic refers to anything hidden: (1) Structurally heterozygous individuals that are not identifiable; (2) A form of polymorphism controlled by recessive genes ('cryptic polymorphism'); (3) Any mutation which is exposed by a sensitizing mutation and otherwise poorly detected (such mutations probably escape detection; (4) Phenotypically very similar species (cryptic species); (5) Cryptic genetic variation refers to the existence of alleles conferring high performance for a trait, in a breed that has low performance for that trait.

Crystal is the part of the ultrasound head that vibrates and changes shape.

CSIRO is an abbreviation for Commonwealth Scientific and Industrial Research Organization.

CSO applies to outsourcing and biotech service providers in the pharmaceutical industry.

CTP is an abbreviation for: (1) cytidine 5'-triphosphate, which is required for RNA synthesis since it is a direct precursor molecule; and (2) Chloroplast transit peptide.

Cultigen is a cultivated plant species with no known wild progenitor.

Cultivar (cv.) is a category of plants that are, firstly, below the level of a sub-species taxonomically, and, secondly, found only in cultivation.

Culture alteration is a term used to indicate a persistent change in the properties of a culture's behavior (e.g., altered morphology, chromosome constitution, virus susceptibility, nutritional requirements, proliferative capacity, etc.).

Culture is a population of plant or animal cells or micro-organisms grown under controlled conditions.

Culture medium is the liquid that covers cells in a culture dish and contains nutrients to nourish and support the cells.

Culture room refers to a room for maintaining cultures and often in a controlled environment of light, temperature and humidity.

Curing is the elimination of a plasmid from its host cell. Many agents which interfere with DNA replication, e.g. ethidium bromide, can cure plasmids from either bacterial or eukaryotic cells.

Current [A] is a rate of flow of electric charge. One ampere current is a flow of one C of charge per second.

Cut refers to make a double-stranded break in DNA, usually with a type II restriction endonuclease.

Cuticle (*L. cuticula*, diminutive of *cutis*, the skin) refers to a layer of cutin or wax on the outer surface of leaves and fruits and that reduces water loss.

Cutting (Noun) refers to a detached plant part that under appropriate cultural conditions can regenerate the complete plant without a sexual process.

Cyanine dye is a non-systematic name of a synthetic dye family belonging to polymethine group. The word cyanin is from the English word "cyan," which conventionally means a shade of blue-green (close to "aqua") and is derived from the Greek "kyanos" which means a somewhat different color: "dark blue."

Cyanosis refers to the bluish color in the skin because of insufficient oxygen.

Cyberknife (by John R. Adler) is a frameless robotic radiosurgery system used for treating benign tumors, malignant tumors and other medical conditions.

Cybrid refers to a cytoplasmic hybrid, originating from the fusion of a cytoplast (the cytoplasm without nucleus) with a whole cell, as in nuclear transfer.

Cyclic adenosine monophosphate (cyclic AMP, cAMP) refers to a "messenger" molecule that regulates many intracellular reactions by transducing signals from extracellular growth factors to cellular metabolic pathways.

Cyclic AMP or cAMP is an abbreviation for cyclic adenosine monophosphate.

Cyclodextrin (cycloamyloses) are a family of compounds made up of sugar molecules bound together in a ring (cyclic oligosaccharides).

Cycloheximide is a molecule that inhibits protein synthesis in eukaryotes, but not in prokaryotes.

Cyclotron is a type of particle accelerator in which charged particles accelerate outwards from the center along a spiral path.

Cy-dyes exhibit excellent spectroscopic performance with high quantum yields, high extinction coefficient, and good stability of fluorescence signal against changes in pH in various buffer conditions.

Cysteine (Cys or C) is an α-amino acid with the chemical formula $HO_2CCH(NH_2)CH_2SH$. It is a semi-essential amino acid, which means that it can be biosynthesized in humans.

Cystic fibrosis is a hereditary disease whose symptoms usually appear shortly after birth. They include faulty digestion, breathing difficulties and respiratory infections due to mucus accumulation, and excessive loss of salt in sweat.

Cystoscopy (cystourethroscopy) is an examination in which a scope, a flexible tube and viewing device, is inserted through the urethra to examine the bladder and urinary tract for structural abnormalities or obstructions, such as tumors or stones.

Cystourethrogram (voiding cystogram) is a specific x-ray series that examine the urinary tract.

Cytidine refers to the (ribo)nucleoside resulting from the combination of the base cytosine (C) and the sugar D-ribose. The corresponding deoxyribonucleoside is called deoxycytidine.

Cytidine triphosphate (cytidine 5'-triphosphate, CTP) is a pyrimidine nucleoside triphosphate [(2R, 3S,4R,5R)-5-(4-amino-2-oxopyrimidin-1-yl)-3,4-dihydroxyoxolan-2-yl] methyl(hydroxy-phosphono-oxyphosphoryl)hydrogen phosphate]. CTP is a substrate in the synthesis of RNA. CTP acts as an inhibitor of the enzyme Aspartate carbamoyltransferase, which is used in pyrimidine biosynthesis.

Cytidylic acid is synonym for **cytidine** monophosphate (CMP), which is a (ribo)nucleotide containing the nucleoside cytidine.

Cytochrome is a class of pigments in plant and animal cells, usually in the mitochondria.

Cytochrome p450 is a highly diversified set (more than 1500 known sequences) of heme-containing **protein**s. In bacteria they are soluble and approximately 400 amino acids long; eukaryotic P450s are larger – about 500 amino acids. In mammals they are critical for drug metabolism, haemostasis, cholesterol biosynthesis and steroidogenesis.

Cytogenetics is a branch of genetics that is concerned with the study of the structure and function of the cell, especially the chromosomes.

Cytokine (Greek *cyto-*, cell; and *-kinos*, movement) are a broad and loose category of small proteins (~5–20 kDa) that are important in cell signaling. Cytokines include chemokines, interferons, interleukins, lymphokines, tumor necrosis factor but generally not hormones or growth factors.

Cytokinesis refers to the cytoplasmic division and other changes exclusive of nuclear division that are a part of mitosis or meiosis.

Cytokinin refers to plant growth regulators (hormones) characterized as substances that induce cell division and cell differentiation (e.g., BA, kinetin, and 2-iP).

Cytology (Greek κύτος, *kytos*, "a hollow"; and -λογία, *-logia*) is the study of cells. Cytology is that branch of life science, which deals with the study of cells in terms of structure, function and chemistry.

Cytolysis refers to cell disintegration.

Cytoplasm (Gr. *kytos*, a hollow vessel + *plasma*, form) refers to the living material of the cell, exclusive of the nucleus, consisting of a complex protein matrix or gel.

Cytoplasmic genes are genes located in cellular organelles such as mitochondria and chloroplasts.

Cytoplasmic inheritance refers to hereditary transmission dependent on the cytoplasm or structures in the cytoplasm rather than the nuclear genes; extrachromosomal inheritance. Thus, plastid characteristics in plants are inherited by a mechanism independent of nuclear genes.

Cytoplasmic male sterility refers to genetic defect due to faulty functioning of mitochondria in pollen development, preventing the formation of viable pollen.

Cytoplasmic organelles refer to discrete sub-cellular structures located in the cytoplasm of cells; these allow division of labour within the cell.

Cytosine (C) is one of the four main bases found in DNA and RNA, along with adenine, guanine, and thymine (uracil in RNA). It is a pyrimidine derivative, with a heterocyclic aromatic ring and two substituents attached.

Cytosol is the fluid portion of the cytoplasm, i.e., the cytoplasm minus its organelles.

Cytotoxic T cell (also known as T_C, cytotoxic T lymphocyte, CTL, T-killer cell, cytolytic T cell, CD8+ T-cells or killer T cell) is a T lymphocyte (a type of white blood cell) that kills cancer cells, cells that are infected (particularly with viruses), or cells that are damaged in other ways.

Cytotoxicity refers to the poisoning of the cell.

Cytotrophoblast can invade the maternal decidua, form columns to attach villi, or fuse to form syncitiotrophoblast cells.

Cytotype is a maternally inherited cellular condition in *Drosophila* that regulates the activity of transposable P elements.

D

D loop is an abbreviation for displacement loop, which is formed when a short stretch of RNA is paired with one strand of DNA, thus displacing the original partner DNA.

dA–dT Tailing (5'-Adenylic acid, thymidylyl-(5'-3')-2'-deoxy-, homopolymer) is synonyms of complementary homopolymeric tailing; dG-dC tailing. It refers to poly-deoxyribo nucleotides made up of deoxyadenine nucleotides and thymine nucleotides. It is desirable for the A tail to be at least as long as oligo(dT) on the surface.

dAb (single domain antibody; plural Dabs) is an antibody with only one (instead of two) protein chain derived from only one of the two domains of the normal antibody structure. Related terms are single-chain antigen binding technology (SCA), biosynthetic antibody binding sites (BABS), minimum recognition units

(MRUs), and complementary determining regions (CDRs).

DAD is an abbreviation for domestic animal diversity.

DAF (DNA amplification fingerprinting) is the enzymatic amplification of arbitrary stretches of DNA, which is directed by very short oligonucleotide primers of arbitrary sequence to generate complex but characteristic DNA fingerprints.

Dalton (Da: John Dalton) is a unit of atomic mass roughly equivalent to the mass of a hydrogen atom: 1.67×10^{-24} g. The kDa or MDa are equal to 1×10^3 and 1×10^6 daltons, respectively.

DAMD is an abbreviation for directed amplification of minisatellite DNA.

Darwinian cloning is a selection of a clone from a large number of essentially random starting points, rather than isolating a natural gene or making a carefully designed artificial one.

dATP is an abbreviation for deoxyadenosine 5'-triphosphate, which is required for DNA synthesis. It is a direct precursor molecule.

Daughterless carp is a carp which only produce male fish. This slows the growth of the population with the aim of reducing overall carp numbers.

dCTP is an abbreviation for deoxycytidine 5'-triphosphate, which is required for DNA synthesis.

ddNTP is an abbreviation for dideoxynucleotide triphosphate.

De novo is a Latin expression meaning "from the beginning," "afresh," "anew," "beginning again." It is used in **Biology and chemistry.** *De novo* may also be a term used to define methods for making predictions about biological features using only a computational model without extrinsic comparison to existing data. In this context, it may be sometimes interchangeable with the Latin term *ab initio*. In bioinformatics, *de novo* is a form of sequencing, as in *de novo* peptide sequencing. *De novo* mutation is neither parent possessed nor transmitted. *De novo* transcriptome assembly is the method of creating a transcriptome without a reference genome.

Deamination refers to hydrolysis changes of a normal base to an atypical base containing a keto group in place of the original amine group. Examples include C → U and A → HX (hypoxanthine), and 5MeC (5-methylcytosine) → T.

Death phase is the final growth phase of cell culture.

Deceleration phase refers to the phase of declining growth rate, following the linear phase and preceding the stationary phase in most batch-suspension cultures.

Decidua basalis is a term given to the uterine endometrium at the site of implantation where signaling transforms the uterine stromal cells (fibroblast-like) into decidual cells. This forms the maternal component of the placenta, the decidualization process gradually spreads through the remainder of the uterus, forming the *decidua parietalis*.

Decidua basalis **reaction** is a term describing the maternal endometrial changes that occur initially at the site of, and following, blastocyst implantation.

Decidua capsularis is a term given to the uterine endometrium, which has been converted to decidua surrounding the conceptus on the smooth chorion side.

Decidua parietalis is a term given to the remainder of the uterine endometrium, away from the site of implantation, that gradually becomes converted to decidua.

Decidual cell refers to the uterine stromal cell (fibroblast-like) that differentiate in response to both steroid hormones (progesterone) and embryonic signals.

Decidualization is a process by which uterine stromal cells differentiate in response to both steroid hormones and embryonic signals into large epitheliod decidual cells.

Dedifferentiation is the process by which plant cells can become unspecialized and start to proliferate by cell division to form a mass of undifferentiated cells (or callus) that may later differentiate again to form either the same cell type or a different one.

Deep vein thrombosis is a blood clot in the deep vein in the calf.

Defective virus is a virus that is unable to reproduce when infecting its host cell, but that can grow in the presence of another virus.

Deficiency is an insufficiency or absence of one or more usable forms of enzymatic, nutritional or environmental requirements, so that development, growth or physiological functions are affected.

Defined refers to: (1) Fixed conditions of medium, environment and protocol for growth; (2) Precisely known and stated elements of a tissue culture medium. Opposite is undefined.

Degeneracy (of the genetic code) refers to the specification of one amino acid by more than one codon. It arises from the inevitable redundancy resulting from 64 triplets in a triplet code (4 x 4 x 4 = 64) encoding only 20 amino acids.

Degeneration refers to: (1) Changes in cells, tissues or organs due to disease; (2) The reduction in size or complete loss of organs during evolution.

Degradation is a gradual wearing down or away.

Degrees of freedom is the total number of items in a data set that are free to vary independently of each other. In testing for Hardy Weinberg proportions, this is the number of possible genotypes minus the number of alleles.

Dehalogenation is the removal of halogen atoms (chlorine, iodine, bromine, fluorine) from molecules, usually during biodegradation.

Dehiscence is the spontaneous and often violent opening of a fruit, seed pod or anther.

Dehydration is a loss of body water component of body fluid.

Dehydrogenase is an enzyme that catalyzes the removal of hydrogen atoms in the biological reaction.

Dehydrogenation is an chemical reaction in which hydrogen is removed from the compound.

De-ionized water is a water, which is free of most inorganic and most organic compounds.

Deletion is a mutation involving the removal of one or more base pairs in a DNA sequence.

Deletion mapping is the use of overlapping deletions to localize the position of an unknown gene on a chromosome or linkage map.

Deliberate release refers to putting a genetically modified organism (GMO) into field trials.

Delta endotoxins (δ-endotoxins, also called Cry and Cyt toxins) are pore-forming toxins produced by bacillus thuringiensis species of bacteria.

Deme is the local conspecific group of individuals that mate at random.

Demineralize is the removal of the mineral contents (salts, ions) from a substance, especially water. Removal methods include distillation, electrodialysis and ion exchange.

Demographic are topics relating to the structure and dynamics of populations.

Demographic stochasticity refers to differences in the dynamics of a population that are the effects of random events on individuals in the population.

Denaturation refers to loss of native configuration of a macro-molecule (protein or nucleic acid) by physical or chemical means, usually accompanied by loss of biological activity. Denatured proteins often unfold their polypeptide chains and express changed properties of solubility.

Denature refers to the induction of structural alterations that disrupt the biological activity of a molecule.

Denatured DNA is double-stranded DNA that has been converted to single strands by breaking the hydrogen bonds linking complementary nucleotide pairs.

Denaturing gradient gel electrophoresis (DGGE) is an electrophoresis method for separating similar sized DNA fragments on the basis of their sequence, by applying across the gel a gradient of increasingly denaturing conditions.

Dendrimer is a polymer having a regular branched structure. Synonymous terms for dendrimer include arborols and cascade molecules.

Dendrite is a branch of a neuron that receives and conducts nerve impulses toward the cell body.

Dendrogram is a tree diagram that serves as a visual representation of the relationships between populations within a species.

Denitrification is a chemical process in which nitrates in the soil are reduced to molecular nitrogen.

Density gradient centrifugation is a high-speed centrifugation in which molecules are separated on the basis of their different densities using a concentration gradient of caesium chloride or sucrose.

Density is the mass of fluid per unit volume. For a compressible fluid flow, the density can vary.

Deoxyadenosine is a deoxyribonucleoside. It is a derivative of the nucleoside adenosine, differing from the latter by the replacement of a hydroxyl group (-OH) by hydrogen (-H) at the 2' position of its ribose sugar moiety. Deoxyadenosine is the DNA nucleoside A, which pairs with deoxythymidine (T) in double-stranded DNA.

Deoxycytidine is a deoxyribonucleoside, a component of deoxyribonucleic acid. It is similar to the ribonucleoside cytidine, but with one hydroxyl group removed from the 2' position.

Deoxyguanosine is composed of the purine nucleoside guanine linked by its N9 nitrogen to the

C1 carbon of deoxyribose. It is similar to guanosine, but with one hydroxyl group removed from the 2' position of the ribose sugar (making it deoxyribose). If a phosphate group is attached at the 5' position, it becomes deoxyguanosine monophosphate.

Deoxyribonuclease (DNase) is any enzyme that catalyzes the hydrolytic cleavage of phosphodiester linkages in the DNA backbone, thus degrading DNA.

Deoxyribonucleic acid (DNA) is the molecule that carries the genetic information in most living organisms. It is a double-stranded helix held together by hydrogen bonds between pairs of nucleotides.

Deoxyribonucleic acid is abbreviated as DNA.

Deoxyribonucleoside is a type of nucleoside including deoxyribose as a component.

Deoxyribonucleotide is the monomer, or single unit, of DNA, or deoxyribonucleic acid. Each deoxyribonucleotide comprises three parts: a nitrogenous base, a deoxyribose sugar, and one phosphate group.

Deoxyribose (2-deoxyribose) is a monosaccharide with idealized formula H-(C=O)-(CH2)-(CHOH)3-H. Its name indicates that it is a deoxy sugar, meaning that it is derived from the sugar ribose by loss of an oxygen atom.

Deoxythymidine (Thymidine can also be labeled deoxyribosylthymine, and thymine deoxyriboside) is a chemical compound, more precisely a pyrimidine deoxynucleoside. Deoxythymidine is the DNA nucleoside T, which pairs with deoxyadenosine (A) in double-stranded DNA.

Depolarization is a process of neutralizing the cell membrane's resting potential.

Depurination refers to loss of a purine base (A or G) to form an apurinic site (AP site).

Derepression is the process of "turning on" the expression of a gene or set of genes whose expression has been repressed (turned off), usually by the displacement of a repressor from a promoter, since, when attached to the DNA, the repressor prevents transcription.

Derivative is used to identify a variant during meristematic cell division.

Derived character is found only in a particular lineage within a larger group.

Dermis refers to the layer of dense connective tissue beneath the epidermis in the skin of vertebrates.

Descending colon is the portion of the large intestine located on the left side of the body.

Desiccant is any compound used to remove moisture or water.

Desiccate refers to dry, exhaust or deprive of water or moisture. The process is desiccation.

Design is a process to plan and delineate with an end in mind and subject to constraints.

Desoxyribonucleic acid is an obsolete spelling of deoxyribonucleic acid (DNA).

Dessicator is an apparatus for drying or depriving of moisture.

Desulphurization (In USA: desulfurization) is a technology for removing sulphur from oil and coal by use of bacteria.

Detergent is a substance which lowers the surface tension of a solution, improving its cleaning properties.

Determinate growth refers to a growth determined and limited in time, as in most floral meristems and leaves.

Determination is a process by which undifferentiated cells in an embryo become committed to develop into specific cell types, such as neurons, fibroblasts or muscle cells.

Determined describes embryonic tissue at a stage when it can develop only as a certain kind of tissue.

Deterministic refers to events that have no random or probabilistic aspects but rather occur in a completely predictable fashion.

Development is the sum total of events that contribute to the progressive elaboration of an organism.

Developmental biology is the study of the process by which organisms grow and develop, including the formation and specialization of cells and tissues, from embryo to adulthood.

Deviation refers to: (1) An alteration from the typical form, function or behavior; (2) A statistical term describing the difference between an actual observation and the mean of all observations.

Dextrin is an intermediate polysaccharide compound resulting from the hydrolysis of starch to maltose by amylase enzymes.

Dextran is a complex, branched glucan (polysaccharide made of many glucose molecules) composed of chains of varying lengths (from 3 to 2000 kilodaltons).

Dextrose is related to glucose.

dG – dC tailing (complementary homopolymeric tailing) is the process of adding complementary nucleotide extensions to DNA molecules, to facilitate the ligation of the two DNA molecules.

DGGE is an abbreviation for denaturing gradient gel electrophoresis.

dGTP is an abbreviation for deoxy-guanosine 5'-triphosphate, which is required for DNA synthesis.

Diabetes is a disease associated with the absence or reduced levels of insulin, a hormone essential for the transport of glucose to cells.

Diagnostic procedure is a test or assay used to determine the presence of a specific sub-stance, organism or nucleic acid sequence alteration, etc.

Diagnostic locus is a locus that is fixed, or nearly fixed for different alleles allowing dif-ferentiation between parental species, populations, or their hybrids.

Diagnostic procedure is a test or assay used to determine the presence of an organism, sub-stance or nucleic acid sequence alteration.

Diakinesis is a stage of meiosis at the end of prophase I, in which the contraction of the chromo-somes is almost at a maximum, pairing configurations are well defined, the nucleolus normally disappears and the nuclear envelope is disrupted.

Dialysis (Greek διάλυσις, dialu-sis meaning *dissolution*: dia, meaning *through*; and lysis, meaning *loosening or splitting*) is a biochemical technique by which large molecules such as proteins in solution are sepa-rated from smaller species such as salts.

Diaphragm is the primary muscle of inspiration; and is a thin, dome-shaped sheet of muscle that inserts into the lower ribs.

Diarrhea refers to an increase in frequency of stools compared to normal, or looser bowel movements than usual.

Diastole refers to the phase of the cardiac cycle in which the heart is relaxed.

Diastolic pressure is the lowest pressure to which blood pres-sure falls between contractions of the ventricles.

Diathermy is the application of high-frequency electrical energy that is used to generate heat in body tissues as a result of the resistance of the tissue to the passage of energy.

Diazotroph is an organism that can fix atmospheric nitrogen.

Dicentric chromosome is a chro-mosome having two active centromeres.

Dichogamy refers to the condi-tion in which the male and the female reproductive organs of a flower mature at different times.

Dicotyledon (dicot) refers to a plant with two cotyledons.

Di-deoxynucleotide (ddNTP, didN) is a synthetic deoxynucleotide that lacks a 3'-hydroxyl group, and is thus unable to form the 3' to 5' phosphodiester bond necessary for chain elongation.

didN is an abbreviation for di-deoxynucleotide.

Dielectric constant (ε) is a ratio of the permittivity of a medium to that of a vacuum. It is also called the relative dielectric constant or relative permittivity.

Dielectric displacement is the magnitude of charge per unit area of the capacitor plate.

Dielectric is any material that is electrically insulating.

Dielectric strength is the magnitude of an electric field necessary to cause significant current passage through a dielectric material.

Differential centrifugation is a method for separating sub-cellular particles according to their sedimentation coefficients, which are roughly proportional to their size.

Differential display is a method to identify mRNAs which are present at different levels in different tissues, or in response to specific treatments.

Differentially permeable refers to a membrane, through which different substances diffuse at different rates.

Differentiation (*L. differre*, to carry different ways) is a process in which unspecialized cells develop structures and functions characteristic of a particular type of cell.

Diffuse axonal injury is an injury, when after a closed brain injury, the shifting and rotation of the brain inside the skull will result in shearing injury to the brain's long connecting nerve fibers or axons.

Diffusion (*L. diffusus*, spread out) is the movement of molecules from a region of higher concentration to a region of lower concentration.

Diffusion coefficient is a constant of proportionality between the diffusion flux and the concentration gradient in **Fick's first law**. Its magnitude is indicative of the rate of atomic diffusion.

Diffusion system is a system, when the drug is either encapsulated in a polymer membrane or suspended within a polymer matrix; water diffuses into the membrane or matrix, the drug dissolves, and finally the dissolved drug diffuses out of the polymer.

Digest refers to treat DNA molecules with one or more restriction endonucleases in order to cleave them into smaller fragments.

Digestion refers to a process describing how the body breaks down food and uses it for energy, cell repair, and growth.

Digestive tract consists of the organs that are involved in digestion; including the mouth, salivary glands, esophagus, stomach, pancreas, liver, gallbladder, small intestine, and large intestine.

Digital rectal exam (DRE) is a procedure in which the physician inserts a gloved finger into the rectum to examine the rectum and the prostate gland for signs of cancer.

Digital refers to systems employing only quantized (discrete) states to convey information.

Dihaploid is an individual, which arises from a doubled haploid.

Dihybrid is an individual that is heterozygous for two pairs of alleles; the progeny of a cross between homozygous parents differing at two loci.

Dilate refers to relax or expand.

Dimensionless number is a quantity which describes a certain physical system and which is a pure number without any physical units. Such a number is typically defined as a product or ratio of quantities which do have units, in such a way that all units cancel. Dimensionless numbers are widely applied in the field of bioengineering, mechanical and chemical engineering. According to the Buckingham π-theorem of dimensional analysis, the functional dependence between a certain number (e.g., n) of variables can be reduced by the number (e.g., k) of independent dimensions occurring in those variables to give a set of p = n − k independent, dimensionless numbers. For the purposes of the experiment, different systems which share the same description by dimensionless numbers are equivalent.

Dimer refers to: (1) A molecule formed by the covalent combination of two monomers; (2) The reversible association of two similar (or nearly similar) molecules.

Dimethyl sulphoxide (dimethyl sulfoxide, DMSO, C_2H_6OS) is a highly hygroscopic liquid and powerful solvent with little odor or color.

Dimorphism is the existence of two distinctly different types of individuals within a species.

Dinucleotide refers to a nucleotide dimer.

Diode is a two-terminal device that conducts current well in one direction and poorly in the other.

Dioecious is a plant species in which male and female flowers form on different plants.

Dip (Dual In-line Package) is a common ceramic or plastic enclosure for an integrated circuit.

Dip pen nanolithography (DPN) is a new AFM- based soft-lithography technique. DPN is a scanning probe nanopatterning technique in which an AFM tip is used to deliver molecules to a surface via a solvent meniscus, which naturally forms in the ambient atmosphere.

Diplochromosome refers to a chromosome arising from an abnormal duplication in which the centromere fails to divide and the daughter chromosomes fail to move apart.

Diploid (Gr. *diploos*, double + *oides*, like) refers to: (1) The status of having two complete sets of chromosomes; (2) An organism or cell with a double set $(2n)$ of chromosomes (most commonly one of paternal origin, and the other of maternal origin), or referring to an individual containing a double set of chromosomes per cell.

Diploid cell is a cell, which contains two sets of chromosomes.

Diplonema (adj.: diplotene) refers to stage in prophase of meiosis I following the pachytene stage, but preceding diakinesis, in which one pair of sister chromatids begin to separate from the other pair.

Diplophase refers to a phase with 2n chromosomes.

Direct current refers to a galvanic current that always flows in the same direction and may flow in either a positive or a negative direction.

Direct embryogenesis is the formation in culture, on the surface of zygotic or somatic embryos or on explant tissues (leaf section, root tip, etc.), of embryoids without an intervening callus phase. Opposite is indirect embryogenesis.

Direct organogenesis is the formation of organs directly on the surface of cultured intact explants. Opposite is indirect organogenesis.

Direct repeat refers to two or more stretches of DNA within a single molecule which have the same nucleotide sequence in the same orientation. Direct repeats may be either adjacent to one another or far apart on the same molecule. For example TATTA...TATTA; ATAAT...ATAAT

Direct reprogramming is the process of taking fully mature, differentiated stem cells and inducing them to become another cell type without going through a stem cell (iPS) state. The transition from one cell type directly to another is referred to as "transdifferentiation." In the case of pancreatic cells, the method employs three developmental factors (different from the ones used to create iPS cells) causing acinar cells to switch directly to a insulin producing beta cells.

Directed amplification of minisatellite DNA (DAMD) refers to a polymerase chain reaction technique used for obtaining molecular markers in the region of minisatellites.

Directed differentiation is the manipulation of stem cell culture conditions to induce differentiation into a particular cell type.

Directed mutagenesis refers to the generation of changes in the nucleotide sequence of a cloned gene by one of several procedures.

Directional cloning is the technique by which a vector and a DNA insert are both digested with two different restriction endonucleases to create non-complementary sticky ends at either end of both molecules.

Directional selection refers to the selective increase in the frequency of an advantageous allele, gene, or phenotypic trait in a population.

Disaccharide refers to a dimer consisting of two covalently linked monosaccharides.

Disarm is the deletion from a plasmid or virus of genes that are pathogenic.

Disassortative mating is a preferential mating of individuals with different phenotypes.

Discontinuous variation refers to the variation where individuals can be classified as belonging to one of a set of discrete, non-overlapping classes. Opposite is continuous variation.

Discordant refers to members of a pair showing different, rather than similar, characteristics.

Discrete generations can be defined by whole integers and in which all individuals will breed only with individuals in their generation.

Disease (*L. dis*, a prefix signifying the opposite + M.E. *aise*, comfort) is the opposite of ease.

Disease free is a plant or animal certified through specific tests as being free of specified pathogens. Disease-free should be interpreted to mean "free from any known diseases" as "new" diseases may yet be discovered to be present.

Disease indexing refers to when disease-indexed organisms have been assayed for the presence of known diseases according to standard testing procedures.

Disease resistance is the genetically determined ability to prevent the reproduction of a pathogen, thereby remaining healthy.

Disinfestation is the elimination or inhibition of the activity of surface-adhering micro-organisms and removal of insects.

Disjunction is a separation of homologous chromosomes during anaphase I of meiosis; separation of sister chromatids during anaphase of mitosis and anaphase II of meiosis.

Disomy (adj.: disomic) refers to the presence of a pair of specific chromosome.

Dispense is a portion out a nutrient medium into containers, such as test tubes, jars, Erlenmeyer flasks, Petri dishes, etc.

Dispersal is the movement of individuals from one genetic population (or birth place) into another.

Disrupter gene is used to enforce the sterility of seed saved from a genetically engineered crop.

Dissecting microscope is a microscope with a magnifying power of about 50x, used as an aid in the manipulation of small objects.

Dissection (*L. dissectio*, a dissecting or being dissected) refers to separation of a tissue by cutting for analysis or observation.

Dissolve refers to passing chemicals into solution.

Distention is a swelling or bloating, usually referring to the abdomen.

Distillation (*L. distillatio*, a distilling process) refers to the process of heating a mixture to separate the more volatile from the less volatile parts, and then cooling and condensing the resulting vapor so as to produce a more nearly pure or refined substance.

Distinct population segment (DPS) is a level of classification under the ESA that allows for legal protection of populations that are distinct, relatively reproductively isolated, and represent a significant evolutionary lineage to the species.

Di-sulphide bond is a chemical bond that stabilizes the three-dimensional structure of proteins, and hence the protein's normal function.

Disulphide bridge is a chemical bond between pairs of sulphur atoms that stabilizes the three-dimensional structure of proteins, and hence the protein's normal function. Synonym is disulphide bond.

Ditype is a tetrad in a fungi that contains two kinds of meiotic products (spores), e.g., 2AB and 2ab.

Diurnal refers to an event that occurs repetitively on a daily basis, generally during daylight hours.

Divergent evolution is the accumulation of differences between groups, which can lead to the formation of new species, usually a result of diffusion of the same species to different and isolated environments which blocks the gene flow among the distinct populations allowing differentiated fixation of characteristics through genetic drift and natural selection.

Diverticulitis occurs when one or more small pouches in the large intestine (called a diverticulum) become irritated or infected. **Diverticulum** is a small pouch in the wall of the large intestine, which usually do not cause a problem unless it becomes irritated.

Dizygotic twins refers to two-egg twins, i.e., a pair of individuals that shared the same uterus at the same time, but which arose from separate and independent fertilization of two ova.

DMD is a Duchenne muscular dystrophy.

DMSO is an abbreviation for dimethyl sulphoxide.

DNA (deoxyribonucleic acid; formerly spelt desoxyribonucleic acid) is a nucleic acid. It consists of a backbone comprised of alternating sugar and phosphate molecules, and each of the sugar molecules is bound to a nucleotide base. Related terms are: b-DNA, cDNA, Complementary DNA or RNA, DNA polymorphism, DNA sequencing, Double-stranded complementary DNA, Duplex DNA, and Z-DNA.

DNA amplification refers to the multiplication of a piece of DNA in a test-tube into many thousands of millions of copies.

DNA amplification fingerprinting (DAF) is an arbitrarily primed polymerase chain reaction technique for obtaining molecular markers using very short (5-8 bp) primers.

DNA bank refers to storage of DNA, which may or may not be the complete genome, but should always be accompanied by inventory information.

DNA barcoding refers to the use of a short gene sequence from a standardized region of the genome that can be used to help discover, characterize, and distinguish species, and to assign unidentified individuals to species.

DNA carrier is a substance or particle that can transfer genes into a cell.

DNA construct refers to a chimeric DNA molecule, carrying all the genetic information necessary for its transgenic expression in a host cell.

DNA delivery system is a generic term for any procedure that transports DNA into a recipient cell.

DNA diagnostics refers to the use of DNA polymorphisms to detect the presence of a specific sequence, which could indicate the presence of a contaminant, of a pathogen, or of a particular allele at a target gene. Most commonly utilizes the polymerase chain reaction.

DNA fingerprint is a description of the genotype of an individual from the pattern of DNA fragments obtained from DNA fingerprinting. Synonym is DNA profile.

DNA fingerprint technique is a method employed to determine differences in amino acid sequences between related proteins.

DNA fingerprinting or DNA profiling is a genetic tool used to compare and contrast DNA sequences using electrophoresis.

DNA helicase (gyrase) is an enzyme that catalyzes the unwinding of the complementary strands of a DNA double helix.

DNA hybridization is the annealing of two single-stranded DNA molecules, possibly of different origin, to form a partial or complete double helix.

DNA ligase is an enzyme that catalyzes a reaction that links two DNA molecules via the formation of a phospho-diester bond between the 3′ hydroxyl and 5′ phosphate of adjacent nucleotides.

DNA micro-array refers to a small glass surface to which has been fixed an array of DNA fragments, each with a defined location.

DNA polymerase is an enzyme that catalyzes the synthesis of double-stranded DNA, using single-stranded DNA as a template.

DNA polymorphism refers to the existence of two or more alternative forms (alleles) of a chromosomal locus that differ in nucleotide sequence or have variable numbers of repeated nucleotide units.

DNA primase is an enzyme that catalyzes the synthesis of short strands of RNA that initiate the synthesis of DNA strands.

DNA probe is a short strand of DNA that locates and binds to its complementary sequence in samples containing single strands of DNA or RNA enabling identification of specific sequences.

DNA repair enzymes are enzymes that catalyze the repair of DNA.

DNA repair is a variety of mechanisms that repair errors (e.g., the incorporation of a non-complementary nucleotide) that occur naturally during DNA replication.

DNA replication is the process whereby DNA makes exact copies itself, under the action of and control of DNA polymerase.

DNA sequencing refers to the procedure for determining the nucleotide sequence of a DNA fragment. Two common methods available: (1) The Maxam Gilbert technique; (2) the Sanger technique (also called the di-deoxy or chain-terminating method).

DNA topo-isomerase is an enzyme that catalyzes the introduction or removal of super-coils in DNA. Synonym is topo-isomerase.

DNA vaccine is a vaccine generated by the injection of specific **DNA** fragments to stimulate an immune response.

DNase (deoxyribonuclease) is an enzyme that catalyzes the cleavage of DNA. DNase I is a digestive enzyme secreted by the pancreas, that degrades DNA into shorter nucleotide fragments.

Dobzhansky Muller incompatibilities are genic interactions between alleles at multiple loc in which alleles that enhance fitness within their parental genetic backgrounds that may reduce fitness in the novel genetic background produced by hybridization.

Docetaxel (generic or under the trade name Taxotere or Docefrez) is a clinically well-established anti-mitotic chemotherapy medication that works by interfering with cell division.

Docking is a three-dimensional molecular structure that is one of the foundations of structure-based drug design. The program AutoDock was originally written in FORTRAN-77 in 1990 by David S. Goodsell here in Arthur J. Olson's laboratory. It performs automated docking of ligands (small molecules like a candidate drug) to their macromolecular targets (usually proteins, sometimes DNA).

Docking programs evaluate lead compounds against target proteins; and these programs are "informed" by structure data. Traditional ligand- docking programs – such as DOCK, MacroModel, GOLD give information about potential ligands for a known protein structure.

Docking studies are computational techniques for the exploration of the possible binding modes of a substrate to a given receptor, enzyme or other binding site.

Dolly refers to the first mammal to be created by cloning a cell from an adult animal, ewe.

Domain is a portion of a protein or DNA molecule that has a discrete function or conformation.

Domestic animal diversity (DAD) refers to the spectrum of genetic differences within each breed, and across all breeds within each domestic animal species, together with the species differences.

Dominance genetic variation refers to the proportion of total genetic variation that can be attributed to the interactions of alleles at a locus in heterozygotes.

Dominance refers to the type of gene action exhibited by a dominant allele.

Dominant (-acting) oncogene is a gene that stimulates cell proliferation and contributes to oncogenesis when present in a single copy.

Dominant is a dominant gene that will almost always be expressed and lead to a specific physical characteristic.

Dominant alleles determine the phenotype displayed in a heterozygote with another (recessive) allele.

Dominant gene is a gene whose phenotype is when it is present in a single copy.

Dominant marker selection refers to selection of cells via a gene encoding a product that enables only the cells that carry the gene to grow under particular conditions.

Dominant oncogene is a gene that stimulates cell proliferation.

Dominant selectable marker gene is a gene that allows the host cell to survive under conditions where it would otherwise die.

Donor is a person who donates living tissue to be used in another body.

Donor junction site is the junction between the 5' end of an exon and the 3' end of an intron.

Donor plant (mother plant) refers to an explant, graft or cutting used as a source of plant material for micro-propagation purposes.

Doping is a process of introducing impurity atoms into a semiconductor to affect its conductivity.

Doppler spectroscopy (the radial-velocity method, or colloquially, the wobble method) is an indirect method for finding extrasolar planets and brown dwarfs from radial-velocity measurements via observation of Doppler shifts in the spectrum of the planet's parent star.

Dormancy (F. *dormir*, from L. *dormire*, to sleep) is an inactive period in the life of an animal or plant during which growth slows or completely ceases.

Dosage compensation is a regulatory mechanism for sex-linked genes, to allow equivalent levels of gene expression from (in mammals) XY or XX genotypes, even though the gene copy number in XX is double that in XY.

Double crossing over refers to two simultaneous reciprocal breakage and reunion events between the same two chromatids.

Double fertilization is a process, unique to flowering plants, in which two male nuclei, which have travelled down the pollen tube, separately fuse with different female nuclei in the embryo sac.

Double helix describes the coiling of the two strands of the double-stranded DNA molecule, resembling a spiral staircase in which the base pairs form the steps and the sugar-phosphate backbones form the rails on each side. One strand runs 3' to 5,' while the complementary one runs 5' to 3.'

Double recessive is an organism homozygous for a recessive allele at each of two loci.

Double stranded complementary DNA (dscDNA) is a duplex DNA molecule copied from a cDNA template.

Double-stranded DNA (dsDNA) refers to two complementary strands of DNA annealed in the form of a double helix. Synonym is duplex DNA.

Doubling time is the period of time required for a quantity to double in size or value.

Douche refers to a current of water directed against the skin surface.

Down promoter mutation is a mutation that decreases the frequency of initiation of transcription.

Down syndrome is an inherited condition due to an extra chromosome 21, either as a third chromosome 21 or attached to chromosome 13, 14 or 15. Also called trisomy 21.

Down-regulate refers to induce genetically a reduction in the level of a gene's expression.

Downstream refers to: (1) In **molecular biology**, the stretch of nucleotides of DNA that lie in the 3′ direction from the site of initiation of transcription; (2) In **chemical engineering**, those phases of a manufacturing process that follow the biotransformation stage.

Downstream processing is a general term for all the things which happen in a biotechnological process after the biology, be it fermentation of a micro-organism or growth of a plant.

Doxil is a pegylated (polyethylene glycol coated) liposome-encapsulated form of doxorubicin formerly made by Ben Venue Laboratories in the United States. It was developed to treat Kaposi's sarcoma, an AIDS-related cancer that causes lesions to grow under the skin, in the lining of the mouth, nose and throat, or in other organs.

Doxorubicin (trade name Adriamycin; liposome-encapsulated trade name Doxil; also known as hydroxydaunorubicin) is a drug used in cancer chemotherapy and derived by chemical semisynthesis from a bacterial species.

DPS is a distinct population segment.

Draft sequence generally refers to sequence that is not yet finished but is of generally high quality. In terms of clone based project, Draft sequence refers to a project in which greater than 90% of the bases are of high quality.

DRAM (Dynamic Random Access Memory) is a memory in which each stored bit must be refreshed periodically.

Drift is a gradual departure of the instrument output from the calibrated value.

Drosophila melanogaster refers to the fruit fly, used for many years as a model for eukaryotic genetics.

Drug delivery refers to a method by which a therapeutic agent is delivered to its site of action. For traditional therapeutic agents this is another name for *formulation*. However, biotechnology has allowed the development of a range of new therapeutic-agent delivery systems, such as liposomes and other encapsulation techniques, and a range of mechanisms that target a therapeutic agent to a particular cell or tissue.

Drug Identification Number (DIN) is a number issued to a drug indicating that it is authorized for sale in Canada.

Dry weight is the weight of tissue obtained following sufficiently prolonged oven-drying at high temperature to remove all water.

DSC is an abbreviation for differential scanning calorimetry, or the differential scanning calorimeter.

dscDNA is an abbreviation for double-stranded complementary DNA.

dsDNA is an abbreviation for double-stranded DNA.

DSP (Digital Signal Processing) is a process by which a sampled and digitized data stream are modified in order to extract relevant information.

dTTP is an abbreviation for deoxythymidine 5'-triphosphate.

Dual culture is a culture made of a plant tissue and one organism.

Duchenne/ Becker muscular dystrophy is the most common and severe form of muscular dystrophy; transmitted as an X-linked trait. X-linked recessive.

Ductility is a measure of a material's ability to undergo plastic deformation before fracture. It is expressed as percent elongation or percent area reduction in a tensile test.

Ductus arteriosus is a vascular connection between the pulomonary trunk and aorta that functions throughout the fetus.

Ductus venosus is a vessel that allows the blood to bypass the liver.

Duodenal ulcer is an open sore in the duodenum (first part of the small intestine).

Duodenum is the portion of the small intestine into which the contents of the stomach first enter.

Duplex DNA is related to double-stranded DNA.

Duplication refers to multiple occurrence of: (1) A DNA sequence within a defined length of DNA; or (2) A specific segment in the same chromosome or genome.

Dwarfism refers to conditions of short stature with adult height under 4'10" as adult, usually with normal intelligence and lifespan. Ehlers Danlos Syndrome connective tissue condition includes problems with tendons, ligaments, skin, bones, cartilage, and membranes surrounding blood vessels and nerves.

Dynamic characteristics is a description of an instrument's behavior between the time a measured quantity changes value and the time the instrument obtains a steady response.

Dynamic error is an error when the output does not precisely follow the transient response of the measured quantity.

Dynamic light scattering (DLS, photon correlation spectroscopy or quasi-elastic light scattering) is a technique that can be used to determine the size distribution profile of small particles in suspension or polymers in solution. It can also be used to probe the behavior of complex fluids such as concentrated polymer solutions.

Dynamic range is a ratio of the largest to the smallest values of a range. It is often expressed in decibels.

Dysphagia is a condition that indicates the difficulty swallowing food or liquid.

Dyspnea is a sensation of difficulty in breathing.

Dystonia is a neurologic condition involving repeated twisting and movement. Three forms are: childhood-autosomal dominant, autosomal recessive, adult-acquired.

E

. . . .

E. coli is an abbreviation for *Escherichia coli*.

EBV is an abbreviation for estimated breeding value.

EC is an abbreviation for enzyme commission number.

Eccrine sweat glands are distributed over the entire surface of the body except at a relatively few locations.

Ecdysone is a steroid hormone in insects that stimulates moulting and metamorphosis.

ECG is an abbreviation for electrocardiogram.

Echocardiogram is an ultrasonic record of the dimension and movement of the hearts and his valves.

Eclosion is an: (1) Emergence of an adult insect from the pupal stage; (2) Initial phase of germination of fungal spores.

Ecocentrism states that all elements of the environment have worth and should be valued and cared for.

Ecological diversity refers to the variety of ecosystems present in a biosphere, the variety of species and ecological processes that occur in different physical settings.

Ecology is the study of the interactions between organisms and their natural environment, both living and non-living.

Economic trait locus (ETL) is a locus influencing a trait that contributes to income.

Ecosystem is the complex of a living community and its environment, functioning as an ecological unit in nature.

Ecosystem services are the products and services humans receive from functioning ecosystems.

Ecotone is the region that encompasses the shift between two biological communities.

Ecotype is a population or a strain of an organism that is adapted to a particular habitat.

Ectoderm is the outermost germ layer of cells derived from the inner cell mass of the blastocyst.

Ectopic refers to anomalous situation or relation, particularly with respect to pregnancy, where the foetus is implanted outside the uterus.

Edema is a swelling due to the buildup of fluid.

Edible vaccine is an edible antigen-containing material, that activates the immune system via gut-associated lymphoid tissues.

EDP refers to Ethylene diamine pyrocatechol.

EDTA is an abbreviation for ethylene-diamine tetra-acetic acid. It is an aminopolycarboxylic acid and a colorless, water-soluble solid. Its conjugate base is named ethylenediaminetetraacetate.

EDV is an abbreviation for essential derivation of varieties.

EEG is an abbreviation for electroencephalogram.

Effective number of alleles is the number of equally frequent alleles that would create the same heterozygosity as observed in the population.

Effective population size is the size of the ideal, panmictic population that would experience the same loss of genetic variation, through genetic drift, as the observed population.

Effector cells refer to cells of the immune system that are responsible for the production of cell-mediated cytotoxicity.

Effector molecule is a molecule that influences the behavior of a regulatory molecule, such as a repressor protein, thereby influencing gene expression.

EGF is an abbreviation for epidermal growth factor (a growth factor in biology); Exponential generating function (a function in mathematics).

Egg refers to: (1) The fertilized ovum (zygote) in egg-laying animals after it emerges from the body; (2) The mature female reproductive cell in animals and plants.

EGS is an abbreviation for external guide sequence. EGS represent a new class of RNA-based gene-targeting agents, consist of a sequence complementary to a target mRNA, and render the target RNA susceptible to degradation by ribonuclease P.

EIA is an abbreviation for enzyme immunoassay. EIA is an assay that uses an enzyme-bound antibody to detect antigen.

Elastic deformation is a nonpermanent deformation that totally recovers upon release of an applied stress.

Elastin is a fibrous protein that is the major constituent of the yellow elastic fibers of animal connective tissue.

Elastomer is a polymeric material that may experience large and reversible elastic deformations.

Electrical breakdown is a condition in which a nominal insulator becomes electrically conducting.

Electrical field is a technique of heating the tissues in shortwave diathermy in which the patient is part of the electrical circuit.

Electrical potential is the difference between charged particles at a higher and lower potential.

Electroblotting refers to the electrophoretic transfer of macromolecules (DNA, RNA or protein) from a gel, in which they have been separated, to a support matrix, such as a nitrocellulose sheet.

Electrocardiogram (EKG) is a record of the electric activity of the heart. The electrocardiogram is a graph that indicates electrical impulses flow through the heart.

Electrocardiograph is an instrument used for the measurement of the electrical activity of the heart.

Electrochemical sensor are biosensors, such as an enzyme electrode, in which a biological process is harnessed to an electrical sensor system.

Electroencephalogram (EEG) is a procedure that records the brain's continuous electrical activity by means of electrodes attached to the scalp.

Electroencephalograph is an instrument for measuring and recording electrical activities from the brain (brain waves).

Electroluminescence describes the emission of visible light by a p-n junction across which a forward-biased voltage is applied.

Electrolyte is a solution through which an electric current can be carried by the motion of ions.

Electrolyte/Insulator/Silicon (EIS) are structures of potentiometric silicon sensors. The best-known EIS is the ion-sensitive field effect transistor (ISFET or CHEMFET) and the light-addressable potentiometric sensor (LAPS).

Electromagnetic radiation refers to the electromagnetic waves, including ultraviolet (UV), X-rays, and gamma radiation.

Electromagnetic spectrum is the range of frequencies and wavelengths associated with radiant energy.

Electromotive force (emf) is the greatest potential difference that can be generated by a particular source of electric current.

Electromyography is a pick-up and amplification of electrical signals generated by the muscle as it contracts.

Electron is an elementary negative particle whose charge is 1.602×10^{-19} coulombs.

Electron level is a set of discrete, quantized energies that are allowed for electrons. In the atomic case, each state is specified by 4 quantum numbers.

Electron microscope is a microscope that uses an electron beam focused by magnetic lenses.

Electrophoresis is a molecular biology technique that is used to resolve complex mixtures of macromolecules into their components. Related terms are: agarose gel electrophoresis, polyacrylamide gel electrophoresis, denaturing gradient gel electrophoresis, capillary electrophoresis, sodium dodecyl sulphate polyacrylamide gel electrophoresis, thermal gel gradient electrophoresis pulsed-field gel electrophoresis, and iso-electric focusing gel.

Electronegative describes elements that tend to gain electrons and form negative ions.

Electroporation refers to: (1) An electrical treatment of cells that induces transient pores, through which DNA can enter the cell; (2) The introduction of DNA or RNA into protoplasts or other cells by the momentary disruption of the cell membrane through exposure to an intense electric field.

Electrostatic or condenser field refers to when the patient is placed between electrodes and becomes a part of a series circuit.

ELISA (enzyme-linked immunosorbent assay) is a sensitive technique for accurately determining specific molecules in a mixed sample. The amount of protein or other antigen in a given sample is determined by means of an enzyme-catalyzed color change, avoiding both the hazards and expense of radioactive techniques.

Elite tree is a phenotypically superior tree in a tree breeding program.

Elongation factors are soluble proteins required for the elongation of polypeptide chains on ribosomes.

ELSI is ethical, legal and social implications of HGP.

EM *is an expectation maximization* algorithm.

Embryo refers to an immature organism in the early stages of development. The word *embryon* derives from Medieval Latin *embryo*, itself from Greek ἔμβρυον (*embruon*), plural ἔμβρυα (*embrua*), lit. "young one," which is the neuter of ἔμβρυος (*embruos*), lit. "growing in," from ἐν (*en*), "in"+ βρύω (*bruō*), "swell, be full"; the proper Latinized form of the Greek term would be *embryum*.

Embryo cloning refers to the creation of identical copies of an embryo by embryo splitting or by nuclear transfer from undifferentiated embryonic cells.

Embryo culture refers to the culture of embryos on nutrient media.

Embryo multiplication and transfer (EMT) refers to the cloning of animal embryos and their subsequent transfer to recipients via artificial inembryonation. Artificial inembryonation is a non-surgical transfer of embryo(s) to a recipient female.

Embryo rescue is a sequence of tissue culture techniques utilized to enable a fertilized immature embryo resulting from an interspecific cross to continue growth and development, until it can be regenerated into an adult plant.

Embryo sac is a large thin-walled space within the ovule of the seed plant in which the egg and, after fertilization, the embryo develop; the mature female gametophyte in angiosperms.

Embryo sexing is the determination of the sex of an embryo prior to birth.

Embryo splitting refers to the splitting of young embryos into several sections, each of which develops into an animal.

Embryo storage refers to the cryogenic preservation of animal embryos, allowing inembryonation or other manipulations long after embryo formation.

Embryo technology is a generic name for any modification of mammalian embryos.

Embryo transfer (ET) refers to a step in the process of assisted reproduction in which embryos are placed into the uterus of a female with the intent to establish a pregnancy. Related terms are: embryo multiplication and transfer, multiple ovulation.

Embryogenesis refers to: (1) (General) Development of an embryo; (2) (In plants) *In vitro* formation of plants from plant tissues, through a pathway closely resembling normal embryogeny from the zygote.

Embryoid bodies are rounded collections of cells that arise when embryonic stem cells are cultured in suspension.

Embryoid refers to an embryo-like body developing *in vitro*, forming a complete, self-contained plantlet with no vascular connection with the callus. Embryoid bodies (EBs) are three-dimensional aggregates of pluripotent stem cells.

Embryonic germ cells are pluripotent stem cells that are derived from early germ cells.

Embryonic stem cell line is a process where embryonic stem cells, which have been cultured under *in vitro* conditions that allow proliferation without differentiation for months to years.

Embryonic stem cells are undifferentiated cells in an embryo that are able to multiply and become differentiated into any type of cell in the body.

EMI refers at electromagnetic interference.

Emission wavelength is the specific wavelength of light emitted by a fluorescent molecule, such as a labeled probe, upon absorption of light at the (higher) excitation wavelength.

Empirical is related to or based upon practical experience, trial and error, direct observation or observation alone, without benefit of scientific method, knowledge or theory.

EMT is an abbreviation for embryo multiplication and transfer.

Emulsifying agent refers to an emulsifier.

Emulsion is a mixture of two or more liquids that are normally immiscible (nonmixable or unblendable).

Encapsidation is the process by which the nucleic acid of a virus is enclosed in a capsid.

Encapsulating agents refers to anything which forms a shell around an enzyme or bacterium, although the agents used are usually polysaccharides such as alginate or agar.

Encapsulation is any method of getting something, usually an enzyme or bacterium, into a small package or capsule while it is still working or alive.

Encode refers to the gene product specified by a particular nucleic acid sequence. ENCODE is also an abbreviation for the Encyclopedia of DNA Elements.

Encopresis is a constipation and intestinal obstruction (blockage) that lead to an involuntary leakage of loose stool.

End Sequence Profiling (ESP) is a method for detecting genome-level variation.

Endangered breed is a breed where the total number of breeding females is between 100 and 1000 or the total number of breeding males is less than or equal to 20 and greater than five; or the overall population size is close to, but slightly above 100 and increasing, and the percentage of pure-bred females is above 80%; or the overall population size is close to, but slightly above, 1000 and decreasing, and the percentage of pure-bred females is below 80%.

Endangered species refers to plant or animal species in immediate danger of extinction because its population numbers have reached a critical level or its habitats have been drastically reduced.

Endemic refers to: (1) Describing a plant or animal species whose distribution is restricted to one or a few localities; (2) Describing a disease or a pest that is always present in an area.

End-labeling refers to the introduction of a radioactive atom at the end of a DNA or RNA molecule.

Endocrine gland is any gland in an animal that manufactures hormones and secretes them directly into the bloodstream to act at distant sites in the body, known as target organs or cells.

Endocrine interference refers to the interference with the normal balance hormones.

Endocrine system helps regulate metabolic activities; consists of ductless glands and tissues that secretes hormones.

Endocytosis is the process by which materials enter a cell without passing through the cell membrane. Related terms are phagocytosis, pinocytosis.

Endoder refers to the internal layer of cells of the gastrula, which will develop into the alimentary canal (gut) and digestive glands of the adult.

Endoderm is the innermost layer of the cells derived from the inner cell mass of the blastocyst; it gives rise to lungs, other respiratory structures, and digestive organs, or generally "the gut."

Endodermis (Gr. *endon*, within + *derma*, skin) refers to the layer of living cells, with various characteristically thickened walls and no intercellular spaces, which surrounds the vascular tissue of certain plants and occurs in nearly all roots and certain stems and leaves. The endodermis separates the cortical cells from cells of the pericycle.

Endogamy is common in many cultures and ethnic groups. Several ethnic religious groups are traditionally more endogamous, although sometimes with the added dimension of requiring marital religious conversion.

Endogenote refers to the part of the bacterial chromosome that is homologous to a genome fragment (exogenote) transferred from the donor to the recipient cell in the formation of a merozygote.

Endogenous (Gr. *endon*, within, + *genos*, race) refers to one that is developed or added from within the cell or organism.

Endomitosis refers to duplication of chromosomes without division of the nucleus, resulting in increased chromosome number within a cell.

Endonuclease is an enzyme that breaks strands of DNA at internal positions.

Endophyte is an organism that lives inside a plant.

Endoplasmic reticulum (Gr. *endon*, within + *plasma*, anything formed; + *L. reticulum*, a small net) refers to cytoplasmic net of membranes, adjacent to the nucleus, made visible by the electron microscope.

Endopolyploidy refers to the net result of endomitoses.

Endoprotease is an enzyme that cleaves the peptide bonds between amino acids within a protein.

Endoreduplication refers to chromosome reproduction during interphase.

Endorphins refer to endogenous opiates whose actions have analgesic properties.

Endoscope is a small, flexible tube with a light and a camera lens at the end, used to examine the inside of the digestive tract. It can also be used to take tissue samples for testing from inside the digestive tract.

Endoscopy means looking inside and typically refers to looking inside the body for medical reasons using an **endoscope**, an instrument used to examine the interior of a hollow organ or cavity of the body. Unlike most other medical imaging devices, endoscopes are inserted directly into the organ. Endoscope can also refer to using a borescope in technical situations where direct line of-sight observation is not feasible.

Endosome is a membrane-bounded compartment inside eukaryotic cells.

Endosperm (Gr. *endon*, within + *sperma*, seed) refers to nutritive tissue that develops in the embryo sac of most angiosperms.

Endosperm mother cell refers to one of the seven cells of the mature embryo sac, containing the two polar nuclei and, after reception of a sperm cell, gives rise to the primary endosperm cell from which the endosperm develops.

Endothelial cells are the delicate lining, only one cell thick, of the organs of circulation.

Endothelium is the thin layer of cells that lines the interior surface of blood vessels and lymphatic vessels, forming an interface between circulating blood or lymph in the lumen and the rest of the vessel wall.

Endovenous laser ablation is a minimally invasive procedure performed in a physician's office or clinic, for the treatment of varicose veins.

Endotoxin is a component of the cell wall of Gram-negative bacteria that elicits, in mammals, an inflammatory response and fever.

Endovascular trophoblast migrates into the maternal spiral arteries.

End-product inhibition refers to the inhibition of an enzyme by a metabolite.

Enema is a liquid placed into the rectum to either clear stool out of the large intestine, or to examine the large intestine with an x-ray (barium enema).

Energy [J] is a capacity for performing work or to cause heat flow. It is measured in Joules.

Energy function is an empirical energy function in which computationally a shape is assigned to a protein sequence based on an empirical energy function.

Enhanced permeability and retention (EPR) **effect** is the property by which certain sizes of molecules (typically liposomes, nanoparticles, and macromolecular drugs) tend to accumulate in tumor tissue much more than they do in normal tissues.

Enhancer element (enhancer sequence) refers to: (1) A sequence found in eukaryotes and certain eukaryotic viruses which can increase transcription of a gene when located (in either orientation) up to several kilobases from the gene concerned; (2) A substance or object that increases a chemical activity or a physiological process; (3) A major or modifier gene that increases a physiological process.

Enkephalin are neurotransmitter proteins that are pain-relieving molecules.

Enkephalinergic neurons are neurons with short axons that release enkephalin.

Enolpyruvyl – shikimate-3-phosphate synthase (EPSP synthase or EPSPS) refers to an enzyme produced by virtually all plants, which is essential for normal metabolism, and for the biosynthesis of aromatic amino acids.

Enteral administration is a feeding or drug administration by the digestion process of a gastrointestinal (GI) tract.

Enteral route refers to the absorption via one or more of the following components of the GI tract: the buccal cavity and sublingually, gastrically, intestinally, and rectally.

Enterotoxin is a bacterial protein that, following release into the intestine; causes cramps, diarrhoea and nausea.

Enthalpy (H) is a property of a system and = E + PV, where, E is the internal energy of the system, P is the pressure of the system, and V is the volume of the system. At constant pressure, the change in enthalpy equals the internal energy.

Enucleated ovum refers to an egg cell from which the nucleus has been removed.

Enucleated refers to having its nucleus removed.

Enuresis is an involuntary discharge of urine usually during sleep at night; bedwetting beyond the age when bladder control should have been established.

Environment refers to the aggregate of all the external conditions and influences affecting the life and development of an organism.

Environmental Protection Agency (EPA) is a U.S. regulatory agency.

Environmental stewardship refers to the view that humans have a duty to manage and care for the whole natural environment.

Environmental stochasticity is a random variation in environmental factors that influence population parameters affecting all individuals in that population.

Enzymatic is an activity of an enzyme which is a substance produced by a living organism and acting as a catalyst to promote a specific biochemical reaction.

Enzyme bioreactor is a reactor in which a chemical conversion reaction that is catalyzed by an enzyme.

Enzyme Commission number (EC number) refers to the systematic labeling which identifies an enzyme in the technical literature. For example, the EC 3.1.21.1 is deoxyribonuclease I.

Enzyme electrode is a type of biosensor, in which an enzyme is immobilized onto the surface of an electrode.

Enzyme immunoassay (EIA) is a method where an enzyme-labeled antibody or antigen is used for the detection and quantification of the antigen-antibody reaction.

Enzyme is a protein that speeds up chemical reactions in the body, e.g., an enzyme in the saliva of the mouth starts the process of breaking down the food.

Enzyme kinetics refers to the quantitative characteristics of enzyme reactions. **Enzyme kinetics** is the study of the chemical reactions that are catalyzed by enzymes

Enzyme Linked Immunosorbent Assay (ELISA) is a blood test used to detect bacteria (H. pylori) that can cause ulcers.

Enzyme stabilization using antibodies is a method of stabilizing enzymes by binding antibodies to them.

Enzyme-Linked Immuno Assays (EIA) measure the amount of a particular substance by virtue of its binding to a specific antibody. Examples of EIA include ELISA and Western blotting.

EPA stands for *Environmental Protection Agency* **of the United States**.

e-PCR (Electronic PCR) is a program that searches a given sequence for the presence of primer pairs. There are currently two versions of this program: Forward e-PCR and reverse e-PCR.

EPD (expected progeny difference) is an evaluation of an animal's genetic worth as a parent.

Epicotyl (Gr. *epi*, upon + *kotyledon*, a cup-shaped hollow) refers to the upper portion of the axis of a plant embryo or seedling, above the cotyledons.

Epidemiology is the study of the spread and control of a disease in a population.

Epidermis (Gr. *epi*, upon + *derma*, skin) refers to: (1) The outmost layer of cells of the body of an animal; (2) The outermost layer of cells covering a plant.

Epididymis is a coiled duct that connects the rete testis to the *ductus deferens*; site of functional maturation of spermatozoa.

Epigenesis describes the developmental process whereby each successive stage of normal development is built up on the foundations created by the preceding stages of development.

Epigenetic is the process by which regulatory proteins can turn genes on or off in a way that can be passed on during cell division.

Epigenetic variation is the non-hereditary and reversible variation; often the result of a change in gene expression due to methylation of DNA.

Epiglottis is a thin, flexible structure that guards the entrance to the larynx, preventing food from entering the airway during swallowing.

Epinasty is a process by which the growth of branches or petioles is abnormally pointing downward.

Epinephrine is hormone produced by the adrenal gland.

Epiphyte is a plant that grows upon another plant, but is neither parasitic on it nor rooted in the ground.

Episome is a genetic extrachromosomal element (e.g., the fertility factor in *Escherichia coli*) which replicates within a cell independently of the chromosome and is able to integrate into the host chromosome.

Epistasis is an interaction between genes at different loci, e.g., one gene suppresses the effect of another gene that is situated at a different locus.

Epistatic genetic variation is the proportion of total genetic variation that can be attributed to the interaction between loci producing a combined effect different from the sum of the effects of the individual loci.

Epitaxial (EPI) is a single-crystal semiconductor layer grown upon a single-crystal substrate having the same crystallographic characteristics as the substrate material.

Epithelium are cells or membrane covering outside of organs.

Epitope is a specific chemical domain on an antigen that stimulates the production of, and is recognized by, an antibody.

Epizootic is a disease simultaneously affecting a large number of animals.

EPROM (Electrically Programmable Read-Only Memory) is a nonvolatile memory device.

EPSPS is an abbreviation for enolpyruvyl-shikimate-3-phosphate synthase.

Equational division refers to a chromosome division in which the two chromatids of each duplicated chromosome separate longitudinally, prior to being incorporated into two daughter nuclei.

Equations in Fluid Mechanics are: Continuity, Euler, Bernoulli, Dynamic and Total Pressure, etc.

Equatorial plate refers to the figure formed by the chromosomes in the center (equatorial plane) of the spindle in mitosis.

Equilibrium density gradient centrifugation is a procedure used to separate macromolecules based on their density (mass per unit volume).

Equilibrium refers to a state of dynamic systems in which there is no net change.

Equimolar refers to identical molar concentrations.

ER is an abbreviation for endoplasmic reticulum.

Erlenmeyer flask is a conical flat-bottomed laboratory flask with a narrow neck.

Erythrocytes is the hemoglobin-containing cell found in the blood of vertebrates.

Erythropoietin is a hormone released from the kidneys and the liver in response to low oxygen concentrations in the blood.

ES cells refer to the embryonic stem cells.

ESA stands for *Endangered Species Act* **of the United States.**

Escherichia coli (*E. coli*) refers to a commensal bacterium inhabiting the colon of many animal species, including human.

E-site is an abbreviation for exit site.

Esophageal atresia is a condition during pregnancy, when the baby's esophagus does not develop properly, and ends before reaching the stomach.

Esophageal manometry is a test that measures the muscle tone in the esophagus.

Esophageal pH monitoring is a test used to monitor the amount of acid in the esophagus, which helps evaluate gastroesophageal reflux disease (GERD).

Esophageal stricture is a narrowing in the esophagus, often caused by irritation from long-term presence of acid in the esophagus with chronic gastroesophageal reflux disease (GERD).

Esophagogastroduodenoscopy (EGD) is a test using an endoscope to look at the inside of the esophagus, stomach, and upper part of the small intestine.

Esophagus is a tube that connects the mouth to the stomach.

Essential requirement refers to the tissue culture plant cell, comprises inorganic salts, including all of the elements necessary for plant metabolism; organic factors (amino acids, vitamins); usually also endogenous plant growth regulators (auxins, cytokinins and often gibberellins); as well as a carbon source (sucrose or glucose).

Essential amino acid is an amino acid that cannot be synthesized by animals and therefore has to be ingested with feed or food.

Essential derivation of varieties (EDV) are genotypes very similar to an originating cultivar.

Essential element is any number of elements required by living organisms to ensure normal growth, development and maintenance.

Essential nutrient is any substance required by living organisms to ensure normal growth, development and maintenance.

Essential requirement refers to when a nutrient is essential when it is mandatory for growth, development and reproduction.

EST (Expressed sequence tag) are single-pass sequences of cDNA clones.

EST is an abbreviation for expressed sequence tag.

Established culture refers to: (1) An aseptic viable explant; (2) A suspension culture subjected to several passages with a constant cell number per unit time.

Estimated breeding value (EBV) is twice the expected progeny difference. The difference is doubled because breeding value is a reflection of all the genes of an animal, in contrast to progeny difference, which is a reflection of a sample half of an animal's genes.

Estrogen are a group of compounds named for their importance in both menstrual and estrous reproductive cycles. They are the primary female sex hormones. Natural estrogens are steroid hormones, while some synthetic ones are non-steroidal. The name comes from the Greek οἶστρος (oistros), literally meaning "gadfly" but figuratively sexual passion or desire, and the suffix *-gen*, meaning "producer of."

Estrogen replacement therapy (ERT) is the use of female hormone estrogen to replace that which the body no longer produces naturally after medical or surgical menopause.

ET is an abbreviation for embryo transfer.

Ethanol (ethyl alcohol, C_2H_6O) is commonly used to disinfest plant tissues, glassware utensils and working surfaces in tissue culture manipulations.

Ethephon (2-chloroethyl phosphonic acid, $ClC_2PO_3H_6$) is a synthetic compound commonly used to treat cultured cells or unripened fruit with ethylene.

Ethics is a branch of philosophy that deals with morality.

Ethidium bromide is a fluorescent dye, which can intercalate between base pairs of double-stranded DNA, and hence is much used to stain DNA in gels.

Ethylene (C_2H_4) is a gaseous plant growth regulator regulating various aspects of vegetative growth, fruit ripening and abscission of plant parts.

Ethylenediamine tetraacetic acid (EDTA) refers to a chelating compound. It is used to keep nutrients, such as iron, bound in a soluble form that leaves them still available to the plant cells *in vitro*. It is also a potent inhibitor of DNase activity and therefore used as an additive for long-term storage of dissolved DNA.

Etiolation is an abnormal increase in stem elongation, accompanied by poor or absent leaf development.

ETL is an abbreviation for economic trait locus.

Euchromatin refers to a chromosomal material that is stained less intensely by certain dyes.

Eugenics (/juːˈdʒɛnɪks/; from Greek *eu*, meaning "good/well," and *-genēs*, meaning "born") is the belief and practice of improving the genetic quality of the human population.

Eukaryote (Gr. *eu*, true + *karyon*, true nucleus) is any organism characterized by having the nucleus enclosed by a membrane. Eukaryotic organisms include animals, plants, fungi and some algae.

Euler equations govern the motion of a compressible, in viscid fluid. They correspond to the Navier-Stokes equations with zero viscosity, although they are usually written in the form shown here because this emphasizes the fact that they directly represent conservation of mass, momentum, and energy.

Euploid is an organism or cell having a chromosome number that is an exact multiple of the monoploid *(n)* number.

Eutrophication is a death of organisms in a lake or pond due to an overabundance of algae that consume all of the dissolved oxygen in the water.

Euvolemia is the state of normal body fluid volume.

Evolution refers to the long-term process through which a population of organisms accumulates genetic changes that enable its members to successfully adapt to environmental conditions and to better exploit food resources.

Evolutionary significant unit (ESU) is a classification of populations that have substantial reproductive isolation which has led to adaptive differences so that the population represents a significant evolutionary component of the species.

Ex situ **conservation of farm animal genetic diversity** refers to all conservation of genetic material *in vivo,* but out of the environment in which it developed, and *in vitro* including, *inter alia,* the cryoconservation of semen, oocytes, embryos, cells or tissues. The *ex situ* conservation and *ex situ* preservation are synonymous.

Ex situ **conservation** is a conservation method which entails the actual removal of germplasm resources (seeds, pollen, sperm, individual organisms) from the original habitat or natural environment.

Ex vitro refers to organisms removed from tissue culture and transplanted.

Ex vivo (testing) is the testing of a substance by exposing it to (excised) living cells (but not to the whole, multicelled organism) in order to ascertain the effect of the substance (for example, pharmaceutical) on the biochemistry of the cell.

Ex vivo **gene therapy** refers to the delivery of a gene or genes to the isolated cells of an individual.

Exact tests is an approach to compute the exact *P*-value for an observed result rather than use an approximation, such as the chi-square distribution.

Excinuclease refers to the endonuclease-containing protein complex that excises a segment of damaged DNA during excision repair.

Excision refers to: (1) The natural or *in vitro* enzymatic removal of a DNA segment from a chromosome or cloning vector; (2) The cutting out and preparation of a tissue, organ, etc., for culture; (3) The removal of adventitious shoots from callus tissue.

Excision repair refers to the DNA repair processes that involve the removal of a damaged or incorrect segment of one strand of double-stranded DNA and its replacement by the synthesis of a new segment using the complementary strand of DNA as template.

Excitation wavelength is the specific wavelength of light required to stimulate a fluorescent molecule, such as a labeled probe, to emit light at the (lower) emission wavelength.

Excrete refers to transport material out of a cell or organism.

Excretion is a discharge from the body from a waste product of metabolism.

Exhalation (or expiration) is the flow of the respiratory current out of the organism. In humans, it is the movement of air out of the bronchial tubes, through the airways, to the external environment during breathing.

Exit site (E site) refers to the ribosome binding site that contains the free tRNA prior to its release.

Exo FISH (Exo Fluorescent *in situ* hybridization) is a technique that utilizes Whole Genome Shotgun (WGS) reads from the pufferfish, *Tetraodan nigroviridis*, to identify potential coding sequences in mammalian genomes based on homology.

Exocrine gland is a gland that excretes its product through a duct that opens onto a free surface.

Exodeoxyribonuclease III (*Escherichia coli exonuclease III, E. coli exonuclease III, endoribonuclease III*) is an enzyme.

Exogamy is the fusion of reproductive cells from distantly related or unrelated organisms, i.e., outbreeding. Exogamy is a social arrangement where marriage is allowed only outside of a social group.

Exogenote refers to chromosomal fragment homologous to an endogenote and donated to a merozygote.

Exogenous (Gr. *exe*, out, + *genos*, race) is produced outside of, originating from, or due to external causes. Opposite is endogenous.

Exogenous DNA is a DNA that has been derived from one organism, and is to be introduced into a cell a different species. Also referred to as foreign DNA or heterologous DNA.

Exogenous refers to one that is produced outside of; originating from, or due to, external causes. Opposite is endogenous.

Exo-III is an abbreviation for exonuclease III.

Exon amplification refers to a procedure that is used to amplify exons.

Exon is a segment of a eukaryotic gene that is transcribed as part of the primary transcript and is retained, after processing, with other exons to form a functional mRNA molecule.

Exon parsing is a process of identifying precisely the 5' and 3' boundaries of genes (the transcription unit) in metazoan genomes.

Exon prediction implies working with eukaryotes.

Exonuclease III (exo III) is an *Escherichia coli* enzyme that removes nucleotides from the 3' hydroxyl ends of double-stranded **DNA**. Synonym is exodeoxyribonuclease III.

Exonuclease is an enzyme that digests DNA or RNA, beginning at the end of a strand.

Exopolysaccharide is a high-molecular-weight polymer that is composed of sugar residues and is secreted by a microorganism into the surrounding environment.

Exotoxin is a toxin released by a bacterium into the medium in which it grows.

Expectation maximization algorithm (EM) is a computational tool in statistics for finding maximum likelihood estimates of parameters in probabilistic models, where the model depends on unobserved variables.

Expected progeny difference (EPD) is the predicted performance of the future offspring of an individual for a particular trait, calculated from measurement(s) of the individual's own performance and/or the performance of one or more of its relatives, for the trait in question and/or for one or more correlated traits.

Expiration is a process of exhaling; giving off carbon dioxide.

Expiratory Reserve Volume is a volume that can be exhaled during forced breathing in addition to tidal volume of 1100 mL.

Explant donor refers to the source plant or mother plant from which is taken the explant used to initiate a culture.

Explant is a portion of a plant aseptically excised and prepared for culture in a nutrient medium.

Explantation is the removal of cells, tissues or organs of animals and plants for observation of their growth and development in appropriate culture media.

Explosion method is a technique for the genetic transformation of cells, in which the transgene is driven into the target (plant) cells by the sudden vaporization (effected by the application of a pulse of high voltage) of a water droplet containing the DNA and gold particles.

Export is the removal of a compound from a cell by active transport.

Express refers to transcribe and translate a gene's message into a peptide product.

Expressed sequence tag (EST) refers to a partially sequenced cDNA clone.

Expression is the process of converting genetic information into RNA and protein for use in the cell. Every gene is not expressed at the same level and at the same time.

Expression library refers to a cDNA library that has been inserted into a bacterial host cell engineered to express transgenes.

Expression system refers to combination of host and vector which provides a genetic context for making a cloned gene function, i.e., produce peptide, in the host cell.

Expression vector is a cloning vector that has been constructed in such a way that, after insertion of a DNA molecule, its coding sequence is properly transcribed and the RNA is translated.

Expressivity refers to the degree of expression of a trait controlled by a particular gene.

Ex-situ conservation is the conservation of components of biological diversity outside their natural habitats.

Extant refers to currently living and not extinct.

Extension refers to the short single-stranded stretch of nucleotides remaining on a double-stranded DNA molecule, following treatment with a restriction endonuclease which makes a staggered cut.

Extension refers to single-stranded DNA region consisting of one or more nucleotides at the end of a strand of duplex DNA.

External guide sequence (EGS) is related to guide sequence.

External respiration is the movement of gases in and out of lungs.

External root sheath is the outermost part of the hair follicle.

Extinct breed is a breed where it is no longer possible to recreate the breed population. Extinction is absolute when there are no breeding males (semen), breeding females (oocytes), nor embryos remaining.

Extinction refers to the irreversible condition of a species or other group of organisms of having no living representatives in the wild, which follows the death of the last surviving individual of that species or group.

Extirpation is the loss of a species or subspecies from a particular area, but not from its entire range.

Extracellular fluid refer to all body fluids other than that contained within the cells; includes plasma and interstitial fluid.

Extracellular matrix is the extracellular part of multicellular structure (e.g., organisms, tissues, biofilms) that typically provides structural and biochemical support to the surrounding cells. Because multicellularity evolved independently in different multicellular lineages, the composition of ECM varies between multicellular structures; however, cell adhesion, cell-to-cell communication and differentiation are common functions of the ECM.

Extrachromosomal inheritance refers to inheritance of traits through DNA that is not connected with the chromosomes but rather to DNA from organelles in the cell. Also called *cytoplasmic inheritance* .

Extrachromosomal refers to nonnuclear DNA in eukaryotes,, present in cytoplasm organelles such as mitochondria and chloroplasts. In prokaryotes, non-chromosomal DNA, i.e., plasmids.

Extrachromosomes refers to when self-replicative genetic elements separate from main chromosome(s) of a cell. This definition usually excludes viruses, but the division is somewhat arbitrary.

Extracorporeal refers to lying outside the body

Extracorporeal shock wave lithotripsy (ESWL) is the use of a machine to send shock waves directly to the kidney stone to break a large stone into smaller stones that will pass through the urinary system.

Extranuclear genes refer to genes residing elsewhere than in the nucleus (e.g., in mitochondria, chloroplasts or plastids).

Extrinsic characterizes doped, rather than pure, semiconductor.

Exude refers to slowly discharge; leak liquid material (exudate such as tannins or oxidized polyphenols) through pores or cuts, or by diffusion into the medium.

Ex-vitro refers to when organisms are removed from tissue culture and transplanted; generally plants to soil or potting mixture.

Ex-vivo gene therapy refers to the delivery of a gene or genes to the isolated cells of an individual, with the intention of alleviating a genetic disorder.

F

F factor (Fertility factor) refers to bacterial plasmid that confers the ability to function as a genetic donor in conjugation. Related term is Hfr (also called an **Hfr strain**), is a bacterium with a conjugative plasmid (often the **F-factor**) integrated into its genomic DNA. Unlike a normal F+ cell, Hfr strains will, upon conjugation with a F− cell, attempt to transfer their entire DNA through the mating bridge, not to be confused with the pilus. This occurs because the F factor has integrated itself via an insertion point in the bacterial chromosome.

F_1 is an abbreviation for first filial generation. It refers to the initial hybrid generation resulting from a cross between two parents. It is related to **Fn**.

F_1, F_2, ..., F_n refer to the subsequent hybrid generations, counting from the F_1. Thus, for example, F_4 describes the progeny of the F_3.

F_2 refers to the second filial generation, produced by crossing two members of the F_1, or by self-pollinating the F_1. The "grand-children" of a given mating.

F_{ab} refers to a product of hydrolysis of an IgG antibody, consisting of the variable region with some of the constant region of a heavy chain, and an entire light chain.

Face-centered cubic (FCC) is a crystal structure that is found in common elemental metals.

Facilitated diffusion refers to the diffusion of a chemical species that is accompanied by chemical reaction involving that species. The net rate of transport is often increased or "facilitated."

Facilitated transport (or Passive-mediated transport) is the process of spontaneous passive transport (as opposed to active transport) of molecules or ions across a biological membrane via specific transmembrane integral proteins.

FACS is an abbreviation for fluorescence-activated cell sorting.

Factor VIII and IX are soluble blood proteins that form part of the cascade of the 12 reactions of blood clotting. Factor VIII deficiency is associated with haemophilia A while factor IX deficiency is associated with haemophilia B.

Factorial mating is a mating scheme in which each male

parent is mated with each female parent.

Facultative anaerobe refers to an organism that will grow under either aerobic or anaerobic conditions.

FAD (flavin adenine dinucleotide) is a co-enzyme important in various biochemical reactions. It comprises of phosphory-lated vitamin B_2 (riboflavin) molecule linked to the nucleo-tide adenine monophosphate (AMP).

False fruit refers to the pseudocarp.

False negative refers to a negative assay result that should have been positive.

False positive refers to a positive assay result that should have been negative.

FAO stands for Food and Agriculture Organization (United Nations).

Farad is a unit of capacitance.

Faraday is a constant representing the charge on one mole of elec-trons = 96,485 Coulombs.

Farm animal genetic resources refer to those animal species that are used, or may be used, for the production of food and agriculture, and the populations within each of them. Within each species, these popula-tions can be classified as wild and feral populations, landra-ces and primary populations,

standardized breeds, selected lines, and any conserved genetic material.

Farmers' privilege refers to rights to hold germplasm, covered by plant variety protection, as a seed source for subsequent seasons. Synonym is farmer-saved seed.

Farmers' rights refer to rights first recognized by Resolution 5 of the 1989 FAO Conference as "rights arising from the past, present and future contributions of farmers in the conservation, improvement and the mak-ing available of plant genetic resources."

Fascicle is related to a vascular bundle.

Father (dad) is a male parent who has raised a child; sup-plied the sperm through sexual intercourse or sperm donation which grew into a child, and/ or donated a body cell which resulted in a clone. The verb "to father" means to procreate or to sire a child from which also derives the noun "father-ing." Fathers determine the sex of their child through a sperm cell which either contains an X chromosome (female), or Y chromosome (male). Related terms of endearment are dad, daddy, pa, papa, poppa, pop, and pops.

Fats are one of three main types of foods, along with proteins and carbohydrates.

F_c refers to a product of hydrolysis of an IgG antibody, consisting of parts of the constant regions of two heavy chains held together by a disulphide bridge, but excluding the antigen-binding regions, and also excluding the light chains.

FCA stands for frequency correspondence analysis.

FDA is an abbreviation for Food and Drug Administration.

FEA stands for Finite Element Analysis.

Fecal fat test assesses how well the body can break down and absorb fat.

Fecal occult blood test is used to check the occult (hidden) blood in a random stool sample.

Fecundity is the potential reproductive capacity of an individual or population (e.g., the number of eggs or young produced by an individual per unit time).

Fed-batch fermentation refers to culture of cells or micro-organisms, where nutrients are added periodically to the bioreactor.

Federal Insecticide, Fungicide, and Rodenticide Act (FIFRA) refers to the act that helps in regulation of pesticides by EPA.

Federal Plant Pest Act (PPA) is implemented by U.S. Department of Agriculture.

Federal Seed Act is implemented by U.S. Department of Agriculture.

Feedback inhibition is the process by which the accumulated end product of a biochemical pathway stops synthesis of that product. Related term is end-product inhibition.

Feeder layer consists of cells that are used in co-culture to maintain pluripotent stem cells.

Femtoengineering involves engineering using mechanisms within a quark.

Feral are domestic or introduced animals living in wild conditions, or plants that have become wild.

Fermentation is the anaerobic breakdown of complex organic substances, especially carbohydrates, by micro-organisms, yielding energy.

Fermentation substrates are materials that are used as food for growing micro-organism.

Fermenter is related to a bioreactor.

Fermi energy is an energy level in a solid at which the probability of finding an electron is 1/2.

Ferroelectric material is a dielectric material such as Rochelle salt and barium titanate with a domain structure containing dipoles, which spontaneously align.

Ferromagnetism is a permanent and large magnetizations that are found in some metals (e.g., Fe, Ni and Co), resulting from the parallel alignment of neighboring magnetic moments.

Fertile refers to capable of breeding and reproduction.

Fertility factor is abbreviated as F factor.

Fertility is the ability to conceive and have offspring.

Fertilization (L. *fertilis*, capable of producing fruit) refers to the union of two gametes from opposite sexes to form a zygote; and involves the fusion of nuclei of gametes (karyogamy) and the fusion of cytoplasm (plasmogamy).

Fertilizer refers to any substance that is added to soil in order to increase its productivity.

FET (Field Effect Transistor) is a semiconductor device whose insulated gate electrode controls current flow.

Fetal alcohol syndrome is a link between excessive alcohol consumption during pregnancy and birth defects.

Fetal tissue is the tissue from the unborn offspring of a human in the post-embryonic period (from eight weeks after fertilization to birth), after major structures have been outlined.

Fetus (*foetus*) is a developing mammal or other viviparous vertebrate after the embryonic stage and before birth. It is also defined as the unborn young of a vertebrate, after developing to its basic form.

Feulgen staining refers to a histochemical stain by which the distribution of DNA in the chromosomes of dividing cell nuclei can be observed.

Feulgen's test is a histochemical test in which the distribution of DNA in the chromosomes of dividing cell nuclei can be observed.

Fever is defined as a temperature 1° or more above the normal 98.6 degrees Fahrenheit (F) or 37 degrees Celsius (C).

FIA is an abbreviation for fluorescence immunoassay.

Fiber is an ingredient in edible plants that aids in digestion.

Fiber optic relates to transmission of information as modulated light in tiny transparent fibers instead of copper wires.

Fibers refer to elongated cells with tapering, pointed ends; the cells interlock to form a strong, rigid tissue; pits in the walls are usually very narrow and not very numerous.

Fibril is a microscopic to sub-microscopic cellulose thread that is part of the cellulose matrix of plant cell walls.

Fibrillation is an apid, uncoordinated contractions of individual heart muscle fibers.

Fibrinogen is a plasma protein that is the soluble precursor of the fibrous protein fibrin.

Fibrinoid layer (Nitabuch's layer) is formed at maternal/ fetal interface during placentation and is thought to prevent excessively deep conceptus implantation.

Fibroblasts are irregularly shaped, branching cells distributed throughout vertebrate connective tissue.

Fibrosis a process by which inflamed tissue becomes scarred.

Fibrous root is a root system in which both primary and lateral roots have approximately equal diameters. Opposite is tap root.

Fick's first law defines a process, where: The diffusion flux is proportional to the concentration gradient for steady-state situations.

Fick's second law defines a process, where: The time rate of change of concentration is proportional to the second derivative of concentration for non-steady-state situations.

Field gene bank is a method of planting plants for the conservation of genes. For this purpose we construct ecosystem artificially. Gene banks are a type of biorepository, which preserve genetic material. There are five types of gene banks: Seed bank, Tissue bank, Cryobank, Pollen bank, and Field gene bank.

FIFRA is an abbreviation for Federal Insecticide, Fungicide, and Rodenticide Act, of EPA.

Filial generation is a generation in a breeding experiment that is successive to a mating between parents of two distinctively different but usually relatively pure genotypes. **Filial generation** is related to F_1, F_2, ..., F_n.

Filler is an inert foreign substance added to a polymer to improve or modify its properties.

Filter bioreactor (mesh bioreactor) refers to cells that are grown on an open mesh of an inert material, which allows the culture medium to flow past it but retains the cells. Synonym is mesh bioreactor.

Filter sterilization is the process of removing microbial contaminants from a liquid by passing through a filter with pores too small to allow the passage of micro-organisms and spores.

Filtration refers to: (1) Separation of solids from liquids by using a porous material that only allows passage of the liquid or of solids smaller than the pore size of the filter. The material passing the filter forms the filtrate; (2) Removal of cell aggregates to obtain a filtrate of single cells that can be utilized as plating inocula. **Biofiltration** is a pollution control technique using living material to capture and biologically degrade process pollutants.

Fingerprint is the pattern of bands produced by a clone when restricted by a particular enzyme, such as Hind III.

Finished Sequence is a clone insert that has been sequenced with an error rate of 0.01%.

Firing is a high temperature heat treatment that increases the density and strength of a ceramic piece.

FISH (Fluorescent *in situ* hybridization) is a process when genomic clones are fluorescently labeled and hybridized to chromosome spreads.

Fisher Wright model is the stochastic model for reproduction in population genetics.

Fission (*L. fissilis*, easily split) is a asexual reproduction involving the division of a single-celled individual into two new single-celled individuals of equal size.

Fistula is an abnormal connection between two organs, or between an organ and the outside of the body.

Fitness is the ability of an individual, or genotype to survive and produce viable offspring. It refers to the proportion of the individual's genes in all the genes contributed to the next generation.

Fitness rebound refers to when following an episode of inbreeding depression, successive generations of breeding may result in a rebound in fitness due to the selective decrease in frequency of deleterious alleles (purging).

Fixation index is the proportional increase of homozygosity through population subdivision. FST is sometimes referred to as the fixation index.

Fixation refers to the situation in which only one allele for a given gene/locus is present in a population.

FLAG is related to affinity tag.

Flagellum (plural: flagella; adj.: flagellate) is a whiplike organelle of locomotion in certain cells; locomotor structures in flagellate protozoa.

Flaming is a technique for sterilizing instruments.

Flanking region refers to the DNA sequences extending either side of a specific sequence.

Flat-band potential refers to Electrolyte/Insulator/Silicon (EIS).

Flavin adenine dinucleotide (FAD) refers to a co-enzyme important in various biochemical reactions. It comprises a phosphorylated vitamin B_2 (riboflavin) molecule linked to AMP, and functions as a hydrogen acceptor in dehydrogenation reactions.

Flip-flop is a binary device whose outputs change value only in response to an input pulse.

Flocculant is a chemical agent that causes small particles to aggregate (flocculate).

Floccule refers to micro-organism aggregate or colloidal particle floating in or on a liquid.

Flow cytometry refers to the automated measurements on large numbers of individual cells or other small biological materials, made as the cells flow one by one in a fluid stream past optical and/or electronic sensors.

Flower refers to the structure in angiosperms (flowering plants) that bears the organs for sexual reproduction.

Fluctuating asymmetry (FA) is asymmetry in which deviations from symmetry are randomly distributed about a mean of zero.

Fludeoxyglucose (fludeoxyglucose F18, fluorodeoxyglucose, ^{18}F-FDG or FDG) is a radiopharmaceutical used in the medical imaging modality positron emission tomography (PET). Chemically, it is 2-deoxy-2-(^{18}F)fluoro-D-glucose, a glucose analog, with the positron-emitting radioactive isotope fluorine-18 substituted for the normal hydroxyl group at the 2' position in the glucose molecule.

Fluid balance is the concept of human homeostasis that the amount of fluid lost from the body is equal to the amount of fluid taken in. The human homeostatic control mechanisms maintain a constant internal environment and ensure that a balance between fluid gain and fluid loss is maintained.

Fluid balance, negative refers to when fluid loss is greater than fluid gain (for example if the patient vomits and has diarrhea), the patient is said to be in negative fluid balance.

Fluid balance, positive refers to when fluid loss is less than fluid gain, the patient is said to be in positive fluid balance that might suggest a problem with either the renal or cardiovascular system.

Fluid bonding refers to unprotected sex in long-term relationships. The relationships can be either monogamous or polyamorous.

Fluidotherapy is an modality of dry heat using a finely divided solid suspended in an air stream with the properties of liquid.

Fluorescence is the emission of light by a substance that has absorbed light or other electromagnetic radiation. It refers to luminescence which persists less than a second after the exciting cause has been removed. If the luminescence persists significantly longer, it is called phosphorescence.

Fluorescence activated cell sorter (FACS) is an indispensable instrument in stem cell research, FACS enables the rapid characterization, counting and isolation of cells suspended in a stream of fluid.

Fluorescence-activated cell sorting (FACS) refers to a flow cytometry method in which targets (cells, individual chromosomes etc.) are labeled with a fluorescent dye, which is excited by a laser beam.

Fluorescence immunoassay (FIA) is an immunoassay based on the use of fluorescence-labeled antibody.

Fluorescence *in situ* hybridization (FISH) refers to the hybridization of cloned, fluorescently labeled DNA or RNA, to intact biological materials, notably chromosome spreads and thin tissue sections.

Fluorescent probe is a probe, which is labeled with a fluorescent dye, so that the signal emitted can be captured by photometric methods.

Fluoroscopy is a process of using an instrument to observe the internal structure of an opaque object (as the living body) by means of x-rays.

FM (Frequency modulation) is an information coding scheme in which the frequency of a steady wave is changed.

Foetus refers to the pre-natal stage of a viviparous animal, between the embryonic stage and birth. Vivipary has two different meanings. In animals, it means development of the embryo inside the body of the mother, eventually leading to live birth, as opposed to laying eggs. In plants, it means reproduction via embryos, such as buds, that develop from the outset without interruption, as opposed to germinating externally from a seed.

Fog refers to fine particles of liquid suspended in the air, such as of water in a fog chamber used for acclimatizing recent *ex vitro* transplants.

Folate is a naturally occurring form of the vitamin, found in food, while folic acid is synthetically produced, and used in fortified foods and supplements.

Fold-back refers to the structure formed when a double-stranded DNA molecule containing an inverted repeat sequence is denatured and then allowed to re-anneal at low DNA concentrations.

Folded genome refers to the condensed intracellular state of the DNA in the nucleoid of a bacterium.

Follicle refers to any enclosing cluster of cells that protects and nourishes a cell or structure within.

Follicle stimulating hormone (FSH) refers to a hormone, secreted by the anterior pituitary gland in mammals, that stimulates the ripening of the specialized structures in the ovary (Graafian follicles) that produce ova in female mammals; and in males, the formation of sperm in the testis.

Food and Drug Administration (FDA) is an U.S. agency responsible for regulation of food products. The major laws under which the agency has regulatory powers include the Food, Drug, and Cosmetic Act; and the Public Health Service Act.

Food biotechnology is the application of biotechnology to the production of food.

Food is any substance, whether processed, semi-processed or raw, which is intended for human consumption.

Food processing enzyme is an enzyme used to control food texture, flavor, appearance and, to a certain extent, nutritional value.

Food, Drug, and Cosmetic Act grants regulatory powers to the FDA.

Foot-candle is an obsolete photometric measure of light intensity. Now superseded by the lux.

Foramen ovale shunts a large portion of the incoming blood from the right atrium into the left atrium. When the baby is born, the foramen ovale closes, forming the separation of the heart into two pumps.

Force field is a set of functions and parametrization used in molecular mechanics calculations.

Forced cloning is the insertion of foreign DNA into a cloning vector in a predetermined orientation.

Forced expiratory flow is an average rate of flow for a specified portion of the forced expiratory volume, usually between 0.0002 and 0.0012 mm^3 (formerly called maximum expiratory flow rate).

Forced midexpiratory flow is an average rate of flow during the middle half of the forced expiratory volume.

Forced vital capacity is the maximum volume of gas that can be expelled as forcefully and rapidly as possible after maximum inspiration.

Foreign DNA refers to a DNA molecule that is incorporated into either a cloning vector or a chromosomal site.

Forensics is the use of scientific methods and techniques, such as genetic fingerprinting, to solve crimes.

Formulation refers to the method by which a therapeutic agent is delivered to its site of action.

Forskolin refers to a medicinal, diterpenoid, compound exclusive to plant roots and used in the preparation of drugs for the treatment of cardiomyopathy, glaucoma and certain cancers.

Fortify refers to add strengthening components or beneficial ingredients to a nutrient medium.

Forward bias is the conducting bias for a p-n junction rectifier that assures electron flow to the n-side of the junction.

Forward mutation refers to a mutation from the wild type to the mutant type. Opposite is reverse mutation.

Fosmid is a cloning system based on the *E. coli* F factor.

Fouling refers to the coating or plugging (by materials or micro-organisms) of equipment, thus preventing it from functioning properly.

Founder animal is an organism that carries a transgene in its germ line and can be used in matings to establish a pure-breeding transgenic line, or one that acts as a breeding stock for transgenic animals.

Founder effect is a loss of genetic variation in a population that was established by a small number of individuals that carry only a fraction of the original genetic diversity from a larger population.

Founder principle refers to the possibility that a new, small, isolated population may be genetically different from the "parent" population, because the founding individuals (being a small, random sample from the large, "parent" population) could be quite different from typical members of the "parent" population.

Four-base cutter (four-base-pair-cutter, four-cutter) refers to a type-II restriction endonuclease that binds (and subsequently cleaves) DNA at sites that contain a sequence of four nucleotide pairs that is uniquely recognised by that enzyme.

Fourier transform infrared spectroscopy (FTIR) is a technique which is used to obtain an infrared spectrum of absorption, emission, photoconductivity or Raman scattering of a solid, liquid or gas. An FTIR spectrometer simultaneously collects spectral data in a wide spectral range. The term *Fourier transform infrared spectroscopy* originates from the fact that a Fourier transform (a mathematical process) is required to convert the raw data into the actual spectrum.

Fractionation refers to the separation in components of a complex mixture of molecules.

Fragile sites refer to a non-staining gap of variable width that usually involves both chromatids and is always at exactly the same point on a specific chromosome derived from an individual or kindred.

Fragile-X syndrome is a X-linked trait; the second most common identifiable cause of genetic mental deficiency.

Fragment refers to a partial structure.

Frameshift mutation refers to a mutation that changes the reading frame of a DNA, either by the insertion or the deletion of nucleotides.

Free energy (G) is a thermodynamic quantity that is a function of the enthalpy (H), the Kelvin temperature (T) and the entropy (S) of a system: $G = H - TS$. At equilibrium, the free energy is at a minimum. Under certain conditions, the change in free energy for a process is equal to the maximum useful work.

Free water refers to the cellular water released into the intercellular spaces when tissue is frozen and thawed. Opposite is bound water.

Free-living conditions refer to natural or greenhouse conditions experienced by plantlets upon transfer from *in vitro* conditions to soil.

Freeze-dry is the removal of water as vapor from frozen material under vacuum. Synonym is lyophilize.

Freeze-drying is the process of drying a tissue or an organ in a frozen state under vacuum. Freeze-drying is the standard way of preserving micro-organisms for long periods of time.

Frequency distribution is a graph showing either the relative or absolute incidence of classes in a population. The classes may be defined by either a discrete or a continuous variable.

Freeze preservation is related to cryobiological preservation. Cryopreservation or cryoconservation is a process where cells, whole tissues, or any other substances susceptible to damage caused by chemical reactivity or time are preserved by cooling to sub-zero temperatures.

Frequency is a number of times per second that a quantity representing a signal, such as a voltage, changes state. Also, the number of waves (cycles) per second that pass a given point in space.

Frequency-dependent selection is a natural selection in which fitness varies as a function of the frequency of a phenotype.

Fresh weight refers to the weight, including the water content, of a specimen. Synonym is wet weight.

Friable is used to describe a crumbling or fragmenting callus.

Froude number (named after William Froude) is the reciprocal of the square root of the Richardson number. The densimetric Froude number is usually preferred by modellers who wish to nondimensionalize a speed preference to the Richardson number which is more commonly encountered when considering stratified shear layers. For example, the leading edge of a gravity current moves with a front Froude number of about unity.

FSH is an abbreviation for follicle stimulating hormone.

Fullerene is a new allotrope of carbon characterized by a closed cage structure consisting of an even number of three coordinate carbon atoms devoid of hydrogen atoms.

Functional genomics refers to the field of research, that aims to determine patterns of gene expression and interaction in the genome, based on the knowledge of extensive or complete genomic sequence of an organism.

Functional food is an ordinary food that has components or ingredients added to give it a specific medical or physiological benefit, other than a purely nutritional effect. Also known as nutraceutical.

Functional gene cloning is related to candidate-gene strategy. The candidate-gene strategy can be defined as the study of genetic determinants of a complex trait based on: (1) generating hypotheses and identifying candidate genes that might have a pathogenic role; (2) identifying variants (SNPs) in or near these genes; and (3) genotyping of variants in a population, followed by statistical methods (linkage or association) to identify correlations with the phenotype.

Functional incontinence is a leakage of urine due to a difficulty reaching a restroom in time because of physical conditions such as arthritis.

Functional magnetic resonance imaging (functional MRI, fMRI) is a functional neuroimaging procedure using MRI technology that measures brain activity by detecting associated changes in blood flow. Other methods of obtaining contrast are arterial spin labeling and diffusion MRI. The procedure is similar to MRI but uses the change in magnetization between oxygen-rich and oxygen-poor blood as its basic measure.

Functional residual capacity is the volume of gas remaining in the lungs at the resisting expiratory level.

Fungicide is an agent, such as a chemical, that kills fungi.

Fungus (plural: fungi) is a multinucleate single-celled or multicellular heterotrophic micro-organisms, including yeasts, moulds, and mushrooms.

Furfural (furfuraldehyde) is used as a solvent and as a raw material for synthetic resin.

Fusarium spp. refers to a group of fungal pathogens of many economic crop species, particularly cereals, where severe infestation leads to losses in both grain yield and quality.

Fusion biopharmaceuticals refer to fusion proteins with pharmaceutical properties.

Fusion gene is related to chimeric gene. The chimeric gene (literally, made of parts from different sources) form through the combination of portions of one or more coding sequences to produce new genes.

Fusion protein is a polypeptide made from a recombinant gene that contains portions of two or more different genes. Also known as hybrid protein or chimeric protein.

Fusion toxin refers to a fusion protein that consists of a toxic protein domain plus a cell receptor binding domain.

Fusogenic agent refers to any chemical or virus, etc., that causes cells to fuse together.

G

. . . .

G is an abbreviation for guanine residue in either DNA or RNA.

G cap refers to the 5'-terminal methylated guanine nucleoside that is present on many eukaryotic mRNAs.

G protein refers to proteins found on the inner surface of the plasma membrane, which bind to the guanine nucleotides, GTP and GDP.

Gadolinium is a chemical element with symbol **Gd** and atomic number 64. It is a silvery-white, malleable and ductile rare-earth metal. It is found in nature only in combined (salt) form.

Gain is a ratio of the amplitude of an output to input signal.

Galactomannan refers to a gum in which the structural chain is made up of D-mannose units with 1-4 linkages. The ratio of galactose to mannose is 1:2.

Gall is a tumorous growth in plants.

Gall bladder stores bile made by the liver; sends bile into the small intestine to help digest fats.

Galvanic corrosion is the preferential corrosion of the more chemically active of two metals electrically coupled and exposed to an electrolyte.

Gamete and embryo storage refers to storage of ova, sperm or fertilized embryos outside their original source.

Gamete is a mature reproductive cell which is capable of fusing with a cell of similar origin but of opposite sex to form a zygote from which a new organism can develop.

Gametic (phase) equilibrium refers to the occurrence of haplotypes (gametes) with a frequency equal to the product of the frequency of the two relevant alleles, e.g., loci A and B are in linkage equilibrium if the frequency of the haplotype (gamete) A_iB_i equals the product of the frequencies of alleles A_i and B_i linkage equilibrium.

Gametic disequilibrium is a nonrandom association of alleles at different loci within a population. Also known as linkage disequilibrium.

Gametoclone refers to a plant regenerated from a tissue culture originating from gametic tissue.

Gametogenesis is the process of the formation of gametes.

Gametophyte refers to that phase of the plant life cycle that bears the gamete producing organs; the cells have *n* chromosomes.

Gametophytic incompatibility is a phenomenon controlled by the complex S locus, in which a pollen grain cannot fertilize an ovule produced by a plant that carries the same S allele as the pollen grain.

Gamma camera (scintillation camera or Anger camera) is a device used to image gamma radiation emitting radioisotopes, a technique known as scintigraphy.

Gamma emission is a quantum transition between two energy levels of a nucleus in which a gamma ray is emitted. It is also known as gamma decay.

Gamma radiation (gamma rays) refers to electromagnetic radiation of extremely high frequency and therefore high energy per photon.

Gamma system consists of nerve fibers that reset the muscle spindle to its adjusted length.

Ganglion is a mass of neuron cell bodies.

Ganglionic blocking agent blocks nerve impulses at autonomic ganglionic synapses.

Gap is a region of the genome for which no sequence is currently available. There are two types of gaps: heterochromatic and euchromatic.

Gapped DNA refers to the double-stranded DNA molecule with one or more internal single-stranded regions.

Gas is an air that collects in the stomach and intestines as a natural result of digesting food. Passed out of the body via the rectum or the mouth.

Gas chromatography (GC) is a common type of chromatography used in analytical chemistry for separating and analyzing compounds that can be vaporized without decomposition.

Gas transfer is the rate at which gases are transferred from gas into solution, an important parameter in fermentation systems because it controls the rate at which the organism can metabolize.

Gastric is related to the stomach.

Gastritis is an inflammation of the lining of the stomach.

Gastroenteritis is an irritation or infection of the stomach and intestines.

Gastroenterologist is a physician whose specialty is related to digestive diseases.

Gastroenterology is a medical specialty that deals with the digestive system.

Gastroesophageal reflux disease (GERD) is a movement of food, fluids, and digestive juices from the stomach back up into the esophagus; causes irritation of the esophagus with acid, resulting in discomfort.

Gastrointestinal tract (Also called the digestive tract) is an internal passageway that begins at the mouth, ends at the anus, and is lined by a mucous membrane; also called digestive tract.

Gastroparesis is a condition related to muscle or nerve damage in the stomach, which causes slow digestion and stomach emptying.

Gastrostomy is a surgically created opening in the stomach and the abdominal muscles. A tube is passed through these openings, into the stomach, to allow for feeding of a person who cannot eat normally.

Gastrula refers to an early animal embryo consisting of two layers of cells; an embryological stage following the blastula.

Gastrulation is the process in which cells proliferate and migrate within the embryo to transform the inner cell mass of the blastocyst stage into an embryo containing all three primary germ layers.

Gate is a circuit whose logical output variables are determined by its inputs.

Gauss is cgs unit to measure magnetic induction.

GC island is a segment of DNA that is rich in GC base pairs and often precedes a transcribed gene in the genomes of vertebrate organisms.

GDP is an abbreviation for guanosine-5'-di-phosphate.

Gel is a lyophilic colloid that has coagulated to a rigid or jelly-like solid.

Gel electrophoresis is a method for separation and analysis of macromolecules (DNA, RNA and proteins) and their fragments, based on their size and charge. Gel electrophoresis uses a gel as an anticonvective medium and/or sieving medium during electrophoresis, the movement of a charged particle in an electrical field. Gels suppress the thermal convection caused by application of the electric field, and can also act as a sieving medium, retarding the passage of molecules; gels can also simply serve to maintain the finished separation, so that a post electrophoresis stain can be applied.

Gel filtration refers to a method of protein or DNA purification, where differences in size are used to separate the components of a complex mixture.

Gel permeation chromatography is a type of size exclusion chromatography (SEC) that separates analytes on the basis of size.

Gelatin is a glutinous, protein-aceous gelling and solidifying agent.

Gelatinization refers to the swelling of starch when added to hot water.

Gelrite (Gelrite™) refers to the brand name of a *Pseudomonas-*derived refined polysaccharide used as a gelling agent and agar substitute.

GEM is an abbreviation for genetically engineered micro-organism. **GEM** is an organism whose genetic material has been altered using genetic engineering techniques. Organisms that have been genetically modified include micro-organisms such as bacteria and yeast, insects, plants, fish, and mammals. GMOs are the source of genetically modified foods, and are also widely used in scientific research and to produce goods other than food.

Gene (Gr. *gen*, race, offspring) refers to the heredity unit transmitted from generation to generation during sexual or asexual reproduction.

Gene (resources) conservation refers to the conservation of species, populations, individuals or parts of individuals, by *in situ* or *ex situ* methods, to provide a diversity of genetic materials for present and future generations.

Gene addition is the addition of a functional copy of a gene to the genome of an organism.

Gene amplification refers to the presence of multiple genes.

Gene bank refers to: (1) The physical location where collections of genetic material in the form of seeds, tissues or reproductive cells of plants or animals are stored; (2) Field gene bank: A facility established for the *ex situ* storage and maintenance, using horticultural techniques, of individual plants; (3) A collection of cloned DNA fragments from a single genome; (4) A population of micro-organisms, each of which carries a DNA molecule that was inserted into a cloning vector. Also called gene library, clone bank, bank, library. This term is sometimes also used to denote all of the vector molecules, each carrying a piece of the chromosomal DNA of an organism, prior to the insertion of these molecules into a population of host cells.

Gene cloning is the process of synthesizing multiple copies of a particular DNA sequence using a bacteria cell or another organism as a host.

Gene conservation (genetic resources conservation) refers to the conservation of species, populations, individuals or parts of individuals, by *in situ* or *ex situ* methods, to provide a diversity of genetic materials for present and future generations.

Gene construct is a functional unit necessary for the transfer or the expression of a gene of interest.

Gene conversion is a process, often associated with recombination, during which one allele is replicated at the expense of another, leading to non-Mendelian segregation ratios.

Gene drop is a simulation of the transmission of alleles in a pedigree. Each founder is assigned two unique alleles, and the alleles are then passed on from parent to offspring.

Gene expression is the process by which a gene produces mRNA and protein, and hence exerts its effect on the phenotype of an organism.

Gene flow is the spread of genes from one breeding population to another (usually) related populations by migration, possibly leading to changes in allele frequency.

Gene frequency (or allele frequency) is the proportion of a particular allele (variant of a gene) among all allele copies being considered. Allele or gene frequency is the percentage of all alleles at a given locus in a population gene pool represented by a particular allele.

Gene genealogies is the tracing of the inherited history of the genes in an individual.

Gene gun (biolistics) was originally designed for plant transformation. It is a device for injecting cells with genetic information. The payload is an elemental particle of a heavy metal coated with plasmid DNA. This technique is often simply referred to as bioballistics or biolistics.

Gene identification is to hone in on otherwise hard to find genes, using marker SNPs. There are two basic approaches to gene identification: by homology and *ab initio* approaches.

Gene imprinting is the differential expression of a single gene according to its parental origin.

Gene insertion is the addition of one or more copies of a normal gene into a defective chromosome.

Gene interaction refers to the modification of the action of one gene by another, non-allelic gene.

Gene knock-in refers to a genetic engineering method that involves the insertion of a protein coding cDNA sequence at a particular locus in an organism's chromosome.

Gene knockout (KO) is a genetic technique in which one of an organism's genes are made inoperative ("knocked out" of the organism). Also known as knockout organisms or simply knockouts.

Gene library is a collection of DNA fragments that is stored and propagated in a population of micro-organisms through the process of molecular cloning. There are different types of DNA libraries: cDNA libraries, genomic libraries, randomized mutant libraries.

Gene linkage is the tendency of genes that are located proximal to each other on a chromosome to be inherited together during meiosis.

Gene machine is related to transposon tagging, which refers to a process in genetic engineering where transposons (transposable elements) are amplified inside a biological cell by a tagging technique.

Gene mapping is also called **genome mapping**. It is the creation of a genetic map assigning DNA fragments to chromosomes. Two different ways of mapping are distinguished: Genetic mapping uses classical genetic techniques, and physical mapping.

Gene modification is the chemical change to a gene's DNA sequence.

Gene parsing methods were initially based on word frequency computation, eventually combined with the detection of splicing consensus motifs. The next generation of software implemented the same basic principles into a simulated neural network architecture. Finally, the last generation of software, based on Hidden Markov Models, added an additional refinement by computing the likelihood of the predicted gene architectures (e.g., favoring human genes with an average of seven coding exons, each 150 nucleotides long) is added.

Gene pool refers to: (1) The sum of all genetic information in a breeding population at a given time; (2) In plant genetic resources, use is made of the terms 'primary,' 'secondary' and 'tertiary' gene pools.

Gene prediction methods are based on compositional signals that are found in the DNA sequence. These methods detect characteristics that are expected to be associated with genes, such as splice sites and coding regions, and then piece this information together to determine the complete or partial sequence of a gene. The two main principles behind state- of-the- art gene prediction software are (1) common statistical regularities; and (2) plain sequence similarity. From an epistemological point of view, those concepts are quite primitive.

Gene probe is a single-stranded DNA or RNA fragment that is used in genetic engineering to search for a particular gene or other DNA sequence.

Gene recognition is used for finding open reading frames, tools of this type also recognize a number of features of genes, such as regulatory regions, splice junctions, transcription and translation stops and starts, GC islands, and poly adenylation sites.

Gene recombination (Genetic recombination) is the process by which two DNA molecules exchange genetic information, resulting in the production of a new combination of alleles.

Gene regulation is the process of controlling the synthesis or suppression of gene products in specific cells or tissues.

Gene replacement refers to the incorporation of a transgene into a chromosome at its normal location by homologous recombination, thus replacing the copy of the gene originally present at the locus.

Gene sequencing (DNA sequencing) is a process in which the individual base nucleotides in an organism's DNA are identified. This technique is used to learn more about the genome of the organism as a whole, and to identify specific areas of interest and concern.

Gene shear technology involves firstly decoding a gene in an organism that produces a harmful protein. A hammerhead ribozyme is then specially designed to match to that gene code. An army of these specific ribozymes are sent into the cells of the organism. These ribozymes will find the matching code of the harmful protein and snip it out of the message, making it impossible to produce the disease-causing protein.

Gene shears (ribozyme: ribonucleic acid enzyme) refer to RNA molecule that is capable of catalyzing specific biochemical reactions, similar to the action of protein enzymes.

Gene silencing is a general term used to describe the epigenetic regulation of gene expression. In particular, this term refers to the ability of a cell to prevent the expression of a certain gene.

Gene splicing involves cutting out part of the DNA in a gene and adding new DNA in its place. The process is entirely chemical with restriction enzymes used as chemical 'scissors.'

Gene stacking is a term that is used in the context of genetically engineered crops, but is not a new idea in plant breeding. Gene stacking is combining desired traits into one line.

Gene targeting is a process of targeting a gene. This is a specific type of transgenesis that targets a particular gene.

Gene technology is the term given to a range of activities concerned with understanding the expression of genes, taking advantage of natural genetic variation, modifying genes and transferring genes to new hosts.

Gene testing are methods that identify the presence, absence or mutation of a particular gene in an individual.

Gene therapy is the proposed treatment of an inherited disease by the transformation of an affected individual with a wild-type copy of the defective gene causing the disorder.

Gene tracking is a method for determining the inheritance of a particular gene in a family. It is used in the diagnosis of genetic diseases, such as cystic fibrosis and Huntington's chorea.

Gene translocation is the movement of a gene fragment from one chromosomal location to another, which often alters or abolishes expression.

Gene translocation is the movement of a gene from one chromosomal location to another.

Gene trapping uses transgenesis to introduce DNA carrying a reporter gene (lacZ or GFP) flanked by various genomic signals (splice donor or acceptor sites, promoters, etc.).

Genera is plural for genus.

Generally Regarded as Safe (GRAS) is a designation given to foods, drugs, and other materials that have been used for a considerable period of time and have a history of not causing illness to humans, even though extensive toxicity testing has not been conducted. More recently, certain host organisms for recombinant DNA experimentation have been given this status.

Generate refers to propagate or (mass) proliferate. The process is generation or regeneration.

Generation time (cell generation time) is a quantity that reflects the average time between two consecutive generations in the lineages of a population, in population biology and demography. In human population, the **generation time** ranges from 20 to 30 years. In Epidemiology, the term refers to the time between receipt of infection by a host and maximal infectivity of that host.

Generative nucleus refers to the shed pollen that is two-celled (in others it is three-celled or has a variable number), in many flowering plants.

Genet refer to the individual(s) descended vegetatively from a single sexually produced zygote, including all entities derived from it.

Genetic assimilation is a process in which phenotypically plastic characters that were originally "acquired" become converted into inherited characters by natural selection. This term also has been applied to the situation in which hybrids are fertile and displace one or both parental taxa through the production of hybrid swarms (i.e., genomic extinction).

Genetic code refers to: (1) The set of 64 nucleotide triplets (codons); (2) The relationships between the nucleotide base-pair triplets of an mRNA molecule and the 20 amino acids.

Genetic complementation refers to when two DNA molecules that are in the same cell together produce a function that neither DNA molecule can supply on its own.

Genetic counseling refers to counseling individuals and prospective parents who are at risk of a particular genetic disease (either themselves or their potential child)..

Genetic disease is a disease caused by an abnormality in the genetic material, which could be at the level of DNA sequence at a locus, or at the level of karyotype.

Genetic disorder is a hereditary condition that results from a defective gene or chromosome.

Genetic distance matrix is a pairwise matrix composed of differentiation between population (or individual) pairs that is calculated using a measure of genetic divergence such as FST.

Genetic distance is a measure of the genetic similarity between any pair of populations. Such distance may be based on phenotypic traits, allele frequencies or DNA sequences.

Genetic distancing is the collection of the data on phenotypic traits, marker allele frequencies or DNA sequences for two or more populations, and estimation of the genetic distances between each pair of populations.

Genetic divergence is the evolutionary change in allele frequencies between reproductively isolation populations.

Genetic diversity refers to the heritable variation within and among populations which is created, enhanced or maintained by evolutionary forces.

Genetic draft is a stochastic process in which selective substitutions at one locus will reduce genetic diversity at neutral linked loci through hitchhiking.

Genetic drift or allelic drift is the change in the frequency of a gene variant (allele) in a population due to random sampling. Genetic drift may cause gene variants to disappear completely and thereby reduce genetic variation.

Genetic engineering is the technique of removing, modifying or adding genes to a DNA molecule to change the information it contains. By changing this information, genetic engineering changes the type or amount of proteins an organism is capable of producing. A broader definition of genetic engineering also includes selective breeding and other means of artificial selection.

Genetic equilibrium is a condition in a group of interbreeding organisms in which the allele frequencies remain constant over time.

Genetic erosion refers to the loss over time of allelic diversity, particularly in farmed organisms, caused by either natural or man-made processes.

Genetic fingerprinting is a technique, in which an individual's DNA is analyzed to reveal the pattern of repetition of particular nucleotide sequences throughout the genome.

Genetic gain refers to the increase in productivity achieved following a change in gene frequency effected by selection.

Genetic heterogeneity refers to the situation in which different mutant genes produce the same phenotype.

Genetic immunization is the delivery to a host organism of a cloned gene that encodes an antigen.

Genetic information refers to the information contained in a nucleotide base sequence in chromosomal DNA or RNA.

Genetic linkage is the tendency of genes that are located proximal to each other on a chromosome to be inherited together during meiosis.

Genetic linkage map is a linear map of the relative positions of genes along a chromosome.

Genetic load is the decrease in the average fitness of individuals in a population due to deleterious genes or heterozygous advantage.

Genetic map refers to the linear array of genes on a chromosome, based on recombination frequencies (linkage map) or physical location (physical or chromosomal map).

Genetic mapping refers to determining the linear order of genes and/or DNA markers along a chromosome.

Genetic marker is a DNA sequence at a unique physical location in the genome, which varies sufficiently between individuals that its pattern of inheritance can be tracked through families and/or it can be used to distinguish among cell types. A marker may or may not be part of a gene.

Genetic modification (GM) is any process that alters the genetic material of living organism. This includes duplicating, deleting or inserting one or more new genes or altering the activities of an existing gene.

Genetic pollution refers to the uncontrolled spread of genetic information (frequently referring to transgenes) into the genomes of organisms in which such genes are not present in nature.

Genetic polymorphism is the simultaneous occurrence in the same locality of two or more discontinuous forms in such proportions that the rarest of them cannot be maintained just by recurrent mutation or immigration. By definition, genetic polymorphism relates to a balance or equilibrium between morphs. According to Cavalli-Sforza & Bodmer, "Genetic polymorphism is the occurrence in the same population of two or more alleles at one locus, each with appreciable frequency," where the minimum frequency is typically taken as 1%. The definition has three parts: a) sympatry: one interbreeding population; b) discrete forms; and c) not maintained just by mutation.

Genetic privacy is the freedom from unauthorized intrusion. Often referred to as the right to be let alone, it protects territorial, bodily, psychological and informational integrity and decision making.

Genetic relatedness is a quantitative estimate of the proportion of genes, shared between the genomes of any two individuals, groups or populations, e.g., r = 0.5 for full siblings and parent offspring pairs.

Genetic rescue is the recovery in the average fitness of individuals through increased gene flow into small populations.

Genetic resources is the genetic material of actual or potential value.

Genetic resources conservation promotes the conservation of genetic diversity and advances the study of conservation genetics through the rapid publication of technical papers and reviews on methodological innovations and improvements, computer programs and genomic resources.

Genetic screening is testing a population for alterations in the activity (mutations) of particular genes.

Genetic selection is the process of selecting genes, cells, clones, etc., within populations or between populations or species.

Genetic stochasticity refers to random changes in the genetic characteristics of populations through genetic drift and binomial sampling of alleles during Mendelian segregation.

Genetic swamping is the loss of locally adapted alleles or genotypes caused by constant immigration and gene flow.

Genetic testing is a laboratory test, done most often on a blood sample, but also on cheek cells, skin cells, bone marrow, amniotic fluid or a placenta sample.

Genetic toxicology is a research field in which genetic samples from a living organism (including humans) are placed on a DNA microarray (gene chip) and tested in a computerized device for the presence of toxic substances from the environment. The study of the pattern of occurrence of such biomarkers in a sample of individuals or a community is called **genetic epidemiology**.

Genetic transformation is a process by which the genetic material carried by an individual cell is altered by the incorporation of foreign (exogenous) DNA into its genome.

Genetic use restriction technology (GURT) refers to a technology applying transgenesis to genetically compromise the fertility or the performance of saved seed of a cultivar or of second generation animals. Two types of GURTs have been patented: variety-level GURT

(V-GURT), and trait-specific GURT (T-GURT).

Genetic variation refers to the differences between individuals attributable to differences in genotype.

Genetically engineered organism (GEO) is an occasional alternative term for genetically modified organism.

Genetically modified organism (GMO) is an organism produced from genetic engineering techniques that allow the transfer of functional genes from one organism to another, including from one species to another.

Genetics (from the Ancient Greek γενετικός *genetikos* meaning "genitive"/"generative," in turn from γένεσις *genesis* meaning "origin"), a field in biology, is the science of genes, heredity, and variation in living organisms. Genetics is the process of trait inheritance from parents to offspring, including the molecular structure and function of genes, gene behavior in the context of a cell or organism (e.g., dominance and epigenetics), gene distribution and variation and change in populations.

Genome refers to: (1) The entire complement of genetic material (genes plus non-coding sequences) present in each cell of an organism, virus or organelle; (2) The complete set of chromosomes (hence of genes) inherited as a unit from one parent. In modern molecular biology and genetics, the **genome** is the genetic material of an organism.

Genomic DNA library (genomic library) is a collection of clones containing the genomic DNA sequences of an organism.

Genomic extinction is the situation in which hybrids are fertile and displace one or both parental taxa through the production of hybrid swarms so that the parental genomes no longer exist even though the parental alleles are still present.

Genomic library is composed of fragments of genomic DNA.

Genomic ratchet is a process where hybridization producing fertile offspring will result in a hybrid swarm over time, even in the presence of outbreeding depression and with relatively few hybrids per generation.

Genomics is the study of the entire genome (chromosomes, genes and DNA) and how different genes interact with each other. Related terms are: bio-informatics, functional genomics and proteomics.

Genotype (from gene + type) refers to: (1) The genetic constitution (gene makeup) of an organism; (2) The pair of alleles at a particular locus, e.g., *Aa* or *aa*.; (3) The sum total of all pairs of alleles at all loci that contribute to the expression of a quantitative trait.

Genus (plural: genera) refers to a group of closely related species, whose perceived relationship is typically based on physical resemblance, now often supplemented with DNA sequence data.

GEO is an abbreviation for genetically engineered organism.

Geotropism (Gr. *ge*, earth + *tropos*, turning) refers to a growth curvature induced by gravity.

Germ refers to: (1) The botanical term for a plant embryo; (2) Colloquial: a disease-causing micro-organism.

Germ cell gene therapy refers to the repair or replacement of a defective gene within the gamete-forming tissues, resulting in a heritable change in an organism's genetic constitution.

Germ cell is a member of a cell lineage (the germ line) leading to the production of gametes. Synonym is germ line cell. Opposite is somatic cell.

Germ layers (primary germ layers) refer to the layers of cells in an animal embryo at the gastrula stage, from which the various organs of the animal's body will be derived.

Germ line cells are cells that produce gametes.

Germ line gene therapy refers to the delivery of a gene or genes to a fertilized egg or an early embryonic cell.

Germ line refers to a lineage of "generative" cells (= germ track) ancestral to the gametes (sperm and egg cells proper) which, during the development of an organism (animal or plant), are set aside as potential gamete-forming tissues.

Germicide refers to any chemical agent used to control or kill any pathogenic and non-pathogenic micro-organisms.

Germinal epithelium refers to: (1) A layer of epithelial cells on the surface of the ovary; (2) The layer of epithelial cells lining the seminiferous tubules of the testis.

Germination refers to: (1) The initial stages in the growth of a seed to form a seedling; (2) The growth of spores (fungal or algal) and pollen grains.

Germline cells are reproductive cells – the egg and the sperm.

Germplasm refers to: (1) The genetic material that forms the physical basis of hereditary and which is transmitted from one generation to the next by means of the germ cells; (2) An individual or clone representing a type, species or culture, that may be held in a repository for agronomic, historic or other reasons.

Gestation The period between conception (fertilization of the egg) to parturition (birth) spent *in utero* by the foetus of viviparous animals.

GFP is an abbreviation for green fluorescent protein.

GH is an abbreviation for growth hormone.

Giardia lamblia is a parasite found in spoiled food or unclean water that can cause diarrhea.

Gibberellins is a class of plant growth regulators which are active in the elongation, enhancement of flower, fruit and leaf size, germination, vernalization and other physiological processes.

Gland is a group of cells or a single cell in animals or plants that is specialized to secrete a specific substance.

Glass is an amorphous solid obtained when silica is mixed with other compounds, heated above its melting point, and then cooled rapidly.

Glass transition temperature (Tg) is the temperature at which,

upon cooling, a noncrystalline ceramic or polymer transforms from a supercooled liquid to a rigid glass.

Glaucous refers to a surface with a waxy, white coating. In most cases, this waxy covering can be rubbed off.

Glaucus is a surface with a waxy, white coating. In most cases, this waxy covering can be rubbed off.

Glial cells surround the neurons and help regulate the biochemical environment within the brain, provide structural supports for neurons and repaired the central nervous systems after injuries.

Globulins are common proteins in blood, eggs and milk, and as a reserve protein in seeds. Globulins are insoluble in water, but soluble in salt solutions. Alpha, beta and gamma globulins can be distinguished in blood serum. Gamma globulins are important in developing immunity to diseases.

Glomerulonephritis is a type of glomerular kidney disease in which the kidneys' filters become inflamed and scarred, and slowly lose their ability to remove wastes and excess fluid from the blood to make urine.

Glomerulosclerosis describes the scarring that occurs within the kidneys in the small balls of tiny blood vessels called the glomeruli. The glomeruli assist the kidneys in filtering urine from the blood.

Glomerulus is the cluster of capillaries at the proximal end of the nephron; the glomerulus is surrounded by the Bowman's capsule.

GLP (Good Laboratory Practice or GMP: Good Manufacturing Practice) refer to codes of practice designed to reduce to a minimum the chance of accidents which could affect a research project or a manufactured product.

Glucocorticoid is a steroid hormone that regulates gene expression in higher animals.

Glucose is a simple sugar made by the body from carbohydrates in food.

Glucose invertase (glucose isomerase) refers to enzymes that catalyze the interconversion of the two sugars, glucose and fructose.

Glucose invertase is an enzyme that catalyzes the hydrolysis of sucrose into its component monosaccharides, glucose and fructose.

Glucose isomerase is an enzyme that catalyzes the interconversion of the two sugars, glucose and fructose.

Glucosinolates is a class of molecules produced in the seeds and green tissue of a range of plants, in particular brassicas.

Glucuronidase is related to beta-glucuronidase.

Gluten is a mixture of two proteins, gliadin and glutenin, occurring in the endosperm of wheat grain.

Glycoalkaloids is a group of modified alkaloids, including solanine and tomatine, having a range of toxic effects in humans and other species.

Glycoform is one of several structures possible for a given glycoprotein, determined by the type and position of attachment of the component oligosaccharide(s).

Glycolysis (Gr. *glycos*, sugar (sweet) + *lysis*, dissolution) is a sequence of reactions that converts glucose into pyruvate, with the concomitant production of ATP.

Glyco nanotechnology can be seen as the synergy between nanotechnology and glycan related biological and medical problems.

Glycoprotein is a protein molecule modified by the addition of one or several oligosaccharide groups.

Glycoprotein remodeling refers to the use of restriction endoglycosidases to enzymatically remove oligosaccharide branches from glycoprotein molecules.

Glycoscience is the study of the complex carbohydrates on the surface of proteins and lipids.

Glyphosate is an active ingredient in some herbicides, killing plants by inhibiting the activity of plant enolpyruvyl-shikimate 3-phosphate synthase.

Glyphosate oxidase is an enzyme which catalyzes the break-down of glyphosate, discovered in a strain of *Pseudomonas* bacteria which were found to produce unusually large amounts of the enzyme.

Glyphosate oxidoreductase is an enzyme from the micro-organism *Ochrobactrum anthropi*, which catalyzes the breakdown of glyphosate.

GM food is an abbreviation for genetically modified food.

GMAC (Genetic Manipulation Advisory Committee) is a government expert advisory committee that provided guidance to the government and industry on the safe and responsible development and use of gene technology in Australia.

GMO is an abbreviation for genetically modified organism.

GMP is an abbreviation for (1) guanosine 5'-monophosphate.

Synonym is guanylic acid; and (2) good manufacturing practice.

Goldman equation (Goldman–Hodgkin–Katz voltage equation) is used in cell membrane physiology to determine the reversal potential across a cell's membrane, taking into account all of the ions that are permeate through that membrane.

Golden rice refers to a biotechnology-derived rice, which contains large amounts of beta carotene (a precursor of vitamin A) in its seeds.

Golgi apparatus is an assembly of vesicles and folded membranes within the cytoplasm of plant and animal cells that stores and transports secretory products (such as enzymes and hormones) and plays a role in formation of a cell wall (when this is present).

Gonad is one of the (usually paired) animal organs that produce reproductive cells (gametes).

Gonad ridge is the area of cells that will develop into the gonads of foetus, within an embryo.

Good laboratory practice (GLP) are written codes of practice designed to reduce to a minimum the chance of procedural or instrument problems, which could adversely affect a research project or other laboratory work.

Good manufacturing practice (GMP) are written odes of practice designed to reduce to a minimum the chance of procedural or instrument/manufacturing plant problems, which could adversely affect a manufactured product.

Good pasture syndrome is a rare, autoimmune disease that can affect the lungs and kidneys.

G-protein coupled receptor is also known as seven-transmembrane domain receptors – 7TM receptors, heptahelical receptors, serpentine receptor, and G protein-linked receptors (GPLR) – constitute a large protein family of receptors that sense molecules outside the cell and activate inside signal transduction pathways and, ultimately, cellular responses. They are called seven-transmembrane receptors because they pass through the cell membrane seven times. There are two principal signal transduction pathways involving the G protein-coupled receptors: the cAMP signal pathway and the phosphatidylinositol signal pathway.

G-proteins refer to the proteins with an important role in relaying signals in mammalian cells. The proteins occur on the inner surface of the plasma membrane and transmit signals from outside the membrane, via transmembrane receptors, to adenylate cyclase, which catalyzes the formation of the second messenger, cyclic AMP, inside the cell.

Graphene is pure carbon in the form of a very thin, nearly transparent sheet, one atom thick.

Graft-versus-host disease refers to the rejection of transplanted organs by the recipient's immune system, due to attack of the recipient's T lymphocytes on the transplanted organ caused by differences in major histocompatibility complex proteins.

Graft chimera is a plant which is a mosaic of two sorts of tissue differing in genetic constitution and assumed to have arisen as the result of a nuclear fusion following grafting.

Graft hybrid refers to an individual formed from graft and stock showing the characteristics of both progenitors.

Graft inoculation test refers to a test based on the use of a suspected viral carrier which is grafted to an indicator plant.

Graft refers to as a: (1) Verb. To place a detached branch or bud (scion) in close cambial contact with a rooted stem (rootstock) in such a manner that scion and rootstock unite to form a single plant; (2) Noun. Colloquial synonym for scion.

Graft union refers to the point at which a scion from one plant is joined to a rootstock from another plant.

Grafting is the process of making a graft.

Grain boundary is the interface separating two adjoining grains having different crystallographic orientations.

Grain growth is the increase in average grain size of a polycrystalline material.

Grain size is the average grain diameter as determined from a random cross section.

Gram staining is a technique to distinguish between two major bacterial groups, based on stain retention by their cell walls. Gram-positive bacteria are stained bright purple, while Gram-negative bacteria are decolorized.

Gram-negative organism refers to any prokaryotic organism that does not retain the first stain (crystal violet) used in Gram's staining technique.

Gram-positive organism refers to any prokaryotic organism that retains the first stain used in the Gram technique, which gives a purple-black color when viewed under a light microscope.

Grana (L. *granum*, a seed) refers to structures within chloroplasts, seen as green granules with the light microscope and as a series of parallel lamellae with the electron microscope.

GRAS is an abbreviation for generally regarded as safe.

Gratuitous inducer is a substance that can induce transcription of a gene or genes, but is not a substrate for the induced enzyme(s).

Gray matter is a nervous tissue in the brain and spinal cord that contains the cell bodies, dendrite, and unmyelinated axons.

Green ceramic body is a ceramic piece, formed as a particulate aggregate, that has been dried but not fired.

Green fluorescent protein (GFP) is a protein derived from a species of jelly fish, that fluoresces when exposed to ultra violet light.

Green revolution refers to a series of research, and development, and technology transfer initiatives, occurring between the 1940s and the late 1960s, that increased agriculture production worldwide.

Gro-luxä refers to a wide-spectrum fluorescent lamp suitable for artificial light for plant growth.

Gro-luxTM is a wide-spectrum fluorescent lamp suitable for plant growth purposes.

Ground is a wire that makes an electrical connection with the earth.

Growth is an increase in cell size or cell number, or both, resulting in an increase in dry weight.

Growth cabinet is a cupboard used for incubating tubes or culture vessels under controlled environmental conditions.

Growth curve is related to the growth phase.

Growth factor refers to any of several chemicals, particularly polypeptides, that have a variety of important roles in the stimulation of new cell growth and cell maintenance.

Growth hormone (GH) is a group of hormones, secreted by the mammalian pituitary gland, that stimulates protein synthesis and growth of the long bones in the legs and arms.

Growth inhibitor is any substance inhibiting the growth of an organism.

Growth phase refers to each of the characteristic periods in the growth curve of a bacterial culture, as indicated by the shape of a graph of viable cell number versus time.

Growth phase curve is a plot of growth parameter versus elapsed; and it refers to the characteristic periods in the growth of a bacterial culture, as indicated by the shape of a graph of viable cell number versus time.

Growth rate is a change in an organism's mass per unit of time.

Growth regulator is a synthetic or natural compound that at low concentrations elicits and controls growth responses in a manner similar to hormones.

Growth retardant is a chemical that selectively interferes with normal hormonal promotion of growth and other physiological processes, but without appreciable toxic effects.

Growth ring refers to one of the rings that can be seen in a cross-section of a woody stem, such as a tree trunk.

Growth substance is any organic substance, other than a nutrient, that is synthesized by plants and regulates growth and development.

GTP is an abbreviation for guanosine 5'-triphosphate, a nucleotide, which is important as a ligand for G proteins and as a direct precursor molecule for RNA synthesis.

Guanine refers to a purine derivative that is one of the major component bases of nucleotides and the nucleic acids, DNA and RNA.

Guanosine is a nucleoside consisting of one guanine molecule linked to a D-ribose sugar molecule.

Guanosine triphosphate (guanosine 5-triphosphate) is abbreviated as GTP.

Guanylic acid is synonym for guanosine monophosphate (GMP), a (ribo)nucleotide containing the nucleoside guanosine. The corresponding deoxyribonucleotide is called deoxyguanylic acid.

Guard cell is a specialized epidermal cell that occurs as a pair around a stoma and controls opening and closing of the stoma through changes in turgor.

GUI (Graphical User Interface) refers to hardware, software, and firmware that produces the display on modern personal computers.

Guide RNA is a RNA molecule that contain sequences that function as a template during RNA editing.

Guide sequence is a RNA molecule (or a part of it), which hybridizes with eukaryotic mRNA and aids in the splicing of intron sequences.

GURT is an abbreviation for genetic use restriction technology.

Gus gene refers to an *E. coli* gene that encodes for production of beta-glucuronidase (GUS).

GUS is an abbreviation for beta-glucuronidase.

Gymnosperm is a class of plant (e.g., conifers) whose ovules and the seeds into which they develop are borne unprotected, rather than enclosed in ovaries, as are those of the flowering plants.

Gynandromorph is an individual in which one part of the body is female and another part is male; a sex mosaic.

Gynodioecy refers to the occurrence of female and hermaphroditic individuals in a population of plants.

Gynogenesis refers to the female parthenogenesis. After fertilization of the ovum, the male nucleus is eliminated and the haploid individual (described as gynogenetic) so produced possesses the maternal genome only.

Gyrase is related to DNA helicase. It refers to a bacterial enzyme that catalyzes the breaking and rejoining of bonds linking adjacent nucleotides in circular DNA to generate supercoiled DNA helices.

H

· ·

h prefix is used to designate the human form of an enzyme. For example, hGH refers to human growth hormone.

H2 blockers refer to the medications used to treat gastroesophageal reflux disease (GERD) that decrease the amount of acid made by the stomach. The stomach lining has sites that react to a chemical normally found in the body called histamine. When histamine attaches to these sites, the stomach produces acid that aids in digestion of food. H2 blockers prevent the stomach from reacting to histamine, thereby decreasing stomach acid.

Habituation refers to the phenomena that, after a number of sub-cultures, cells can grow, without the addition of specific factors, such as no longer needing exogenous growth regulators in the tissue culture medium.

HAC is an abbreviation for human artificial chromosome. It is a microchromosome that can act as a new chromosome in a population of human cells.

Haematopoietic stem cells make all the blood cells in the body.

Haemoglobin refers to the conjugated protein compound containing iron, located in erythrocytes of vertebrates; important in the transportation of oxygen to the cells of the body.

Haemolymph is the mixture of blood and other fluids in the body cavity of an invertebrate.

Haemophilia is an inherited disease that is due to a deficiency or lack of certain compounds, such as factor VIII or IX, in the blood.

Hair is a single or multicellular, absorptive (root hair) or secretory (glandular hair) and sometimes only a superficial outgrowth (covering hair) of the epidermal cells.

Hair follicle are keratinized filaments that developed from an invagination of the epidermis. They are present over almost the entire body, being absent from the sides and palmar surfaces of the hands, from the sides and plantar surfaces of the feet, from the lips, and from the region around the urogenital orifices.

Hair pin loop is a region in one strand of a polynucleotide which may under appropriate conditions fold back on itself and form a limited segment of double-stranded DNA with a loop at one end.

Hairy root culture is a fairly recent development in plant culture, consisting of highly branched roots of a plant. hairs.

Hairy root disease is a disease of broad-leaved plants, where a proliferation of root-like tissue is formed from the stem.

Half life (t$_{1/2}$) is the amount of time required for a quantity to fall to half its value as measured at the beginning of the time period. While the term "half-life" can be used to describe any quantity which follows an exponential decay, it is most often used within the context of nuclear physics and nuclear chemistry.

Hall effect is a phenomenon whereby a force is applied to a moving electron or hole by a magnetic field that is applied perpendicular to the direction of motion.

Halophyte is a plant that can tolerate a high concentration of salt in the growing medium.

Halothane is a volatile anaesthetic. Its IUPAC name is 2-bromo-2-chloro-1,1,1-trifluoroethane. It is the only inhalational anesthetic agent containing a bromine atom.

Hanging droplet technique is a method in which a drop of bacterial suspension, preferably in mid-logarithmic phase, is enclosed in an air-tight chamber prepared in a special depression slide (having a concave depression in the center) or assembled from modeling clay (plasticine), which is a soft, malleable, and nonhardening material available at toy or hobby stores.

Haploid (Gr. *haploos*, single + *oides*, like) is a cell or organism containing only one representative from each of the pairs of homologous chromosomes found in the normal diploid cell.

Haploid cell is a cell containing only one set, or half the usual (diploid) number, of chromosomes.

Haplotype (haploid genotype) is a set of closely linked genetic markers present on one chromosome that tend to be inherited together. Haplotype may also refer to a set of single nucleotide polymorphisms (SNPs) on a single chromatid that are statistically associated with one another.

Haplotype characterization is the characterization of SNPs by coherent packages (SNPs that are usually transmitted together).

Haplozygous *is synonym* of hemizygous. It refers to the condition in which genes are present only once in the genotype and not in pairs.

Hapten is a small molecule, which by itself is not an antigen, but which as a part of a larger structure when linked to a carrier protein.

Haptoglobin is a serum protein that interacts with haemoglobin during recycling of the iron molecule of haemoglobin. Synonym is alpha globulin.

Hardening off refers to adapting plants to outdoor conditions by gradually withholding water, lowering the temperature, increasing light intensity, or reducing the nutrient supply.

Hardness is a measure of a material's resistance to deformation by surface indentation or by abrasion. There are various scales in use to express hardness. The Mohs scale is qualitative and somewhat arbitrary and ranges from 1 on the soft end for talc to 10 for diamond. Quantitative scales are the Rockwell (HR),

Hardy-Weinberg equilibrium states that allele and genotype frequencies in a population will remain constant from generation to generation in the absence of other evolutionary influences.

HardyWeinberg principle is the principle that allele and genotype frequencies will reach equilibrium, defined by the binomial distribution, in one generation and remain constant in large random mating populations that experience no migration, selection, mutation, or nonrandom mating.

HardyWeinberg proportion is a state in which a population's genotypic proportions equal those expected with the binomial distribution.

Harvesting refers to the collection of cells from cell cultures or of organs from donors for the purpose of transplantation.

Hazardous means dangerous.

He$_{2+}$ or 4-2He$_{2+}$ indicates a Helium ion with a +2 charge (missing its two electrons). If the ion gains electrons from its environment, the alpha particle can be written as a normal (electrically neutral) Helium atom 4-2He.

Health product encompasses products subject to the Food and Drugs Act, and are managed along the following broad categories: Biologics; Pharmaceuticals; Medical devices; and Natural Health Products.

Health surveillance is the ongoing, systematic use of routinely collected health data to guide public health action in a timely fashion.

Heart assist device is a mechanical device that is surgically implanted to ease the workload of the heart.

Heart attack refers to a death of, or damage to, part of the heart muscle due to an insufficient blood supply.

Heart block refers to a delay or interference of the conduction mechanism whereby impulses do not go through all or a major part of the myocardium.

Heart beat refers to one complete contraction of the heart.

Heart burn is a burning feeling in the chest or area just above the stomach, caused by acid moving up the esophagus from the stomach.

Heat capacity (*Cv* at constant volume and *Cp* at constant pressure) is the quantity of heat required to produce a unit temperature rise per mole of material.

Heat cramps refer to painful and often incapacitating cramps in muscles.

Heat exchanger is used to alter temperature of blood by principle of conduction.

Heat exhaustion refers to weakness, lassitude, dizziness, visual disturbance, feeling of intense thirst and heat, nausea, vomiting, palpitations, tingling and numbness of extremities after exposure to a hot environment.

Heat hyper pyrexia refers to the rise in body temperature with moist skin and mental dysfunction, caused by exposure to an extremely hot environment.

Heat shock protein (HSP) is a class of protein chaperones which are typically overexpressed as a response to heat stress. Two such proteins – HSP 90 and HSP 70 – have a role in ensuring that crucial proteins are folded into the correct conformation. Synonym is stress protein.

Heat strain refers to the physiological and behavioral response of the body as a result of heat exposure.

Heat stroke is an acute illness caused by overexposure to heat.

Heat therapy is synonym for thermotherapy.

Heat Transfer is a process by which energy in the form of heat is exchanged between bodies or parts of the same body at different temperatures.

Helicobacter pylori (*H. pylori*) is a bacteria found in the stomach that can damage the lining of the stomach and upper small intestine, leading to ulcer formation.

Helix is any structure with a spiral shape. The Watson and Crick model of DNA is in the form of a double helix.

Helminth is a class of parasitic worms, especially those which are internal parasites of man and animals.

Helper cells refers to T cells that respond to an antigen displayed by a macrophage by stimulating B and T lymphocytes to develop into antibody-producing plasma cells and killer T cells, respectively.

Helper plasmid is a plasmid that provides a function or functions to another plasmid in the same cell.

Helper virus is a virus that provides a function or functions to another virus in the same cell.

Hematocrit refers to the amount of blood that is occupied by red blood cells.

Hematomas result when small blood vessels are broken by the injury.

Hematopoietic cell transplantation is the transplantation of hematopoietic stem cells with blood-forming potential.

Hematopoietic stem cells (HSCs) are stem cells that give rise to all blood cell types. They may be multipotent, oligopotent or unipotent and lead to the generation of erythrocytes (red blood cells), macrophages (the "clean up crew" of the immune system), neutrophils, basophils, eosinophils and dendritic cells (immunological "helper" cells), megakaryocytes and platelets (involved in blood clotting), as well as the immune effector T-cells, B-cells, Natural Killer cells.

Hematuria is the presence of red blood cells (RBCs) in the urine.

Heme is an iron complex.

Hemicellulose (Gr. *hemi*, half + cellulose) is any cellulose-like carbohydrate, but with differing chemical composition.

Hemizygous denotes the presence of only one copy of an allele due a locus being in a haploid genome, on a sex chromosome, or only one copy of the locus being present in an aneuploid organism.

Hemodilution is an increase in the volume of blood plasma resulting in reduced concentration of red blood cells.

Hemodynamics explains how the dimensions and configuration of the vasculature, combined with the fluidity of the blood, determine the way blood is distributed to and within organs.

Hemoglobin brightens in color when saturated with oxygen (oxyhemoglobin) and darkens when oxygen is removed (deoxyhemogloblin). It is a biomolecule composed of four myoglobine-like units (proteins plus heme) that can bind and transport four oxygen molecules in the blood. It is also written as haemoglobin.

Hemolymph is also written as haemolymph.

Hemolysis refers to the damage to blood cells

Hemolytic uremic syndrome is a rare kidney disorder that mostly affects children under the age of 10. It is often characterized by damage to the lining of blood vessel

Hemophilia is an x-linked recessive genetic disease, caused by a mutation in the gene for clotting factor VIII (hemophilia-A) or clotting factor IX (hemophilia-B), which leads to abnormal blood clotting.

Hemostasis is a cessation of bleeding; Arrest of bleeding or of circulation. When equilibrium within the body is maintained, homeostasis is said to occur. The human body maintains a steady internal environment for the proper functioning of the body. Maintaining a constant internal environment requires the body to make many adjustments. Adjustments within the body are referred to as regulation of homeostasis. Homeostatic regulation is comprised of three parts: a receptor, a control center and an effector.

Hemotrophic nutrition is a term used to describe in late placenta development the transfer of blood-borne nutrition from maternal to embryo/fetus compared to early histiotrophic nutrition.

HEMT stands for high electron mobility transistor.

Henry (H) is a unit of inductance. One henry (H) is the inductance of a closed circuit in which an electromotive force of 1 volt is produced when the electric current in the circuit varies uniformly at the rate of 1 ampere per second.

Henry's law describes he amount of gas dissolved in a solution is directly proportional to the pressure of the gas above the solution.

HEPA filter is an abbreviation for high efficiency particulate air filter. It is a type of air filter. Filters meeting the HEPA standard have many applications, including use in medical facilities, automobiles, aircraft and homes.

Heparin is an acid occurring in tissues, mostly in the liver.

Hepatic pertains to the liver.

Hepatitis A is a form of infectious hepatitis caused by the hepatitis A virus.

Hepatitis B is a form of infectious hepatitis caused by the hepatitis B virus.

Hepatitis C is a form of infectious hepatitis caused by the hepatitis C virus.

Hepatitis D is a form of infectious hepatitis caused by the hepatitis (Delta) virus.

Hepatitis E is a form of infectious hepatitis caused by the hepatitis E virus.

Hepatitis G is the newest form of infectious hepatitis.

Hepatitis is an inflammation of the liver that sometimes causes permanent damage; caused by viruses, drugs, alcohol, or parasites.

Hepatocyte is any of the polygonal epithelial parenchymatous cells of the liver that secrete bile called also *hepatic cell, liver cell.*

HER2 is a member of the epidermal growth factor receptor (EGFR/ERBB) family.

Herbicide is a substance that is toxic to plants; the active ingredient in agrochemicals intended to kill specific unwanted plants, especially weeds.

Herbicide resistance is the ability of a plant to remain unaffected by the application of a herbicide.

Herceptin (Trastuzumab, Herclon) is a monoclonal antibody that interferes with the HER2/neu receptor. Its main use is to treat certain breast cancers.

Hereditary disorder is a pathological condition due to changes in individual genes, or groups of genes or in sections of chromosomes or whole chromosomes.

Heredity is the transfer of genetic information from parents to children.

Heritability is the proportion of total phenotypic variation within a population that is due to individual genetic variation (*H*B; broad sense heritability). Heritability is more commonly referred to as the proportion of phenotypic variation within a population that is due to additive genetic variation (*H*N; narrow sense heritability).

Hermaphrodite refers to: (1) An animal that has both male and female reproductive organs, or a mixture of male and female attributes; (2) A plant whose flowers contain both **stamen** and carpels. Synonym is intersex.

Hernia is a section of intestine or other internal organ that pushes through an opening in an abdominal muscle.

Herniorrhaphy refers to an operation that is done to repair a hernia.

Hertz is a unit of frequency and is equal to one cycle per second.

Heteroallele is a gene having mutations at two or more different sites.

Heteroalleles are mutations that are functionally allelic but structurally non-allelic; mutations at different sites in a gene.

Heterochromatin refers to the regions of chromosomes that remain contracted during interphase and therefore stain more intensely in cytological preparations. Opposite is euchromatin.

Heteroduplex analysis is the use of the electrophoretic mobility of heteroduplex DNA to estimate the degree of non-homology between the sequences of the two strands.

Heteroduplex is a double-stranded DNA molecule or DNA-RNA hybrid, where each strand is of a different origin, and consequently containing one or more mismatched (non-complementary) base pairs.

Heterogametic refers to producing unlike gametes with regard to the sex chromosomes. In mammals, the XY male is heterogametic, and the XX female is homogametic.

Heterogametic sex refers to producing unlike gametes with regard to the sex chromosomes.

Heterogeneity is the state of being heterogeneous. It is the nature of opposition, or contrariety of qualities.

Heterogeneous nuclear RNA (hnRNA) refers to large RNA molecules, which are found in the nucleus of a eukaryotic cell and the precursors of mRNA and other RNA molecules.

Heterokaryon refers to the cell with two or more different nuclei as a result of cell fusion. Opposite is homokaryon.

Heterologous is a mixed or divergent cell population or of a divergent origin.

Heterologous probe is a DNA probe that is derived from one species and used to screen for a similar DNA sequence from another species.

Heterologous protein (Foreign protein) is a protein that differs from any protein normally found in the organism in question.

Heteroplasmy refers to a condition in which two genetically different organelles are present in the same cell. Opposite is homoplasmy.

Heteroploid refers to a term given to a cell culture when the cells comprising the culture possess nuclei containing chromosome numbers other than the diploid number.

Heteropyknosis (adj.: heteropyknotic) is the property of certain chromosomes, or of their parts, to remain more dense during the cell cycle and to stain more intensely than other chromosomes or parts.

Heterosis (Gr. *heteros*, different + *osis*, suffix for "a state of") refers to a case when hybrid progeny have higher fitness than either of the parental organisms. Also called hybrid vigor.

Heterotroph is an organism non capable of self-nourishment utilizing carbon dioxide or carbonates as the sole source of carbon and obtaining energy from radiant energy or from the oxidation of inorganic elements, or compounds such as iron, sulphur, hydrogen, ammonium and nitrites. Opposite is autotroph.

Heterozygosity is a measure of genetic variation that accounts for either the observed, or expected proportion of individuals in a population that are heterozygotes.

Heterozygote (adj.: heterozygous; Gr. *heteros*, different + *zygon*, yoke) refers to an individual that has different alleles at the same locus in its two homologous chromosomes.

Heterozygous advantage is a situation where heterozygous genotypes are more fit than homozygous genotypes.

Heterozygous disadvantage is a situation, where heterozygous genotypes are less fit than homozygous genotypes. Also called underdominance.

Heterozygous refers to two different forms of a particular gene that are nor same, one inherited from each parent.

Hetrotrophic refers to xenogeneic transplantation

HFC is heterozygosity fitness correlation.

Hfr refers to high-frequency recombination strain of *Escherichia coli*; in these strains, the F factor (plasmid) is integrated into the bacterial chromosome.

hGH is an abbreviation for human growth hormone.

Hidden Markov Models searches a protein sequence database for homologues and is a powerful tool for discovering the structure and function of a sequence. Amongst the algorithms and tools available for this task, Hidden Markov model (HMM) improves dependent scores to characterize and build a model for an entire family of sequences. HMMs have been used to analyze proteins using two complementary strategiesa: Pfam (to find which of the families it matches); for the second approach, HMM for a family is used to search a primary sequence database to identify additional members of the family.

High efficiency particulate air filter is abbreviated as HEPA filter.

High frequency recombination refers to strain of *Escherichia coli*.

High-performance liquid chromatography (HPLC; formerly referred to as high-pressure liquid chromatography) is a technique in analytic chemistry used to separate the components in a mixture, to identify each component, and to quantify each component. It relies on pumps to pass a pressurized liquid solvent containing the sample mixture through a column filled with a solid adsorbent material. HPLC has been used for medical, legal, research, and manufacturing purposes.

High throughput screening is technology that employs automation and robotics to conduct hundreds or thousands of biological assay experiments within a short period of time.

High voltage current refers to a current in which the wave form has an amplitude of greater than 150 volts with a relatively short pulse duration of less than 100 sec.

Hill Robertson effect is an effect, where selection at one locus will reduce the effective population size of linked loci; increasing the chance of genetic drift forming negative genetic associations.

Hirschsprung's disease is caused by malformation of a baby's large intestine during pregnancy.

Histiotrophic nutrition is a term used to describe in early placenta development the initial transfer of nutrition from maternal to embryo (histiotrophic nutrition) compared to later blood-borne nutrition (hemotrophic nutrition). Histotroph is the nutritional material accumulated in spaces between the maternal and fetal tissues, derived from the maternal endometrium and the uterine glands.

Histocompatible is a tissue or organ from a donor that will not be rejected by the recipient. Rejection is caused because the immune system of the recipient that sees the transplanted organ or tissue as foreign and tries to destroy it.

Histocompatibility refers to the degree to which tissue from one organism is tolerated by the immune system of another organism.

Histocompatibility complex (histocompatibility system) refers to the collection of genes coding for peptides present on the surface of nucleated cells.

Histoglobulin refers to peptides present on the surface of nucleate cells (responsible for the differences between genetically non-identical individuals) that cause rejection of tissue grafts between such individuals.

Histology (compound of the Greek words: ἱστός *histos* "tissue," and -λογία *-logia* "science") is the study of the microscopic anatomy of cells and tissues of plants and animals. The trained scientists, who perform the preparation of histological sections, are *histotechnicians,* histology technicians (HT), histology technologists (HTL), medical scientists, medical laboratory technicians, or biomedical scientists/engineers. Their field of study is called *histotechnology.*

Histone is a group of water-soluble proteins rich in basic amino acids, closely associated with DNA in plant and animal chromatin.

Hitchhiking is the increase in frequency of a selectively neutral allele through gametic disequilibrium with a beneficial allele that selection increases in frequency in a population.

HIV is an abbreviation for human immunodeficiency virus. It is a retrovirus that causes AIDS in humans.

HLA is an abbreviation for human-leukocyte-antigen system.

hnRNA is an abbreviation for heterogeneous nuclear RNA.

Hofbauer cells are found within placental villi connective tissue.

Hogness box is synonym for TATA box.

Hollow fiber is a tube of a porous material, having an internal diameter of a fraction of a millimeter, and so its ratio of surface area to internal volume is very large.

Holoenzyme is an active, complex enzyme consisting of an apoenzyme and a coenzyme.

Holometabolous is an insect that undergoes complete metamorphosis to the adult from a morphologically distinct larval stage.

Homeobox is a highly conserved 180 bp DNA sequence that controls body part-, organ- or tissue- specific gene expression, most particularly involved in segmentation in animals, but also in a variety of other eukaryotes.

Homeodomain is a domain in a protein that is encoded for by a homeobox.

Homeotic genes refers to genes that act in concert to determine fundamental patterns of development.

Homeotic mutation is a mutation that causes a body part to develop in an inappropriate position in an organism, such as the mutation in *Drosophila melanogaster*.

Homoallele is one of the number of otherwise identical alleles, which differ at the same site in their sequence.

Homoalleles are inherited as strict alternatives; but heteroalleles could through recombination create a genotype containing a 'double' variant.

Homodimer refers to a protein comprising two identical polypeptide chains, or a dimer of identical residues.

Homoduplex DNA refers to a double-stranded fully complementary DNA molecule.

Homoeologous refers to chromosomes which are descended from a common progenitor, but which have evolved to be no longer fully homologous. Homoeologous chromosomes have similar gene content to one another, but are structurally altered in subtle ways to inhibit, and sometimes completely prevent their pairing with one another at meiosis.

Homogametic refers to producing similar gametes with regard to the sex chromosomes. In mammals, the female is homogametic (XX), and the male is heterogametic (XY).

Homogametic sex refers to producing similar gametes with regard to the sex chromosomes.

Homogeneous assay does not require a separation step to remove free antigen from bound antigen and relies upon the fact that the function of the label is modified upon binding, leading to a change in signal intensity.

Homogenotization is a genetic technique used to replace one copy of a gene, or other DNA sequence within a genome, with an altered copy of that sequence.

Homokaryon refers to a cell with two or more identical nuclei as a result of fusion. Opposite is heterokaryon.

Homologous is similar or uniform, often used in the context of genes and DNA sequences. In the context of stem cells, the term homologous recombination is a technique used to disable a gene in embryonic stem cells.

Homologous chromosomes are chromosomes that occur in pairs and are generally similar in size and shape: one comes from the male parent and the other from the female.

Homologous recombination is a technique used to inactivate a gene and determine its function in a living animal. The process of homologous recombination is more efficient in embryonic stem cells than in other cell types.

Homology is the degree of identity between individuals, or characters. Although sequence determination is the ultimate test of homology, yet useful estimates can be provided by either DNA-DNA or DNA-RNA hybridization.

Homomultimer refers to a protein consisting of a number of identical subunits.

Homoplasmy refers to the condition in which all copies of an organelle in a cell are genetically identical. Opposite is heteroplasmy.

Homoplasy is an independent evolution or origin of similar traits, or gene sequences.

Homopolymer refers to a polymer (nucleic acid, polypeptide, etc.), which contains only one kind of residue (e.g., the polynucleotide GGGGGGGGG...).

Homopolymeric tailing is a technique where a nucleic acid homopolymer (a long chain of the same nucleotide over and over) is attached to the end of a piece of DNA as a part of DNA cloning.

Homozygosity is a measure of the proportion of individuals in a population that are homozygous. Opposite is heterozygosity.

Homozygote (adj.: homozygous; Gr. *homos*, one and the same + *zygon*, yoke) is an individual that has two copies of the same allele at a particular locus in its two homologous chromosomes.

Homozygous refers to two forms of a particular gene that are the same, one inherited from each parent.

Hormone (Gr. *hormaein*, to excite) is a specific organic product, produced in one part of a plant or animal body, and transported to another part, where it promotes, inhibits or modifies a biological process.

Host genomics refers to the genetic makeup of a person (host or patient).

Host is an animal or plant on which, or in which, a parasite lives.

Host specific toxin is a metabolite produced by a pathogen, which has a host specificity equivalent to that of the pathogen.

Hot spot (molecular biology) is a particular area of DNA, which is especially prone to spontaneous mutations or recombinations.

House keeping genes are those genes that are expressed in all cells because they provide functions needed for sustenance of all cell types.

HSP (heat shock protein) is a group of proteins induced by heat shock. The most prominent members of this group are a class of functionally related proteins involved in the folding and unfolding of other proteins.

HTGS (high throughput genome sequence) is a term to distinguish all genomic sequences generated in a high-throughput manner. In order to release data more rapidly, it became standard for all sequence centers to submit unfinished sequence into public repositories. Keywords associated with HTGS are: HTGS_phase 0; HTGS_phase 1; HTGS_phase 2; HTGS_phase 3; HTGS_draft; HTGS_full-top; HTGS_activefin; HTGS_cancelled.

http (Hypertext transfer protocol) is a transfer protocol used on the WWW.

HUGO is a Human Genome Organization.

Human artificial chromosome (HAC) is analogous to yeast artificial chromosome, which would allow for the cloning of very large fragments of DNA, and their transfer into human cells for the purpose of gene therapy.

Human clone is defined as "an embryo that, as a result of the manipulation of human reproductive material or an *in vitro* embryo, contains a diploid set of chromosomes obtained from a single – living or deceased – human being, foetus or embryo."

Human Computer Interaction (HCI) is the study of how humans interact with computers, and how to design computer systems that are usable, easy, quick, and productive for humans to use. HCI is multidisciplinary and is based on system design/software engineering, psychology, cognitive science, ergonomics, sociology, engineering, business, graphic design, and technical writing.

Human embryonic stem cell (hESC) is a stem cell that is derived from the inner cell mass of a blastocyst and can differentiate into several tissue types in a dish. Human embryonic stem cells are harder to grow than mouse embryonic stem cells.

Human genetics is the study of how traits are passed on in families and how genes are involved in health and disease.

Human Genome Project is coordinated by the National Institutes of Health (NIH) and the Department of Energy (DOE) to determine the entire nucleotide sequence of the human chromosomes.

Human growth hormone (hGH, somatotrophin) is a protein produced in the pituitary gland that stimulates the liver to produce somatomedins, which stimulate growth of bone and muscle.

Human health is a state of complete physical, mental and social well-being and not merely the absence of disease or infirmity.

Human serum albumin are soluble blood proteins that make up about 55% of plasma proteins.

Human leukocyte antigen system is abbreviated as HLA.

Human umbilical vein endothelial cells (HUVEC) are cells derived from the endothelium of veins from the umbilical cord. They are used as a laboratory model system for the study of the function and pathology of endothelial cells.

Humoral immune response (Humoral immunity) is also called the antibody-mediated immune system. It refers to immunity that is mediated by macromolecules, as opposed to cell-mediated immunity, found in extracellular fluids such as secreted antibodies, complement proteins and certain antimicrobial peptides. Humoral immunity is so named because it involves substances found in the humours, or body fluids.

Huntington disease is an inherited disease due to a defective gene on the short arm of chromosome (4) It results in loss of motor control and mental deterioration.

Hup$^+$ is an abbreviation for hydrogen-uptake positive.

HW is an abbreviation for Hardy Weinberg.

Hybrid arrested translation is a method used to identify the proteins encoded by a cloned DNA sequence.

Hybrid cell is related to synkaryon that refers to the nucleus of a fertilized egg immediately after the male and female nuclei have fused.

Hybrid dysgenesis is a syndrome of abnormal germ-line traits, including mutation, chromosome breakage, and sterility, which results from activity of transposable elements.

Hybrid refers to a mixed origin or composition. A hybrid is the offspring of genetically dissimilar parents.

Hybrid released translation is a method used to identify the gene product of a cloned gene.

Hybrid seed refers to: (1) Seed produced by crossing genetically dissimilar parents; (2) In plant breeding, it is used for seed produced by specific crosses of selected pure lines.

Hybrid selection is the process of choosing individuals possessing desired characteristics from among a hybrid population.

Hybrid sink is the situation where immigration of locally unfit genotypes produces hybrids with low fitness that reduces local density and thereby increases the immigration rate.

Hybrid swarm is a population of individuals that are all hybrids by varying numbers of generations of backcrossing with parental types and matings among hybrids.

Hybrid vigor refers to the extent to which a hybrid individual outperforms both its parents with respect to one or many traits. Synonym is heterosis.

Hybrid zone is an area of sympatry between two genetically distinct populations where hybridization occurs without forming a hybrid swarm in either parental population beyond the area of co-occurrence.

Hybridization refers to: (1) Interbreeding of species, races, varieties and so on, among plants or animals; (2) The production of offspring of genetically different parents, normally from sexual reproduction, but also asexually by the fusion of protoplasts or by transformation; (3) The pairing of two polynucleotide strands, often from different sources, by hydrogen bonding between complementary nucleotides.

Hybridoma is a hybrid cell, derived from a B (antibody producing) lymphocyte fused to a tumor cell, which grows indefinitely in tissue culture and is selected for the secretion of the specific antibody produced by that B cell.

Hydatiform mole is a uterine tumor with "grape-like" placenta appearance without enclosed embryo formation, arises mainly from a haploid sperm fertilizing an egg without a female pronucleus.

Hydrate is a compound formed by the incorporation of water.

Hydraulics is a branch of science and engineering concerned with the use of liquids to perform mechanical tasks.

Hydrocephalus is an abnormal accumulation of body fluids within the skull.

Hydrochloric acid is an acid made by the stomach that breaks down proteins in the foods we eat.

Hydrocollator is a synthetic hot (170°F) or cold (0°F) gel that is used as an adjunctive modality to stimulate tissue temperature rise or tissue temperature lowering.

Hydrocortisone is an antinflamatory steroid.

Hydrodynamics is the fluid dynamics applied to liquids, such as water, alcohol, and oil.

Hydrogel is a network of polymer chains that are hydrophilic, sometimes found as a colloidal gel in which water is the dispersion medium.

Hydrogen bond is a relatively weak bond formed between a hydrogen atom (which is covalently bound to a nitrogen or oxygen atom) and a nitrogen or oxygen atom with an unshared electron pair.

Hydrogen breath test is a test that measures the amount of hydrogen in the breath, and helps diagnose lactose intolerance. If the body is unable to digest lactose properly, it will make excess amounts of hydrogen.

Hydrogen-uptake positive (Hup⁺) describes a micro-organism that is capable of assimilating (or taking up) hydrogen gas.

Hydrogymnastics refers to the exercises using the buoyant properties of immersion in water.

Hydrolysis is a chemical reaction in which water is added across a covalent bond, often cleaving the molecule into two.

Hydronephrosis is a condition that occurs as a result of urine accumulation in the upper urinary tract.

Hydrophilic literally means 'water-loving.' This can describe a molecule or part of a molecule that has an affinity for water, or a substance that readily absorbs or dissolves in water.

Hydrophilic molecule is one that has a tendency to interact with or be dissolved by water and other polar substances.

Hydrophobic literally means 'water hating.' This describes a molecule or part of a molecule that prefers to be in an environment where there is no water. It means repelling, tending not to combine with, or incapable of dissolving in water.

Hydrophobic interaction refers to an interaction between a hydrophobic ('water-hating') part of a molecule and an aqueous environment.

Hydrophobic molecules tend to be non-polar and, thus,

prefer other neutral molecules and non-polar solvents. Hydrophobic molecules in water often cluster together, forming micelles. Water on hydrophobic surfaces will exhibit a high contact angle.

Hydroponics refers to the growing of plants without soil.

Hydrostatic pressure is the pressure exerted by a liquid at rest.

Hydrotherapy refers to cryotherapy and thermotherapy techniques that use water as the medium for heat transfer.

Hygromycin refers to an antibiotic used as selective agent in bacterial and transgenic plant cell cultures.

Hyperactive describes a situation in which a body tissue is especially likely to have an exaggerated reaction to a particular situation.

Hyperbilirubinemia refers to too much bilirubin in the bloodstream, due to liver problems.

Hyperkalemia refers to the levels of plasma K above 8 mEq/L.

Hyperploid is a genetic condition in which a chromosome or a segment of a chromosome is over-represented in the genotype. Opposite is hypoploid.

Hypersensitive response refers to: (1) A specific reaction of a plant to attack by a pathogen; (2) The abnormal response of an animal to the presence of a particular antigen.

Hypersensitive site are regions in the DNA that are highly susceptible to digestion with restriction endonucleases.

Hypertension is a condition of a development of the elevated blood pressure when the body's blood vessels narrow, causing the heart to pump harder than normal to push blood through the narrowed openings.

Hyperthermia refers to exceptionally high fever especially when induced artificially for therapeutic purposes. It is a condition of elevated body temperature due to failed thermoregulation that occurs when a body produces or absorbs more heat than it dissipates.

Hypertonic is a solution with an osmotic potential greater than that of living cells. Treatment with such solutions leads to the plasmolysis of cells. Opposite is hypotonic.

Hypervariable region refers to the parts of both the heavy and light chains of an antibody molecule that enable it to bind to a specific site on an antigen.

Hypervariable segment is a region of a protein that varies considerably between strains or individuals.

Hyperventilation is an excessive rate and depth of respiration leading to abnormal loss of carbon dioxide from the blood.

Hypochlorite refers to the aqueous solutions of sodium hypochlorite, potassium hypochlorite or calcium hypochlorite, which are oxidizing agents and used for disinfecting surfaces and surface-sterilizing tissues, and for bleaching.

Hypocotyl (Gr. *hypo*, under + *kotyledon*, a cup-shaped hollow) is a portion of an embryo or seedling below the cotyledons, which is a transitional area between stem and root.

Hypodermis refers to layer of connective tissue, looser than that of the dermis.

Hypomorph is a mutation that reduces but does not completely abolish gene expression.

Hypoplastic is an effective and reduced growth or development (e.g., dwarfing and stunting in plants) resulting from an abnormal condition, for example disease or nutritional stress.

Hypoploid is a genetic condition in which a chromosome or segment of a chromosome is underrepresented in the genotype. Opposite is hyperploid.

Hypotension refers to a condition with abnormally low blood pressure.

Hypothalamic peptides are generated in the vertebrate forebrain and concerned with regulating the body's physiological state.

Hypothalamus is a part of the brain that regulates the pituitary gland, the autonomic system, emotional responses, body temperature, water balance, and appetite; located below the thalamus.

Hypothermia refers to a body temperature below normal.

Hypothesis (Gr. *hypothesis*, foundation) refers to something provisionally adopted to explain certain facts and to guide in the investigation of other facts. Once proven by rigorous scientific investigation, it becomes a theory or a law.

Hypotonic refers to when the osmotic potential is less than that of living cells. Cells placed in a hypotonic solution will absorb water and display swelling and turgidity. Opposite is hypertonic.

Hysteresis is a difference in the output when a specific input value is approached first with an increasing and then with a decreasing input. This phenomenon occurs in ferroelectric materials and results in irreversible loss of energy through heat dissipation.

I

I/E region is an abbreviation for integration-excision region.

I/O (Input/output) is a transfer of information between computer and peripherals such as keyboard or printer.

IC refers to an integrated circuit.

Ichthyosis is any of several hereditary or congenital skin conditions; skin of affected individuals has a dry, scaly appearance.

ICSI is an abbreviation for intracytoplasmic sperm injection.

Ideal (or Perfect) fluid is an inviscid, incompressible fluid.

Ideal Gas Law is for a perfect or ideal gas. It states that the change in density is directly related to the change in temperature and pressure.

Identical by descent are alleles that are identical copies of the same allele from a common ancestor.

Identical twin is related to monozygotic twin.

Idiogram is related to karyogram. An ideogram or ideograph (from Greek ιδέα idéa "idea" + γράφω gráphō "to write") is a graphic symbol that represents an idea or concept.

Idiotype refers to an identifying property or characteristic of an item or system.

IgA/Igd/IgG/IgE/IgM are referred to the antibody class.

IGS ia an abbreviation for internal guide sequence.

Ileum refers to the lower end of the small intestine.

Illuminate refers to brighten with light. Illumination is an absolute requirement for tissue cultures. Fluorescent lights are commonly employed.

Imaginal disc is a mass of cells in the larvae of *Drosophila melanogaster* and other holometabolous insects that gives rise to particular adult organs, such as antennae, eyes or wings.

Imbibition (L. *imbibere*, to drink) refers to: (1) The absorption of liquids or vapors into the ultramicroscopic spaces or pores found in materials; (2) The initial water uptake by seeds starting germination.

Immediate early gene is a viral gene that is expressed promptly after infection.

Immobilized cells are cells that are entrapped in matrices such as alginate, polyacrylamide and agarose designed for use in membrane and filter bioreactors.

Immortalization refers to the genetic transformation of a cell type into a cell line, which can proliferate indefinitely.

Immortalizing oncogene refers to a gene that upon transfection enables a primary cell to grow indefinitely in culture.

Immune response is the reaction of the body to substances that are foreign or treated as foreign. The response is in a variety of forms, from the recognition of antigens in the body, the production of antibodies against the foreign substance and the response of lymphocytes.

Immune system is a network of molecules, cells and organs that work together to protect the body against infection and disease.

Immunity is the lack of susceptibility of an animal or plant to infection by a particular pathogen, or to the harmful effects of their toxins.

Immunization refers to the production of immunity in an individual by artificial means. Active immunization involves the introduction, either orally or by infection, of specially treated bacteria, viruses or their toxins so as

to stimulate the production of antibodies.

Immuno refers to therapies and/ or treatments that stimulate the immune system.

Immunoaffinity chromatography is a purification technique in which an antibody is bound to a matrix and is subsequently used to bind and separate a protein from a complex mixture.

Immunoassay (immunodiagnostics) refers to an assay system which detects proteins by using an antibody specific to that protein.

Immunochemical control refers to the use of immune agents to combat infections.

Immunocontraception is a method (not currently used) of reducing fertility of a pest species by controlling or preventing conception and pregnancy.

Immunodeficiency refers to an innate, acquired, or induced inability to develop a normal immune response.

Immunodiagnostics is related to immunoassay.

Immunogen is related to antigen.

Immunogenicity is the ability to elicit an immune response.

Immunoglobulin refers to a group of proteins (globulins) in the body that act as antibodies.

Immunoprophylaxis refers to the process of active or passive immunization. Related to: adoptive immunization, passive immunity.

Immunosensor is a biosensor having an antibody as biological part.

Immunosuppression is the suppression of the immune response.

Immunosuppressor is a substance, an agent or a condition that prevents or diminishes the immune response.

Immunotherapy refers to the use of an antibody or a fusion protein containing the antigen binding site of antibody to cure a disease or enhance the well-being of a patient. Synonym is immunochemical control.

Immunotoxicity is the toxicity of a therapeutic agent because it could cause immune reactions or allergy.

Immunotoxin refers to protein drugs consisting of an antibody joined to a toxin molecule.

Immunotyping is the process of screening patients specimens to identify the specific viral antigen on antigen presenting cells or detecting specific viral antibodies.

Impedance is a complex ratio of a force-like quantity (force, pressure, voltage, temperature, or electric field) to a corresponding related velocity-like quantity (velocity, volume velocity, current, heat flow, or magnetic field strength).

Impeller is an agitator that is used for mixing the contents of a bioreactor.

Imprinting is a suppression or silencing of genes depending on which parent they were received from.

in silico **biology** refers to advances in genomics and proteomics have greatly improved our knowledge of the components of biological systems at the molecular level. The next logical step is to try to understand how these components interact well enough to model those biological systems *in silico*.

in silico means literally "in the computer." It can be used to screen out compounds which are not druggable.

in silico **modeling** is a modeling of biological pathways and other biological processes for drug discovery and development.

in silico **proteomics** is a prediction of protein structure and function.

in situ (Latin for "in place") refers to the natural place or in the original place. (1) Experimental treatments performed on cells or tissue rather than on extracts from them. (2) Assays or manipulations performed with intact tissues.

in situ **colony** (*in situ* plaque hybridization) is a procedure for screening colonies or plaques growing on plates or membranes for the presence of specific DNA sequences by the hybridization of nucleic acid probes to the DNA molecules present in these colonies or plaques.

In situ **colony hybridization** is a procedure for screening bacterial colonies or plaques growing on plates or membranes for the presence of specific DNA sequences by the hybridization of nucleic acid probes to the DNA molecules present in these colonies or plaques. Synonym is in situ plaque hybridization.

in situ **conservation of farm animal genetic diversity** refers to all measures to maintain live animal breeding populations, including those involved in active breeding programs in the agro-ecosystem where they either developed or are now normally found, together with husbandry activities that are undertaken to ensure the continued contribution of these resources to sustainable food and agricultural production, now and in the future.

in situ **conservation** is a method that attempts to preserve the integrity of genetic resources by conserving them within the evolutionary dynamic ecosystems of their original habitat or natural environment.

In situ **plaque hybridization** is related to in situ colony hybridization.

In vitro (L. for "in glass") refers to living in test tubes, outside the organism or in an artificial environment, typically in glass vessels in which cultured cells, tissues, organs or whole plants may reside.

In vitro **embryo production** (IVEP) refers to the combination of ovum pickup, *in vitro* maturation of ova, and *in vitro* fertilization.

In vitro **fertilization (IVF) is** a procedure where an egg cell (the oocyte) and sperm cells are brought together in a laboratory dish (i.e., *in vitro*), so that a sperm cell can fertilize the egg.

In vitro **maturation** (IVM) refers to culture of immature ova in the laboratory, usually until they are ready for *in vitro* fertilization.

In vitro **mutagenesis** is related to directed mutagenesis.

In vitro **translation** (cell-free translation) refers to the synthesis of proteins in the test-tube from purified mRNA molecules using cell extracts containing ribosomal sub-units, the necessary protein factors, tRNA molecules and aminoacyl tRNA synthetases, ATP, GTP, amino acids and an enzyme system for re-generating the nucleoside triphosphates.

In vivo refers to experiments conducted using a whole, living organism. *In vivo* experimentation is often necessary to confirm hypotheses that can not be thoroughly tested in the artificial environment of laboratory glassware.

In vivo **gene therapy** refers to the delivery of a gene or genes to a tissue or organ of a complete living individual to alleviate a genetic disorder.

Inactivated agent refers to a virus, bacterium or other organism that has been treated to prevent it from causing a disease.

Inbred line refers to the product of inbreeding, i.e., the mating of individuals that have ancestors in common; in plants and laboratory animals, it refers to populations resulting from at least 6 generations of selfing or 20 generations of brother-sister mating, that are for all practical purposes, completely homozygous.

Inbreeding refers to matings between individuals that have one or more ancestors in common, the extreme condition being self-fertilization, which occurs naturally in many plants and some primitive animals. Synonym is endogamy.

Inbreeding coefficient is a measure of the level of inbreeding in a population that determines the probability that an individual possesses two alleles at a locus that are identical by decent.

Inbreeding depression refers to reduction in vigor, yield, etc., of a population that is commonly seen as the level of inbreeding increases.

Inbreeding effect is a process when inbreeding eventually will occur in panmictic small populations due to the individuals becoming increasingly related through time.

Inbreeding effective number (*N*eI) is the size of the ideal panmictic population that loses heterozygosity at the same rate as the observed population.

Inclusion body refers to the protein that is overproduced in a recombinant bacterium and forms a crystalline array inside the bacterial cell.

Incompatibility refers to: (1) Selectively-restricted mating competence, which limits fertilization, such as lack of proper functions by otherwise normal pollen grains or certain pistils, a condition that may be caused by a variety of factors; (2) A physiological interaction resulting in graft rejection or failure; (3) A function of a related group of plasmids.

Incompatibility group refers to a classification scheme indicating which plasmids can co-exist within a single cell. Plasmids must belong to different incompatibility groups to co-exist within the same cell.

Incomplete digest is also named as partial digest.

Incomplete dominance is a condition where a heterozygous offspring has a phenotype that is distinctly different from, and intermediate to, the parental phenotypes.

Incomplete penetrance is the gene for a condition is present, but not obviously expressed in all individuals in a family with the gene.

Incompressible fluid is one whose density is constant everywhere. All fluids behave incompressibly (to within 5%) when their maximum velocities are below Mach 0.3.

Incubation (L. *incubare*, to lie on) refers to: (1) The hatching of eggs by means of heat, either natural or artificial; (2) Period between infection and appearance of symptoms induced by parasitic organisms; (3) The culture of cells and organisms.

Incubator is an apparatus in which environmental conditions (light, photoperiod, temperature, humidity, etc.) are fully controlled, and used for hatching eggs, multiplying microorganisms, culturing plants, etc. *cf* culture room; growth cabinet.

Indehiscent Describing a fruit or fruiting body that does not open to release its seeds or spores when ripe.

Independent assortment refers to the random distribution of alleles (from different loci) to the gametes that occurs when genes are located in different chromosomes or far apart on large chromosomes.

Indeterminate growth refers to unlimited growth potential for a definite or indefinite period. Some apical meristems can produce unrestricted numbers of lateral organs.

Indifferent or dispersive electrode is a large electrode used to spread out electrical charge.

Indigestion refers to the feeling of nausea, bloating, gas, and/or heartburn caused by poor digestion.

Indirect embryogenesis refers to embryo formation on callus tissues derived from zygotic or somatic embryos, seedling plant or other tissues in culture. *cf* direct embryogenesis.

Indirect organogenesis refers to plant organ formation on callus tissues derived from explants. Opposite is direct organogenesis.

Indocyanine green (ICG) is a cyanine dye used in medical diagnostics. It is used for determining cardiac output, hepatic function, and liver blood flow, and for ophthalmic angiography.

Induced pluripotent stem cells are somatic cells induced to a pluripotent embryonic stem cell-like state. The process of creating these cells, often referred to as "reprogramming" involves introducing a combination of three to four genes for transcription factors, delivered by retroviruses, into the somatic cell.

Inducer is a low-molecular-weight compound or a physical agent that is bound by a repressor so as to produce a complex that can no longer bind to the operator; thus, the presence of the inducer turns on the expression of the gene(s) controlled by the operator.

Inducible is a gene or gene product whose transcription or synthesis is increased by exposure of the cells to an inducer or to a condition, e.g., heat. Opposite is constitutive.

Inducible enzyme is an enzyme that is synthesized only in the presence of the substrate that acts as an inducer.

Inducible gene is a gene that is expressed only in the presence of a specific metabolite, the inducer.

Inducible promoter refers to the activation of a promoter in response to either the presence of a particular compound, i.e., the inducer, or to a defined external condition, e.g., elevated temperature.

Inductance [Henry, H] is the property of an electric circuit which tends to oppose the change in current in the circuit. One henry (H) is the inductance of a closed circuit in which an electromotive force of 1 volt is produced when the electric current in the circuit varies uniformly at the rate of 1 ampere per second.

Induction (*L. inducere*, to lead in) refers to the process of causing to occur; process whereby a cell or tissue influences neighboring cells or tissues.

Induction electrode refers to an electrical current that is passed through a coil that in turn gives off eddy currents of electromagnetic energy. This energy is absorbed by the tissues and heating occurs as a result of the resistance of the tissues.

Induction media refers to the: (1) Media used to induce the formation of organs or other structures; (2) Media causing variation or mutation in the tissues exposed to it.

Inductor is an energy storage circuit component consisting of a coil of wire and possibly a magnetic material.

Inembryonation is related to artificial inembryonation.

Inert is a support structure that makes no chemical contribution, and whose only function is support. Physiologically it is a neutral or immobile unit.

Infection refers to the invasion of any living organisms by disease-causing micro-organisms, which proceed to establish themselves, multiply, and produce various clinical signs in their host.

Infectious agent refers to a proliferating virus, bacterium or parasite that causes a disease in plants or animals.

Inferior vena cava refers to the large vein returning blood from the legs and abdomen to the heart.

Infertile means incapable of initiating, sustaining, or supporting reproduction. Alternatively, not fertilized and therefore incapable of growing and developing.

Infiltrate refers to force the passage of liquid into tissue pores or space, such as by applying a vacuum, then releasing it, during the disinfectation procedure.

Inflammatory arthritis is characterized by the tendency to cause inflammation in joints and tendons.

Inflammatory bowel disease (IBD) cause irritation and ulcers in the intestinal tract.

Inflorescence (L. *inflorescere*, to begin to bloom) refers to: (1) The way the flowers are arranged on the stalk of a plant; (2) The flowers of a plant collectively; (3) The flowering or blooming process.

Informed consent is a term describing the responsibility of doctors or researchers to ensure that patients or people have an understanding of the relevant facts regarding their care or participation in research.

Infrared is the portion of the electromagnetic spectrum associated with thermal changes; located adjacent to the red portion of the visible light spectrum.

Inguinal hernia is a part of the small intestine that pushes through an opening in the abdominal muscle, causing a bulge underneath the skin in the groin area.

Inhalation (inspiration) is the flow of air into an organism. It is a vital process for all human life. In humans, it is the movement of air from the external environment, through the airways, and into the alveoli.

Inheritance refers to the transmission of particular characteristics and/or genes from generation to generation.

Inherited are characteristics that come from one's ancestors and are transmitted from parents to offspring through genes. The traits will therefore be present at birth.

Inhibitor refers to (1) Any substance or object that retards a chemical reaction; (2) A metabolite or modifier gene that interferes with a reaction or with the expression of another gene.

Initial (Noun) refers to the cells in a meristem that remain permanently meristematic and form tissues of particular structure and function.

Initiation codon (initiator) refers to the codon AUG which specifies the first amino acid methionine of a polypeptide chain. In bacteria, the initiation codon is either AUG, which is translated as *n*-formyl methionine or, rarely, GUG (valine). In eukaryotes, the initiation codon is always AUG and is translated as methionine. The term is also used to describe the corresponding sequence in DNA, namely ATG.

Initiation factor refers to soluble protein required for the initiation of translation.

Initiation refers to: (1) Early steps or stages of a tissue culture process (culture growth, organogenesis, embryogenesis); (2) Early stages of biosynthesis.

Inner cell mass (ICM) is a small group of cells attached to the wall of the blastocyst (the embryo at a very early stage of development that looks like a hollow ball).

Inoculate refers to deliberately introduce, in contrast to contamination: (1) In bacteriology, tissue culture, etc., placing an inoculum into (or onto) medium to initiate a culture; (2) In immunology, to carry out immunization; (3) In plant pathology, application of pathogen spores etc. on to plants under conditions where infection should result in the absence of resistance.

Inoculation cabinet is a small cabinet for inoculation (of tissue or micro-organism cultures) operations, often with a current of sterile air to carry contaminants away from the work area.

Inoculum (plural: inocula) refers to: (1) A small piece of tissue cut from callus, or an explant from a tissue or organ, or a small amount of cell material from a suspension culture, transferred into fresh medium for continued growth of the culture; (2) Microbial spores or parts (such as mycelium). (3) Vaccine.

Inorganic refers to: (1) Chemicals that are not organic, that is, not manufactured within living organisms; (2) Any chemical compound not based on carbon chains or rings (except oxides, sulphides of carbon and metallic carbides that are also inorganic).

Inorganic chemistry is the study of the synthesis and behavior of inorganic and organometallic compounds.

Inorganic compound is a chemical compound that generally is not derived from living processes; compounds that do not contain carbon.

Inositol (hexahydroxycyclohexane: $C_6H_6(OH)_6$) refers to: (1) A cyclic acid that is constituent of certain cell phosphoglycerides; (2) A water-soluble nutrient frequently referred to as a "vitamin" in plant tissue culture. Also it acts as a growth factor in some animals and micro-organisms.

Inositol is a cyclic acid (hexahydroxycyclohexane) that is a constituent of certain cell phosphoglycerides. It is a nutrient frequently referred to as a "vitamin" in plant tissue culture.

Inositol lipid is a membrane-anchored phospholipid that transduces hormonal signals by stimulating the release of any of several chemical messengers.

Inotropic agent is any of a class of agents affecting the force of muscle contraction, particularly a drug affecting the force of cardiac contraction; positive inotropic agents increase, and negative inotropic agents decrease the force of cardiac muscle contraction.

Input traits are traits that are introduced into crop plants with the aim of lowering the cost of production and improving the performance of the crop in the field.

Insecticide is a substance that kills insects.

Insensible fluid loss is a loss that cannot be easily measured.

Some sources say insensible losses account for 500 to 650 ml/day (0.5 to 0.6 qt.) of water in adults. This loss includes fluid lost through perspiration and as water vapor in expired air.

Insert refers to: (1) To incorporate a DNA molecule into a cloning vector; also used as a noun to describe such a DNA molecule; (2) To introduce a gene or gene construct into a new genomic site or into a new genome.

Insertion element is a DNA sequences that are found in bacteria capable of genome insertion.

Insertion mutation refers to changes in the base sequence of a DNA molecule resulting from the random integration of DNA from another source.

Insertion point (in lithography context) is an adaptation of a new lithography technique that is referred to as the insertion point of that technique.

Insertion sequence is related to insertion element.

Insertion site refers to: (1) A unique restriction site in a vector DNA molecule into which foreign DNA can readily be inserted. This is achieved by treating both the vector and the insert with the relevant restriction endonuclease and then ligating the two different molecules, both having the same sticky ends: Synonym is cloning site; (2) The position of integration of a transposon.

Inspiration is a process of inhaling; taking in oxygen.

Inspiratory capacity is the maximal volume of gas that can be inspired from the resting expiratory level.

Inspiratory reserve volume refers to maximal volume of gas that can be inspired from the end-inspiratory position. Volume that can be inhaled during forced breathing in addition to tidal volume = 3000 mL.

Instability is a lack of consistent phenotype, usually as a result of uncontrolled genetic changes. These may be due to transposon activity, or in cell lines, to changes in karyotype.

Insulator is a material that conducts electricity very poorly.

Insulin is a peptide hormone secreted by the islets of Langerhans of the pancreas that regulates the level of sugar in the blood.

Integrated circuit (IC) is a semiconductor circuit, typically on a very small silicon chip, containing microfabricated transistors, diodes, resistors, capacitors, etc.

Integrating vector is a vector that is designed to integrate cloned DNA into the host's chromosomal DNA.

Integration refers to the recombination process which inserts a small DNA molecule (usually by homologous recombination) into a larger one. If the molecules are circular, integration involves only a single crossing-over; if linear, then two crossings-over are required. A well known example is the integration of lambda phage DNA into the *E. coli* genome.

Integration-excision region refers to the portion of bacteriophage lambda DNA that enables bacteriophage lambda DNA to be inserted into a specific site in the *E. coli* chromosome and excised from this site.

Integument is one of the layers that enclosed the ovule, and is the precursor of the seed coat.

Integumentary system performs a number of functions related to its location on the surface of the body.

Intellectual property (IP) refers to the content of the human intellect, or the result of intellectual effort, which is considered to be unique and original and have value in the marketplace, and therefore requires legal protection and ownership.

Intellectual property rights (IPR) refers to the legal framework, which includes patenting and plant variety protection, by which inventors control the commercial application of their work.

Intensifying screen is a plastic sheet impregnated with a rare-earth compound, such as calcium tungstate, which absorbs ß radiation and emits light.

Interaction is an effect that cannot be explained by the additive action of contributing factors; a departure from strict additivity.

Intercalary (L. *intercalare*, to insert) refers to meristematic tissue or growth not restricted to the apex of an organ, i.e., growth at nodes.

Intercalary growth is a pattern of stem elongation typical of grasses.

Intercalary refers to: (1) Meristematic tissue or growth not restricted to the apex of an organ, i.e., growth at nodes; (2) Referring to internal segments of a chromosomes (i.e., not at the ends).

Intercalating agent is a chemical capable of inserting between adjacent base pairs in a double-stranded nucleic acid. A prominent example is ethidium bromide.

Intercellular space is a pore space between cells, especially typical of leaf tissues.

Interfascicular cambium refers to the cambium that arises between vascular bundles.

Interference refers to crossing over at one point that alters the chance of another crossing-over nearby; detected by studying the pattern of crossings-over with three or more linked genes. Interference is positive or negative depending on whether the chance of another crossing-over nearby is reduced or increased, respectively.

Interferon (IFN) is a protein first recognized in animals for its action in inhibiting viral replication and inducing resistance in host cells. The interferons (IFNs) are a highly conserved family of multi-functional, species-specific, secreted proteins originally classified on the basis of cellular origin including: leucocyte IFN (alpha), fibroblastic IFN (beta) and immune IFN (gamma).

Intergeneric cross refers to a hybrid made between parents belonging to two different genera.

Intergeneric is a cross between two different genera.

Intergenic regions refer to non-coding DNA located between genes; this comprises a variable but considerable proportion of all eukaryotic genomic DNA, and its function is largely unknown.

Intergenic spacer (IGS) refers to non-coding DNA separating tandemly arranged copies of a repeated gene sequence (typically ribosomal DNA).

Interleukin is a group of proteins that transmit signals between immune cells and are necessary for mounting normal immune responses. *See*: cytokine.

Internal guide sequence is an abbreviation IGS.

Internal transcribed spacer (ITS) refers to non-coding regions separating the individual components of the ribosomal DNA units.

International Society for Pharmaceutical Engineering is abbreviated as ISPE.

International Treaty on Plant Genetic Resources for Food and Agriculture refers to the international treaty resulting from the revision of the International Undertaking on Plant Genetic Resources.

International Undertaking on Plant Genetic Resources refers to the first comprehensive voluntary, international agreement (adopted in 1983) dealing with plant genetic resources for food and agriculture.

Internet is a worldwide digital communication network in which packets of information travel between the senders and recipients.

Internode (L. *inter*, between + *nodus*, a knot) refers to the region of a stem between two successive nodes.

Interphase refers to the stage in the cell cycle when the cell is not dividing; the stage follows telophase of one division and extends to the beginning of prophase in the next division.

Intersex is a synonym of hermaphrodite. It refers to an organism displaying a mixture of male and female attributes.

Inter-simple sequence repeat (ISSR) refers to a PCR-based molecular marker assay of genomic sequence lying between adjacent microsatellites.

Interspecific cross refers to a hybrid made between parents belonging to two different species.

Interspecific refers to between two different species, e.g., an interspecific cross is a cross between two species.

Interstitial cystitis is a complex, chronic disorder characterized by an inflamed or irritated bladder wall.

Interstitial diffusion is a mechanism that causes atomic motion from interstitial site to interstitial site.

Interstitial trophoblast invades the decidual stroma.

Intestinal flora refers to the normal bacteria, yeast, and fungi found in the intestines that aid in digestion.

Intestinal mucosa is the lining of the intestines, through which nutrients and water are absorbed into the body.

Intestine is a digestive organ in the abdomen, also known as either the large or small bowel. The small intestine removes nutrients from food to be used for energy, while the large intestine absorbs water from the digested food and processes it into stool. Intestine absorbs nutrients from the foods we eat.

Intolerance refers to an allergy to a food or other substance.

Intracardiac vent is a device used to decompress cardiac chambers and to aspirate air and blood that is returned to cardiotomy reservoir.

Intracellular (L. *intra*, within + cell) refers to that occurs within a cell.

Intracellular fluid refers to the cytosol.

Intracranial pressure will cause internal or external herniation.

Intracytoplasmic sperm injection (ICSI) refers to the injection, using micromanipulation, of a single sperm into the cytoplasm of a mature oocyte.

Intrageneric cross refers to a hybrid made between parents belonging to two species in the same genus.

Intrageneric refers to within a genus, such as a hybrid resulting from a cross between species within one genus.

Intragenic complementation refers to complementation that occurs between two mutant alleles of a gene; common only when the product of the gene functions as a homomultimer.

Intramuscular (IM) injection is the injection of a substance directly into a muscle. In medicine, it is one of several alternative methods for the administration of medications (see route of administration).

Intraoperative magnetic resonance imaging (iMRI) refers to an operating room configuration that enables surgeons to image the patient via an MRI scanner while the patient is undergoing surgery, particularly brain surgery.

Intraspecific cross refers to a hybrid made between parents belonging to the same species.

Intrathoracic refers to: within the thoracic cavities

Intravenous pyelogram (IVP) is a series of x-rays of the kidney, ureters, and bladder with the injection of a contrast dye into the vein to detect tumors, abnormalities, kidney stones, or any obstructions, and to assess renal blood flow.

Intravenous therapy (IV therapy) is the infusion of liquid substances directly into a vein. The word intravenous simply means "within vein." Therapies administered intravenously are often called specialty pharmaceuticals. Intravenous therapy may be used to correct electrolyte imbalances, to deliver medications, for blood transfusion or as fluid replacement to correct, for example, dehydration.

Intravital microscopy is a technique to observe biological systems *in vivo* at high resolution through an attached window preparation. It is a type of microscopy.

Introduction is the placement, or escape, of a species or individual into a novel habitat. Often introductions are used in conservation to aid genetic rescue of isolated populations.

Introgression refers to the introduction of new alleles or gene(s) into a population from an exotic source, usually another species.

Intron (intervening sequence) is a segment of DNA sequence of a eukaryotic gene, not represented in the mature (final) mRNA transcript, because it is spliced out of the primary transcript before it can be translated; a process known as intron splicing.

Intussusception is a disorder in which the intestine folds into itself in a telescope fashion, causing obstruction (blockage).

Invariant refers to that is constant, unchanging; usually referring to the portion of a molecule that is the same across species.

Invasive procedure is one that penetrates or breaks the skin or enters a body cavity.

Invasive species is an introduced alien species that is likely to cause harm to the natural ecosystem, the economy, or human health.

Invasiveness is the ability of a plant to spread beyond its introduction site and become established in new locations where it may provide a deleterious effect on organisms already existing there.

Inversion refers to a chromosome re-arrangement, which involves the re-orientation of a segment so that the order of a linear array of genes within it is reversed.

Inverted repeat refers to two regions of a nucleic acid molecule which have the same nucleotide sequence but in an inverted orientation, such as: 5' GCACTTG... ...CAAGTGC 3'; 3' CGTGAAC... ...GTTCACG 5'. Because they contain exactly the same message when read in either direction, inverted repeats are said to be palindromes.

Inviscid is not viscous.

Invivo refers to one that is occurring in a living organism.

Ion is a charged particle.

Ion channel is an integral protein within a cell membrane, through which selective ion transport occurs.

Ion gel is a relatively new material where an ionic conducting liquid is immobilized inside a polymer matrix.

Ionic bond is a coulombic interatomic bond existing between two adjacent and oppositely charged ions, characteristic of salts. Also called *electrovalent bond*.

Ionic bonding is a type of chemical bond that involves the electrostatic attraction between oppositely charged ions. These ions represent atoms that have lost one or more electrons (known as cations) and atoms that have gained one or more electrons (known as an anions).

Ionic gelation (IG) is a method quite frequently used for synthesis of chitosan (CS) microparticles (MPs) and nanoparticles (NPs).

Ionic interactions are highly sensitive to changes in pH. As the pH drops, H^+ bind to the carboxyl groups (COO^-) of aspartic acid (Asp) and glutamic acid (Glu), neutralizing their negative charge; and H^+ bind to the unoccupied pair of electrons on the N atom of the amino (NH_2) groups of lysine (Lys) and arginine (Arg) giving them a positive charge.

Ionizing radiation refers to the portion of the electromagnetic spectrum that results in the production of positive and negative charges (ion pairs) in molecules. X-rays and gamma rays are examples of ionizing radiation.

Ionophore is a macro-organic molecule capable of specifically solubilizing an inorganic ion of suitable size in organic mediums.

IP relates to the biotechnology industry.

IPO (Initial public offering) is a commonly used term when referring to new companies and their entry into the stock exchange.

IPR is an abbreviation for intellectual property rights.

IPTG is an abbreviation for isopropyl-3-D-thiogalactopyranoside. A synthetic inducer of beta-galactosidase activity in many bacteria.

Iris refers to the pigmented portion of the eye.

Irradiation (L. *in*, into + *radius*, ray) refers to: (1) Exposure to any form of radiation; (2) Treatment with a ray, such as ultraviolet rays, etc.

Irrotational fluid flow is one whose streamlines never loop back on themselves. Typically, only inviscid fluids can be irrotational. Of course, a uniform viscid fluid flow without boundaries is also irrotational, but this is a special (and boring!) case.

IS element is an abbreviation for insertion sequence element. A short (800–1400 nucleotide pairs) DNA sequence found in bacteria that is capable of transposing to a new genomic location; DNA sequences contained within an IS element can be transposed along with the IS itself.

ISE (Ion Selective Electrode) refers to ions in solution are quantified by measuring the change in voltage between a sensing membrane (the ion selective membrane) and the solution. This potential is measured at zero current with respect to a reference electrode which is also in contact with the solution. The potential measured is proportional to the logarithm of the analyte concentration.

ISFET (Ion Sensitive Field Effect Transistor) is a logical extension of ISE's. These can be conceptualized by imagining that the lead from an ion-selective electrode, attached via a cable to a FET in the high impedance input stage of a voltmeter, is made shorter until no lead exists and the selective membrane is attached directly to the FET.

Island model of migration is a model in which a population is subdivided into a series of demes, of size N, that randomly exchange migrants at a given rate, m.

Islets of Langerhans is also called Islands of Langerhans, irregularly shaped patches of endocrine tissue located within the pancreas of most vertebrates. The normal human pancreas contains about 1,000,000 islets.

ISO Certification is related to explanation of the ISO standards and QMS certification.

Isoallele refers to multiple similar copies of a gene, usually located at independent positions in the genome, which encode similar gene products and produce the same, or a very similar phenotype.

Isochromosome is a metacentric chromosome produced during mitosis or meiosis when the centromere splits transversely instead of longitudinally; the arms of such chromosome are equal in length and genetically identical, however, the loci are positioned in reverse sequence in the two arms.

Isodiametric is commonly used to describe cells with equal diameters.

Isoelectric focusing gel (IEF gel) is a variant of gel electrophoresis, in which macromolecules (usually proteins) are separated on the basis of differing isoelectric point, rather than on the basis of size.

Isoform is a member of a family of closely related proteins that have some amino acid sequences in common and some different.

Isogamy refers to sexual reproduction involving the production fusion of gametes that are similar in size and structure.

Isogenic is a group of individuals that possesses the same genotype, irrespective of their being homozygous or heterozygous.

Isogenic stock refers to strains of organisms that are genetically nearly identical, except with respect to identified genes. Generally produced by repeated backcrossing, or by transformation.

Isolating mechanism refers to any of the biological properties of organisms that prevent interbreeding (and therefore exchange of genetic material) between members of different species that inhabit the same geographical area.

Isolation by distance is the case where genetic differentiation is greater than the further individuals (or populations) are from each other because gene flow decreases as geographic distance increases.

Isolation medium refers to an optimum medium suitable for explant survival, growth and development.

Isomer refers to: (1) Structural isomers have the same chemical formula but different structures; e.g., leucine and isoleucine; (2) Stereoisomers are different topological forms of an otherwise single chemical structure, due to changes in bond configurations about some axis or plane of symmetry; eg, D- and L-glucose or cis- and trans-cinnamic acid.

Isomerase refers to any of a class of enzymes that catalyze the re-arrangement of the atoms within a molecule, thereby converting one isomer into another.

Isomorphous refers to having the same structure. In the phase diagram sense, isomorphicity means having the same crystal structure or complete solid solubility for all compositions.

Isothermal is a process at a constant temperature. In an isothermal process heat is, if necessary, supplied or removed from the system at just the right rate to maintain constant temperature.

Isotonic (iso-osmotic) are solutions having the same osmotic potential; the same molar concentration.

Isotope refers to one of two or more forms of an element that have the same number of protons (atomic number) but differing numbers of neutrons (mass numbers).

Isotropic refers to having identical values of a property in all crystallographic directions.

Isozyme refers to the variant of a particular enzyme. In general, all the isozymes of a particular enzyme have the same function and sometimes the same activity, but they differ in amino-acid sequence.

ISSR are inter-simple sequence repeat markers that use similar PCR methods as PINE fragments, but have primers based on simple sequence repeats of microsatellites.

ITS is an abbreviation for internal transcribed spacer.

IUCN stands for World Conservation Union, formerly called *International Union for Conservation of Nature*.

Iuxtaglomerular apparatus refers to a structure in the kidney that secretes rennin in response to a decrease in blood pressure.

IVEP is an abbreviation for *in vitro* embryo production.

IVF is an abbreviation for *in vitro* fertilization.

IVM is an abbreviation for *in vitro* maturation.

J

. . . .

J is an abbreviation for joining segment.

J/m² or J m⁻² (joules per square meter) is a unit of light measurement.

Jaundice refers to a yellow color of the skin and eyes that is caused by too much bilirubin in the blood stream due to liver problems.

Jejunum is the middle section of the small intestine.

Jiffy pots are pots made from wood pulp and peat, commonly used for transplanting tissue-culture-derived plants into soil medium.

JIVET (or JIVT) is an abbreviation for juvenile *in vitro* embryo technology.

Joining segment (J). A small DNA segment that links genes in order to yield a functional gene encoding an immunoglobulin.

Joule (SI unit: J) is the amount of energy needed to apply a force of 1 newton over a distance of 1 meter.

Juvenile hormone is a hormone secreted by insects from a pair of endocrine glands close to the brain. Its function is to inhibit metamorphosis so maintaining the larval features.

Juvenile *in vitro* embryo technology (JIVT or JIVET) is a technology involving collection of immature ova from young animals, *in vitro* maturation of those ova, *in vitro* fertilization, and then transfer of the resultant embryos into recipient females. It is a method of achieving rapid generation turnover.

Juvenility refers to an early phase of development in which an organism is juvenile and incapable of sexual reproduction.

K

Kanamycin is an antibiotic of the aminoglycoside family that inhibits translation by binding to the ribosomes.

kanr is an abbreviation for Kanamycin-resistance gene.

Kappa chain is one of two classes of antibody light chains. The other is a lambda chain.

Karyogamy refers to the fusion of nuclei or nuclear material that occurs during sexual reproduction.

Karyokinesis refers to the division of a cell nucleus.

Karyotype is a picture of an individual's chromosomes as seen under a microscope. The chromosomes can be identified by their unique banding patterns and arranged in order of size (1 is the largest and 22 is the smallest). The karyotype is a test to look for major changes in the chromosomes, such as a change in the number or the structure.

Karyotyping (traditional) is a laboratory technique to view all of the human chromosomes at one time in black and white. It is useful for observing the number, size and shape of the chromosomes. **Spectral karyotyping** paints each pair of chromosomes in a different fluorescent color.

kb is an abbreviation for kilobase (of single-stranded nucleic acid).

kbp is an abbreviation for kilobase pairs (of double-stranded DNA).

k_{cat}/K_m is the catalytic efficiency of an enzyme-catalyzed reaction. The greater the value of k_{cat}/K_m, the more rapidly and efficiently the substrate is converted into product.

k_{cat} is the catalytic rate constant that characterizes an enzyme-catalyzed reaction. The larger the k_{cat} value, the faster the conversion of substrate into product.

K_d is an abbreviation for dissociation constant. It describes the strength of binding (or affinity) between molecules and their ligands.

kDa is an abbreviation for kiloDalton. It is a unit of molecular mass equal to 1000 Dalton.

Keratin is an insoluble protein that is a constituent of intermediate filaments in animal cells and is found in the epidermis of vertebrates: in nails, feathers, hair, horns, scales and claws.

Kidney refers to either of two bean-shaped excretory organs that filter wastes (especially urea) from the blood and excrete them and water in urine; urine passes out of the kidney through ureters to the bladder.

Kidney stone is a solid piece of material that forms from crystallization of excreted substances in the urine.

Kidney transplantation is a procedure that places a healthy kidney from one person into a recipient's body.

Killer T cells are T cells that carry T-cell receptors and kill the cells displaying the recognized antigens.

Kilobase (kb) refers to length of single-stranded nucleic acid composed of 1000 bases. One kilobase of single-stranded DNA has a mass of about 330 kiloDalton. One kilobase of double-stranded DNA has a mass of about 660 kiloDalton.

Kilobase pairs (kbp) is a length of double-stranded DNA composed of 1000 base pairs.

Kilobyte (kB) is 1,000 bytes of information.

Kilocalorie (kcal) is equal to 1000 calories.

Kilodalton (kDa) is a unit of atomic mass equal to 1,000 daltons.

Kilohertz (kHz) is one thousand cycles per second.

Kilojoule (kJ) is equal to 1,000 J; and 1 kcal = 4.184 kJ.

Kinase is an enzyme that can transfer a phosphate from a high energy phosphate such as ATP, to an organic molecule.

Kinetic molecular theory is a model with an assumption that an ideal gas is composed of tiny particles (molecules) in constant motion.

Kinetics refers to dynamic processes involving motion.

Kinetin (Gr. *kinetikos*, causing motion) is one of the cytokinins, a group of growth regulators that characteristically promote cell division in plants.

kinetochore refers to the structure at the centromere of eukaryotic chromosomes. The kinetochore consists of inner and outer electron dense plates and a central zone containing repetitive DNA elements.

Kinetosome refers to the granular cytoplasmic structure which forms the base of a cilium or flagellum. Synonym is basal body.

Kinin is a substance promoting cell division. In plant systems, the prefix cyto- has been added (cytokinin) to distinguish it from kinin in animal systems.

Klenow fragment is a product of proteolytic digestion of the DNA polymerase I from *E. coli*; it has both polymerase and 3'-exonuclease activities but not 5'-exonuclease activity.

Klinefelter syndrome is an endocrine condition caused by a an extra X-chromosome (47,XXY); characterized by the lack of normal sexual development and testosterone, leading to infertility and adjustment problems if not detected and treated early.

K_m is a dissociation constant that characterizes the binding of an enzyme to a substrate. It is also called the Michaelis constant.

Knockout refers to an animal resulting from an embryonic stem cell in which a normal functional gene has been replaced by a non-functional form of the gene.

L

Label is a compound or atom that is attached to, or incorporated into, another molecule in order to allow detection of the latter's presence: tag.

Label or marker detects the bound antibody-antigen complex. Several markers have been established for use in immunoassays, such as: Particles (e.g., latex, gold particles, erythrocytes); Metal and dye sols (e.g., Au, Palanil®; Luminous Red G); Chemiluminescent and bioluminescent compounds (e.g., Luciferase/luciferins, Luminol and derivatives, Acridinium esters, Peroxidase); Electrochemical active species (ions, redox species, ionophores); Fluorophores (e.g., Dansyl chloride DANS, Rare earth metal chelates, Umbelliferones); Chromophores; Enzymes (e.g., Alkaline phosphatase, β-D-Galactosidase, Peroxidase), substrates, cofactors; Liposomes; and Iodine-125, tritium, 14C, 75Se, 57Co.

Labeling refers to the process of replacing a stable atom in a compound with a radioactive isotope of the same element to enable it to be detected by autoradiography or other techniques.

Lac repressor (lacI) is a DNA-binding protein, which inhibits the expression of genes coding for proteins involved in the metabolism of lactose in bacteria. These genes are repressed when lactose is not available to the cell, ensuring that the bacterium only invests energy in the production of machinery necessary for uptake and utilization of lactose when lactose is present.

Lacerations is the tearing of frontal and temporal lobes or blood vessels caused by brain rotating across ridges inside skull.

Lactase is an enzyme in the small intestine needed to digest lactose, a sugar found in milk and milk products.

Lactase deficiency is a lack of an enzyme made by the small

intestine called lactase, which prevents the body from digesting lactose (a sugar found in milk and milk products) properly.

Lactoferrin is a breast milk protein that promotes infant growth.

Lactose is a disaccharide sugar produced in milk, composed of one unit each of glucose and galactose.

Lactose tolerance test is a test that checks the body's ability to digest lactose (a sugar found in milk and milk products).

Lag phase refers to: (1) A state of apparent inactivity preceding a response; called also a latent phase; (2) The initial growth phase, during which cell number remains relatively constant, prior to rapid growth; (3) The first of five growth phases of most batch-propagated cell suspension cultures, being the phase in which inoculated cells in fresh medium adapt to the new environment and prepare to divide.

Lagging strand refers to the strand of DNA that is synthesized discontinuously during replication (because DNA synthesis can proceed only in the 5′ to 3′ direction).

Lambda chain is one of two classes of antibody light chains. Opposite is a kappa chain.

Lambda phage (also called, Enterobacteria phage λ, coliphage λ) is a bacterial virus, or bacteriophage, that infects the bacterial species *Escherichia coli* (*E. coli*). This virus has a temperate lifecycle that allows it to either reside within the genome of its host through lysogeny or enter into a lytic phase (during which it kills and lyses the cell to produce offspring).

Lamella (L. diminutive of *lamina*, a thin blade; plural: lamellae) is a double-membrane structure, plate or vesicle that is formed by two membranes lying parallel to each other.

Lamina (*L. lamina*, a thin plate) refers to expanded part of a leaf.

Lamina propria is a highly vascular layer of connective tissue under the basement membrane lining a layer of epithelium.

Laminar (non-turbulent) is an organized flow field that can be described with streamlines. In order for laminar flow to be permissible, the viscous stresses must dominate over the inertia stresses.

Laminar air-flow cabinet refers to the cabinet designed for cell or tissue culture manipulations requiring a sterile environment. Synonym is laminar air-flow hood.

Laminarin is a storage polysaccharide of the brown algae.

Lampbrush chromosomes refer to large diplotene chromosomes present in oocyte nuclei, and particularly conspicuous in amphibians. These chromosomes have extended regions called loops, which are active sites of transcription.

Lanagement unit is a local population that is managed as a unit due to its demographic independence.

Landrace refers to an early, cultivated form of a crop species (in plant genetic resources,), evolved from a wild population, and generally composed of a heterogeneous mixture of genotypes.

Landscape genetics is the study of the interaction between landscape or environmental features and population genetics, such as gene flow.

Laparoscope is a tube with a camera lens attached that looks inside organs to check for abnormalities.

Laparoscopy (Greek λαπάρα, *lapara* meaning flank, side, and σκοπέω, *skopeo* meaning to see) is a procedure that uses a tube with a light and a camera lens at the end (laparoscope) to examine organs, check for abnormalities, or perform

minimally invasive surger-
ies. Laparoscopy is a surgery,
which avoids making large
incisions. During the procedure, tissue samples may also
be taken for examination and
testing.

Laplace's equation describes the
behavior of gravitational, electric, and fluid potentials.

Large intestine (Also called the
colon) is the last section of the
digestive tract, from the cecum
to the rectum; absorbs water
from digested food and processes it into stool.

Larginal overdominance refers
to greater fitness of heterozygous genotypes, which are
not the most fit in any single
environment, due to an organism's interactions with multiple
environments that each favor
different alleles.

Larynx is the organ at the upper
end of the trachea that contains
the vocal cords.

Laser (Light Amplification by
the Stimulated Emission of
Radiation) is a quantum device
that produces coherent light.

Laser ablation refers to when a
pulsed laser vaporizes a graph-
ite target in a high-temperature
reactor while an inert gas is
bled into the chamber.

Laser ablation therapy has
become a popular choice

for treating a broad range of
illnesses. It provides precision control a laser provides;
the blood loss is minimum;
and recovery time is faster
than with traditional surgical
techniques.

Laser trimming is a method for
adjusting the value of thin- or
thick-film resistors by using
a computer-controlled laser
system.

Latch probability (MP) is the
probability of sampling an
individual with an identical
multilocus genotype to the one
already sampled ("in hand").

Latent agent is a pathogen, usually
a virus, present in a host organism without producing any
symptoms.

Latent bud refers to an inactive
bud not held back by rest or
dormant period, but which may
start growth if stimulated.

Latent phase is related to lag phase.

Lateral bud is related to axillary
bud.

Lateral meristem is a meristem
giving rise to secondary plant
tissues, such as the vascular
and cork cambia. The term is
sometimes used to refer to an
axillary meristem.

Laternal effects refer to the influence of the genotype or phenotype of the mother on the
phenotype of the offspring.

Lawn refers to a uniform and uninterrupted laver of bacterial growth, in which individual colonies cannot be observed.

Layering is a technique for vegetative propagation, in which new plants produce adventitious roots before being severed from the parent plant.

LCD (Liquid Crystal Display) is display device employing light source and electrically alterable optically active thin film.

LCR is an abbreviation for ligase chain reaction , which is a method of DNA amplification. While the better-known PCR carries out the amplification by polymerizing nucleotides, LCR instead amplifies the nucleic acid used as the probe.

LD$_{50}$ (lethal dose$_{50\%}$) is the amount of a chemical required to kill 50% of the test population. The higher the LD$_{50}$, the lower the presumed toxicity of the chemical.

LE (lethal equivalent) is the number of deleterious alleles in an individual whose cumulative effect is the same as that of a single lethal allele. For example, four alleles each of which would be lethal 25% of the time (or to 25% of their bearers), and are equivalent to one lethal allele.

Lead compound is a chemical that has demonstrated promising biological activity in preliminary assays.

Leader peptide is an enzyme and catalyzes the cleavage of hydrophobic, N-terminal signal or leader sequences. It is also named as Signal peptidase-I. Other related terms are: leader peptidase I, signal proteinase, *Escherichia coli* leader peptidase, eukaryotic signal peptidase, eukaryotic signal proteinase, leader peptidase, leader peptide hydrolase, leader proteinase, signal peptidase, pilin leader peptidase, SPC, prokaryotic signal peptidase, prokaryotic leader peptidase, HOSP, prokaryotic signal proteinase, propeptidase, PuIO prepilin peptidase, signal peptide hydrolase, signal peptide peptidase, signalase, bacterial leader peptidase 1.

Leader sequence is a variable length sequence of nucleotides at the 5' end of an mRNA molecule that precedes the AUG initiation codon where translation begins and is not itself translated into protein.

Leading strand is the strand of DNA that is synthesized continuously during replication.

Leaf blade refers to usually flattened portion of the leaf.

Leaf bud cutting refers to a cutting that includes a short section of stem with attached leaf.

Leaf margin refers to the edge of a leaf.

Leaf primordium (L. *primordium*, a beginning) is a lateral outgrowth from the apical meristem, which will become a leaf when fully developed and expanded.

Leaf roll is a symptom of some virus diseases, characterized by curling of the leaves. It can also occur as a response to water stress.

Leaf scar refers to the mark left on a stem after leaf abscission.

Leaflet is an expanded leaf-like part of a compound leaf.

Leakage is a loss of all or parts of a useful agent, as of the electric current that flows through an insulator or the magnetic flux that passes outside useful flux circuits.

Leaky mutant refers to a mutant in which the gene product still retains some biological activity.

Lectin is any of a group of proteins, derived from plants, that can bind to specific oligosaccharides on the surface of cells, causing cells to clump together.

LED (Light emitting diode) is a semiconducting diode that produces visible or infrared radiation.

Legume is a member of the pea family that possesses root nodules containing nitrogen-fixing bacterias.

Lek is a specific area, where the males of a population (that exhibits female sexual selection) will congregate and display for females.

Leptonema (adj.: leptotene) is a stage in meiosis immediately preceding synapsis, in which the chromosomes appear as single, fine, threadlike structures (but they are really double because DNA replication has already taken place).

Leptotene (adjective for leptonema) is an early stage of prophase in meiosis in which the replicated chromosomes contract and become visible as long filaments well separated from one another.

Lethal allele (lethal gene) is a mutant form of a gene that eventually results in the death of an organism if expressed in the phenotype.

Lethal gene (Lethal alleles) cause an organism to die only when present in homozygous condition. The gene involved is considered an essential gene.

Lethal mutation prevents the organism (or embryo) from performing vital functions.

Leukaemia is an increase in the number of ineffective and immature white blood cells, causing a weakened immune system. This leaves the body susceptible to infection.

Leukocyte refers to white blood cells, up to 0.02 mm in diameter, of which there are normally 4–11 million per milliliter of human blood.

Lewis acid is an electron-pair acceptor.

Lewis base is an electron-pair donor.

Library is a collection of DNA fragments that have been cloned into a vector that can propagate in *E. coli*.

Life cycle refers to the sequence of events from a given developmental stage in one generation to the same stage in the following generation.

Lifetime is a length of time the sensor can be used before its performance changes.

Ligand is a small molecule (e.g., activators, substrates and inhibitors of enzyme activity) bound to a protein by non-covalent forces; an ion or a molecule that binds to another chemical entity to form a larger complex.

Ligase is an enzyme that is used to join fragments of DNA together, for example in gene splicing.

Ligase (DNA ligase, from the Latin verb *ligāre*—"to bind" or "to glue together") is an enzyme

that can catalyze the joining of two large molecules by forming a new chemical bond, usually with accompanying hydrolysis of a small chemical group dependent to one of the larger molecules or the enzyme catalyzing the linking together of two compounds, e.g., enzymes that catalyze joining of C-O, C-S, C-N, etc. In general, a ligase catalyzes the following reaction: Ab + C → A–C + b or Ab + cD → A–D + b + c, where the lowercase letters signify the small, dependent groups.

Ligase chain reaction (LCR) refers to a technique for the detection and amplification of target DNA sequences.

Ligate or ligation refers to the joining of two linear double-stranded DNA fragments by the formation of phosphodiester bonds.

Ligation refers to the joining of two linear nucleic acid molecules by the formation of phospho-diester bonds. In cloning experiments, a restriction fragment is often ligated to a linearized vector molecule using T4 DNA ligase.

Lignification (L. *lignum*, wood + *facere*, to make) refers to the impregnation of a cell wall with lignin.

Lignin (L. *lignum*, wood) is a group of high-molecular-weight amorphous materials, comprising polymers of phenylpropanoid compounds, giving strength to certain tissues.

Lignocellulose is the combination of lignin, hemicellulose and cellulose that forms the structural framework of plant cell walls.

Likelihood statistics is an approach for parameter estimation and hypothesis testing that involves building a model (i.e., a likelihood function) and the use of the raw data (not a summary statistic), which often provides more precision and accuracy than frequentist statistic approaches (method of moments).

Limbic system is the region of the brain that wraps around the brain stem and lies beneath the cerebrum.

Limit of detection is a smallest measurable input. This differs from resolution, which defines the smallest measurable change in input.

LINE is an abbreviation for long interspersed nuclear element. It refers to a group of non-LTR retrotransposons, which are widespread in the genome of many eukaryotes.

Lineage is a group of individuals, related by common descent, e.g., an *in vitro* cell line derived from a single cell.

Lineage sorting is a process where different gene lineages within an ancestral taxon are lost by drift or replaced by unique lineages evolving in different derived taxa.

Linear phase refers to when cell numbers constantly increases. The linear phase is located between the exponential growth and the deceleration phases.

Linearity is a degree to which the calibration curve of a device conforms to a straight line.

Linearized vector refers to a covalently closed circular DNA vector (typically a plasmid), which has been opened by restriction digestion to convert it to a linear molecule.

LINEs (long interspersed nuclear elements) refers to families of long (average length = 6 500 bp), moderately repetitive (about 10,000 copies). LINEs are cDNA copies of functional genes present in the same genome; also known as processed pseudo-genes.

Linkage disequilibrium is related to gametic phase disequilibrium.

Linkage equilibrium is the non-random association of alleles at two or more loci, that descend from single, ancestral chromosomes. Linkage disequilibrium is the occurrence of some combinations of alleles or genetic markers in a population more often or less often than would be expected from a random formation of haplotypes from alleles based on their frequencies. It is a second order phenomenon derived from linkage, which is the presence of two or more loci on a chromosome with limited recombination between them.

Linkage map is a linear or circular diagram that shows the relative positions of genes on a chromosome as determined by recombination fraction.

Linkage mapping measures meiotic recombination using polymorphic markers to produce the relative order of markers with respect to each other. Distance between markers is measured in centiMorgans (cM). A centiMorgan is equivalent to 1% cross-over rate.

Linked gene or **linked marker** refers to a gene or marker that is linked to another gene or marker.

Linker refers to a synthetic double-stranded oligonucleotide that carries the recognition sequence for one or more restriction endonucleases.

Lipases are enzymes which break down lipids into their component fatty acid and "head group" moieties.

Lipids are water-insoluble (fat) biomolecules that are highly soluble in organic solvents such as chloroform. Lipids serve as "fuel" molecules in organisms, highly concentrated energy stores, "signalling" molecules; and are basic components of cell membranes.

Lipofection is the delivery into eukaryotic cells of DNA, RNA or other compounds that have been encapsulated in an artificial phospholipid vesicle.

Lipophilic is the ability to dissolve or attach to lipids.

Lipopolysaccharide (LPS) refers to a compound containing lipid bound to a polysaccharide; often a component of microbial cells walls.

Liposomes are microscopic vesicles developed in a laboratory environment. Each liposome has an outer wall comprised of lipids similar and sometimes identical to those which compose the cell wall, allowing liposomes to interact directly with cells.

Liquefaction refers to enzymatic digestion of gelatinized starch to form lower molecular weight polysaccharides.

Liquid media is a media without a solidifying agent.

Liquid medium refers to a culture solution, without a solidifying agent, for *in vitro* cell growth.

Liquid membrane refers to thin films made up of liquids (as opposed to solids) which are stable in another liquid (usually water). Thus the liquid must not dissolve in the water, but nevertheless must be prevented from collapsing into small droplets.

Liquid nitrogen is a nitrogen gas condensed to a liquid with a boiling point of about −196°C. Commonly used as a medium for long-term storage of biological materials.

Litmus is a pH indicator paper (range 4.5–8.3) impregnated with an extracted lichen pigment. It turns red in acidic and blue in alkaline solutions. However, the use of litmus paper as an indicator is not a precise method of pH measurement.

Litmus paper is a pH indicator paper. It turns red in acidic and blue in alkaline solutions.

Live recombinant vaccine refers to a vaccine created by the expression of a pathogen antigen in a non-pathogenic organism.

Live vaccine refers to a living, non-virulent form of a micro-organism or virus that is used to elicit an antibody response that will protect the inoculated organism against infection by a virulent form of the micro-organism or virus.

Liver function tests are blood tests that indicate how well the liver is working.

Liver is a digestive organ located on the right side of the abdomen, under the ribs. It has many important functions, including storing and helping make blood, making bile (which aids in the digestion of fats in the food we eat), processing medicines and removing toxins from the bloodstream, and changing food and fats stored in our bodies into energy.

Living modified organism (LMO) is any living organism that possesses a novel combination of genetic material. A living organism is a biological entity that can transfer or replicate genetic material.

LMO is an abbreviation for living modified organism.

Lobectomy is the removal of an entire lobe of the lung.

Local adaptation refers to greater fitness of individuals in their local habitats due to natural selection.

Local scale is the spatial scale at which individuals routinely interact with their environment.

Locus (plural: loci) is a site on a chromosome.

Lod score refers to yhe logarithm of the odds of linkage between two loci. It is calculated from pedigree data, as the log (to base 10) of the ratio of the probability of the observed pedigree assuming linkage with a specified recombination fraction to the probability of the observed pedigree assuming no linkage, i.e., recombination fraction = 0.5. A lod score of +3.00 (which is odds of 1000:1) or greater is regarded as acceptable evidence for linkage; the -2.00 (which is the log of 1:100) or less indicates that no linkage exists.

Log of odds ratio (LOD) refers to logarithmic of the odds ratio. The odds ratio is the odds of an event occurring in one group to the odds of it occurring in another group. For example, if 80% of the individuals in a population are *Aa* and 20% are *AA*, then the odds of *Aa* over *AA* is four; there are four (4.0) times as many *Aa* as *AA* genotypes. The natural log of this ratio is often computed because it is convenient to work with statistically.

Log phase is an abbreviation for logarithmic phase.

Logarithmic phase (log or exponential growth phase) refers to the steepest slope of the growth curve.

Long day plant refers to plants requiring a period of short nights before the switch from vegetative to reproductive growth can be initiated.

Long interspersed nuclear element (LINE) refers to families of common DNA elements, of average length 6.5 kb, which are dispersed at numerous locations within the genome. The human genome contains over 500,000 LINEs (representing ca. 16% of the genome).

Long template is a DNA strand that is synthesized during the polymerase chain reaction and has a primer sequence at one end but is extended beyond the site that is complementary to the second primer at the other end.

Long terminal repeat (LTR) refers to a characteristic sequence of nucleotides that occurs at each end of a retrovirus element that has become integrated into the host genome. Involved in the integration process.

Long-day plant refers to plant requiring short nights before flowering is initiated.

Long-term self-renewal is the ability of stem cells to replicate themselves by dividing into the same non-specialized cell type over long periods (many months to years) depending on the specific type of stem cell.

Loop bioreactors are fermenters in which the fermenting material is cycled between a bulk tank and a smaller tank or loop of pipes.

Low voltage current is a current in which the waveform has an amplitude of less than 150 volts.

Lower esophageal sphincter is a muscle at the top portion of the stomach relaxes to allow food to pass from the esophagus to the stomach when we eat, and closes to keep food from moving back into the esophagus from the stomach.

Lower GI series is a study that looks at the rectum, the large intestine, and the lower part of the small intestine. A fluid called barium that shows up well on x-rays is given into the rectum as an enema. X-rays of the abdomen shows strictures (narrowed areas), obstructions (blockages), and other problems.

LPS is an abbreviation for lipopolysaccharide.

LTR is an abbreviation for long terminal repeat.

Luciferase is a generic term for the class of oxidative enzymes used in bioluminescence and is distinct from a photoprotein.

Lumen is the cavity of a tubular organ or instrument.

Luminescence is emission of light by a substance not resulting from heat; it is thus a form of cold body radiation. It can be caused by chemical reactions, electrical energy, subatomic motions, or stress on a crystal. This distinguishes luminescence from incandescence, which is light emitted by a substance as a result of heating.

Lung is an internal respiratory organ that functions in gas exchange; enables a person to breathe air.

Lung capacity is the amount of gas contained n the lung at the end of maximal inspiration.

Lung volume is the amount of air the lungs hold.

Luteinizing hormone is a pituitary hormone which causes growth of the yellow body of the **ovary** and also stimulates activity of the interstitial cells of the testis.

Lux (SI unit: lx) is the unit of measurement for illuminance (i.e., the amount of illumination) impinging upon a surface. One lx is the illuminance impinging upon a surface of 1 m², each point of which is at a distance of 1 m away from a uniform point source of light of 1 cd (candela). It supersedes the foot-candle.

Luxury consumption refers to nutrient absorption by an organism in excess of that required for optimum growth and productivity.

Lyase is any of a class of enzymes that catalyze either the cleavage of a double bond and the addition of new groups to a substrate, or the formation of a double bond.

Lymph is the fluid content of the lymphatic vessels, similar in composition to the interstitial fluid.

Lymphatic system is a network of lymphatic vessels that play a central role in the body's defense system.

Lymphocyte refers to white blood cells that are important components of the immune system of vertebrates.

Lymphokine refers to proteins that are released by lymphocytes to act on other cells involved in the immune response. The term includes interleukins and interferons.

Lymphoma refers to cancer originating in the lymph nodes, spleen and other lympho-reticular sites.

Lyophilize refers to freeze-dry (blood plasma or other biological substances). Freeze-drying, also known as lyophilization, lyophilization, or cryodesiccation, is a dehydration process typically used to preserve a perishable material or make the material more convenient for transport.

Lysis (Gr. *lysis*, a losing) refers to the destruction or breakage of cells either by viruses or by chemical or physical treatment.

Lysogen is a bacterial cell whose chromosome contains integrated viral DNA.

Lysogenic is a type or phase of the virus life cycle during which the virus integrates into the host chromosome of the infected cell, often remaining essentially dormant for some period of time.

Lysogenic bacteria (singular: Lysogenic bacterium) refer to bacteria harboring temperate (non-virulent, lysogenic) bacteriophages.

Lysogeny is a condition in which a bacteriophage genome (prophage) survives within a host bacterium, either as part of the host chromosome or as part of an extrachromosomal element, and does not initiate lytic functions.

Lysosome is a membrane-bound sac within the cytoplasm of animal cells that contains enzymes responsible for the digestion of material in food vacuoles, the dissolution of foreign particles entering the cell and, on the death of the cell, the breaking down of all cell structures.

Lytic cycle are the steps in viral production that lead to cell lysis. The lytic cycle is one of the two cycles of viral reproduction, the other being the lysogenic cycle. The lytic cycle results in the destruction of the infected cell and its membrane.

Lytic refers to a phase of the virus life cycle during which the virus replicates within the host cell, releasing a new generation of viruses when the infected cell undergoes lysis.

M

M13 strand refers to the single-stranded DNA molecule that is present in the infective form of bacteriophage M13.

MAAP is an abbreviation for multiple arbitrary amplicon profiling, which is a collective term for a number of related polymerase chain reaction techniques, all of which use arbitrary primers, and which generate a number of distinct amplification products.

mAb is an abbreviation for monoclonal antibody, which refers to a monospecific antibodies that are the same because they are made by identical immune cells that are all clones of a unique parent cell, in contrast to polyclonal antibodies which are made from several different immune cells.

Macerate refers to disintegrate tissue to disrupt cells. Commonly achieved via mechanical shearing, plasmolysis or enzymatic cell wall degradation.

Mach number is the relative velocity of a fluid compared to its sonic velocity. Mach number less than 1 correspond to sub-sonic velocities; and Mach number > 1 correspond to super-sonic velocities.

Macromolecule is a molecule of large molecular weight, such as proteins, nucleic acids and polysaccharides.

Macronutrient (Gr. *makros*, large + L. *nutrire*, to nourish) is for growth media: an essential element normally required in concentrations > 0.5 millimole/l.

Macrophages are large, white blood cells that ingest foreign substances and display on their surfaces antigens produced from the foreign substances, to be recognized by other cells of the immune system.

Macropropagation refers to production of plant clones from growing parts.

Macrospore (megaspore) are a type of spore that is present in heterosporous plants. These plants have two spore types, megaspores and microspores. Generally speaking, the megaspore, or large spore, germinates into a female gametophyte, which produces egg cells.

Mad cow disease refers to bovine spongiform encephalopathy. It is a fatal neurodegenerative disease (encephalopathy) in cattle that causes a spongy degeneration in the brain and spinal cord. BSE has a long incubation period, about 30 months to 8 years, usually affecting adult cattle at a peak age onset of four to five years, all breeds being equally susceptible. In humans, it is known as new variant Creutzfeldt–Jakob disease (vCJD or nvCJD).

MADS box is a highly conserved DNA sequence motif found in a large family of transcription factors, most of which play important roles in developmental processes.

Magenta is a type of plastic container frequently used for plant micropropagation and tissue culture.

Magnetic field is a technique of heating the tissues in shortwave diathermy in which the patient is not part of the electrical circuit.

Magnetic field strength (H, A/m) is a magnetic field produced by a current, independent of the presence of magnetic material. The units of H are ampere-turns per meter, or just amperes per meter.

Magnetic flux density or magnetic induction (designated by B) is a magnetic field produced in a substance by an external magnetic field. The units of B are tesla (T). One tesla is the magnetic flux density given by a magnetic flux of 1 weber per square meter. One henry (H) is the inductance of a closed circuit in which an electromotive force of 1 volt is produced when the electric current in the circuit varies uniformly at the rate of 1 ampere per second. The magnetic field strength and flux density are related according to: $B = \mu H$, where μ is the permeability (see under permeability).

Magnetic resonance imaging (MRI) is a diagnostic procedure that uses a combination of large magnets, radiofrequencies, and a computer to produce detailed images of organs and structures within the body.

Magnetic susceptibility (χ_m) is a proportionality constant between the magnetization M and the magnetic field strength H. The magnetic susceptibility is unitless.

Magnetization (M) is a total magnetic moment per unit volume of material. Also, a measure of the contribution to the magnetic flux by some material within an H field. The magnitude of M is proportional to the applied field as: $M = \chi_m \times H$, where, χ_m is the magnetic susceptibility.

Magnetostrictive material is a material that changes dimension in the presence of a magnetic field or generates a magnetic field when mechanically deformed.

Major histocompatibility antigen is a cell-surface macromolecule that allows the immune system to distinguish foreign or "nonself" from "self."

Major histocompatibility complex is group of genes that control several aspects of the immune response. They code for markers located on the surface of all body cells and are recognized by the body as 'self' (belonging to the body).

Malabsorption is a state arising from abnormality in absorption of food nutrients across the gastrointestinal (GI) tract.

Malabsorption syndrome is a condition related to abnormality in absorption of food nutrients across the GI tract

Malignant refers to having the properties of cancerous growth.

Malignant tumors of the brain are more common that benign ones; the most frequent of all are the liomas, which arise form the neuroglial cells.

Malnutrition is a situation caused by eating a poorly balanced diet, or by not eating enough food to meet the body's needs.

Malt extract is a mixture of organic compounds prepared from malt, used as a culture medium additive.

Malting is a process of generating starch-degrading enzymes in grain by allowing it to germinate in a humid atmosphere. *See* brewing.

Mammalian is the group of vertebrates that have: Internal development of the embryo; mammary glands that can produce milk; live-born young with a body covering of hair or fur; a four-chambered heart; a well-developed cerebral cortex; the ability to maintain a constant body temperature; and a permanent set of teeth.

Mammary glands are the milk-producing organs of female mammals, which provide food for the young.

Mammary tumors are tumors of the milk glands.

Management of farm animal genetic resources refer to the sum total of technical, policy and logistical operations involved in understanding (characterization), using and developing (utilization), maintaining (conservation), accessing, and sharing the benefits of animal genetic resources.

Mannitol ($C_6H_{14}O_6$) is a sugar alcohol widely distributed in plants.

Mannose ($C_6H_{12}O_6$) is a hexose component of many polysaccharides and mannitol.

Map distance is the standard measure of distance between loci, expressed in centiMorgans (cM). It is estimated from recombination fraction via a mapping function. For small recombination fractions, map distance equals the percentage of recombination (recombination frequency) betwen two genes. One % recombination = 1 cM. Sometimes called a map unit.

Map refers to: (1) verb: to determine the relative positions of loci (genes or DNA sequences) on a chromosome; (2) noun: a diagram showing the relative position of, and distances between, loci on a chromosome.

Map unit refers to one centiMorgan (1 cM). It is defined as the distance between chromosome positions (also termed, loci or markers) for which the expected average number of intervening chromosomal crossovers in a single generation is 0.01.

Mapping refers to the construction of a localized (around a gene), or broad-based (whole genome) genetic map.

Mapping function is a mathematical expression relating observed recombination fraction to map distance expressed in centiMorgans.

Megabase cloning is the cloning of very large DNA fragments.

Marfan syndrome is an autosomal dominant condition of connective tissue; affects the skeletal, ocular and cardiovascular systems.

Mariculture is a specialized branch of aquaculture involving the cultivation of marine organisms for food and other products in the open ocean, an enclosed section of the ocean, or in tanks, ponds or raceways which are filled with seawater.

Marine biotechnology explores and uses marine bioresources as the target for or origin of technological applications, which are used for the production of products and services.

Marker gene is a gene of known function or known location, used for marker-assisted selection or genetic studies.

Marker is an identifiable DNA sequence that facilitates the study of inheritance of a trait or a gene.

Marker-assisted introgression refers to the use of DNA markers to increase the speed and efficiency of introgression of a new gene or genes into a population.

Marker-assisted selection (MAS) refers to the use of DNA markers to improve response to selection in a population.

Marker peptide is a portion of fusion protein that facilitates its identification or purification.

Marsupial is a mammal whose distinguishing features include the birth of young at an early foetal stage of development, and generally, a pouch (marsupium) in which further development of the foetus occurs.

Martensite is a metastable iron phase supersaturated in carbon that is the product of a diffusionless (athermal) transformation from austenite.

MAS is an abbreviation for marker-assisted selection.

Mask is a pattern on glass, like a photographic negative, for producing integrated-circuit elements on semiconductor wafer.

Mass selection refers to (as practiced in plant and animal breeding) the selection of a number of individuals, on the basis of their individual phenotypes, to interbreed to form the next generation.

Mass spectroscopy is an analytical technique that produces spectra of the masses of the atoms or molecules comprising a sample of material.

Mass transfer is a process when a system contains two or more components whose concentrations vary from point to point, there is a natural tendency for mass to be transferred, minimizing the concentration differences within the system and moving it towards a equilibrium. It is the transport of one component from a region of higher concentration to that of a lower concentration.

Massive chronic intervillositis (MCI) refers to maternal blood-filled space that is filled with CD68-positive histiocytes and an increase in fibrin, occurring more commonly in the first trimester.

Mate pair is the sequence obtained from opposite ends of a particular clone are referred to as mate pairs.

Maternal effect refers to an effect attributable to a genetic contribution of the female parent of the individual being evaluated.

Maternal inheritance refers to an inheritance controlled by extrachromosomal (cytoplasmic) factors that are transmitted through the egg.

Matric potential is a water potential component, always of negative value, resulting from capillary, imbibitional and adsorptive forces. *See* pressure potential.

Matrigel is the trade name for a gelatinous protein mixture secreted by Engelbreth-Holm-Swarm (EHS) mouse sarcoma cells and marketed by Corning Life Sciences and by Trevigen, Inc., under the name Cultrex BME.

Maturation is the formation of gametes or spores.

Maximal voluntary ventilation is the volume of air that a subject can breathes with maximal effort over a given time interval.

Maximum likelihood estimate (MLE) is a method of parameter estimation that obtains the parameter value that maximizes the likelihood of the observed data.

Maximum likelihood is a statistical method of determining which of two or more competing alternative hypotheses (such as alternative phylogenetic trees) yields the best fit to the data.

MCM (Multi Chip Module) is the interconnection of two or more semiconductor chips in semiconductor – type package.

MCMC (Markov chain Monte Carlo) is a tool or algorithm for sampling from probability distributions based on constructing a Markov chain. Sometimes called a random walk Monte Carlo method.

MCS is an abbreviation for multiple cloning site.

MDA is an abbreviation for multiple drop array.

MDS (Multidimensional scaling) is a statistical graphing technique used to represent genetic distances between samples in two or three dimensions, and thereby visualizing similarities and differences between different groups or samples.

Mean blood pressure is an average blood pressure, taking account of the rise and fall that occurs with each heartbeat. It is often estimated by multiplying the diastolic pressure by two, adding the systolic pressure, and then dividing this sum by three.

Mean is the arithmetic average; the sum of all measurements or values in a sample divided by the sample size.

Measurand is a physical quantity, condition, or property that is to be measured.

Mechanical effects refer to ultrasonic effects that involve movement as a result of vibratory motion.

Mechanical valves are artificial valves made from metal, plastic, and/or pyrolytic carbon.

Mechatronics is a synergistic combination of precision mechanical engineering with electronic control.

Meckel's diverticulum is a condition that occurs as a baby is developing during pregnancy, in which a small pouch (or sac) forms

Meconium myonecrosis is a prolonged meconium exposure leads to toxic death of myocytes of placental vessels (umbilical cord or chorionic plate).

Media refers to the culture medium.

Median is the central value above and below which there are an equal number of measurements.

Medical Imaging combines knowledge of a unique physical phenomenon (sound, radiation, magnetism, etc.) with high speed electronic data processing, analysis and display to generate an image.

Medium (plural: media) refers to: (1) In plant tissue culture, a term for the liquid or solid formulation upon which plant cells, tissues or organs develop; (2) In general terms, a substrate for plant growth, such as nutrient solution, soil, sand, etc., e.g., potting medium.

Medium formulation refers to (in tissue culture) the particular constituents for the culture medium, commonly comprising macro- and micro-elements, vitamins, plant hormones, and a carbohydrate source.

Mega yeast artificial chromosome refers to a yeast artificial chromosome (YAC), which can carry particularly large inserts (up to 1Mbp) – standard YACs typically carry inserts of up to 500 kbp.

Megabase (Mb) is a length of DNA consisting of 10^6 base pairs (if double-stranded) or 10^6 bases (if single-stranded). 1 Mb = 10^3 kb = 10^6 bp.

Megabase cloning is the cloning of very large DNA fragments.

Megabyte (MB) = 220 (= 1,048,576) = about one million bytes of information.

Megadalton (MDa) refers to 1 MDa = 10^6 daltons.

Megagametophyte refers to the female gametophyte; the plant that develops from a megaspore.

Megahertz (Mhz) = One million cycles per second.

Megaspore (macrospore) is a haploid (n) spore developing into a female gametophyte in heterosporous plants.

Meiosis (Gr. *meioun*, to make smaller) refers to the special cell division process by which the chromosome number of a reproductive cell becomes reduced to half (n) the diploid (2n) or somatic number.

Meiotic analysis refers to the use of patterns of chromosome pairing at meiotic prophase and metaphase to detect relationships between chromosomes, from which can be deduced the relationship between the parents of the organism studied..

Meiotic drive refers to any mechanism that causes alleles to be recovered unequally in the gametes of a heterozygotc.

Meiotic product refers to one of the (usually four) cells formed by the two meiotic divisions (Meiosis).

Melanin is a pigment that is typically produced by specialized epidermal cells called melanocytes.

Melanoma is a type of cancer that begins in the melanocytes (the skin cells that produce pigments).

Melting temperature (T_m) is the temperature at which a double-stranded DNA or RNA molecule denatures into separate single strands. The T_m is characteristic of each DNA species and gives an indication of its base composition.

Membrane bioreactor refers to a vessel in which cells are cultured on or behind a permeable membrane, which allows the diffusion of nutrients to the cells, but retains the cells themselves. A variation is the hollow-fiber reactor.

Membrane is a thin layer of tissue that covers a surface or divides a space or organ.

Membrane oxygenator uses a semipermeable membrane through which oxygen diffuses into, and carbon dioxide diffuses out of, desaturated blood; no direct blood-gas interface exists, preferred method for long bypass runs.

Memory cell refers to long-lived B cells and T cells that mediate rapid secondary immune responses to a previously encountered antigen.

MEMS (MicroElectro Mechanical Systems) is an application area in itself than a manufacturing or fabrication technique that enables other application areas. Many authors use MEMS as shorthand to imply a number of particular application areas. As it is used in this book, MEMS is a "top- down" fabrication technology that is especially useful for integrating mechanical and electrical systems together on the same chip. MEMS techniques can be extended in the future to also help integrate biological and chemical components on the same chip.

Mendelian inheritance is a hereditary process where genetic traits are passed from parents to offspring and are explained in terms of chromosomes separating, independent assortment of genes and the homologous exchange of segments of DNA. There are three modes of Mendelian inheritance: autosomal dominant, autosomal recessive and X-linked inheritance.

Mendelian population refers to a natural, interbreeding unit of sexually reproducing plants or animals sharing a common **gene** pool.

Mendelian segregation is the random separation of paired alleles (or chromosomes) into different gametes.

Mendelism is a theory of heredity that forms the basis of classical genetics, which was proposed by Gregor Mendel in 1866.

Mendel's Laws are two laws summarizing Gregor Mendel's theory of inheritance. The **Law of Segregation** states that each hereditary characteristic is controlled by two 'factors' (now called alleles), which segregate and pass into separate germ cells. The **Law of Independent Assortment** states that pairs of 'factors' segregate independently of each other when germ cells are formed.

Meninges are the three membranes that envelop the brain and spinal cord: the dura madre, arachnoid, and pia madre.

Meningitis is an infection of the cerebrospinal fluid.

Mericlinal refers to a chimera with tissue of one genotype partly surrounded by that of another genotype.

Mericloning is a propagation method using shoot tips in culture to proliferate multiple buds, which can then be separated, rooted and planted out.

Meristele is the vascular cylinder tissue in the stem.

Meristem (Gr. *meristos*, divisible) refers to an undifferentiated but determined tissue, the cells of which are capable of active cell division and differentiation into specialized and permanent tissue such as shoots and roots.

Meristem culture refers to a tissue culture containing meristematic dome tissue without adjacent leaf primordia or stem tissue. The term may also imply the culture of meristemoidal regions of plants, or meristematic growth in culture.

Meristem tip culture refers to the cultures derived from meristem tip explants. Used widely to achieve virus elimination and axillary shoot proliferation, less commonly for callus production.

Meristem tip refers to an explant comprising the meristem (meristematic dome) and usually one pair of leaf primordia. Also refers to explants originating from apical meristem tip or lateral or axillary meristem tip.

Meristemoid refers to a localized group of callus cells, characterized by their accumulation of starch, RNA and protein, and giving rise to adventitious shoots or roots.

Meristic character is a trait of an organism that can be counted using integers (e.g., fin rays or ribs).

Merozygote is a partial zygote produced by a process of partial genetic exchange, such as transformation in bacteria.

Mesemchymal stem cell is also known as bone marrow stromal cells. Mesenchymal stem cells are rare cells, mainly found in the bone marrow, that can give rise to a large number of tissue types such as bone, cartilage (the lining of joints), fat tissue, and connective tissue (tissue that is in between organs and structures in the body).

Mesh bioreactor is also called a filter bioreactor. It refers to a vessel in which cells are cultured on or behind a permeable membrane, which allows the diffusion of nutrients to the cells, but retains the cells themselves. A variation is the hollow-fiber reactor.

Mesoderm is the middle of three germ layers that gives rise later in development to such tissues as muscle, bone, and blood.

Mesophile is a micro-organism able to grow in the temperature range 20–50°C; optimal growth often occurs at about 37°C.

Mesophyll (Gr. *mesos*, middle + *phyllon*, leaf) refers to leaf parenchyma tissue occurring between epidermal layers.

Messenger RNA (mRNA) refers to the RNA molecule that specifies the amino acid sequence of a protein. It is the intermediary molecule between DNA and ribosomes. mRNA takes encoded specifications from the cell's DNA, processes the message and takes it to the ribosomes, where the amino acids are assembled into polypeptides, which are then folded into proteins.

Metabolic acidosis refers to the levels of plasma K below 2 mEq/L. It is a type of acidosis caused by the inability to excrete hydrogen ions, the production of numerous fixed and/or organic acids, or a severe bicarbonate loss [all by H2].

Metabolic cell is a cell that is not dividing.

Metabolic rate is a rate of energy (heat) production of the body which varies with the level of activity.

Metabolism (Gr. *metobolos*, to change) is the biochemical process by which nutritive material is built up into living matter, or aids in building living matter, or by which complex substances and food are broken down into simple substances.

Metabolite refers to: (1) A low-molecular-weight biological compound that is usually synthesized by an enzyme; (2) A compound that is essential for a metabolic process.

Metabolome is the quantitative complement of all the low molecular weight molecules present in cells in a particular physiological or developmental state.

Metabonomics and metabolomics are very similar terms. **Metabonomics** deals with integrated, multicellular and biological systems, including communicating extracellular environments. **Metabolomics** deals with simple cell systems and, at least in terms of published data, mainly intracellular metabolite concentrations.

Metacentric chromosome is a chromosome with the centromere near the middle and, consequently, two arms of about equal length.

Metacentric is a chromosome in which the centromere is centrally located.

Metagenomics refers to the research using DNA

sequencing technologies to sample the structure and function of the genomes of organisms inhabiting a common environment.

Metal affinity chromatography refers to a chromatographic technique, in which a compound interacting with a specific metal ion can be captured by immobilizing the relevant ion on the column's solid matrix.

Metal nanoshells is a new type of nanoparticle composed of a semiconductor or dielectric core coated with an ultrathin conductive layer. By adjusting the relative core and shell thicknesses, metal nanoshells can be fabricated that will absorb or scatter light at any wavelength across the entire visible and infrared range of the electromagnetic spectrum.

Metalloenzyme is an enzyme which requires the presence of a metal in order to be catalytically active.

Metallothionein is a protective protein that binds heavy metals such as cadmium and lead.

Metaphase (Gr. *meta*, after + *phasis*, appearance) refers to the stage of mitosis or meiosis (following prophase and preceding anaphase) during which the chromosomes, or at least the

kinetochores, lie in the central plane of the spindle.

Metapopulation is a collection of spatially divided subpopulations that experience a certain degree of gene flow among them.

Metapopulation scale is the spatial scale at which individuals migrate between local subpopulations, often across habitat that is unsuitable for colonization.

Metastasis refers to the spread of cancer cells to previously unaffected organs.

Methionine is a sulphur-containing amino acid.

Methylation refers to the addition of a methyl group ($-CH_3$) to a molecule, most commonly in the context of DNA where cytosine and, less often, adenine residues can be modified in this way, sometimes resulting in a change in transcription.

Methylmalonic acidemia is a group of conditions characterized by the inability to metabolize methylmalonic acid or by a defect in the metabolism of Vitamin B12.

MHC is an abbreviation for major histocompatibility complex.

Micelle (plural: micelles or micellae) is an aggregate of surfactant molecules dispersed in a liquid colloid. A typical micelle in aqueous solution forms an aggregate with the hydrophilic "head" regions in contact with surrounding solvent, sequestering the hydrophobic single-tail regions in the micelle centre. This phase is caused by the packing behavior of single-tail lipids in a bilayer.

Michaelis constant is abbreviated as K_m.

Microalgal culture refers to the culture in bioreactors of microalgae (including seaweeds).

Micro-array refers to a large set of cloned DNA molecules immobilized as a compact and orderly pattern of sub-microliter spots onto a solid matrix (typically a glass slide). It is used to analyse patterns of gene expression, presence of markers, or nucleotide sequence.

Microbe is a microscopic organism such as a virus, bacterium, fungus or protozoan (single-celled organism). **Microbe** is a general term for a micro-organism.

Microbial genetics is the study of genetics in microorganisms.

Microbial mat refers to a layered microbial populations, usually growing on the surface of a solid medium or on a membrane.

Microbiology is the study of microorganisms and how they interact with the environment and other organisms.

Microbody (Gr. *mikros*, small + body) is a cellular organelle always bound by a single membrane, frequently spherical, from 20 to 60 nm in diameter, containing a variety of enzymes.

Microbubbles are very small encapsulated gas bubbles (diameters of micrometers) that can be used in diagnostic and therapeutic applications. Upon exposure to sufficiently intense ultrasound, microbubbles will cavitate, rupture, disappear, release gas content, etc.

Micro-carriers are small particles used as a support material for cells, and particularly mammalian cells, which are too fragile to be pumped and stirred as bacterial cells are in a large-scale culture.

Microchemistry is the development and use of techniques and equipment to study or perform chemical reactions, with small quantities of materials, frequently less than a milligram or a milliliter.

Microchromosomes are small chromosomes found in many bird species which, unlike heterochromosomes, carry functional genes.

micro-CT can have excellent spatial resolution, up to 6 μm when combined with contrast agents.

However, the radiation dose needed to achieve this resolution is lethal to small animals, and a 50 μm spatial resolution is a better representation of the limits of micro-CT. It is also decent in terms of image acquisition times, which can be in the range of minutes for small animals. In addition, micro-CT is excellent for bone imaging.

Microdroplet array (multiple drop array, MDA; hanging droplet technique) is used to evaluate large numbers of media modifications, employing small quantities of medium into which are placed small numbers of cells.

Microelectronics is a subfield of electronics. The microelectronics relates to the study and manufacture (or microfabrication) of very small electronic designs and components. Usually, but not always, this means micrometre-scale or smaller. These devices are typically made from semiconductor materials.

Micro-element is an element required in very small quantities.

Microemulsions are clear, thermodynamically stable, isotropic liquid mixtures of oil, water and surfactant, frequently in combination with a cosurfactant. The three basic types of microemulsions are direct (oil dispersed in water, o/w), reversed (water dispersed in oil, w/o) and bicontinuous.

Micro-encapsulation is a process of enclosing a substance in very small sealed capsules from which material is released by heat, solution or other means.

Microengineering is the process of fabrication of miniature structures of micrometre scales and smaller. Historically, the earliest microfabrication processes were used for integrated circuit fabrication, also known as "semiconductor manufacturing" or "semiconductor device fabrication."

Micro-environment (Gr. *mikros*, small + O.F. *environ*, about) refers to the environment close enough to the surface of a living or non-living object to be influenced by it.

Microfabrication includes techniques used to manufacture integrated circuits (ICs), discrete microelectronic devices, MEMS devices such as sensors and actuators, and various electro-optic devices. University of Louisville Microfabrication Lab is currently serving as a center for research activity in the areas of micromachined sensors and actuators, electro-optic devices, special-purpose microelectronic devices, planar waveguides, chemical transducers, microstrip and microgap radiation detectors, micromachined nozzles, and micromachined ink- jet printheads.

Microfibrils (Gr. *mikros*, small + *fibrils*, diminutive of fiber) are exceedingly small fibers visible only at the high magnification of the electron microscope.

Microfluidics is a multidisciplinary field intersecting engineering, physics, chemistry, biochemistry, nanotechnology, and biotechnology, with practical applications to the design of systems in which small volumes of fluids will be handled. Microfluidics is used in the development of inkjet printheads, DNA chips, lab-on-a-chip technology, micro-propulsion, and micro-thermal technologies. It deals with the behavior, precise control and manipulation of fluids that are geometrically constrained to a small, typically sub-millimeter, scale.

Microgametophyte refers to the male gametophyte that develops from the microspores of heterosporous plants. The pollen grains of gymno-sperms and angiosperms are microgametophytes.

Micrograft is a small graft of tissue containing a single hair taken from a thin strip of tissue in the patient's donor area which is located in the back of the head.

Microinjection is the insertion of a substance into a cell through a microelectrode. Typical applications include the injection of drugs, histochemical markers (such as horseradish peroxidase or lucifer yellow) and RNA or DNA in molecular biological studies.

Micro-isolating system refers to the mechanical separation of single cells or protoplasts thus allowing them to proliferate individually.

Micromachining are techniques for fabricating MEMS.

Micromanipulation is the technique or practice of manipulating cells or tissues

Micromaterials refer to materials of μ size.

micro-MRI has good spatial resolution, up to 100 μm and even 25 μm in very high strength magnetic fields. It also has excellent contrast resolution to distinguish between normal and pathological tissue. Micro-MRI can be used in a wide variety of applications, including anatomical, functional, and molecular imaging.

Micron (micrometer; Gr. *mikros*, small) is a unit of distance: 10^{-6} m $= 0.001$ mm $= 1$ μm.

Micronucleus refers to a nucleus, distinct from and smaller than the main nucleus, but lying within the same cell.

Micronutrient (Gr. *mikros*, small + *L. nutrire*, to nourish) is an essential element normally required in concentrations < 0.5 millimole/liter.

Microorganism is an organism that is visible only under a microscope, such as protozoa, bacteria, fungi and viruses.

Microparticles refer to applications that include calibration of flow cytometers, particle and hematology analyzers, confocal laser scanning microscopes and zetapotential measuring instruments; flow measurements in gases and liquids like Laser Doppler Anemometry (LDA); Particle Dynamics Analysis (PDA), and Particle Image Velocimetry (PIV); medical diagnostics; separation phases for chromatography; support for immobilized enzymes; spacer in liquid crystal displays (LCD's); peptide synthesis; cell separation; tracers in environmental science; model systems in medicine, biochemistry, colloid chemistry, and aerosol research.

Micro-PAT can be described as an imaging modality that is applicable in a wide variety of functions. It combines the high sensitivity of optical imaging with the high spatial resolution of ultrasound imaging. For this reason, it can not only image structure, but also separate between different tissue types, study hemodynamic responses, and even track molecular contrast agents conjugated to specific biological molecules. Furthermore, it is non-invasive and can be quickly performed, making it ideal for longitudinal studies of the same animal.

Micro-PET refers to using a beta emitter such as Fluorine-18 or Copper-61, so that tomographic images can be collected using the Siemens Inveon system which provides the highest commercially available 3-dimensional image resolution. The LSO-based system provides greater than 10% sensitivity in the center of the field of view and is expected to deliver less than 1.4 mm FWHM spatial resolution in images reconstructed using the filtered back projection algorithm.

Microphone is a device that produces voltage or current in response to a sound wave.

Microplast refers to vesicle produced by subdivision and fragmentation of protoplasts or thin-walled cells.

Microprocessor is a chip containing the logical elements for performing calculations and carrying out stored instructions.

Microprojectile bombardment (biolistics) is a method by which foreign substances such as DNA are introduced into living cells and tissues via high-velocity microprojectiles.

Micropropagation refers to one that is miniaturized *in vitro* multiplication and/or regeneration of plant material under aseptic and controlled environmental conditions on specially prepared media that contain substances necessary for growth; used for three general types of tissue: excised embryos (= embryo culture); shoot-tips (= meristem culture or mericloning); and pieces of tissue that range from bits of stems to roots.

Micropyle refers to: (1) A small opening in the surface of a plant ovule through which the pollen tube passes prior to fertilization; (2) A small pore in some animal cells or tissues.

Microsatellite refers to a segment of DNA characterized by a variable number of copies (typically 5–50) of a sequence of around 5 or fewer bases (called a repeat unit).

Microshock is an electrical shock that is imperceptible because of a leakage of current of less than 1 mamp.

Micro-SPECT has very good sensitivity and only nanograms of molecular probes are needed. Furthermore, by using different energy radioisotopes conjugated to different molecular targets, micro-SPECT has the advantage over micro-PET in being able to image several molecular events simultaneously.

Microspheres are small spherical particles, with diameters in the micrometer range (typically 1 μm to 1000 μm (1 mm)). Microspheres are sometimes referred to as microparticles. Microspheres can be manufactured from various natural and synthetic materials.

Microspore is the smaller of the two kinds of meiospores produced by heterosporous plants in the course of microsporogenesis; in seed plants, microspores give rise to the pollen grain, the male gametophyte.

Microstructures refer to the fabrication, characterization and conceptual understanding of synthetic microstructures in many different material systems including silicon, III-V and II-VI semiconductors, metals, ceramics and organics.

Microsystem is a microscale machine that can sense information from the environment and act accordingly. Outside the U.S., it can also refer to microelectromechanical systems (MEMS).

MicroTAS (micro Total Analysis Systemsor, uTAS) research offer many advantages over traditional analysis systems. Low power consumption and small reaction volumes, faster analysis, ultrasensitive detection, and minimal human intervention are key parameters in the development of micro- TAS. The development of MicroTAS is linked to the design of liquid handling micro-devices.

Microtransponder approach to treating tinnitus is based on decades of scientific research. The innovative approach pairs a well-known treatment called vagus nerve stimulation with listening to specific tones.

Microtuber refers to the miniature tuber, produced in tissue culture, which is readily regenerable into a normal tuberous plant.

Microtubules is a minute filament in living cells that is composed of the protein tubulin and occurs singly, in pairs, triplets or bundles.

Micro-ultrasound is the only real-time imaging modality per se, capturing data at up to 1000 frames per second. This means that not only is it more than capable of visualizing blood flow *in vivo*, it can even be used to study high speed events such as blood flow and cardiac function in mice.

Micturition relates to urination.

Middle lamella (L. *lamella*, a thin plate or scale) refers to original thin membrane separating two adjacent protoplasts and remaining as a distinct cementing layer between adjacent cell walls.

Mid-parent value refers to the average of the phenotypic measure, with respect to a given trait, of the two parents used to generate the population being analyzed.

Migration is the movement of individuals from one generically distinct population to another resulting in gene flow.

Miller indices is a set of 3 integers (4 for hexagonal) that designate crystallographic planes, as determined from reciprocals of fractional axial intercepts.

Mineralization is the conversion of organic compounds into inorganic (mineral) ones. For example, the conversion of ethanol into carbon dioxide and water.

Miniaturization has been desirable for many technologies for overall cost reduction. Important to remember that building space is often the least available and most expensive component of an overall laboratory budget.

Miniaturization of DNA diagnostics are the key in sample preparation and assay. To reduce the size of samples by a factor of 10 or greater, barriers in microfluidics, micromachining, robotics, microchemistry, nucleic acid chemistry, and surface chemistry must be overcome. To implement miniaturized protocols accurately and efficiently, substantial automation of the process will be required. In the development of miniaturized systems, it is essential that the system can be adapted for high levels of parallelization.

Minimally invasive procedure (MIP) is similar to newer surgical techniques, such as *Minimally Invasive Surgery* (MIS). It requires a simulated learning environment facilitated by intensive cadaveric training. MIP is any procedure (surgical or otherwise) that is less invasive than open surgery used for the same purpose. A minimally invasive procedure typically involves use of arthoscopic (for joints and the spine) or laparoscopic devices and remote-control manipulation of instruments with indirect observation of the surgical field through an endoscope or large scale display panel, and is carried out through the skin or through a body cavity or anatomical opening.

Minimum effective cell density is the inoculum density below which the culture fails to give reproducible cell growth. The minimum density is a function of the tissue (species, explant, cell line) and the culture phase of the inoculum suspension. Minimum density decreases inversely to the aggregate size and division rate of the stock culture.

Minimum inoculum size refers to the critical volume of inoculum required to initiate culture growth, due to the diffusive loss of cell materials into the medium. The subsequent culture growth cycle is dependent on the inoculum size, which is determined by the volume of medium and size of the culture vessel.

Minimum viable population (MVP) is the minimum

population size at which a population is likely to persist over some defined period of time.

Mini-prep is a small-scale (mini-) preparation of plasmid or phage DNA. It is used to analyse DNA in a cloning vector after a cloning experiment.

Minisatellite is a form of VNTR in which the repeat units typically range from 10 to 100 bases. They are usually detected by Southern hybridization, using a probe comprising a clone of the repeat unit.

Minitubers are small tubers (5–15 mm in diameter) formed on shoot cultures or cuttings of tuber-forming crops, such as potato.

MIPS (Millions of Instructions per second) is a measure of computing power.

Mismatch refers to the occurrence of a non-complementary pairs of bases in a double helix of DNA, e.g., A:C, G:T.

Mismatch repair refers to DNA repair processes that correct mismatched base pairs.

Missense mutation is a mutation that changes a codon for one amino acid into a codon specifying another amino acid.

Mist propagation is the application of fine droplets of water to maintain humidity around plantlets or cuttings, which have not yet developed effective roots.

Mitochondria is the cell organelles responsible for energy production.

Mitochondrial DNA (mtDNA) refers to the genetic material of the mitochondria – the organelle that generates energy for the cell. The DNA in mitochondria is different from that in the nucleus.

Mitochondrion (Gr. *mitos*, thread + *chondrion*, a grain; plural: mitochondria) refers to organelle possessing its own DNA which appear in all eukaryotic cells (and never in prokaryotic cells) and produce adenosine triphosphate as an energy source for the cell via oxidative phosphorylation.

Mitogen is a substance that can cause cells to initiate mitosis.

Mitosis (Gr. *mitos*, a thread; adj.: mitotic; plural: mitoses) refers to misjunction of replicated chromosomes and division of the cytoplasm to produce two genetically identical daughter cells.

Mitral valve controls blood flow between the heart's left atrium (upper chamber) and left ventricle (lower chamber).

Mixed bud is a bud containing both rudimentary leaves and flowers.

Mixoploid are cells with variable (euploid, aneuploid) chromosome numbers. Mosaics or chimeras differ in chromosome number as a result of a variety of mitotic irregularities.

Mobility (electron, and hole) is the proportionality constant between the carrier drift velocity and applied electric field.

Mobilization refers to: (1) The transfer between bacteria of a non-conjugative plasmid by a conjugative plasmid; (2) The transfer between bacteria of chromosomal genes by a conjugative plasmid.

Mobilizing functions are the genes on a plasmid that give it the ability to facilitate the transfer of either a non-conjugative or a conjugative plasmid from one bacterium to another.

Modal class is the class having the greatest frequency.

Mode refers to the class having the greatest frequency, in a frequency distribution.

Model refers to: (1) A mathematical description of a biological phenomenon; (2) A simplified biological system used to test hypotheses (e.g., *Arabidopsis thaliana* as a model plant).

Modern biotechnology is a term to refer to biotechnological techniques for the manipulation of genetic material and the fusion of cells beyond normal breeding barriers. The most obvious example is genetic engineering to create genetically modified/ engineered organisms (GMOs/ GEOs) through "transgenic technology" involving the insertion or deletion of genes.

Modification refers to: (1) Enzymatic methylation of a restriction enzyme DNA recognition site; (2) Specific nucleotide changes in DNA or RNA molecules.

Modifier (modifying gene) is a gene that affects the expression of some other gene.

Modifying gene is a gene that affects the expression of some other gene.

Modulation refers to any alteration in the magnitude or any variation in the duration of an electrical current.

Modulus of elasticity (E) is a ratio of stress to strain when deformation is totally elastic. Also called Young's modulus.

MOEMS (Micro Optical Electro Mechanical systems) integrate waveguides or other optical features into the body of the silicon chip, in addition to mechanical and electrical components,.

MOET is an abbreviation for multiple ovulation and embryo transfer. MOET is a way to produce an animal of certain genetic qualities faster. The multiple ovulation part is where the female animal is manipulated hormonally to produce more than the usual number of eggs during ovulation.

Molality or molal concentration is the amount of substance per unit mass of solvent or mol. kg^{-1}.

Molarity is a concentration in a liquid solution, in terms of the number of moles of a solute dissolved in 106 mm^3 (103 cm^3) of solution in $mol.l^{-1}$.

Molding (plastics) is shaping of a plastic material by forcing it, under pressure at a high temperature, into a mold cavity.

Mole (M) is an amount of substance that has a weight in grams numerically equal to the molecular weight of the substance. Also called gram molecular weight. A mole contains 6.023×10^{23} molecules or atoms of a substance.

Molecular biology is the study of living processes at the molecular level. **Molecular biology** (mə'lɛkjʊlər) is the branch of biology that deals with the molecular basis of biological activity. This field overlaps with other areas of biology and chemistry, particularly genetics and biochemistry. Molecular biology chiefly concerns itself with understanding the interactions between the various systems of a cell, including the interactions between the different types of DNA, RNA and protein biosynthesis as well as learning how these interactions are regulated.

Molecular chaperone are proteins that assist the non-covalent folding or unfolding and the assembly or disassembly of other macromolecular structures.

Molecular clock is the observation that mutations sometimes accumulate at relatively constant rates, thereby allowing researchers to estimate the time since two species diverged (TMRCA).

Molecular cloning is the biological amplification of a DNA sequence via the mitotic division of a host cell into which it has been transformed or transfected.

Molecular computing is a generic term for any computational scheme, which uses individual atoms or molecules as a means of solving computational problems. Molecular computing is most frequently associated with DNA computing, because that has made the most progress, but it can also refer to quantum computing or molecular logic gates.

Molecular dynamics is a simulation procedure consisting of the computation of the motion of atoms in a molecule or of individual atoms or molecules in solids, liquids and gases, according to Newton's laws of motion. The forces acting on the atoms, required to simulate their motions, are generally calculated using molecular mechanics force fields.

Molecular electronics offers the tantalizing prospect of eventually building circuits with critical dimensions of a few nanometers. Some basic devices utilizing molecules have been demonstrated, including tunnel junctions with negative differential resistance, rectifiers and electrically configurable switches that have been used in simple electronic memory and logic circuits.

Molecular genetics is the branch of genetics that studies the molecular structure and function of genes, or that (more generally) uses molecular markers to test hypotheses.

Molecular graphics is a technique for the visualization and manipulation of molecules on a graphical display device.

Molecular imaging originated from the field of radiopharmacology due to the need to better understand the fundamental molecular pathways inside organisms in a noninvasive manner.

Molecular marker refers to a genetic marker which is assayed at the DNA level.

Molecular mechanics is the calculation of molecular conformational geometries and energies using a combination of empirical force fields. The method implies transferability of the potential functions within a network of similar molecules.

Molecular modeling is a technique for the investigation of molecular structures and properties using computational chemistry and graphical visualization techniques in order to provide a plausible three-dimensional representation under a given set of circumstances. *Molecular Modeling* includes subjects, such as: computer-aided molecular design, rational drug design, *de novo* ligand design and receptor modeling, application of computational and modeling methods in the field of medical chemistry, protein and peptide modeling, quantum chemistry, application of semi empirical, DFT and *ab initio* calculations, prediction of bio-logical activities (QSAR) and physico- chemical properties (QSPR), molecular mechanics/dynamics simulation of poly-mers and biopolymers, genetic algorithms and neural nets, modeling of catalysts, advanced materials, and stationary phases in separation science, enhanced desktop computational tools for the life sciences visualiza-tion, classification and handling of chemical data. **Molecular modeling** applications use falls into two broad categories: interactive visualization and computational analyses. Three of the most prominent uses of modern molecular model-ing applications are structure analysis, homology modeling, and docking. Molecular model-ing software includes AMBER, DOCK, MODELER, RasMol and many other programs.

Molecular models are models used experimentally or theoreti-cally to study molecular shape, electronic properties, or inter-actions; includes analogous molecules, computer gener-ated graphics, and mechanical structures.

Molecular mutations refer to changes to the genetic mate-rial of a cell, including single nucleotide changes, deletions,

and insertions of nucleotides as well as recombinations and inversions of DNA sequences.

Molecular nanoscience is an emerging interdisciplinary field that combines the study of molecular/biomolecular systems with the science and technology of nanoscale struc-tures and systems.

Molecular pathway is a series of interactions between genes, proteins, and other biological factors that leads to a biologi-cal effect, such as the differ-entiation of one cell type to another, the division of a cell, the production of antibodies, cell homing, the formation of tissue and body patterns during development, the secretion of factors such as insulin, etc.

Molecular pharming (biopharm-ing) is also known as molecular farming or pharming. It refers to the use of genetic engineer-ing to insert genes that code for useful pharmaceuticals into host animals or plants that would otherwise not express those genes, thus creating a genetically modified organism (GMO).

Molecular robotics is a new family of six degree of freedom posi-tional devices.

Molecule (L. diminutive of *moles,* a little mass) is a unit of matter, the smallest portion of an element or a compound that retains chemical identity with the substance in mass.

Monitoring are activities conducted to measure levels, concentrations or quantities of material and the use of these measurement results to evaluate potential exposures and doses, and to determine existing environmental conditions, pollutant levels (rates) and effects on species in the environment.

Monoclonal antibodies are produced by injecting animals to elicit a response from lymphocytes to produce antibodies.

Monocotyledon (monocot) refers to a flowering plant whose embryo has one cotyledon. Examples are cereals (corn, wheat, rice etc.), banana, and lily.

Monoculture is the agricultural practice of cultivating a single crop on a whole farm or area.

Monoecious is a plant in which male and female organs are found on the same plant but in different flowers (for example maize).

Monogastric animal is a non-ruminant animal with a simple stomach.

Monogenic refers to one that is controlled by a single gene, as opposed to multigenic.

Monohybrid (Gr. *monos*, solitary + *L. hybrida*, a mongrel) is the offspring of two homozygous parents that differ from one another by the alleles present at only one locus.

Monohybrid cross is a cross between parents differing in only one trait or in which only one trait is being considered.

Monohybrid heterozygous refers to one gene.

Monokine is a generic name for proteins that are released by monocytes to act on other cells involved in the immune response. It is a sub-class of cytokines.

Monolayer is a single layer of cells growing on a surface.

Monolignols refers to the building blocks of lignin that undergo polymerization.

Monomer refers to a small molecule (in the biological sciences typically individual amino acids, nucleotides or monosaccharides) that can combine with identical or similar others to form a larger, more complex molecule called a polymer.

Monomorphic refers to the presence of only one allele at a locus, or the presence of common allele at a high frequency (>95% or 99%) in a population.

Monophyletic describes any group of organisms that are assumed to have originated from the same ancestor.

Monophyly is the presence of a monophyletic group.

Monosaccharide refers to a simple sugar (e.g., glucose, fructose).

Monosomic (noun: monosomy) describes a diploid organism lacking one chromosome (2n – 1) of its proper (disomic) complement; a form of aneuploidy.

Monosomy refers to when one chromosome of a pair is missing. In humans, this would result in a total of 45 chromosomes. An example of monosomy is 45, X, also known as Turner syndrome.

Monotypic is a taxonomic group that encompasses only one taxonomic representative. The reptile family that contains tuatara (Sphenodontia) is currently monotypic.

Mono-unsaturated are molecules, such as fats, with only one double bond in their chemical structure.

Mono-unsaturates are oils containing mono-unsaturated fatty acids (i.e., where one -CH_2-CH_2- group in the hydrocarbon chain is replaced by -CH=CH-).

Monozygotic twin refers to one of a pair of twins derived from a single fertilized egg. Synonym is identical twin.

Monte Carlo technique is a simulation procedure consisting of randomly sampling the conformational space of a molecule.

Moore's law states that the number of transistors per computer chip will double roughly every two years.

Moratorium is a temporary prohibition or suspension of an activity.

Morbidity is a state of being diseased. The number of sick persons or cases of disease in relationship to a specific population.

Morphogen is a substance that stimulates the development of form or structure in an organism.

Morphogenesis is the development, through growth and differentiation, of form and structure in an organism.

Morphogenic response is the effect on the developmental history of a plant or its parts exposed to a given set of growth conditions or to a change in the environment.

Morphology (Gr. *morphe*, form + *logos*, discourse) refers to: (1) The science of studying form and its development; (2) General: Shape, form, external structure or arrangement.

MOS (metal oxide semiconductor) is an integrated circuit containing n-channel and p-channel, MOSFETs.

Mosaic is an organism or part of an organism that is composed of cells with different origin.

Mosaicism is the presence of two or more cell populations that have a different genetic or chromosomal makeup in a single individual or tissue.

MOSFET (Metal Oxide Semiconductor Field Effect Transistor) is a device where gate electrode potential controls current flow

Mother plant refers to the donor plant.

Motif refers to a conserved sequence of nucleotides or amino acids that can be associated with some function of, respectively, a length of DNA or a protein.

Motility is the movement of food through the digestive tract, aided by contractions of muscles in the stomach and intestines known as peristalsis.

Movable genetic element (MGE) are a type of DNA that can move around within the genome. They include: Transposons (also called transposable elements); Retrotransposons; DNA transposons; Insertion sequences; Plasmids; Bacteriophage elements, like Mu, which integrates randomly into the genome; Group II introns; and Group I introns. The total of all mobile genetic elements in a genome may be referred to as the mobilome.

mRNA (messenger RNA) refers to the RNA transcript of a protein-encoding gene.

MRUs is an abbreviation for minimum recognition units.

MSD stands for multiple factor sex determination.

mtDNA stands for mitochondrial DNA.

MU stands for management unit.

Mucus is a thick, jelly-like substance made by the intestines and other organs of the body (such as the nose), that helps coat and protect the lining of the organ.

Multi-copy describes plasmids which replicate to produce many plasmid molecules per host genome.

Multifactorial is a characteristic influenced in its expression by many factors, both genetic and environmental.

Multigene family is a set of genes (not necessarily mapping to the same genomic location) that are related in nucleotide sequence and/or that produce polypeptides with similar amino acid sequences. Sequence similarity does not always result in functional similarity.

Multigenic refers to a trait controlled by several genes, as opposed to monogenic. Synonym is polygenic.

Multi-locus probe is a probe that hybridizes to a number of different sites in the genome of an organism.

Multimer (multimeric) is a protein made up of more than one peptide chain.

Multiple alleles refers to the existence of more than two alleles at a locus in a population.

Multiple arbitrary amplicon profiling refers to a collective term for a number of related polymerase chain reaction techniques, all of which use arbitrary primers, and which generate a number of distinct amplification products.

Multiple cloning site (MCS), also called a **polylinker**, is a short segment of DNA which contains many (up to ~20) restriction sites – a standard feature of engineered plasmids.

Multiple drop array is abbreviated as MDA.

Multiple ovulation and embryo transfer (MOET) is a technology by which a single female that usually produces only one or two offspring can produce a litter of offspring.

Multiplex refers to: (1) The simultaneous amplification of a number of amplicons in a single polymerase chain reaction, achieved by including more than one set of primers in the reaction mix; (2) The inheritance pattern of alleles in autopolyploids.

Multipotent is the potential to make a few cell types in the body.

Multipotent stem cells are stem cells whose progeny are of multiple differentiated cell types, but all within a particular tissue, organ, or physiological system. For example, blood-forming (hematopoietic) stem cells are single multipotent cells that can produce all cell types that are normal components of the blood.

Multivalent vaccine is a single vaccine that is designed to elicit an immune response either to more than one infectious agent or to several different epitopes of a molecule.

Muscular dystrophy is a group of hereditary diseases that cause progressive muscle wastage due to defects in the biochemistry of a muscle tissue. The most common type is Duchenne muscular dystrophy, which is due to a deffective gene on the X chromosome.

Mutable gene refers to a gene which has an unusually high rate of mutation.

Mutagen is any agent or process that can cause mutations.

Mutagenesis is the natural or intentional formation of mutations in a genome.

Mutagenisis is the formation or development of a mutation.

Mutant is an organism or an allele that differs from the wild type because it carries one or more genetic changes in its DNA. A mutant organism may carry mutated gene(s) (= gene mutation); mutated chromosome(s) (= chromosome mutation); or mutated genome(s) (= genome mutation).

Mutation is a change in the DNA sequence that can interfere with protein production. A mutation can arise in a germ cell and be passed on to an individual's children, who will then carry it in every cell of their body. A mutation can also arise in one cell in the body, such as a skin or heart cell. Mutations like these can lead to cancer if they interrupt the cell cycle.

Mutation pressure is a constant mutation rate that adds mutant genes to a population; repeated occurrences of mutations in a population.

Mutational melt down is the process by which a small population accumulates deleterious mutations, which leads to loss of fitness and decline of the population size, which leads to further accumulation of deleterious mutations.

Mutualism (symbiosis) is the way two organisms of different species exist in a relationship in which each individual benefits. Similar interactions within a species are known as co-operation. Mutualism is a type of symbiosis. Symbiosis is a broad category, defined to include relationships that are mutualistic, parasitic, or commensal.

MUX is a device for combining several signals or data streams into a single flow.

Mycelium (plural: mycelia) is a threadlike filament making up the vegetative portion of thallus fungi.

Mycoprotein (also known as fungal protein) is defined in the *Oxford English Dictionary* as "the albuminoid which is the principal constituent of the protoplasm of the cell." "Myco" is from the Greek word for "fungus." Mycoprotein means protein from fungi.

Mycorrhiza (Gr. *mykos*, fungus + *riza*, root) is fungi that form an association with or have a symbiotic relationship with roots of more developed plants.

Mycorrhizae are fungi that form symbiotic relationships with roots of more developed plants.

Mycotoxin is a toxic substance of fungal origin, such as aflatoxin.

Myelin sheath is the white fatty material that forms a sheath around the axons of certain nerve cells, which are then called myelinated nerve fibers.

Myeloma (from Greek *myelo-*, marrow), also known as plasma cell myeloma or Kahler's disease (after Otto Kahler), is a cancer of plasma cells, a type of white blood cell normally responsible for producing antibodies. In multiple myeloma, collections of abnormal plasma cells accumulate in the bone marrow, where they interfere with the production of normal blood cells.

Myo inositol (Inositol or cyclohexane-1,2,3,4,5,6-hexol) is a chemical compound with formula $C_6H_{12}O_6$ or $(-CHOH-)_6$, a sixfold alcohol (polyol) of cyclohexane. It exists in nine possible stereoisomers, of which the most prominent form, widely occurring in nature, is *cis*-1,2,3,5-*trans*-4,6-cyclohexanehexol, or *myo*-inositol (former names: meso-inositol or i-inositol).

Myoepithelial cells move secretions out of the gland by contraction.

Myotonic dystrophy is a combination of progressive weakening of the muscles and muscle spasms or rigidity, with difficulty relaxing a contracted muscle; inherited as an autosomal dominant trait.

Myxoma is a virus that causes myxomatosis in rabbits. It is carried by mosquitos and fleas.

Myxomatosis is a disease of rabbits caused by the myxoma virus. It is an early form of biological control.

N

N_2 free nitrogen gas is used as a cryopreservant, in the liquid form.

N50 is the contig/scaffold length at which the bases in a given assembly reside. This provides a measure of continuity. For instance, a scaffold N50 of 15 Mb means that at least half of the bases in the assembly are in a contig that is at least 15 Mb.

Naked bud is a bud not protected by bud scales.

NAND (NOT-AND) is a logic gate whose output is the negation of that of the AND gate.

Nano refers to 10^{-9}.

Nanoarray is related to **microarray that refers to a** glass slide or bead on which are deposited biomolecules or other material in a regular micro-scale pattern to enable automated simultaneous multiple assays of target substances or activities.

Nanobarcode is an alternative tagging or monitoring device that works more like the UPC code, but on the nano-scale. One type of nanobarcode is a nanoparticle consisting of metallic stripes, where variations in the striping provide the method of encoding information.

Nanobiology is the study of biology for nano materials. Many fundamental biological functions are carried out by molecular machineries that have the sizes of 1–100 nm.

Nanobioprocessors are implantable nano scale processors that can integrate with biological pathways and modify biological processes.

Nanobiotechnology is an emerging area of scientific and technological opportunity. It is the interface between biotechnology and inorganic nanoengineering. Nanobiotechnology applies the tools and processes of nano/ microfabrication to build devices for studying biosystems.. Nanobiotechnology Center (NBTC), a National Science Foundation, Science and Technology Center at Cornell University is characterized by its highly interdisciplinary nature and features a close collaboration between life scientists, physical scientists, and engineers. According to NanoBioNexus New volume 1, *nanobiotechnology is an application of nano science in the life sciences. These may include, but are not limited to, therapeutics, medical devices/ implants, biosensors, and tools for the development of drugs.*

Nanochemistry is related with the production and the reactions of nanoparticles and their compounds. It is concerned with the unique properties associated with assemblies of atoms or molecules on a scale between that of the individual building blocks and the bulk material (from 1 to 1000 nm), At this level, quantum effects can be significant, and also new ways of carrying out chemical reactions become possible.

Nanochip is the fabrication of of transistors made from carbon nanotube.

Nanocircle is a term coined by Stanford University scientists who have synthesized a molecule of DNA that is capable of shutting off specific genes in living bacteria. The new nanometer- size molecule might one day give researchers the ability to target harmful genes that cause cancer and other diseases in humans.

Nanocluster or **nanocrystal** is a fragment of solid comprising somewhere between a few atoms to a few tens of thousands of atoms. Nanoclusters are therefore a novel state of matter with properties that are neither those of a bulk crystal nor those of individual atoms and molecules.

Nanocomposites have broadened significantly to encompass a large variety of systems such as one-dimensional, two-dimensional, three-dimensional and amorphous materials, made of distinctly dissimilar components and mixed at the nanometer scale.

Nanocomputer is a computer whose fundamental components measure only a few nanometers in size. State of the art current computer components are no smaller than about 350 nm.

Nanocrystal has a diameter of between 1 and 10 nm and may contain as few as a hundred or as many as tens of thousands of atoms.

Nanodetectors are detectors of nano size.

Nanodevices may lead to computer chips with billions of transistors, instead of millions – which is the typical range in today's semiconductor technology.

Nanoelectronics in Japan and Korea tends to focus on next generation semiconductor devices and single electron devices (SED).

Nanoelectrospray is an electrospray ionization at a very low solvent flow rate, typically hundreds of nanoliters per minute of sample solution or lower, often without the use of an external solvent delivery system.

Nanoengineering represents the extension of the engineering fields into the nano-scale realm (nanofabrication, nanodevices, etc.). It uses chemistry to construct nanostructures and their composites, then focus the attention on the electronic, optical, and transport properties of these nanostructures and the macroscopic films and materials that can be constructed from them. This includes common frontier of chemistry, condensed matter physics, optics, genetics and bioengineering.

Nanofabrication methods can be divided into two categories: top–down methods, which carve out or add aggregates of molecules to a surface, and bottom-up methods, which assemble atoms or molecules into nanostructures. It is a abrication on the nanotechnology scale.

Nanofibers are fibers of nano size. Using a proprietary process, eSpin is able to produce minute fibers which are 10 to 100 times smaller in diameter than what is possible with conventional textile technology.

Nanofiltration is a pressure driven separation process. The filtration process takes place on a selective separation layer formed by an organic semipermeable membrane. The driving force of the separation process is the pressure difference between the feed (retentate) and the filtrate (permeate) side at the separation layer of the membrane.

Nanofluidics uses nanotechnology to build microscopic silicon devices with features comparable in size to DNA, proteins and other biological molecules – to count molecules, analyze them, separate them, perhaps even work with them one at a time.

Nanofoam is a class of nanostructured, porous materials, foams, containing a significant population of pores with diameters less than 100 nm. Aerogels are one example of nanofoam.

Nanoharvesting agents are being designed to act as molecular mops for these [cancer] biomarkers and can be directly queried by mass spectrometry.

Nanoimaging presents a diverse collection of microscopy techniques and methodologies that provides guidance to successfully image cellular molecular complexes at nanometer spatial resolution.

Nanoimprinting uses an imprinting technique to create nanostructures. With photolithography successfully producing sub-100–nm structures, attention has now turned to 50-nm structures. Nanoimprinting is also called soft lithography: A technique that is very simple in concept, and totally analogous to traditional mould- or form-based printing technology, but that uses moulds (masters) with nanoscale features.

Nanolabels are labels that are designed for general use to highlight existing connections to nanoscale science, engineering, or technology.

Nanomachines can be manufactured with an approach that involves using biological molecules – such as DNA, RNA, enzymes and proteins – to synthesize and duplicate useful devices (this might be termed the bottom–up approach). The other major approach involves miniaturizing today's microfabrication tools by stages, ultimately to work at the nanoscale (the top–down approach).

Nanomanufacturing is high-volume, high-rate, integrated assembly of nano-elements into commercial products. This involves controlling position, orientation, and interconnectivity of the nano-elements.

Nanomaterial research is a field that takes a materials science-based approach on nanotechnology. It studies materials with morphological features on the nanoscale, and especially those that have special properties stemming from their nanoscale dimensions.

Nanomaterials are materials possessing grain sizes on the order of a billionth of a meter. They manifest extremely fascinating and useful properties, which can be exploited for a variety of structured and non-structured applications. Nano particles are generally described as a minute particle of a few nanometers in size (one nanometer is one-billionth of a meter). This area combines nanotechnology and many applications of nano-structured materials.

Nanomedicine or Molecular Medicine is a broad field, where physical, chemical, biological and medical techniques are used to describe molecular structures and mechanisms, identify fundamental molecular and genetic errors of disease, and to develop molecular interventions to correct them.

Nanometals are metal particles of diameter in the range of a few nanometers or thin films in the same thickness range, are interesting not only because of their special mechanic properties but also because of other physical and chemical properties, sometimes totally different from those of coarse-grained metals.

Nanometer is a unit of length = 1 $\times$ 10^{-9} m = One millionth of a millimeter = 1 millimicron = 10 angstroms.

Nanomotor is a motor of nano size. At University of Florida, chemistry professor Weihong Tan has made a "nanomotor" from a single DNA molecule. According to him, the motor is so small that hundreds of thousands can fit on the head of a pin, curls up and extends like an inchworm.

Nanonewtons are forces 1 billion times smaller than the force required to hold an apple against Earth's gravity. Nanonewton forces are estimated with atomic force microscopes and instruments that measure the properties of ultrathin coatings like those used on computer hard drives or turbine blades.

Nanoparticle sensors are tiny particles on the order of a single atom that will recognize compounds.

Nanoparticles include nanoclusters, [nano]-layers, [nano]-tubes. Two- and three-dimensional structures in the size range between the dimensions of molecules and 50 nm (or in a broader sense, submicron sizes as a function of materials and targeted phenomena), are seen as tailored precursors for building up functional nanostructures.

Nanophotonics can provide high bandwidth, high speed and ultra-small optoelectronic components.

Nanophysics uses various experimental techniques to examine the physical properties of objects in the nanoscale size range, that is, a little bit larger than the size of atoms.

Nanoplumbing is a stretching of DNA inside nanofluidic channels with a diameter of about 100 nm.

Nanopore is a membrane with very small channels (a few nanometers in diameter), that separates two solutions.

Nanopositioning is a method of controlling motion on the nanometer scale.

Nanoprecipitation (solvent displacement method) refers to a method of forming nanocapsules by creating a colloidal suspension between two separate phases. The organic phase consists of a solution and a mixture of organic solvents. The aqueous phase consists of a mixture of non-solvents that forms a surface film.

Nanoprism is a nanoparticle with a new shape that can be a useful tool in the race to detect biological threats. The nanoprism, which resembles a tiny Dorito, exhibits unusual optical properties that can be used to improve biodetectors, allowing them to test for a far greater number of biological warfare agents or diseases at one time.

Nanoscale refers to 1 to 100 billionths of a meter. At the nanoscale, physics, chemistry, biology, materials science, and engineering converge toward the same principles and tools. The nanoscale is not just another step toward miniaturization, but a qualitatively new scale.

Nanoscience is a study of phenomena and manipulation of materials at atomic, molecular and macromolecular scales, where properties differ significantly from those at a larger scale. Nanoscience is primarily the extension of existing sciences into the realms of the extremely small (nanomaterials, nano-chemistry, nanobio, nanophysics, etc.).

Nanosensors are any biological, chemical, or surgical sensory points used to convey information about nanoparticles to the macroscopic world.

Nanoshells are procedures that target cancer cells while leaving normal cells untouched, patients controlling the release of medicine in their bodies with an infrared light, and medical test results produced in seconds rather than days. At Rice University, the research focuses on nanoshells, a new type of nanoparticle invented by Naomi Halas. Nanoshells are layered nanoparticles whose ability to manipulate light and color can be designed into the nanoparticle by varying the thickness of the nanoparticle's layers.

Nanospheres are self-assembling nanospheres that fit inside each other like Russian dolls are one form of a broad range of submicroscopic spheres. The durable silica spheres, which range in size from 2 to 50 nanometers, form in a few seconds, are small enough to be introduced into the body, and have uniform pores that could enable controlled release of drugs. The spheres can absorb organic and inorganic substances including small particles of iron, which means they can be controlled by magnets and the contents released as needed.

Nanostructures may be considered as small, familiar, or large, depending on the view point of the disciplines concerned. To chemists, nanostructures are molecular assemblies of atoms numbering from 10^3 to 10^9 and of molecular weights of 10^4 to 10^{10} Daltons. To molecular biologists, nanostructures have the size of familiar objects from proteins to viruses and cellular organelles. But to material scientists and electrical engineers, nanostructures are the current limit of microfabrication and thus are rather small. Nanostructures are complex systems, which evidently lie at the interface between solid-state physics, supramolecular chemistry, and molecular biology.

Nanotechnology (NT) is the science and technology of precisely structuring and controlling matter on the nanometer scale. From the Latin *nanus* = "dwarf," so it literally means "dwarf technology." The word was originally coined by Norio Taniguchi in 1974, to refer to high precision machining. Advancements in nanotechnology can be classified in two categories: Information Technology or Molecular Electronics. According to Royal Academy of Engineering Nanotechnology and Nanoscience, nanotechnology includes *the production and application of structures, devices and systems by controlling shape and size at nanometer scale.* According to Bioengineering Nanotechnology Initiative

– NIH, *NT is the development and use of techniques to study physical phenomena and construct structures in the nanoscale size range or smaller.* It is a technology to create functional materials, devices and systems through control of matter at the scale of 1 to 100 nanometers, and the exploitation of novel properties and phenomena at the same scale.

Nanotechnology and Biomedical Engineering are sometimes loosely used interchangeably in general; however, NT more typically denotes specific products which use "biological systems, living organisms, or derivatives thereof." Even some complex "medical devices" can reasonably be deemed "biotechnology" depending on the degree to which such elements are central to the principle of operation. Biologics/biopharmaceuticals (e.g., vaccines, stored blood product), genetic engineering, and various agricultural applications are some major classes of biotechnology. Pharmaceuticals are related to bionanotechnology in two indirect ways: (1) certain major types (e.g., biologics) fall under both categories; and (2) together they essentially comprise the "*non*-medical-device" set of BME applications.

Nanotoxicology is the study of the toxicity of nanomaterials. Because of quantum size effects and large surface area to volume ratio, nanomaterials have unique properties compared with their larger counterparts. Nanotoxicology is a branch of bionanoscience, which deals with the study and application of toxicity of nanomaterials.

Nanotubes are nanometer-sized tubes composed of various substances including carbon (carbon nanotubes), boron nitride, or nickel vanadate. Nanotube is a one dimensional fullerene with a cylindrical shape.

Nanowire is a nanostructure, with the diameter of the order of a nanometer (10^{-9} meters). It can also be defined as the ratio of the length to width being greater than 20. Alternatively, nanowires can be defined as structures that have a thickness or diameter constrained to tens of nanometers or less and an unconstrained length.

Narrow sense heritability (HN) is the amount of individual phenotypic variation that is due to additive genetic variation.

Narrow-host-range plasmid refers to plasmid that can replicate in one, or at most a few, different bacterial species.

Narrow-sense heritability refers to the proportion of the phenotypic variance that is due to variation in breeding values; the proportion of the phenotypic variance that is due to additive genetic variance.

National Center of Microelectronics Barcelona – Spain belongs to the Spanish Research Council (Consejo Superior de Investigaciones Científicas – CSIC) since its foundation in 1985, together with the institutes in Madrid, (IMM-CSIC) and Seville (IMSE-CSIC). The main activity of IMB-CNM is basic and applied research and development in micro- and nano-electronics based *in silico*n technology. Its mission is to improve the knowledge in these fields and to contribute to the implementation of solutions based in these technologies in industrial products. The R&D activities of IMB-CNM are complemented with training of researchers and with support and technology transfer to companies of the main industrial fields in its environment.

National Institutions of Health (NIH) is a US Govt. non-regulatory agency which oversights of research activities that the agency funds.

National Nanotechnology Initiative is a US federal government agencies participating include the National Science Foundation, the Department of Defense, the National Institute of Health, NASA, and NIST.

National Science Foundation (NSF) is a US Govt. non-regulatory agency which has oversight of biotechnology research activities that the agency funds.

Native organisms are those that have not been recently introduced into an ecosystem.

Native protein is the naturally occurring form of a protein.

Native species is a species that was not introduced and historically, or currently, occurs in a given ecosystem.

Natural catastrophes are natural events causing great damage to populations and that increase their probability of extinction.

Natural selection is differential contribution of genotypes to the next generation due to differences in survival and reproduction.

Nausea is a feeling of needing to vomit (throw-up).

Navier-Stokes equations refers to when the motion of a nonturbulent, Newtonian fluid is governed by the Navier-Stokes equation. The equation can be used to model turbulent flow, where the fluid parameters are interpreted as time-averaged values.

NCA (Nested clade analysis) is a statistical approach to describe how genetic variation is distributed spatially within a species' geographic range. This method uses a haplotype tree to define a nested series of branches (clades), thereby allowing a nested analysis of the spatial distribution of genetic variation, often with the goal of resolving between past fragmentation, colonization, or range expansion events.

NDP is an abbreviation for ribonucleoside diphosphate.

Near infrared Spectroscopy (NIRs) is a spectroscopic method that uses the near-infrared region of the electromagnetic spectrum (from about 800 nm to 2500 nm).

Nearly exact test is a method of using a nearly exact P-value to test if the observed test statistic deviates from the expected value under the null hypothesis, For example, a test of whether populations are in HardyWeinberg proportions by comparing the observed chi-squared value to the chi-squared values of random computer permutations of genotypes from the population's allele frequencies.

Necrosis (Greek νέκρωσις, death; νεκρός, dead) is a form of cell injury that results in the premature death of cells in living tissue by autolysis.

Necrotizing enterocolitis is a situation that may affect underweight or premature infants, and occurs when part of the intestine is damaged or destroyed by a bacterial infection.

Negative autogenous regulation refers to the inhibition of the expression of a gene or set of co-ordinately regulated genes by the product of the gene or the product of one of the genes. Synonym: negative self-regulation.

Negative control system is a mechanism in which the regulatory protein(s) is required to turn off gene expression.

Negative selection is the selection of individuals that do not possess a certain character.

Negative self-regulation is related to negative autogenous regulation.

Neighborhood is the area in a continuously distributed population that call be considered panmictic.

Nematode refers to a slender, unsegmented worms, often parasitic. Also known as eelworm, especially when phytoparasitic.

NEMS (Nano Electro Mechanical Systems) are machines, sensors, computers and electronics that are on the nanoscale.

Neo-formation is related to organogenesis. It is a a new and abnormal growth of tissue; tumor; neoplasm.

Neomycin phosphotransferase II (npt-II) refers to an enzyme which detoxifies the antibiotic neomycin, used as a marker gene to select for successfully transformed cells in plant transgenesis.

Neoplasm refers to localized cell multiplication. Generally it designates a collection of cells which have undergone genetic transformation, forming a tumor.

Neo^r refers to a neomycin-resistance gene. Related terms are: antibiotic resistance marker gene, neomycin phosphotransferase II, selectable marker.

Neoteny refers to the retention of juvenile body characters in the adult state, or the occurrence of adult characters in the juvenile state.

Nephrectomy is a surgical procedure to remove the kidney; the most common treatment for kidney cancer.

Nephritis is the inflammation of the kidneys.

Nephrolithotomy is a small cut is made in the patient's back and a narrow tunnel is made through the skin to the stone inside the kidney. The physician can remove the stone through this tunnel.

Nephrology is the medical specialty concerned with diseases of the kidneys.

Nephron is the functional, microscopic unit of the kidney.

Nephrotic syndrome is a condition characterized by high levels of protein in the urine, low levels of protein in the blood, tissue swelling, and high cholesterol.

Nernst equation relates the potential of an electrochemical cell to the concentrations of the cell components: $E = Eo + (RT/zF)$ [ln $(C1/C2)$], where, z = the charge exchanged at the electrode and $C1$ and $C2$ concentrations of two electro-active compounds.

Nerve is a bundle of axons or dendrites wrapped in connective tissue that conveys impulses between the central nervous system and some other part of the body.

Nervous system consists of specialized cells (neurons, or nerve cells) that conduct stimuli from a sensory receptor through a neuron network to the site (e.g., a gland or muscle) where the response occurs.

Nervous tissue is a type of tissue specialized for conducting electrochemical impulses.

Neural stem cell is a type of stem cell that resides in the brain, which can make new nerve cells (called neurons) and other cells that support nerve cells (called glia). In the adult, neural stem cells can be found in very specific and very small areas of the brain where replacement of nerve cells is seen.

Neuroblastoma is a fetal malignancy that leads to an enlarged placenta, with tumor cells in the fetal circulation and rarely in the chorionic villi.

Neurofibromatosis is one of the most common single gene conditions affecting the human nervous system; in most cases, "cafe au lait" spots, are the only symptom; inherited as an autosomal dominant trait, with 50% being new mutations.

Neurogenic bladder (Also called neuropathic bladder) is a bladder disorder that can be caused by a tumor or other condition of the nervous system.

Neuromuscular blocking agent is a compound that causes paralysis of skeletal muscle by blocking neural transmission at the neuromuscular junction.

Neurons are nerve cells, the principal functional units of the nervous system. A neuron consists of a cell body and its processes – an axon and one or more dendrites. Neurons transmit information to other neurons or cells by releasing neurotransmitters at synapses.

Neurotransmission refers to transmission of information between neutrons.

Neurotransmitter is a chemical messenger used by neurons to transmit impulses across the synapse.

Neutral allele is an allele that is not under selection because it does not affect fitness.

Neutral mutation is a mutation that changes the nucleotide sequence of a gene but has negligible effect on the fitness of the organism.

Neutral theory refers to the theory that much of evolution has been primarily due to random drift of neutral mutations.

Neutrophil refers to a type of leukocyte involved in the early inflammatory response.

Newtonian fluid is a fluid, when the viscosity is constant applied to shear force.

NFT is an abbreviation for nutrient film technique. NFT is a hydroponic technique wherein a very shallow stream of water containing all the dissolved nutrients required for plant growth is re-circulated past the bare roots of plants in a watertight gully, also known as channels.

Nick refers to break (or a break in) a phosphodiester bond in one of the strands of a double-stranded DNA molecule.

Nick translation is a procedure for labeling DNA. A DNA fragment is treated with DNase to produce single-stranded nicks. The nick is moved along the DNA molecule in the presence of labeled deoxyribonucleoside triphosphates by the concerted action of the 5'≥3' exonuclease and 5'≥3' polymerase activities of *E. coli* DNA polymerase I.

Nicked circle (relaxed circle) refers to when during extraction of plasmid DNA from the bacterial cell, one strand of the DNA becomes nicked. This relaxes the torsional strain needed to maintain supercoiling, producing the familiar form of plasmid.

Nif gene cluster is a group of bacterial genes responsible for the biological fixation of atmospheric nitrogen.

NIH refers to National Institutes of Health.

Nissen fundoplication is an operation that helps to treat gastroesophageal reflux disease (GERD). The fundus (top of the stomach) is pulled around the esophagus and sewn. This helps prevent food from moving back from the stomach into the esophagus by creating a muscular band at the top of the stomach, which becomes tighter as the stomach fills up.

Nitabuch's layer (fibrinoid layer) is formed at maternal/fetal interface during placentation and is thought to act to prevent excessively deep conceptus implantation.

Nitrate refers to the form of nitrogen that can be used directly by plants; a major component of inorganic fertilizers.

Nitrification is a chemical process in which nitrogen in plant and animal wastes and dead remains is oxidized, first to nitrites and then to nitrates.

Nitrite is a salt or ester of nitrous acid.

Nitrocellulose (cellulose nitrate) is a nitrated derivative of cellulose. It is made into membrane filters of defined porosity, used to immobilize DNA, RNA or protein, which can then be probed with a labeled sequence or antibody.

Nitrogen assimilation refers to the incorporation of nitrogen into the cells of living organisms.

Nitrogen fixation is the conversion of atmospheric nitrogen gas to oxidized forms that can be assimilated by plants, particularly by blue-green algae and some genera of bacteria (e.g., *Rhizobium* spp.; *Azotobacter* spp.). An important source of nitrogen in unfertilized soils.

Nitrogenous bases are the purines (adenine and guanine) and pyrimidines (thymine, cytosine and uracil) that form DNA and RNA molecules.

NMP is an abbreviation for ribonucleoside monophosphate.

NO is an abbreviation for nucleolar organizer.

NOAA stands for National Oceanic Atmospheric Administration, USA.

Nod box is a DNA sequence that controls the transcriptional regulation of *Rhizobium* nodulation genes.

Nodal culture refers to the culture of a lateral bud and a section of adjacent stem tissue.

Node (L. *nodus*, a knot) refers to slightly enlarged portion of the stem where leaves and buds arise and where branches originate.

Nodular describes a pebbly (rough) texture of a callus.

Nodulation is the formation of nodules by symbiotic bacteria on the roots of plants.

Nodule is the enlargement or swelling on roots of nitrogen-fixing plants.

Non-repetitive DNA/RNA refers to a nucleotide sequence which does not include a significant proportion of repetitive sequences of nucleotides.

Non-additive genetic variation refers to the proportion of the total genetic variation in a population that does not respond to simple mass selection and that causes specific pairwise crosses to depart from performance values predicted by the breeding values of the parents.

Non-autonomous refers to a term referring to biological units that cannot function by themselves; such units require the assistance of another unit, or "helper." Opposite is autonomous.

Non-coding strand is related to antisense DNA. The non-coding strand contains anti codons.

Noncovalent bonding holds the two strands of the DNA double helix together (hydrogen bonds). It folds polypeptides into such secondary structures as the alpha helix and the beta conformation. It enables enzymes to bind to their substrate.

Non-disjunction is the failure of disjunction or separation of homologous chromosomes or chromatids in mitosis or meiosis, resulting in too many chromosomes in some daughter cells and too few in others.

Non-histone chromosomal protein refers to all of the proteins except the histones, in chromosomes.

Nonindigenous species are species that are present in a given ecosystem that were introduced and did not historically occur in that ecosystem.

Non-Newtonian fluid is a fluid, when the viscosity changes with the applied shear force.

Nonsense mutation is a mutation which converts an amino-acid-specifying codon into a stop codon, e.g., a change from UAU (tyr) to UAG (amber) would lead to the premature termination of a polypeptide chain at the place where a tyrosine was inserted in the wild-type.

Nonsteroidal antiinflammatory agent is a drug to reduce inflammation.

Non-target organism is an organism which is affected by a treatment (e.g., pesticide application) for which it was not the intended recipient.

Non-template strand refers to the non-transcribed strand of DNA. a.k.a. sense strand or coding strand. It will have the same sequence as the RNA transcript, except that T is present at positions where U is present in the RNA transcript.

Non-virulent agent is related to an attenuated vaccine. Virulent (chiefly medicine, of a disease or disease-causing agent) refers to the one that is highly infectious, malignant or deadly.

Noonan syndrome is a condition characterized by short stature and ovarian or testicular dysfunction, mental deficiency, and lesions of the heart.

NOR (NOT-OR) is a logic gate whose output is the negation of that of the OR gate.

Norepinephrine is a neurotransmitter that may enhance pain. When it is inhibited, analgesia is increased. Increased levels in the central nervous system decrease analgesia.

Northern analysis is a technique for transferring electrophoretically resolved RNA segments from an agarose gel to a nitrocellulose filter paper sheet via capillary action.

Northern blot is a cellulose or nylon membrane to which RNA molecules have been attached by capillary action. The transferred RNA is hybridized to single-stranded DNA probes.

Northern blotting is similar to Southern blotting, except that RNA is transferred onto a matrix and the presence of a specific RNA molecule is detected by DNA-RNA hybridization.

Northern hybridization is the hybridization of a labeled DNA probe to RNA fragments that have been transferred from an agarose gel to a nitrocellulose filter.

NOT is a logic gate whose output is binary 1 when its input is 0, and whose output is a 0 when its input is a 1.

Notice of Compliance refers to: Once a product submission has been reviewed, assessed and deemed by Health Canada to meet the Food and Drug Regulations, it is given a Notice of Compliance.

Novel food refers to: a substance, including a microorganism, that does not have a history of safe use as a food; a food that has been manufactured, prepared, preserved or packaged by a process that has not been applied before to that food, and causes the food to undergo a major change; or a food that is derived from a plant, animal or microorganism that has been genetically modified.

Novel trait in a plant is a plant with characteristics not normally found in that species in which the new characteristic has been created through specific genetic manipulation, transformation, mutation, etc.

NPT-II is an abbreviation for neomycin phosphotransferase II. It confers resistance to kanamycin and neomycin in bacteria and Geneticin in mammalian cells.

NSF is an abbreviation for National Science Foundation.

NTP is an abbreviation for ribonucleoside triphosphate.

n-type characterizes a semiconductor containing predominantly mobile electrons.

Nucellar embryo is an embryo which has developed vegetatively from somatic tissue surrounding the embryo sac, rather than by fertilization of the egg cell.

Nucellus (L. *nucella*, a small nut) is a tissue composing the chief part of the young ovule in which the embryo sac develops; megasporangium.

Nuclear DNA (nDNA) is a DNA that forms chromosomes in the cell nucleus of eukaryotes.

Nuclear gene is a gene located on a chromosome in the nucleus of a eukaryotic cell.

Nuclear magnetic resonance (NMR) is a physical phenomenon in which nuclei in a magnetic field absorb and re-emit electromagnetic radiation. This energy is at a specific resonance frequency which depends on the strength of the magnetic field and the magnetic properties of the isotope of the atoms; in practical applications, the frequency is similar to VHF and UHF television broadcasts (60–1000 MHz). NMR allows the observation of specific quantum mechanical magnetic properties of the atomic nucleus.

Nuclear reactor is a device to initiate and control a sustained nuclear chain reaction.

Nuclear transfer is a technique in which an egg has its original nucleus removed and exchanged for the nucleus of a donor cell.

Nuclear transfer technology is a method of cloning a living organism. The process involves removing the nucleus of an egg cell and replacing it with a nucleus from any cell of the organism being cloned.

Nuclease is a class of largely bacterial enzymes that degrade DNA or RNA molecules by catalyzing the cleavage of the phosphodiester bonds that link adjacent nucleotides. For deoxyribonuclease (DNAse) the substrate is DNA, for ribonuclease (RNAse) the substrate is RNA, and for

S1 nuclease, the substrate is single-stranded DNA or RNA. Endonucleases cleave at internal sites in the substrate molecule, while exonucleases progressively cleave from the end of the substrate molecule.

Nucleation is the initial stage in a phase transformation, evidenced by the formation of small particles (nuclei) of the new phase, which are capable of growing.

Nucleic acid probe is related to DNA probe.

Nucleic acid is a macromolecule composed of phosphoric acid, pentose sugar, and organic bases. The two nucleic acids, deoxyribonucleic acid (DNA) and ribonucleic acid (RNA), are made up of long chains of molecules called nucleotides.

Nuclein is the term used by Friedrich Miescher to describe the nuclear material he discovered in 1869, which today is known as DNA.

Nucleo-cytoplasmic ratio is the ratio of nuclear to cytoplasmic volume. This ratio is high in meristematic cells and low in differentiated cells.

Nucleolar organizer (NO; nucleolar organizer region, NOR) is a chromosomal segment containing genes that encode ribosomal RNA; located at the secondary constriction of some chromosomes.

Nucleolar organizer region (NOR) refers to a chromosomal segment containing a large array of genes that encode ribosomal RNA; located at the secondary constriction of specific chromosomes.

Nucleolar zone is any chromosome region, irrespective of whether or not it is a secondary-constriction, that is associated with the formation of the nucleolus during telophase.

Nucleolus (*L. nucleolus*, a small nucleus) is an RNA-rich intranuclear organelle in the nucleus of eukaryotic cells, produced by a nucleolar organizer. It represents the storage place for ribosomes and ribosome precursors. The nucleolus consists primarily of ribosomal precursor RNA, ribosomal RNA, their associated proteins, and some, perhaps all, of the enzymatic equipment (RNA polymerase, RNA methylase, RNA cleavage enzymes) required for synthesis, conversion and assembly of ribosomes. Subsequently the ribosomes are transported to the cytoplasm.

Nucleoplasm refers to the non-staining or slightly chromophilic, liquid or semi-liquid, ground substance of the inter-phase nucleus and which fills the nuclear space around the chromosomes and the nucleoli.

Nucleoprotein is a conjugated protein composed of nucleic acid and protein; the material of which the chromosomes are made.

Nucleoside analog (or Nucleoside analogue) is a synthetic molecule that resembles a naturally occurring nucleoside, but that lacks a bond site needed to link it to an adjacent nucleotide.

Nucleoside is a base (purine or pyrimidine) that is covalently linked to a 5-carbon (pentose) sugar. When the sugar is ribose, the nucleoside is a ribonucleoside; when it is deoxyribose, the nucleoside is a deoxyribonucleoside. Adenine, guanine and cytosine occur in both DNA and RNA; thymine occurs in DNA; and uracil in RNA. They are the building blocks of DNA and RNA.

Nucleosome are spherical sub-units of eukaryotic chromatin that are composed of a core particle consisting of an octamer of histones (two molecules each of histones H_{2a}, H_{2b}, H_3 and H_4) and 146 nucleotide pairs.

Nucleotide is a nucleoside with one or more phosphate groups linked at the 3'- or 5'-hydroxyl of a pentose sugar. When the

sugar is ribose, the nucleotide is a ribonucleotide; when it is 2-deoxyribose, the nucleotide is a deoxyribonucleotide. RNA and DNA are polymers of, respectively, ribonucleoside 5'-monophosphates and deoxyribonucleoside 5'-monophosphates. Nucleotides containing the bases adenine, guanine and cytosine (A, G, C) occur in both DNA and RNA; thymine (T) occurs only in DNA, and uracil (U) only in RNA. Ribonucleoside mono-, di-, and triphosphates for which a specific base is not assigned are abbreviated NMP, NDP, and NTP, while deoxyribonucleoside mono-, di-, and tri-phosphates are abbreviated dNMP, dNDP, and dNTP. Otherwise, the "N" is replaced by the base letter abbreviation.

Nucleus (*L. nucleus*, kernel) is a dense protoplasmic-membrane-bound region of a eukaryotic cell that contains the chromosomes separated from the cytoplasm by a membrane; present in all eukaryotic cells except mature sieve-tube elements.

Null allele is an allele that is not detectable either due to a failure to produce a functional product or a mutation in a primer site that precludes amplification during PCR analysis.

Nullisomy (adj.: nullisomic) is an otherwise diploid cell or organism lacking both members of a chromosome pair (chromosome formula 2n – 2).

Nurse culture refers to planting a cell from a suspension culture on a raft of filter paper above a callus tissue piece (nurse tissue).

Nusselt number is a dimensionless number which measures the enhancement of heat transfer from a surface which occurs in a 'real' situation, compared to the heat transfer that would be measured if only conduction could occur. Typically it is used to measure the enhancement of heat transfer when convection takes place.

Nutraceutical is a product isolated or purified from food that is generally sold in medicinal forms not usually associated with food.

Nutrient cycle refers to the passage of a nutrient or element through an ecosystem, including its assimilation and release by various organisms and its transformation into various organic or inorganic chemical forms.

Nutrient deficiency is an absence or insufficiency of some factor needed for normal growth and development.

Nutrient film technique (NFT) is a hydroponic technique used to grow plants. NFT delivers a thin film of water or nutrient solution either continuously or through on-off cycles.

Nutrient gradient refers to a diffusion gradient of nutrients and gases that develops in tissues where only a portion of the tissue is in contact with the medium.

Nutrient medium is a solid, semi-solid or liquid combination of: major and minor salts; an energy source (sucrose); vitamins; plant growth regulators; and occasionally other defined or undefined supplements. Often made from stock solutions, then sterilized by autoclaving or filtering through a micropore filter.

O

Obstruction is a blockage in the digestive tract that prevents the forward movement of foods and liquids as they are digested.

Occipital lobes refer to the posterior areas of the cerebrum; interpret visual stimuli from the retina of the eye.

Occult blood refers to the blood in the stool that is not visible to the naked eye.

Occupational Safety and Health Administration (OSHA) is a U.S. agency responsible for regulation of biotechnology. The major law under which the agency has regulatory powers is the Occupational Safety and Health Act.

Ocham's razor is the principle that the least complicated explanation (most parsimonious hypothesis) generally should be accepted to explain the data at hand.

Octoploid is a cell or organism with eight sets of chromosomes, i.e., chromosome number $2n = 8x$.

OEIS (Opto Electronic Integrated Systems) refers to the development of methodologies for self-assembly of micron scale objects with the objective of fabricating optoelectronic integrated systems (OEIS). In this process, DNA-modified optoelectronic components may be transported to the surface of a microelectrode using electrophoresis whereupon sequence specific oligonucleotide interactions direct component localization and binding.

Oestrogen (estrogen) is a group of female sex hormones which control the development of sexual characteristics and control oestrus.

Oestrous cycle (from oestrus) is the cycle of reproductive activity shown by most sexually mature non-pregnant female mammals.

Oestrus (adj.: oestrous) is the period of sexual excitement and acceptance of the male.

Office of the Gene Technology Regulator (OGTR) is a part of the Australian Department of Health and Ageing that assists the Gene Technology Regulator, a statutory office holder, to administer the Gene Technology Act 2000 (the Act). The objective of the Act is to protect the health and safety of people and to protect the environment by identifying risks posed by, or resulting from, gene technology and by managing those risks through regulating certain dealings with genetically modified organisms.

Offset is a young plant produced at the base of a mature plant.

Offshoot is a short, usually horizontal, stem produced near the crown of a plant.

Offspring (progeny) are new individual organisms that result from the process of sexual or asexual reproduction.

Ohm (Ω) is a unit of resistance. One ohm is the electrical resistance between two points of a conductor when a constant potential difference of 1 volt, applied to these points, produces a current of 1 ampere in the conductor.

Ohmmeter measures electrical resistance.

Okazaki fragment indicates that since DNA can be replicated in only one direction, i.e., nucleotides can be added only at the 3′ end, only one of the two strands of a double helix can be replicated continuously. The other strand is replicated in small segments (Okazaki fragments) that are subsequently joined together by DNA ligase.

OLA is an abbreviation for oligonucleotide ligation assay. OLA is a rapid, sensitive, and specific method for the detection of known single nucleotide polymorphisms (SNPs). This method is based on the joining of two adjacent oligonucleotide probes (Capture and Reporter Oligos) using a DNA ligase while they are annealed to a complementary DNA target (e.g., PCR product).

Oligodendrocyte is a supporting cell that provides insulation to nerve cells by forming a myelin sheath (a fatty layer) around axons.

Oligohydramnios is a condition that occurs in pregnant women when the amniotic fluid is <5 cm.

Oligomer is a molecule formed from a small number of monomers.

Oligonucleotide is a short molecule (usually 6 to 100 nucleotides) of single-stranded DNA.

Oligonucleotide ligation assay (OLA) is a diagnostic technique for determining the presence or absence of a single nucleotide polymorphism within a target DNA sequence, often indicating whether the gene is wild type (normal) or mutant (usually defective).

Oligonucleotide-directed mutagenesis (site-specific mutagenesis or oligonucleotide-directed site-specific mutagenesis) involves the incorporation of a mutant oligonucleotide into one strand of plasmid DNA. After DNA replication of this heteroduplex plasmid, any progeny plasmids that arise by replication of the wild-type strand will be homozygous for the wild-type allele, and any plasmids that arise by replication of the mutant strand will be homozygous for the mutant allele. The trick to oligonucleotide-directed mutagenesis is to identify the desired mutants. There are many different methods of oligonucleotide-directed mutagenesis: (1) methods which destroy the wild-type template strand of DNA and thus favor replication of the mutant strand; and (2) methods which simply allow the wild-type and mutant strands of DNA an equal chance of replicating. The *dut ung* method is an example of the first approach, and the *mutS* method is an example of the second approach.

Oligopotent progenitor cells are progenitor cells that can produce more than one type of mature cell. An example is the myeloid progenitor cell, which can give rise to mature blood cells, including blood granulocytes, monocytes, red blood cells, platelets, basophiles, eosinophiles and dendritic cells, but not T lymphocytes, B lymphocytes, or natural killer cells.

Oligosaccharide refers to carbohydrate consisting of several linked monosaccharide units.

Oncogene is a gene that causes cells to grow in an uncontrolled manner, i.e., that causes cancer. Oncogenes are mutant forms of normal functional genes (called proto-oncogenes) that have a role in normal cell proliferation.

Oncogenes are found in tumors and in retroviruses.

Oncogenesis is the progression of cytological, genetic, and cellular changes that culminate in a malignant tumor.

Oncogenic are genes that are responsible for the transformation of normal cells into tumor cells.

Oncology is the study of tumors.

Onco-mouse refers to a mouse that has been genetically modified to incorporate an oncogene, which acts as an animal model for studies of human cancer.

Oncovirus is a virus associated with cancer.

Ontogeny is the developmental life history of an organism.

Oocyte refers to egg mother cell; it undergoes two meiotic divisions (oogenesis) to form the egg cell. The primary oocyte is before completion of the first meiotic division; the secondary oocyte is after completion of the first meiotic division.

Oogenesis is the formation and growth of the egg or ovum in an animal ovary.

Oogonium (plural: oogonia) refers to: (1) A germ cell of the female animal, that gives rise to oocytes by mitotic division; (2) The female sex organ of algae and fungi.

Oosphere refers to the non-motile female gamete in plants and some algae.

Oospore (Gr. *oion*, an egg + spore) is a resistant spore developing from a zygote, resulting from the fusion of heterogametes in certain algae and fungi.

Opal stop codon is related to stop codon (or termination codon) is a nucleotide triplet within messenger RNA that signals a termination of translation a, in the genetic code.

Op-amp (operational amplifier) is a semiconductor amplifier characterized by high gain and high internal resistance.

Open continuous culture is a continuous culture system, in which inflow of fresh medium is balanced by outflow of a corresponding volume of spent medium plus cells. In the steady state, the rate of cell wash-out equals the rate of formation of new cells in the system.

Open pollination is a pollination by wind, insects or other natural mechanisms.

Open reading frame (ORF) is a sequence of nucleotides in a DNA molecule that has the potential to encode a peptide or protein: it starts with a start triplet (ATG), is followed by a string of triplets each of which encodes an amino acid, and ends with a stop triplet (TAA, TAG or TGA). This term is often used when, after the sequence of a DNA fragment has been determined, the function of the encoded protein is not known. The existence of open reading frames is usually inferred from the DNA (rather than the RNA) sequence.

Opening pressure of CSF is measured when the patient is in the lateral decubitus position with the legs and neck in a neutral position.

Operational definition is an operation or procedure that can be carried out to define or delimit something.

Operator is the region of DNA that is upstream from a gene or genes and to which one or more regulatory proteins (repressor or activator) binds to control the expression of the gene(s).

Operon refers to a functionally integrated genetic unit for the control of gene expression in bacteria. It consists of one or more genes that encode one or more polypeptide(s) and the adjacent site (promoter and operator) that controls their expression by regulating the transcription of the structural genes.

Opine is the condensation product of an amino acid with either a keto-acid or a sugar, produced by the crown gall tissues. Opine synthesis is a unique characteristic of tumor cells.

Opsonin (Greek *opsonin*, to prepare for eating), is any molecule that enhances phagocytosis by marking an antigen for an immune response (i.e., causes the phagocyte to "relish" the marked cell). However, the term is usually used in reference to molecules that act as a binding enhancer for the process of phagocytosis, especially antibodies, which coat the negatively charged molecules on the membrane. Molecules that activate the complement system are also considered opsonins.

Optical mapping is the light microscope based technique in which images of single DNA molecules undergoing restriction enzyme digest are recorded and used for the construction of physical maps of large pieces of DNA.

Optical trapping is also called an optical tweezer.

Optical tweezers (single-beam gradient force trap) are scientific instruments that use a highly focused laser beam to provide an attractive or repulsive force (typically on the order of piconewtons), depending on the refractive index mismatch to physically hold and move microscopic dielectric objects.

Optoelectronics is the study and application of electronic devices that source, detect and control light, usually considered a sub-field of photonics. In this context, *light* often includes invisible forms of radiation such as gamma rays, X-rays, ultraviolet and infrared, in addition to visible light. Optoelectronic devices are electrical-to-optical or optical-to-electrical transducers, or instruments that use such devices in their operation.

OPU is an abbreviation for ovum pickup.

OR is a logic gate whose output is a binary 1 if any of its inputs is a 1; zero otherwise.

Ordinate is the vertical axis of a graph. Opposite is absciss or abscissa.

ORF is an abbreviation for open reading frame.

Organ culture is the aseptic culture of complete living organs of animals and plants outside the body in a suitable culture medium.

Organ of corti is the structure within the inner ear that contains receptor cells that sense sound vibrations.

Organ refers to a tissue or group of tissues that constitute a morphologically and functionally distinct part of an organism.

Organ system is a body system; an organized group of tissues and organs that work together to perform a specialized set of functions.

Organellar gene refers to the genes located on organelles outside the nucleus.

Organelle is a structure within a cell that performs a particular function. Examples include mitochondria, endoplasmic reticulum, vacuoles, chloroplasts and lysosomes. Organelles are like smaller versions of the organs in your body, each performing a particular function to keep the whole cell alive.

Organic chemistry is a chemistry subdiscipline involving the scientific study of the structure, properties, and reactions of organic compounds and organic materials, i.e., matter in its various forms that contain carbon atoms.

Organic complex is a chemically undefined compound added to nutrient media to stimulate growth, e.g., coconut milk; yeast extract; casein hydrolysate.

Organic co-solvent is a compound to dissolve some neutral organic substances, such as in media preparation. Organic co-solvents include alcohols (usually ethanol), acetone and dimethylsulphoxide (DMSO).

Organic evolution is the process by which changes in the genetic composition of populations of organisms occur in response to environmental changes.

Organic refers to compounds containing carbon, many of which have been in some manner associated with living organisms.

Organism is an individual living system, such as animal, plant or micro-organism, that is capable of reproduction, growth and maintenance.

Organized growth refers to the development under tissue culture conditions of organized explants (meristem tips or shoot tips, floral buds or organ primordia).

Organized tissue is composed of regularly differentiated cells.

Organizer is an inductor; a chemical substance in a living system that determines the fate in development of certain cells or groups of cells.

Organogenesis is the initiation of adventitious or *de novo* shoots or roots from callus, meristem or suspension cultures.

Organoid is an organ-like structure produced in culture, such as leaves, roots or callus.

Organoleptic refers to having an effect on one of the organs of sense, such as taste or smell.

Origin of replication is the nucleotide sequence at which DNA synthesis (replication) is initiated.

Orphan gene is a gene or DNA sequence whose function is not known.

Orphan receptor is a receptor for which a cellular function or ligand has yet to be identified.

Ortet is the plant from which a clone is obtained.

Orthologous refers to homologous genes/gene products that have evolved divergently *between* species; many rice genes have orthologues in other cereal genomes, because of the common ancestry of cereal species.

Orthopaedic Bioengineering is the specialty where methods of engineering and computational mechanics have been applied for the understanding of the function of bones, joints and muscles, and for the design of artificial joint replacements. Orthopaedic bioengineers analyze the friction, lubrication and wear characteristics of natural and artificial joints; they perform stress analysis of the musculoskeletal system; and they develop artificial biomaterials (biologic and synthetic) for replacement of bones, cartilages, ligaments, tendons, meniscus and intervertebral discs. They often perform gait and motion analyses for sports performance and patient outcome following surgical procedures. Orthopaedic bioengineers also pursue fundamental studies on cellular function, and mechano-signal transduction.

Oscillator is a circuit that produces an alternating voltage (current) when supplied by a steady (DC) energy source.

OSHA is an abbreviation for Occupational Safety and Health Administration.

Osmic acid *(osmium tetroxide, OsO_4)* is a fixing agent commonly used to prepare tissue samples for electron microscopy.

Osmolarity is the total molar concentration of the solutes. Osmolarity affects the osmotic potential of solution or nutrient medium.

Osmosis is a process that differs primarily from ordinary diffusion in that only the solvent, rather than the solute, is able to penetrate the separating membrane. Usually, the solute molecules that cause the osmotic driving force are unable to pass through the membrane because of their size. In ordinary diffusion, all species are able to penetrate the membrane, although the permeability may be different for each species. Here, water moves through permeable membranes. Water moves from an area of high/low water potential to one of low/high water potential until a dynamic equilibrium is reached.

Osmotic potential refers to a change in the energy state of solvent brought about by dissolving a substance in the solvent-water in the biological sciences. The potential of aqueous solutions is always negative compared to pure water. Solvent flows from higher to lower osmotic potential solutions by diffusion or osmosis.

Osmotic pressure (π) is the pressure that must be applied to a solution to stop osmosis: $\pi =$ MRT.

Osmoticum is an agent, such as PEG, mannitol, glucose or sucrose, employed to maintain the osmotic potential of a nutrient medium equivalent to that of the cultured cells (isotonic). Because of this osmotic equilibrium, cells are not damaged *in vitro*.

Osteoarthritis is the most common form of arthritis is osteoarthritis (OA), also known as degenerative joint disease.

Osteogenesis imperfecta is a condition also known as brittle bone disease; characterized by a triangular shaped face with yellowish brown teeth, short stature and stunted growth, scoliosis, high pitched voice, excessive sweating and loose joints.

Ostomy is an operation that is done when there is damage to a section of intestine. It creates an opening in the wall of the abdomen, and brings a portion of intestine through the opening so stool can leave the body.

Otoliths refers to small calcium carbonate crystals in the saccule and utricule of the inner ear; sense gravity and are important in static equilibrium.

Outbreeding is a mating system characterized by the breeding of genetically unrelated or dissimilar individuals.

Outbreeding depression is the relative reduction in the fitness of hybrids compared to parental types.

Outflow is the volume of growing cells that is removed from a bioreactor during a continuous fermentation process.

Outlier loci refer to loci that may be under selection (or linked to loci under selection) that are detected because they fall outside the range of expected variation for a given summary statistic (e.g., extremely high or low *F*ST compared to most "neutral" loci in a sample).

Output traits are raits produced in GM crops, which are beneficial or of direct value to the consumer.

Outsourcing refers to: Not every biotech startup has the resources to carry out every step in the research, development, processing and marketing chain. Often at least one or more steps need outsourced.

Ovary refers to: (1) Enlarged basal portion of the pistil of a plant flower that contains the ovules; (2) The reproduction organ in female animals in which eggs are produced.

Overdominance is a condition in which heterozygotes are superior (on some scale of measurement) to either of the associated homozygotes.

Overflow incontinence refers to leakage of urine that occurs when the quantity of urine produced exceeds the bladder's capacity to hold it.

Overlapping generations is a breeding system where sexual maturity does not occur at a specific age, or where individuals breed more than once, causing individuals from different brood years to interbreed in a given year.

Overlapping reading frames refers to start triplets in different reading frames generate different polypeptides from the same DNA sequence.

Ovulation refers to the release of mammalian egg(s) from the ovary.

Ovule (L. *ovulum*, diminutive of *ovum*, egg) is the part of the reproductive organs in seed plants that consists of the nucellus, the embryo sac and integuments.

Ovum (L. *ovum*, egg; plural: ova) refers to: (1) A gamete of female animals, produced by the ovary. (2) The oosphere in plants.

Ovum pickup (OPU) refers to the non-surgical collection of ova from a female.

Oxidation is the removal of one or more electrons from an atom, ion or molecule.

Oxidative phosphorylation is the enzymatic addition of a phosphate to ADP to make ATP, coupled to electron transport from a substrate to molecular oxygen.

Oxidizing agent is a substance capable of accepting electrons from another substance, thereby oxidizing the second substance and itself becoming reduced.

Oxygen dissociation curve depicts the percentage saturation of hemoglobin with oxygen, as a function of certain variables such as oxygen concentration, carbon dioxide concentration, or pH.

Oxygenation refers to the saturation or combination with oxygen, as the aeration of the blood in the lungs.

Oxygenator performs gas exchange functions; provides oxygen, removes carbon dioxide; contains an arterial reservoir.

Oxygen-electrode-based sensor is a sensor in which an oxygen electrode is coated with a biological material which generates or absorbs oxygen. When the biological coating is active, the amount of oxygen next to the electrode changes and the signal from the electrode changes.

Oxyhemoglobin refers to hemoglobin that has combined with oxygen.

P

· ·

p denotes the shorter of the two chromosome arms, e.g., human 14p is the short arm of human chromosome 14.

P element is a *Drosophila* transposon.

P_1 is a symbol for the parental generation or parents of a given individual.

P_1, P_2 are generational symbols for the two parents of a given individual.

p53 gene is also known as tumor protein p53, p53, cellular tumor antigen p53, phosphoprotein p53, or tumor suppressor p53.

It is a protein that is encoded by the *TP53* gene. The p53 protein is crucial in multicellular organisms, where it regulates the cell cycle and, thus, functions as a tumor suppressor, preventing cancer. As such, p53 has been described as "the guardian of the genome" because of its role in conserving stability by preventing genome mutation. Hence *TP53* is classified as a tumor suppressor gene. The name p53 is in reference to its apparent molecular mass: SDS-PAGE analysis indicates that it is a 53-kilodalton (kDa) protein. However, based on calculations from its amino acid residues, p53's mass is actually only 43.7 kDa. This difference is due to the high number of proline residues in the protein; these slow its migration on SDS-PAGE, thus making it appear heavier than it actually is. It is a human tumor-suppressor transcription factor gene, damage or mutation to which is believed to be responsible for up to 60% of all human cancer tumors. If, in spite of the presence of p53 protein, a cell begins to divide uncontrollably following damage to its DNA, the p53 gene acts to prevent tumors by triggering apoptosis.

Pacemaker (of the heart) refers to a sinoatrial node.

Pachynema (adj.: pachytene) is a mid-prophase stage in meiosis, immediately following zygonema and preceding diplonema. In microscopic preparations, the chromosomes are visible as long, paired threads.

Package is a protective enclosure for a chip or a sensor, typically made of plastic or ceramic.

Packaging cell line is a cell line that is designed to produce viral particles that do not contain nucleic acid. After transfection of these cells with a full-size viral genome, fully infective viral particles are assembled and released.

Packed cell volume (PCV) is the volume of cells in a set volume of culture expressed as a percentage of that set volume after sedimentation (packing) by means of low speed centrifugation.

Paclitaxel is a mitotic inhibitor used in cancer chemotherapy. Paclitaxel is an anti-cancer drug, also known as Taxol and Onxol. The drug is first line treatment for ovarian, breast, lung, and colon cancer. It was originally extracted from the Pacific Yew tree, Taxus brevifolia.

Paclitaxel total synthesis in organic chemistry is a major ongoing research effort in the total synthesis of paclitaxel.

Pad electrodes is a capacitor type electrode used with shortwave diathermy.

PAGE is an abbreviation for polyacrylamide gel electrophoresis.

Paired End Mapping (PEM) is a method for detecting genome-level variation. Paired ends from size-selected sheared genomic DNA fragments are subjected to high-throughput sequencing and then mapped onto a reference genome (*in silico*).

Pairing (synapsis) is the pairing of homologous chromosomes during the prophase of the first meiotic division, when crossing over occurs.

Pair-rule gene is a gene that influences the formation of body segments in *Drosophila.*

Paleontology is the study of the fossil record of past geological periods and of the phylogenetic relationships between ancient and contemporary plant and animal species.

Palindrome (Gr. *palindromos*, running back again) is a segment of double-stranded DNA, in which the order of bases, read 5'-3' in one strand, is the same as that in the complementary antiparallel strand, also read 5'-3.' If the sequence is written in the normal convention, on two lines with paired bases shown one above the other, the base order on one strand runs in the opposite direction to that on the complementary strand.

Palindromic sequence is a segment of duplex DNA whose 5'-to-3' sequence is identical on each DNA strand. The sequence is the same when one strand is read left-to-right and the other strand is read right-to-left. Typically, recognition sites for type II restriction endonucleases are palindromes.

Palisade parenchyma refers to elongated cells found just beneath the upper epidermis of leaves, and containing many chloroplasts.

Palpitation is a sensation of rapid heartbeats.

pAMP is an abbreviation for Ampicillin-resistant plasmid.

Pancreas is large gland located in the abdominal cavity. The pancreas produces pancreatic juice containing digestive enzymes; also serves as an endocrine gland; secreting the hormone insulin and glucagons.

Panicle (L. *panicula*, a tuft) is an inflorescence, the main axis of which is branched; the branches bear loose racemose flower clusters.

Panicle culture refers to an aseptic culture of immature panicle explants to induce microspore germination and development.

Panmictic is a population that is randomly mating.

Panmictic population is a population in which mating occurs at random.

Panmixis is a random mating in a population.

Papilla (A small nipple) is like projection or elevation, such as the papilla at the base of each hair follicle.

Papillary layer consists of loose connective tissue. It is located immediately under the epidermis and is separated from it by the basal lamina. The papillary layer is a relatively thin layer extending into the dermal papillae and ridges. It contains blood vessels that serve, but do not enter, the epidermis.

Par gene is one of a class of genes required for faithful plasmid segregation at cell division. Initially, par loci were identified on plasmids, but have also been found on bacterial chromosomes.

PAR is an abbreviation for photosynthetically active radiation. It designates the spectral range (wave band) of solar radiation from 400 to 700 nanometers that photosynthetic organisms are able to use in the process of photosynthesis. This spectral region corresponds more or less with the range of light visible to the human eye.

Paracentric inversion is an inversion that is entirely within one arm of a chromosome and does not include the centromere.

Paraffin (wax) is a translucent, white, solid hydrocarbon with a low melting point. Paraffin is used as an embedding medium to support tissue for sectioning for light microscopy observation.

Paraffin bath is a combined paraffin and mineral oil immersion technique in which the paraffin sub stance is heated to 126°F for conductive heat gains; commonly used on the hands and feet for distal temperature gains in blood flow and temperature.

Parafilm is a stretchable film based on paraffin wax; used to seal tubes and Petri dishes. Parafilm™ is a proprietary name, which is applied colloquially to similar products.

Parahormone is a substance with hormone-like properties that is not a secretory product (e.g., ethylene; carbon dioxide).

Parallel evolution refers to the development of different organisms along similar evolutionary paths due to similar selection pressures acting on them.

Paralogous refers to homologous genes/gene products that have duplicated and evolved divergently *within* a species. e.g., beta- and gamma- globulin genes.

Paramagnetism is a relatively weak form of magnetism resulting from the independent alignment of atomic dipoles (magnetic) with an applied magnetic field. Also a type of induced magnetism, associated with unpaired electrons, that causes a substance to be zapped into the inducing magnetic field.

Parameter refers to a value or measurement that varies with circumstances, and is used as a reference to quantify a situation or a process.

Paraphyletic is a clade that does not include all of the descendants from the most recent common ancestor taxon.

Parasexual cycle refers to a sexual cycle involving changes in chromosome number but differing in time and place from the usual sexual cycle; occurring in those fungi in which the normal cycle is suppressed or apparently absent.

Parasexual hybridization refers to the hybridization by induced fusion of cells (protoplasts) from two contrasting genotypes for production of hybrids or cybrids which contain various mixtures of nuclear and/ or cytoplasmic genomes, respectively.

Parasite (Gr. *parasites*, one who eats at the table of another) is an organism deriving its food from the living body of another organism.

Parasitism is the close association of two or more dissimilar organisms where the association is harmful to at least one.

Parasporal crystal refers to tightly packaged insect pro-toxin molecules that are produced by strains of *Bacillus thuringiensis* during the formation of resting spores.

Parathyroid gland is a small, pea-sized glands closely adjacent to the thyroid gland; their secretion regulates calcium and phosphate metabolism.

Parathyroid hormone is a hormone secreted by the parathyroid glands; regulates calcium and phosphate metabolism.

Paratope is synonym for antibody binding site. It is the part of an antibody which recognizes an antigen, the antigen-binding site of an antibody. It is a small region (of 15–22 amino acids) of the antibody's Fv region and contains parts of the antibody's heavy and light chains. The part of the antigen to which the paratope binds is called an epitope. This can be mimicked by a mimotope.

Parenchyma is the bulk of a substance. In animals, a parenchyma comprises the functional parts of an organ and in plants parenchyma is the ground tissue of nonwoody structures.

Parenchymal cells make up the bulk of the soft parts of plants.

Parenchymatous describes spherical and undifferentiated cells with primary cell walls, capable of both cell division and differentiation.

Parentage analysis is the assessment of the maternity and/or paternity of a given individual.

Parenteral is a method administering the medicine by some means other than oral intake, particularly intravenously or by injection.

Parenteral nutrition is a means of providing protein, fats, carbohydrates, fluid, and vitamins to the body through a special solution given through a vein into the bloodstream.

Parsimony is the principle that the preferred phylogeny of an organism is the one that requires the fewest evolutionary changes; the simplest explanation.

Parthenocarpy (Gr. *parthenos*, virgin + *karpos*, fruit) is the development of fruit without fertilization.

Parthenogenesis (Gr. *parthenos*, virgin + *genesis*, origin) is the production of an embryo from an unfertilized egg.

Partial digest refers to the incomplete reaction of a restriction enzyme with DNA, such that only a proportion of the target sites are cleaved. Partial digests are often performed to give an overlapping collection of

DNA fragments for use in the construction of a gene bank. Synonym is incomplete digest. Opposite is complete digest.

Partial nephrectomy is a surgery to remove the kidney; only the part of the kidney that contains the tumor is removed.

Particle radiation refers to gamma particles (positively charged) and beta (ß) particles (negatively charged), electrons, protons and neutrons.

Parts per million (ppm) refers to units of any given substance per one million equivalent units, such as the weight units of solute per million weight units of solution (i.e., 1 ppm = 1 mg/l).

Parturition is the process of giving birth.

Passage is the process in which cells are disassociated, washed, and seeded into new culture vessels after a round of cell growth and proliferation.

Passage number is the number of times the cells in the culture have been sub-cultured. In descriptions of this process, the dilution ratio of the cells should be stated so that the relative cultural "age" can be ascertained.

Passage time is the interval between successive sub-cultures.

Passive immunity refers to: (1) Natural acquisition of antibodies by the foetus or neonate (newborn) from the mother; (2) The artificial introduction of specific antibodies by the injection of serum from an immune animal. In both cases, temporary protection is conferred on the recipient.

Passive transport is a movement of biochemicals and other atomic or molecular substances across cell membranes. Unlike active transport, it does not require an input of chemical energy, being driven by the growth of entropy of the system. The rate of passive transport depends on the permeability of the cell membrane, which, in turn, depends on the organization and characteristics of the membrane lipids and proteins. The four main kinds of passive transport are: diffusion, facilitated diffusion, filtration and osmosis.

Pat gene refers to a gene obtained from *Streptomyces* sp. encoding resistance to glufosinate-ammonium containing herbicides. Used widely as a means of transgenically inducing herbicide resistance in crop plants. Synonym is bar gene.

Patent is a legal permission to hold exclusive right – for a defined period of time – to manufacture, use or sell an invention.

Paternal refers to that is pertaining to the father.

Pathogen (Gr. *pathos*, suffering + *genesis*, beginning) is an organism or agent that causes disease. For example, bacteria, viruses, parasites and fungi.

Pathogen free refers to one that is free from disease-causing organisms (bacteria, fungi, viruses, etc.).

Pathogenesis related protein (PR protein) refers to one of a group of proteins that are characteristically highly expressed as part of a plant's response to pathogen attack.

Pathogenicity is the ability to cause disease.

Pathotoxin is a very dilute substance synthesized and released by some pathogen, and which interacts with the host metabolism.

Pathovar refers to strain of a plant-attacking bacterium or fungus that can be differentiated from others by their interaction with specific host cultivars.

PBR is an abbreviation for plant breeders' rights. PBR also known as plant variety rights (PVR), are rights granted to the breeder of a new variety of plant that give the breeder exclusive control over the propagating material (including seed, cuttings, divisions, tissue culture) and harvested material (cut flowers, fruit, foliage) of a new variety for a number of years.

pBR322 is a plasmid and was one of the first widely used *E. coli* cloning vectors. The p stands for "plasmid," and BR for "Bolivar" and "Rodriguez." pBR322 is 4361 base pairs in length and contains the replicon of plasmid pMB1, the *amp*^{R} gene, encoding the ampicillin resistance protein (source plasmid RSF2124) and the *tet*^{R} gene, encoding the tetracycline resistance protein (source plasmid pSC101). The plasmid has unique restriction sites for more than forty restriction enzymes. The 11 of these 40 sites lie within the *tet*^{R} gene. There are 2 sites for restriction enzymes HindIII and ClaI within the promoter of the *tet*^{R} gene. There are 6 key restriction sites inside the *amp*^{R} gene. The *ori* site in this plasmid is pMB1 (a close relative of ColE1). The circular sequence is numbered such that 0 is the middle of the unique EcoRI site and the count increases through the tet gene. The ampicillin resistance gene is penicillin beta-lactamase. Promoters P1 and P3 are for the beta-lactamase gene. P3 is the natural promoter, and P1 is artificially created by the ligation of two different DNA fragments to create pBR322. P2 is in the same region as P1, but it is on the opposite strand and initiates transcription in the direction of the tetracycline resistance gene.

PCA stands for principal component analysis.

PCoA stands for principle coordinates analysis.

PCR is an abbreviation for polymerase chain reaction.

PCR-RFLP is an alternative term for cleaved amplified polymorphic sequence.

PCV is an abbreviation for packed cell volume.

pdf stands for probability density function.

PE stands for paternity exclusion.

Pectin refers to a group of naturally occurring complex polysaccharides, containing galacturonic acid, found in plant cell walls, where their function is to cement cells together.

Pectinase (Gr. *pektos*, congealed) is an enzyme catalyzing the hydrolysis of pectins.

Pediatric gastroenterologist is a physician who treats infants and children with diseases of the digestive system.

Pedicel (L. *pediculus*, a little foot) is a stalk or stem of the individual flowers of an inflorescence.

Pedigree is a table, chart or diagram recording the ancestry of an individual.

Peduncle (L. *pedunculus*, a late form of *pediculus*, a little foot) is a stalk or stem of a flower that is born singly; the main stem of an inflorescence.

PEG is an abbreviation for polyethylene glycol that is a polyether compound with many applications from industrial manufacturing to medicine. The structure of PEG is H-(O-CH$_2$-CH$_2$)$_n$-OH. PEG is also known as polyethylene oxide (PEO) or polyoxyethylene (POE), depending on its molecular weight.

Penetrance is the percentage of individuals in population that show a particular phenotype among those capable of showing it, i.e., among those that have the genotype normally associated with that phenotype.

Peptic is related to the stomach and the upper part of the small intestine (duodenum).

Peptic ulcer is a sore in the lining of the esophagus, stomach, or duodenum (beginning of the small intestine); often caused by a bacteria called Helicobacter pylori.

Peptidase refers to an enzyme that catalyzes the hydrolysis of a peptide bond.

Peptide is a sequence of amino acids linked by peptide bonds; a breakdown or build-up unit in protein metabolism. Typically used to describe low molecular weight species.

Peptide bond is the chemical bond holding amino acid residues together in peptides and proteins. The (CO-NH) bond is formed by the condensation, with loss of a water molecule, between the carboxyl group of one amino acid and the amino (-NH$_2$) of the next amino acid.

Peptide expression library refers to a collection of peptide molecules, produced by recombinant cells, in which the amino acid sequences are varied.

Peptide nanotubes are nanotubes are formed from cyclic peptides. Alternating D and L linkages create planar rings that self-assemble by stacking into nanotubes. They can form pores through cell membrane causing damage.

Peptide nucleic acid (PNA) is a synthetic oligonucleotide analogue, in which the sugar backbone is replaced by a peptide chain, upon which the nucleoside residues are strung. Probes made from PNA appear to have greater specificity than those made from DNA.

Peptide vaccine is a short chain of amino acids that can induce antibodies against a specific infectious agent.

Peptidyl transferase is an enzyme bound tightly to the large sub-unit of the ribosome that catalyzes the formation of peptide bonds between amino acids during translation.

Peptidyl-tRNA binding site (P-site) refers to the site on a ribosome that hosts the tRNA to which the next amino acid for the growing polypeptide chain is attached.

Percutaneous is affected through the skin. Applying a medicated ointment by friction, or removal or injection by needle.

Perennial (L. *perennis*, lasting years) is a plant that grows more or less indefinitely from year to year and, once mature, usually produces seed each year.

Perfect fluid is defined as a fluid with zero viscosity (i.e., inviscid).

Perforation is a hole in the wall of an organ.

Perfusion is passing of a fluid through spaces. Supplying an organ or tissue with nutrients and oxygen by injecting blood or a suitable fluid into an artery.

Pericentric inversion is a chromosomal inversion that includes the centromere because the breaks were on opposite chromosomal arms.

Periclinal chimera refers to: (1) Genotypically or cytoplasmically different tissues arranged in concentric layers; (2) A chimera in which one or more layers of tissue derived from one graft member enclose the central tissue derived from the other member of the graft.

Periclinal is the plane of cell wall orientation or cell division parallel to the surface of the organ.

Pericycle (Gr. *peri*, around + *kyklos*, circle) is a region of the plant bounded externally by the endodermis and internally by the phloem.

Periplasm is the space (periplasmic space) between the cell (cytoplasmic) membrane of a bacterium or fungus and the outer membrane or cell wall.

Peristalsis are wave-like movements of food forward through

Peritonitis is an infection inside the abdominal cavity.

Permeability [μ] is a constant in the equation for the relationship between magnetic induction and magnetic field: $B = \mu H$ for free space, and $\mu_0 = 1.26 \times 10^{-6}$ H/m.

Permeable (L. *permeabilis*, that which can be penetrated) refers to a membrane, cell or cell system through which substances may diffuse.

Permitivity [ε] relates polarization charge and electric field; for free space, $\varepsilon_0 = 8.85 \times 10^{-12}$ F/m.

Persistence is the ability of an organism to remain in a particular setting for a period of time after it is introduced.

Persistent refers to chemicals with a long inactivation or degradation time, such as some pesticides. Persistent substances can become dangerously concentrated in the tissues of organisms at the top end of a food chain.

Persistent right umbilical vein (PRUV) is a placental cord abnormality associated with fetal abnormalities and poor neonatal prognosis.

Personalized medicine is a medical model that proposes the customization of healthcare using molecular analysis – with medical decisions, practices, and/or products being tailored to the individual patient. In this model, diagnostic testing is often employed for selecting appropriate and optimal therapies based on the context of a patient's genetic content.

PERV is an abbreviation for porcine endogenous retrovirus. PERV are remnants of ancient viral infections, found in the genomes of most, if not all, mammalian species. Integrated into the chromosomal DNA, they are vertically transferred through inheritance. Due to the many deletions and mutations they accumulate over time, they usually are not infectious in the host species, however the virus may become infectious in another species.

Pessary is a device placed in the vagina to hold the bladder in place or to treat a prolapsed uterus.

Pesticide is a toxic chemical product that kills harmful organisms (e.g., insecticides, fungicide, weedicides, rodenticides).

Petal is one of the parts of the flower that make up the corolla.

Petiole (L. *petiolus*, a little foot or leg) is the stalk of leaf.

Petite mutant is a respiration-deficient yeast mutant that produces small colonies when grown on glucose-containing medium.

Petri dish is a flat round dish with a matching lid, made of glass or plastic material, and used for culturing organisms.

Peyronie's disease is a plaque, or hard lump, that forms on the erection tissue of the penis. The plaque often begins as an inflammation that may develop into a fibrous tissue.

PFGE is an abbreviation for pulsed-field gel electrophoresis. PFGE is a technique used for the separation of large deoxyribonucleic acid (DNA) molecules by applying to a gel matrix an electric field that periodically changes direction.

PG is an abbreviation for polygalacturonase. PG is an enzyme produced in plants which is involved in the ripening process, and by some bacteria and fungi which are involved in the rotting process.

pH is a measure of acidity and alkalinity. This is equal to the log of the reciprocal of the hydrogen ion concentration of a solution, expressed in grams per liter. A reading of 7 is neutral (e.g., pure water), whereas below 7 is acid and above 7 is alkaline.

Phage is an abbreviation for bacteriophage. It is a virus that infects and replicates within bacteria. The term is derived from 'bacteria' and the Greek φαγεῖν *phagein* "to devour."

Phagemids refers to cloning vectors that contain components derived from both phage and plasmid DNA.

Phagocytosis (Greek φαγεῖν, phagein meaning to devour; κύτος, kytos meaning cell; and -*osis* meaning process) is the process by which a cell engulfs a solid particle to form an internal vesicle known as a phagosome. Phagocytosis was revealed by Élie Metchnikoff in 1882. Phagocytosis is a specific form of endocytosis involving the vesicular internalization of solids such as bacteria, and is therefore distinct from other forms of endocytosis such as the vesicular internalization of various liquids (pinocytosis).

Pharmaceutical agent is also called therapeutic agent. Agent is something capable of producing an effect. It is any chemical substance formulated or compounded as single active ingredient or in combination of other pharmacologically active substance, it may be in a separate but packed in a single unit pack as combination product intended for internal, or external or for use in the medical diagnosis, cure, treatment, or prevention of disease.

Pharmaceutical is a medical drug.

Pharmacodynamics is the study of how drugs achieve their therapeutic effect.

Pharmacogenomics is the study of variability in the expression of individual genes that relate to disease susceptibility and drug response at the cellular, tissue, individual and population level. A major objective of pharmacogenomics is the development of innovative classes of targeted drugs and vaccines designed to affect highly specific processes in the body while minimizing side effects. A related area is biopharmaceuticals, whereby transgenic techniques are used to insert therapeutic properties, including vaccines, into foods, potentially replacing pills and syringe injections.

Pharmacokinetics refers to the quantitative measurement of how drugs move around the body, and the processes which control their absorption, distribution, metabolism, and excretion.

Pharming is the process of farming GM plants or animals to be used as living pharmaceutical factories.

Pharynx is a part of the digestive tract. It is bounded anteriorly by the mouth and nasal cavities and posteriorly by the esophagus and larynx; the throat region.

Phase change is the developmental change from one maturation state to another.

Phase shift is a time difference between the input and output signals.

Phase state refers to the coupling or repulsion of two linked genes.

Phase transformation is a change in the number and/or character of the phases that make up the microstructure of an alloy.

PHB is an abbreviation for polyhydroxybutyrate.

pH-electrode-based sensor is a sensor in which a standard electrochemical pH electrode is coated with a biological material. Many biological processes raise or lower pH, and the changes can be detected by the pH electrode.

Phenetics is a taxonomic classification solely based on overall similarity (usually of phenotypic traits), regardless of genealogy.

Phenocopy is an organism whose phenotype (but not genotype) has been changed by the environment to resemble the phenotype usually associated with a mutant organism.

Phenogram is a branching diagram or tree that is based on estimates of overall similarity between taxa derived from a suite of characters.

Phenolic oxidation refers to many plant species that contain phenolic compounds, which blacken through oxidation. The process is initiated after plants are wounded. Phenolic oxidation may lead to growth inhibition or, in severe cases, to tissue necrosis and death. Antioxidants are incorporated into the sterilizing solution or isolation medium to prevent or reduce oxidative browning.

Phenolics are compounds with hydroxyl group(s) attached to the benzene ring, forming esters, ethers and salts.

Phenols (phenolics) are compounds with hydroxyl group(s) attached to the benzene ring, forming esters, ethers and salts. Phenolic substances are produced from newly explanted tissues, oxidizing to form colored compounds visible in nutrient media.

Phenomics is the study of an overall organism and how the characteristics or traits of an organism that we can see (its phenotype) fits with the information we know about its genes (genomics) and proteins (proteomics).

Phenotype is the description of the characteristics of a cell, a tissue or an animal; as black and white fur of a mouse are two phenotypes that can be found.

Phenotypic plasticity is the variation in the phenotype of individuals with similar genotypes due to differences in environmental factors during development. For example, cod in areas with red algae develop a reddish color.

Phenylketonuria (PKU) is a hereditary disorder that results in reduced production of the liver enzyme phenylalanine hydroxylase. This substance is involved in the breakdown of phenylalanine in food to tyrosine. Without a modified diet, affected infants will develop severe, irreversible brain damage.

Pheromone is a hormone-like substance that is secreted by an organism into the environment as a specific signal to another organism, usually of the same species.

Philopatry is a characteristic of reproduction of organisms where individuals faithfully home to natal sites. Individuals exhibiting philopatry are philopatric.

Phloem is a specialized vascular plant tissue for the transport of assimilates (generally sugars) from the point of synthesis (in the leaf) to other parts of the plant. It consists of sieve tubes, companion cells, phloem parenchyma and fibers.

Phonon is a single quantum of vibrational or elastic energy.

Phosphatase is an class of enzymes that catalyze the hydrolysis of esters of phosphoric acid, removing a phosphate group from an organic compound.

Phosphatase is an enzyme that hydrolyzes esters of phosphoric acid, removing a phosphate group.

Phospho-diester bond is a bond in which a phosphate group joins adjacent carbons through ester linkages. A condensation reaction between adjacent nucleotides results in a phospho-diester bond between 3′ and 5′ carbons in DNA and RNA.

Phospholipase A2 is an enzyme which degrades type A2 phospholipids.

Phospholipid is a class of lipid molecules in which a phosphate group is linked to glycerol and two fatty acyl groups. A major component of biological membranes.

Phosphorescence is luminescence that occurs at times greater than on the order of a second after an electron excitation event.

Phosphorolysis refers to the cleavage of a bond by orthophosphate; analogous to hydrolysis referring to cleavage by water.

Phosphorylation is the addition of a phosphate group to a compound.

Photoacoustic imaging is a hybrid biomedical imaging modality, and is developed based on the photoacoustic effect. In photoacoustic imaging, non-ionizing laser pulses are delivered into biological tissues (when radio frequency pulses are used, the technology is referred to as thermoacoustic imaging). Some of the delivered energy will be absorbed and converted into heat, leading to transient thermoelastic expansion and thus wideband (e.g., MHz) ultrasonic emission. The generated ultrasonic waves are then detected by ultrasonic transducers to form images. It is known that optical absorption is closely associated with physiological properties, such as hemoglobin concentration and oxygen saturation. As a result, the magnitude of the ultrasonic emission (i.e., photoacoustic signal), which is proportional to the local energy deposition, reveals physiologically specific optical absorption contrast. 2D or 3D images of the targeted areas can then be formed.

Photo-bioreactor is the bioreactor dependent on sunlight, which is taken up by its content of plant material, usually algae.

Photodiode is a semiconductor diode that produces voltage (current) in response to illumination (see also "phototransitor").

Photodynamic therapy (photo-chemotherapy, PDT) is a form of phototherapy using nontoxic light-sensitive compounds that are exposed selectively to light, whereupon they become toxic to targeted malignant and other diseased cells. PDT has proven ability to kill microbial cells, including bacteria, fungi and viruses.

PHOTOFRIN® is contraindicated in patients with porphyria.

Photomicrograph is a foto taken with a microscope.

Photon is a quantum of light; the energy of a photon is proportional to its frequency: $E = hv$, where E is energy; h is Planck's constant = 6.62×10^{-27} erg-second; and v is the frequency.

Photoperiod (Gr. *photos*, light + period) is the length of day or period of daily illumination provided or required by plants for reaching the reproductive stage.

Photoperiodism refers to the photoperiod required by a plant to switch from the vegetative to the reproductive stage.

Photophosphorylation is the formation of ATP from ADP and inorganic phosphate using light energy in photosynthesis.

Photoreactivation is a light dependent DNA repair process.

Photosensitizer is a substance that sensitizes an organism, cell, or tissue to light; an agent used in photodynamic therapy which, when absorbed by CA cells and exposed to light, is activated, killing cancer cells

Photosynthate refers to the carbohydrates and other compounds produced in photosynthesis.

Photosynthesis is a chemical process by which green plants synthesize organic compounds from carbon dioxide and water in the presence of sunlight.

Photosynthetic efficiency is the efficiency of converting light energy into organic compounds.

Photosynthetic photon flux (PPF) refers to a measure of the intensity of light utilized by plants for photosynthetic activity.

Photosynthetic refers to one that is able to use sunlight energy to convert atmospheric carbon dioxide into organic compounds. Nearly all plants, most algae and some bacteria are photosynthetic.

Photosynthetically active radiation (PAR) is radiant energy captured by the photosynthetic system in the light reactions, usually taken to be the wavelengths between 400 and 700 nm.

Photothermal therapy refers to efforts to use electromagnetic radiation (most often in infrared wavelengths) for the treatment of various medical conditions, including cancer. This approach is an extension of photodynamic therapy, in which a photosensitizer is excited with specific band light.

Phototransistor is a transistor that, when powered, produces amplified voltage (current) in response to illumination.

Phototropism (Gr. *photos*, light + *tropos*, turning) is a growth curvature in which light is the stimulus.

Phylogenetic refers to evolutionary relationships between taxa or gene lineages.

Phylogenetic species concept (PSC) refers to states when a species is a discrete lineage or recognizable monophyletic group.

Phylogeny is a diagram illustrating the deduced evolutionary history of populations of related organisms.

Phylogeography is the assessment of the geographic distributions of the taxa of a phylogeny to understand the evolutionary history (e.g., origin and spread) of a given taxon.

Physical map is a map showing physical locations on a DNA sequence, such as restriction sites and sequence-tagged sites. Also a diagram of a chromosome or a karyotype, showing the location of loci (genes and markers).

Phyto- (Prefix) is related with plants. A combining form meaning "plant," used in the formation of compound words.

Phytochemical refers to molecules characteristically found in plants.

Phytochrome refers to a pigment, found in the cytoplasm of green plants, which can exist in two forms: Pr (biologically inactive) and Pfr (biologically active).

Phytohormone is a substance that stimulates growth or other processes in plants. They include auxins, abscissic acid, cytokinins, gibberellins and ethylene. Phytohormones are chemical messengers that may pass through cells, tissues and organs and stimulate biochemical, physiological and morphological responses.

Phytoparasite (adj.: phytoparasitic) refers to parasite on plants.

Phytopathogen is an organism that causes disease in plants.

Phytoremediation refers to the use of plants actively to remove contaminants or pollutants from either soils (e.g., polluted fields) or water resources (e.g., polluted lakes).

Phytosanitary refers to plant health, including quarantine.

Phytostat refers to the name adopted by Tulecke in 1965 for an apparatus designed for the semi-continuous chemostat culture of plant cells.

Phytosterol is one of a group of biologically active phytochemicals present in the seeds of certain plants.

PI refers to probability of identity.

Picoengineering involves engineering at the level of subatomic particles (e.g., electrons). Ray Kurzweil, in his book, "*When Computers Exceed Human Intelligence by Penguin,*" defines picoengineering as age of spiritual machines.

Picomole is one trillionth of a mole.

Piconewton is trillionth of a Newton.

Piconewtons is a plural form of *piconewton*. It is an SI unit of force equal to 10^{-12} newtons.

Piezoelectric material is a ferroelectric material in which an electrical potential difference is created due to mechanical deformation, or conversely,

in which the application of a voltage causes dimensional changes in the material.

Pigment are compounds that are colored by the light they absorb. Light absorption is exploited by plants both as a means of energy capture (photosynthesis) and as a signaling mechanism (phytochrome).

PINEs (Paired interspersed nuclear elements) refer to the use of PCR primes that bind one end of a transposable element (along with a few adjacent single-copy nucleotides), to generate DNA markers for studies in population genetics (e.g., hybridization or admixture).

Pinhole embraces a wide variety of oxide defects and is used in a broad sense today. Listed in this category are cracks caused by thermal contraction after oxidation or by handling, and regions of oxide with low dielectric strength caused by dust particles, inadequate masking, contamination, or poor resist adhesion.

Pinocytosis is used primarily for the absorption of extracellular fluids (ECF). In contrast to phagocytosis, it generates very small amounts of ATP from the wastes of alternative substances such as lipids (fat). Unlike receptor-mediated endocytosis, pinocytosis is nonspecific in the substances that it transports.

Pin-out is a diagram showing for electronic components the relations between connecting pins and internal components.

Pipelines apply to new drug development in the biotechnology industry and how they are used to describe the research and development progress and potential value of biotech companies.

Pipette is a widely used device for accurate dispensing of small volumes of liquids.

Pistil (L. *pistillum*, a pestle) is the central organ of the flower, typically consisting of ovary, style and stigma. The pistil is usually referred to as the female part of a perfect flower.

Pitot Tube can be used to measure the velocity of a fluid.

Pitting is a form of very localized corrosion wherein small pits or holes form, usually in a vertical direction.

Pixel (Picture Element) is a smallest element of an image, such as a dot on a computer monitor screen.

pK value is a measure of the strength of an acid on a logarithmic scale. The pK value is given by $\log_{10} (1/K_a)$, where, K_a is the acid dissociation constant pK values often are used to compare the strengths of different acids.

PKU refers to phenylketonuria: An enzyme deficiency condition characterized by the inability to convert one amino acid, phenylalanine, to another, tyrosine, resulting in mental deficiency. plasmid double-stranded, circular, bacterial DNA into which a fragment of DNA from another organism can be inserted.

Placenta (Greek, plakuos = flat cake) refers to the developmental organ formed from maternal and fetal contributions in animals with placental development. In human, the placenta at term is a discoid shape "flat cake" shape: 20 cm diameter, 3 cm thick and weighs 500-600 gm. Placenta are classified by the number of layers between maternal and fetal blood (Haemochorial, Endotheliochorial and Epitheliochorial) and shape (Discoid, Zonary, Cotyledenary and Diffuse). The placenta has many different functions including metabolism, transport and endocrine.

Placenta accrete is the abnormal placental adherence, either in whole or in part of the placenta with absence of decidua basalis, leading to retention as an after-birth to the underlying uterine wall. The incidence of placenta accreta also significantly increases in women with previous cesarean section compared to those without a prior surgical delivery.

Placenta increta occurs when the placenta attaches deep into the uterine wall and penetrates into the uterine muscle, but does not penetrate the uterine serosa. Placenta increta accounts for approximately 15-17% of all cases.

Placenta percreta refers to when placental villi penetrate myometrium and through to uterine serosa.

Placenta previa refers to a condition that occurs in approximately 1 in 200 to 250 pregnancies. In the third trimester and at term, abnormal bleeding can require caesarian delivery and can also lead to "Abruptio Placenta." Ultrasound screening programs during 1st and early 2nd trimester pregnancies now include placental localization. Diagnosis can also be made by transvaginal ultrasound.

Placental arteries (umbilical arteries) are blood vessels. In placental animals, the blood vessels which develop within the placental cord carrying relatively deoxygenated blood from the embryo/fetus to the placenta. In humans, there are two placental arteries continuous with the paired internal iliac arteries (hypogastric arteries) arising off the dorsal aortas. At birth, this vessel regresses and form the remnant medial umbilical ligament.

Placental cord (umbilical cord) is the structure connecting the embryo/fetus to the placenta. It is initially extra-embryonic mesoderm forming the connecting stalk within which the placental blood vessels (arteries and veins) form. In human placental cords the placental blood vessels are initially paired, later in development only a single placental vein remains with a pair of placental arteries. This structure also contains the allantois, an extension from the hindgut cloaca then urogenital sinus. Blood collected from the placental cord following delivery is a source of cord blood stem cells.)

Placental diameter is measured in the transverse section by calculating the maximum dimensions of the chorionic surface.

Placental growth factor (PlGF) of the vascular endothelial growth factor (VEGF) family, released from the placental trophoblast cells and other sources that stimulates blood vessel growth.

Placental malaria is a malarial infection of the placenta by sequestration of the infected red blood cells.

Placental membranes (chorionic membrane; amniotic membrane) are general terms to describe the membrane bound extra-embryonic fluid-filled cavities surrounding the embryo then fetus. In humans, the amniotic membrane and chorionic membrane fuse.

Placental thickness is measured at its mid-portion from the chorionic plate to the basilar plate, on a longitudinal plane (less than 4 cm at term). It excludes any abnormalities (fibroids, myometrial contractions, or venous lakes).

Placental vein (umbilical vein) are blood vessels. In placental animals, the blood vessels which develop within the placental cord carrying relatively oxygenated blood from the placenta to the embryo/fetus. In humans, there are initially two placental veins which fuse to form a single vein. The presence of paired veins in the placental cord can be indicative of developmental abnormalities.

Placental volume is measured by a range of different methods and calculations, more recently with three-dimensional ultrasound.

Placentophagia is a term used to describe the maternal ingestion of after birth materials (placental membranes and amniotic fluid) that can occur following mammalian parturition (birth).

Plant breeders' rights (PBR) refers to legal protection of a new plant variety granted to the breeder or his successor in title.

Plant cell culture refers to *in vitro* growth of plant cells.

Plant cell immobilization is an entrapment of plant cells in gel matrices; the cells are suspended in small drops of the material, which then is set or allowed to harden to make little carriers. Materials such as alginates, agar or polyacrylamide can be used.

Plant genetic resources (PGR) refers to reproductive or vegetative propagating material of: (1) cultivated varieties (cultivars) in current use and newly developed varieties; (2) obsolete cultivars; (3) primitive cultivars (landraces); (4) wild and weed species, near relatives of cultivated varieties; and (5) special genetic stocks (including elite and current breeder's lines and mutants).

Plant genetics is the study of genetics in plants.

Plant growth regulator is an organic compound, either

natural or synthetic, and other than a nutrient, that modifies or controls one or more specific physiological processes within a plant.

Plant hormone refers to a plant growth regulator.

Plant molecular farming (PMF) is a technique that involves using genetically modified plants to produce substances that the plants typically do not produce naturally, such as industrial compounds or therapeutics.

Plant Pest Act (PPA) of 1957 (P.L. 85-36) prohibited the movement of pests from a foreign country into or through the United States unless authorized by United States Department of Agriculture (USDA).

Plant variety protection (PVP) is synonym for plant breeders' rights.

Plant Variety Protection Act (PVPA) was signed into federal law on December 24, 1970. The law allows the developer and owner of a new distinct seed variety to obtain property rights protection if the variety meets certain requirements. The act was amended by US Congress in 1994.

Plantibody refers to an antibody expressed transgenically in an engineered plant.

Plantlet is a small rooted shoot regenerated from cell culture following embryogenesis or organogenesis. Plantlets can normally develop into normal plants when transplanted to soil.

Plaque is a clear spot on an otherwise opaque culture plate of bacteria or cultured bacteria cells, showing where cells have been lysed by viral infection.

Plasma cells are antibody-producing white blood cells derived from B lymphocytes.

Plasma clearance is the sum of all the drug elimination processes of the body.

Plasma is the fluid portion of the blood in which is suspended the white and red blood cells. Contains 8-9% solids, of which 85% is composed of the proteins fibrinogen, albumin, and globulin. The essential function of plasma is the maintenance of blood pressure and the transport of nutrients and waste.

Plasmalemma (Gr. *plasma*, anything formed + *lema*, a husk or shell of a fruit) is a delicate cytoplasmic double membrane found on the outside of the protoplast, adjacent to the cell wall.

Plasmid is a circular DNA molecule, capable of autonomous replication, which typically carries one or more genes encoding antibiotic resistance proteins. Plasmids can transfer genes between bacteria and are important tools of transformation for genetic engineers.

Plasmodesma (Gr. *plasma*, something formed + *demos*, a bond, a band; plural: plasmodesmata) is a fine protoplasmic thread passing through the plant cell wall that separates two protoplasts.

Plasmolysis (Gr. *plasma*, something formed + *lysis*, a loosening) is a separation of the cytoplasm from the cell wall, due to removal of water from the protoplast.

Plastic deformation is a permanent or nonrecoverable deformation, accompanied by permanent atomic displacement.

Plasticity is the ability of adult-derived stem cells to be capable of developing into cells types outside of the tissue of origin (for example, human blood stem cells have been shown to differentiate into liver cells.

Plasticizer is a low molecular weight polymer additive that enhances flexibility and workability and reduces stiffness and brittleness.

Plastid (Gr. *plastis*, a builder) is a cytoplasmic body found in the cells of plants and some protozoa. Chloroplastids, for example, produce chlorophyll that is involved in photosynthesis.

Plastoquinone is a quinone which is one of a group of compounds involved in the transport of electrons in photosynthesis in chloroplasts.

Plate refers to: (1) Verb: To distribute a thin film of something. Hence micro-organisms or plant cells are plated onto nutrient agar; (2) Noun: Refers to the two segments of a Petri dish or a similar-shaped item.

Platelets are small packets of cytoplasm that contain enzymes important in the clotting response; manufactured in bone marrow by megakaryocytes.

Platform technology is the technology that has a common starting point but diverges once it is put into actual practice.

Plating efficiency is the percentage of inoculated cells which give rise to cell colonies when seeded into culture vessels.

Pleiotropy (adj.: pleiotropic) is the situation in which a particular gene has an effect on several different traits.

Plethysmography records the changes in the volume of a body part as modified by the circulation of the blood in it.

Pleura is a membrane that covers the outside of the lung.

Plicae circulares are macroscopically visible, crescent-shaped folds of the mucosa and submucosa. Plicae circulares extend around one-half to two-thirds of the circumference of the lumen of the small intestine.

Ploidy is the number of complete sets of chromosomes per cell, e.g., one set: haploid, two sets: diploid, etc. Normally a gamete (sperm or egg) carries a full set of chromosomes that includes a single copy of each chromosome, as aneuploidy generally leads to severe genetic disease in the offspring. The haploid number (n) is the number of chromosomes in a gamete.

Plumule (*L. plumula*, a small feather) is the first bud of an embryo or that portion of the young shoot above the cotyledons.

Pluripotent (*L. plurimus*, meaning very many, and potens, meaning having power) refers to a stem cell that has the potential to differentiate into any of the three germ layers: endoderm (interior stomach lining, gastrointestinal tract, the lungs), mesoderm (muscle, bone, blood, urogenital), or ectoderm (epidermal tissues and nervous system). However, cell pluripotency is a continuum, ranging from the completely pluripotent cell that can form every cell of the embryo proper, e.g., embryonic stem cells and iPSCs, to the incompletely or partially pluripotent cell that can form cells of all three germ layers but that may not exhibit all the characteristics of completely pluripotent cells.

PNA is an abbreviation for peptide nucleic acid. PNA is an artificially synthesized polymer similar to DNA or RNA. The name is somewhat of a misnomer as PNA is not an acid.

Pneumonectomy is a surgical procedure for removal of an entire lung.

Point defect is a crystalline defect associated with one or several atomic sites.

Point mutation is a change in DNA at a specific site in a chromosome. Includes nucleotide substitutions and the insertion or deletion of one or a few nucleotide pairs. *See* mutation.

Poisson distribution is a probability distribution, with identical mean and variance, that characterizes discrete events occurring independently of one another in time, when the mean probability of that event on any one trial is very small.

Poisson's ratio (v) is a negative ratio of lateral and axial strains that result from an applied axial stress.

Polar Body is a structure produced when an early egg cell, or oogonium, undergoes meiosis. In the first meiosis, the oogonium divides its chromosomes evenly between the two cells but divides its cytoplasm unequally. One cell retains most of the cytoplasm, while the other gets almost none, leaving it very small. This smaller cell is called the first polar body. The first polar body usually degenerates. The ovum, or larger cell, then divides again, producing a second polar body with half the amount of chromosomes but almost no cytoplasm. The second polar body splits off and remains adjacent to the large cell, or oocyte, until it (the second polar body) degenerates. Only one large functional oocyte, or egg, is produced at the end of meiosis.

Polar mutation is a mutation that influences the functioning of genes that are downstream from the site of mutagenesis but are in the same transcription unit.

Polar nuclei are two centrally located nuclei in the embryo sac that unite with a second sperm cell in a triple fusion. In certain seeds, the product of this fusion develops into the endosperm.

Polar transport is the directed movement within plants of compounds (usually endogenous plant growth regulators) mostly in one direction; polar transport overcomes the tendency for diffusion in all directions.

Polarity (Gr. *pol*, an axis) is the observed differentiation of an organism, tissue or cell into parts having opposed or contrasted properties or form.

Polarization (P) is the total electric dipole moment per unit volume of dielectric material.

Pole cells refers to a group of cells in the posterior of *Drosophila* embryos that are precursors to the adult germ line.

Pollen (L. *pollen*, fine flour) is the mass of germinated microspores or partially developed male gametophytes of seed plants.

Pollen culture refers to the *in vitro* culture and germination of pollen grains.

Pollen grain is a microspore that is produced in the pollen sac of angiosperms or the microsporangium of gymnosperms. Unicellular, with variable shape and size, and usually ovoid from 25 to 250 μm.

Pollination is the transfer of pollen from anther to stigma in the process of fertilization in angiosperms; transfer of pollen from male to female cone in the process of fertilization in gymnosperms.

Poly (A) polymerase is an enzyme that catalyzes the addition of adenine residues to the 3' end of pre-mRNAs to form the poly-(A) tail.

Poly-(A) tail is related to polyadenylation. It is the addition of a poly(A) tail to a primary transcript RNA. The poly(A) tail consists of multiple adenosine monophosphates; in other words, it is a stretch of RNA that has only adenine bases.

Polyacrylamide gel electrophoresis (PAGE) describes a technique widely used in biochemistry, forensics, genetics, molecular biology and biotechnology to separate biological macromolecule, usually proteins or nucleic acids, according to their electrophoretic mobility.

Polyacrylamide gels refer to incorrectly as acrylamide gels. These gels are made by cross-linking acrylamide with *N,N'*-methylene-*bis*-acrylamide. Polyacrylamide gels are used for the electrophoretic separation of proteins, DNA and RNA molecules. Polyacrylamide beads are also used as molecular sieves in gel chromatography, marketed as Bio-gel™.

Polyadenylation refers to post-transcriptional addition of a polyadenylic acid tail to the 3' end of eukaryotic mRNAs. Also called poly-(A) tailing. The adenine-rich 3' terminal segments is called a poly (A) tail.

Polyanhydrides are a class of biodegradable polymers characterized by anhydride bonds that connect repeat units of the polymer backbone chain.

Polycistronic refers to a single mRNA that contains the information necessary for the production of more than one polypeptide. Particularly characteristic of prokaryotic mRNAs.

Polyclonal antibodies are produced by an animal's white blood cells (lymphocytes, specifically) in response to an antigen. This response occurs naturally or can purposely be created by injecting an animal, such as a rabbit or goat, with a specific antigen.

Polycloning site refers to: In recombinant DNA technology, an artificially synthesized nucleotide sequence incorporated in a plasmid that contains multiple cleavage sites for different restriction enzymes enabling a choice of the most appropriate restriction enzyme for cloning. Called also restriction site bank or polylinker site.

Polycystic kidney disease (PKD) is a group of conditions characterized by fluid filled sacs that slowly develop in both kidneys, eventually resulting in kidney malfunction.

Polyembryony refers to the production of more than one embryo from a single egg cell (in animals) or from a range of embryogenic cell types (in plants). These embryos are genetically identical to one another.

Polyester is a category of polymers which contain the ester functional group in their main chain. Although there are many polyesters, the term "polyester" as a specific material most commonly refers to polyethylene terephthalate (PET).

Polyethylene glycol (PEG, carbowax, $HOCH_2(CH_2OCH_2)XCH_2OH$) is a polymer that is available in a range of molecular weights from *ca* 1000 to *ca* 6000. PEG 4000 and PEG 6000 are commonly used to promote cell or protoplast fusion, and

to facilitate DNA uptake in the transformation of organisms such as yeast. PEG is also used to concentrate solutions by withdrawing water from them.

Polygalacturonase (PG) is an enzyme which catalyzes the breakdown of pectin. A tomato engineered to contain an antisense-PG gene succeeded in delaying the onset of softening, by inhibiting the expression of PG.

Polygene is one of a number of genes, each of small effect, which together act to determine the phenotype of a quantitative trait. The result is continuous variation in the trait and a seemingly non-Mendelian mode of inheritance.

Polygenic refers to one that is controlled by many genes of small effect.

Polyhydramnios is a condition that occurs in pregnant women when the amniotic fluid is > 25 cm.

Polyhydroxybutyrate (PHB) refers to a biopolymer, with physical properties similar to polystyrene, originally discovered in the bacterium *Alcaligenes eutropus*.

Polylactic-*co*-glycolic acid (PLGA) is a copolymer, which is used in a host of Food and Drug Administration (FDA) approved therapeutic devices, owing to its biodegradability and biocompatibility. PLGA is synthesized by means of random ring-opening co-polymerization of two different monomers, the cyclic dimers (1,4-dioxane-2,5-diones) of glycolic acid and lactic acid.

Polylinker is a synthetic segment of DNA, designed to include a number of different restriction endonuclease sites. When ligated to a DNA fragment that is to be cloned, this enables a wide choice of restriction endonucleases to be used for the cloning process. Synonym is multiple cloning site (MCS).

Polymer refers to a macromolecule synthesized by the chemical joining of many identical or similar monomers. For example, amino acids, monosaccharides and nucleotides give rise to proteins, polysaccharides and nucleic acids respectively. Water is eliminated between the monomers as they link to form chains. The individual monomer units condensed within a chain are often referred to as residues, a term which is also employed for the bases incorporated in polynucleotides.

Polymerase (DNA) synthesizes a double-stranded DNA molecule using a primer and DNA as a template.

Polymerase chain reaction (PCR) is a technique to replicate a desired segment of DNA. PCR starts with primers that flank the desired target fragment of DNA. The DNA strands are first separated with heat, and then cooled allowing the primers bind to their target sites. Polymerase then makes each single strand into a double strand, starting from the primer. This cycle is repeated multiple times creating a 106 increase in the gene product after 20 cycles and a 109 increase over 30 cycles.

Polymer chemistry is a multidisciplinary science that deals with the chemical synthesis and chemical properties of polymers.

Polymeric degradation (biodegradation) is a process in which, the drug is contained within a polymer membrane or matrix.

Polymerization is a chemical union of two or more molecules of the same kind such as glucose or nucleotides to form a new compound (starch or nucleic acid) having the same elements in the same proportions but a higher molecular weight and different physical properties.

Polymery refers to the phenomenon whereby a number of genes at different loci (which may be polygenes) can act together to produce a single effect.

Polymorphic is the presence of more than one allele at a locus. Generally defined as having the most common allele at a frequency less than 95% or 99%.

Polymorphism is the presence of more than one allele at a locus. Polymorphism is also used as a measure of the proportion of loci in a population that are genetically variable or polymorphic (*P*).

Polynucleotide is a chain of nucleotides in which each nucleotide is linked by a single phosphodiester bond to the next nucleotide in the chain. They can be double- or single-stranded. The term is used to describe DNA or RNA.

Polypeptide is a peptide containing anywhere between 10 and 100 molecules of amino acids. Peptides can either be small proteins or part of a protein. A polypeptide is the ultimate expression product of a gene, and is folded into a functional protein after it has been assembled.

Polyphyletic is a group of taxa classified together that have descended from different ancestor taxa (i.e., taxa that do not all share the same recent common ancestor).

Polyploid (Gr. *polys*, many + *ploid*, fold) is the tissue or cells with more than two complete sets of chromosomes, that results from chromosome replication without nuclear division or from union of gametes with different number of chromosome sets, hence triploid (3*x*), tetraploid (4*x*), pentaploid (5*x*), hexaploid (6*x*), heptaploid (7*x*), octoploid (8*x*)).

Polyploidy refers to an organism, tissue or cells having more than two complete sets of chromosomes. Many crop plants are polyploid, including bread wheat (hexaploid, 6x), cotton and alfalfa (tetraploid, 4x), and banana (triploid, 3x).

Polysaccharide (Gr. *polys*, many + Gr. *sakcharon*, sugar) are long-chain molecules, such as starch and cellulose, composed of multiple units of a monosaccharide.

Polysaccharide capsule is a polysaccharide layer that lies outside the cell envelope of bacteria, and is thus deemed part of the outer envelope of a bacterial cell. It is a well organized layer, not easily washed off, and it can be the cause of various diseases.

Polysilicon is a polycrystalline silicon used as conductor in

Polysome is a multi-ribosomal structure representing a linear array of ribosomes held together by mRNA.

Polyspermy is the entry of several sperm into the egg during fertilization, although only one sperm nucleus actually fuses with the egg nucleus.

Polytene chromosome refers to giant chromosomes produced by interphase replication without division, and consisting of many identical chromatids arranged side by side.

Polyunsaturated fat is a fat that has more than one double bond in the molecule.

Polyunsaturates are oils in which some of the carbon-carbon bonds are not fully hydrogenated, i.e., of the form -CH=CH-, rather than -CH$_2$-CH$_2$-.

Polyvalent vaccine is a recombinant organism into which antigenic determinants have been cloned from a number of different disease-causing organisms, and used as a vaccine.

Polyvinyl alcohol (PVOH, PVA, or PVAl) is water soluble synthetic polymer (not to be confused with polyvinyl acetate).

Polyvinyl pyrrolidone (PVP) is an occasional constituent of plant tissue culture isolation media. PVP is of variable molecular weight and of general formula $(C_6H_9NO)_n$. Its antioxidant properties are used to prevent oxidative browning of excised plant tissues.

Population density is the number of cells or individuals per unit. The unit could be an area or volume of medium.

Population genetics is the branch of genetics that deals with frequencies of alleles and genotypes in breeding populations.

Population is defined group of interbreeding organisms.

Population viability analysis (PVA) is the general term for the application of models that account for multiple threats facing the persistence of a population to access the likelihood of the population's persistence over a given period of time.

Population viability is the probability that enough individuals in a population will survive to reproductive age to prevent extirpation of the population.

Porcine endogenous retrovirus (PERV) is the provirus of a porcine retrovirus. With the increasing interest in the use of pig organs for xenotransplantation to humans, there has been concern that PERVs could be activated after transplantation, creating an infection in the human recipient.

Porosity or void fraction is a measure of the void (i.e., "empty") spaces in a material, and is a fraction of the volume of voids over the total volume, between 0 and 1, or as a percentage between 0 and 100%.

Portal hypertension refers to a high blood pressure in the portal vein that carries blood to the liver.

Portal vein refers to the large vein that carries blood to the liver from the spleen and intestines.

Position effect is the situation in which a change in phenotype results from the change of the position of a gene or group of genes.

Positional assembly is a molecular manufacturing to add positional control to chemical synthesis.

Positional candidate gene is a gene known to be located in the same region as a DNA marker that has been shown to be linked to a single-locus trait or to a QTL, and whose deduced function suggests that it could be the source of genetic variation in the trait in question.

Positional cloning is the identification of a gene based on its physical location in the genome.

Positive and negative predictive values (**PPV** and **NPV**, respectively) are the proportions of positive and negative results in statistics and diagnostic tests that are true positive and true negative results. The PPV and NPV describe the performance of a diagnostic test or other statistical measure. A high result can be interpreted as indicating the accuracy of such a statistic. The PPV can be derived using Bayes' theorem (**Bayes' law** or **Bayes' rule**).

Positive control system is a mechanism in which a regulatory protein(s) is required to turn on gene expression.

Positive selectable marker is related to dominant selectable marker.

Positive selection is a method by which cells that carry a DNA insert integrated at a specific chromosomal location can be selected, since this integration confers a predictable phenotype.

Positron or antielectron is the antiparticle or the antimatter counterpart of the electron. The positron has an electric charge of +1e, a spin of ½, and has the same mass as an electron. When a low-energy positron collides with a low-energy electron, annihilation occurs, resulting in the production of two or more gamma ray photons.

Positrons may be generated by positron emission radioactive decay (through weak interactions), or by pair production from a sufficiently energetic photon.

Positron emission tomography (PET) is a nuclear medicine, functional imaging technique that produces a three-dimensional image of functional processes in the body. The system detects pairs of gamma rays emitted indirectly by a positron-emitting radionuclide (tracer), which is introduced into the body on a biologically active molecule. Three-dimensional images of tracer concentration within the body are then constructed by computer analysis.

Post-implantation embryo is implanted embryos in the early stages of development until the establishment of the body plan of a developed organism with identifiable tissues and organs.

Post-replication repair is a recombination-dependent mechanism for repairing damaged DNA.

Post-translational modification refers to the addition of specific chemical residues to a protein after it has been translated.

Potassium channel blocking agent is any of a class of antiarrhythmic agents that inhibit the movement of potassium ions through the potassium channels, thus prolonging repolarization of the cell membrane.

Potentiometric device monitors the voltage between a sensing electrode and a reference electrode. A high input impedance voltmeter is used to minimize current flow. The voltage typically is proportional to the logarithm of the analyte concentration.

Potentiometric refers to: (1) An instrument for measuring an unknown voltage by comparison to a standard voltage; (2) A three-terminal resistor with an adjustable center connection, widely used for volume control in radio and television receivers. Also called *pot*.

Pouch is a specialized collection bag worn over an ostomy to collect stool.

Power [W] is a product of voltage and current in a component; also it refers to the field of electric energy supply.

Power of a statistical test is the probability that it correctly rejects the null hypothesis when the null hypothesis is false.

PPFD is an abbreviation for photosynthetic photon flux density. PPFD is a unit of measure to express the light quantum in photons of solar energy (specific to wavelength) related specifically to photosynthesis and is measured with a quantum meter in units called micro-moles. The PPF) is expressed as micro-moles of photons per meter square per second.

ppm is an abbreviation for parts per million.

PR protein is an abbreviation for pathogenesis related protein, which are proteins produced in plants in the event of a pathogen attack. They are induced as part of systemic acquired resistance. Infections activate genes that produce PR proteins.

Prader-Willi syndrome is a condition characterized by obesity and insatiable appetite, mental deficiency, small genitals, and short stature. May be deletion of #15 chromosome.

Prandtl Number is a dimensionless number approximating the ratio of momentum diffusivity and thermal diffusivity.

Precautionary principle refers to the approach whereby any possible risk associated with the introduction of a new technology is avoided, until a full understanding of its impact on health, environment etc. is available. Particularly applied to the release of genetically modified organisms, since

unlike many technologies, these cannot be recalled if problems arise.

Precision is the degree of reproducibility among several independent measurements of the same true value under specified conditions.

Preclinical imaging is the visualization of living animals for research purposes, such as drug development. These days, many manufacturers provide multi-modal systems combining the advantages of anatomical modalities such as CT and MR with the functional imaging of PET and SPECT. As in the clinical market, common combinations are SPECT/CT, PET/CT and PET/MR.

Precocious germination is the premature germination of the embryo, prior to completion of embryogenesis.

Predator is animal that kills another animal for food.

Predisposition is to have a tendency or inclination towards something in advance.

Preeclampsia refers to abnormal state of pregnancy characterized by hypertension and fluid retention and albuminuria.

Pre-filter is a coarse filter used to screen out large particles before air is forced through a much finer filter.

Pre-implantation embryos refer to fertilized eggs (zygotes) and all of the developmental stages up to, but not beyond, the blasto-cyst stage.

Preimplantation means that the embryo has not yet implanted in the wall of the uterus. Human embryonic stem cells are derived from preimplantation-stage embryos fertilized outside a woman's body (*in vitro*).

Premature ejaculation (PE) is the inability to maintain an erec-tion long enough for mutual satisfaction.

pre-mRNA (Precursor mRNA) is an immature single strand of messenger ribonucleic acid (mRNA). Pre-mRNA is synthe-sized from a DNA template in the cell nucleus by transcrip-tion. Pre-mRNA comprises the bulk of heterogeneous nuclear RNA (hnRNA). The term hnRNA is often used as a synonym for pre-mRNA. In the strict sense, hnRNA may include nuclear RNA transcripts that do not end up as cytoplas-mic mRNA. Once pre-mRNA has been completely processed, it is termed "mature messenger RNA," "mature mRNA," or simply "mRNA."

Prenatal refers to existing or occurring before birth: prenatal medical care.

Pressure coefficient is equal 1 at most for incompressible flows. For compressible flows, the coefficient is greater than 1.

Pressure potential is the pressure generated within a cell. It is the difference between the osmotic potential within the cell and the water potential of the external environment, provided the cell volume is constant.

Presymptomatic diagnosis is a diagnosis of a genetic condi-tion before the appearance of symptoms.

Pre-transplant refers to stage III in tissue culture micropropa-gation; the rooting, hardening stage prior to transfer to soil.

Preventive immunization refers to an infection with an antigen to elicit an antibody response that will protect the organ-ism against future infections. Synonym is vaccination.

Priapism is an inflammation of the penis.

Pribnow box refers to the consensus sequence near the mRNA start-point of prokary-otic genes.

Primary (L. *primus*, first) refers to first in order of time or development.

Primary antibody is the antibody that binds to the target mol-ecule, in an ELISA or other immunological assay.

Primary cell is a cell or cell line taken directly from a living organism, which is not immortalized.

Primary cell wall is the cell wall layer formed during cell expansion. Plant cells possessing only primary walls may divide or undergo differentiation.

Primary culture is a culture started from cells, tissues or organs taken directly from organisms. A primary culture may be regarded as such until it is sub-cultured for the first time. It then becomes a cell line.

Primary germ layers refers to a layer of cells that form during embryogenesis. The three germ layers in vertebrates are particularly pronounced; however, all eumetazoans, (animals more complex than the sponge) produce two or three primary germ layers. Animals with radial symmetry, like cnidarians, produce two germ layers (the ectoderm and endoderm) making them diploblastic. Animals with bilateral symmetry produce a third layer (mesoderm), between these two layers. making them triploblastic. Germ layers eventually give rise to all of an animal's tissues and organs through the process of organogenesis.

Primary growth refers to: (1) Apical meristem-derived growth; the tissues of a young plant; (2) Explant growth during the initial culture period.

Primary immune response is the immune response that occurs during the first encounter of a mammal with a given antigen.

Primary meristem is meristem of the shoot or root tip giving rise to the primary plant body.

Primary refers to first in order of time or development.

Primary structure is the linear sequence of residues making up a polymer such as a nucleic acid, polysaccharide or protein.

Primary tissue is a tissue that has differentiated from a primary meristem.

Primary transcript is the RNA molecule produced by transcription prior to any post-transcriptional modifications; also is called a pre-mRNA in eukaryotes.

Primer is a small oligonucleotide (typically 18-22 base pairs long) that anneals to a specific single-stranded DNA sequence to serve as a starting point for DNA replication (e.g., extension by polymerase during PCR).

Primer DNA polymerase is a DNA polymerase that provides primers for the DNA polymerization. Unlike RNA polymerase, DNA polymerase is unable to initiate the *de novo* synthesis of a polynucleotide chain. DNA polymerase can only add nucleotides to a free 3′ hydroxyl group at the end of a pre-existing chain. A short oligonucleotide, known as a primer, is therefore needed to supply such a hydroxyl group for the initiation of DNA synthesis.

Primer walking is a method for sequencing long (>1 kb) cloned pieces of DNA. The initial sequencing reaction reveals the sequence of the first few hundred nucleotides of the cloned DNA. On the basis of these data, a primer containing about 20 nucleotides and complementary to a sequence near the end of sequenced DNA is synthesized, and is then used for sequencing the next few hundred nucleotides of the cloned DNA. This procedure is repeated until the complete nucleotide sequence of the cloned DNA is determined.

Primordial germ cells are the precursors of reproductive cells within the embryo. They are detectable in an embryo after four weeks of development and will develop into either sperm or eggs.

Primordium is a group of cells which gives rise to an organ.

Primosome is a protein-replication complex that catalyzes the initiation of synthesis of Okazaki fragments during discontinuous replication of DNA. It involves DNA primase and DNA helicase activities.

Printed circuit board (PCB) is a selectively metallized insulating sheet for supporting and interconnecting circuit components.

Prion (PrP) is an infectious agent composed of protein in a misfolded form. This is the central idea of the Prion Hypothesis. Prions are responsible for the transmissible spongiform encephalopathies in a variety of mammals, including bovine spongiform encephalopathy (BSE, also known as "mad cow disease") in cattle. In humans, prions cause Creutzfeldt-Jakob Disease (CJD), variant Creutzfeldt-Jakob Disease (vCJD), Gerstmann–Sträussler–Scheinker syndrome, Fatal Familial Insomnia and kuru.

Private allele is an allele present in only one of many populations sampled.

Probability is the certainty of an event occurring. The observed probability of an event, *r*, will approach the true probability as the number of trials, n, approaches infinity.

Probability of identity (PI) is the probability that two unrelated (randomly sampled) individuals would have an identical genotype. This probability becomes very small if many highly polymorphic loci are considered.

Proband is an individual in a family who brought the family to medical attention.

Probe refers to: (1) For diagnostic tests, the agent that is used to detect the presence of a molecule in a sample; (2) A DNA or RNA sequence labeled or marked with a radioactive isotope or that is used to detect the presence of a complementary sequence by hybridization with a nucleic acid sample.

Probe DNA is a labeled DNA molecule used to detect complementary-sequence nucleic acid molecules by molecular hybridization. To localize the probe DNA sequence and reveal the complementary hybridization sequence, autoradiography or fluorescence is used.

Procambium (L. *pro*, before + cambium) is a primary meristem that gives rise to primary vascular tissues and, in most woody plants, to the vascular cambium.

Procaryote (Prokaryotic, adj.) is a group of organisms whose cells lack a membrane-bound nucleus (karyon). Those organisms whose cells have a well defined membrane bound nucleus and organelles are called eukaryotes. The word *prokaryote* comes from the Greek (πρό-, *pro-*: before" and καρυόν: *karyon*, "nut or kernel"). Prokaryotes do not have a membrane bound nucleus, mitochondria, or any other membrane-bound organelles. In other words, all their intracellular water-soluble components (proteins, DNA and metabolites) are located together in the same volume enclosed by the cell membrane (cell capsule), rather than in separate cellular compartments.

Process intensification is an engineering term applied to process design, to improve sustainability, in the bioproducts manufacturing sector.

Processed food is any food product that has undergone physical or chemical treatment resulting in a substantial change in the original state of the food.

Processed pseudo-gene is a copy of a functional gene which has no promoter, no introns and which, consequently, is not itself transcribed. Pseudo-genes are thought to originate from the integration into the genome of cDNA copies synthesized from mRNA molecules by reverse transcriptase. Pseudogenes therefore have a poly (dA) sequence at their 5′ ends.

Prodrug is a medication that is administered in an inactive or less than fully active form, and then it becomes converted to its active form through a normal metabolic process, such as hydrolysis of an ester form of the drug. A prodrug is a precursor chemical compound of a drug. Instead of administering a drug, a prodrug might be used instead to improve how a medicine is absorbed, distributed, metabolized, and excreted (ADME).

Product rule is a statistical rule that states that the probability of n_i independent events occurring is equal to the product of the probability of each n independent event.

Production environment refers to all input-output relationships, over time, at a particular location. The relationships will include biological, climatic, economic, social, cultural and political factors, which combine to determine the productive potential of a particular livestock enterprise.

Production trait refers to characteristics of animals, such as the quantity or quality of the milk, meat, fiber, eggs, draught, etc., they (or their progeny) produce, which contribute directly to the value of the animals for the farmer, and that are identifiable or measurable at the individual level.

Productivity is the amount of product that is produced within a given period of time from a specified quantity of resource.

Pro-embryo (L. *pro*, before + *embryon*, embryo) is a group of cells arising from the division of the fertilized egg cell or somatic embryo before those cells which are to become the embryo are recognizable.

Progenitor cell is progenitor cell, often confused with stem cell, is an early descendant of a stem cell that can only differentiate, but it cannot renew itself anymore. In contrast, a stem cell can renew itself (make more stem cells by cell division) or it can differentiate (divide and with each cell division evolve more and more into different types of cells). A progenitor cell is often more limited in the kinds of cells it can become than a stem cell. In scientific terms, it is said that progenitor cells are more differentiated than stem cells.

Progeny (*plural:* **progenies**) is synonym of offspring. It refers to: a person who comes from a particular parent or family; or the child or descendant of someone; or the young of an animal or plant; and something that is the product of something else.

Progeny testing is the practice of ascertaining the genotype of an individual from its offspring, such as by mating it to other individuals and examining the progeny.

Progestational agent is any of group of hormones secreted by the *corpus luteum* and placenta and, in small amounts, by the adrenal cortex, including progesterone; they induce the formation of a secretory endometrium.

Progesterone is a hormone produced primarily by the corpus luteum of the ovary, but also by the placenta, that prepares the inner lining of the uterus for implantation of a fertilized egg cell.

Prognosis refers to the prediction of the course and probable outcome of a disease.

Programed cell death (PCD) is death of a cell in any form, mediated by an intracellular program. PCD is carried out in a regulated process, which usually confers advantage during an organism's life-cycle.

Prokaryote (L. *pro*, before + Gr. *karyon*, a nut, referring in modern biology to the nucleus) is a member of a large group of organisms, including bacteria and blue-green algae, which do not have the DNA separated from the cytoplasm by a membrane in their cells. The DNA is usually in one long strand. Prokaryotes do not undergo meiosis and do not have functional organelles such as mitochondria and chloroplasts.

Prolactin is a hormone, produced by the anterior pituitary gland, that stimulates and controls lactation in mammals.

Prolapse refers to a condition, when part of the body (for instance, a section of intestine) slips from its normal position.

Proliferation (L. *proles*, offspring + *ferre*, to bear) refers to the increase by frequent and repeated reproduction; growth by cell division.

Pro-meristem refers to the embryonic meristem that is the source of organ initials or foundation cells.

Promoter refers to: (1) A short DNA sequence, usually upstream of (5' to) the relevant coding sequence, to which RNA polymerase binds before initiating transcription. This binding aligns the RNA polymerase so that transcription will initiate at a specific site; (2) A chemical substance that enhances the transformation of benign cells into cancerous cells.

Promoter sequence is a regulatory DNA sequence that initiates the expression of a gene.

Pro-nuclear micro-injection is the initial method of transgenesis in animals, involving injecting many copies of a particular gene into one of the two pro-nuclei of a fertilized ova. Characterized by a very low success rate. As animal cloning becomes more common, pro-nuclear micro-injection will be replaced by micro-injection of cloned genes into a culture of cloned embryos produced by nuclear transfer, which can be tested for expression of the transgene before transfer to recipient females.

Pronucleus refers to either of the two haploid gamete nuclei just prior to their fusion in the fertilized ovum.

Proofreading refers to the scanning of newly-synthesized DNA for structural defects, such as mis-matched base pairs. It is a functional activity of most DNA polymerases.

Propagation (L. *propagare*, to propagate) refers to the multiplication of plants by numerous types of vegetative material.

Propagule is any structure capable of giving rise to a new plant by asexual or sexual reproduction, including bulbils, leafbuds, etc.

Propagule pressure is a measure of the introduction of nonindigenous individuals that includes the number of individuals (or propagules) introduced and the number of introductions.

Prophase (Gr. *pro*, before + *phasis*, appearance) is an early stage in nuclear division, characterized by the shortening and thickening of the chromosomes and their movement to the metaphase plate. It occurs between interphase and metaphase. During this phase, the centriole divides and the two daughter centrioles move apart. Each sister DNA strand from interphase replication becomes coiled, and the chromosome is longitudinally double except in the region of the centromere. Each partially separated chromosome is called a chromatid. The two chromatids of a chromosome are sister chromatids.

Proportion of admixture is the proportion of alleles in a hybrid swarm that come from each of the parental taxa.

Prostatalgia refers to a pain in the prostate gland.

Prostate is a sex gland in men. It is about the size of a walnut, and surrounds the neck of the bladder and urethra, the tube that carries urine from the bladder. It is partly muscular and partly glandular, with ducts opening into the prostatic portion of the urethra. It is made up of three lobes: a center lobe with one lobe on each side.

Prostate specific antigen (PSA) is a blood test used to help detect prostate cancer by measuring a substance called prostate-specific antigen produced by the prostate.

Prostate specific membrane antigen (PSMA; Glutamate carboxypeptidase II, N-acetyl-L-aspartyl-L-glutamate peptidase I, or NAAG peptidase) is an enzyme that in humans is encoded by the *FOLH1* (folate hydrolase 1) gene. Human GCPII contains 750 amino acids and weighs approximately 84 kDa.

Prostatectomy is surgical procedure for the partial or complete removal of the prostate.

Prostatism is any condition of the prostate that causes interference with the flow of urine from the bladder.

Prostatitis is an inflamed condition of the prostate gland that may be accompanied by discomfort, pain, frequent urination, infrequent urination, and, sometimes, fever.

Protamine is a class of small basic proteins that replace the histones in the chromosomes of some sperm cells.

Protease is an enzyme that catalyzes the hydrolysis of proteins, cleaving the peptide bonds that link amino acids in protein molecules. Synonym is peptidase.

Protein is a macromolecule composed of one to several polypeptides. Each polypeptide consists of a chain of amino acids linked together by covalent (peptide) bonds. They are naturally-occurring complex organic substances (egg albumen, meat) composed essentially of carbon, hydrogen, oxygen and nitrogen, plus sulphur or phosphorus, which are so associated as to form submicroscopic chains, spirals or plates and to which are attached other atoms and groups of atoms in a variety of ways.

Protein-bound paclitaxel is an injectable formulation of paclitaxel, a mitotic inhibitor drug used in the treatment of breast cancer.

Protein crystallization is the production of a pure preparation of a protein. In this form, the three-dimensional structure of the molecule can be determined.

Protein engineering is the study of techniques of molecular biology, especially molecular cloning, recombinant protein expression, and site-directed mutagenesis can be employed to engineer and produce proteins with increased propensity to crystallize. Frequently, problematic cysteine residues can be replaced by alanine to avoid disulfide-mediated aggregation, and residues such as lysine, glutamate, and glutamine can be changed to alanine to reduce intrinsic protein flexibility, which can hinder crystallization.

Protein kinase is an enzyme that catalyzes the addition of a phosphate group(s) to a protein molecule at the sites of serine, threonine or tyrosine residues.

Protein metabolic step refers to one step in the chain of reactions that take place in an organism and dictate the composition of that organism.

Protein sequencing is the process of determining the amino acid sequence of a protein. Usually achieved following initially partial hydrolysis of the protein into smaller peptides by enzymatic digestion.

Protein synthesis is the creation of proteins from their constituent amino acids, in accordance with the genetic information carried in the DNA of the chromosomes.

Proteinaceous infectious particle (prion) is a proposed pathogen composed only of protein with no detectable nucleic acid. A prion in the Scrapie form (PrPSc) is an infectious agent composed of protein in a misfolded form. This is the central idea of the Prion Hypothesis.

Proteinuria refers to large amounts of protein in the urine.

Proteolysis refers to the enzymatic degradation of a protein.

Proteolytic is the ability to break down protein molecules.

Proteome refers to the complete complement of proteins made by a given species in all its tissues and growth stages.

Proteomics is the study of the protein products of genes, protein-protein interactions and protein sub-cellular localization. Examples could include engineering of new systems to sequence proteins or study protein interactions with other proteins or DNA, developing faster and cheaper detectors, such as high-density capillaries or high throughput mass spectrometers, and developing centers with expertise and accountability for protein analysis, such as 2D protein databases.

Proteus syndrome is a condition characterized by distorted asymmetric growth of the body and enlarged head, enlarged feet, multiple nevi on the skin; mode of inheritance is unknown.

Protoclone refers to the generated plant derived from protoplast culture or a single colony derived from protoplasts in culture.

Protocol (Gr. *protokollon*, first leaf glued to a manuscript) is the step-by-step experimental procedure to describe or solve a scientific problem, or the defined steps of a specific procedure.

Protocorm are seeds that contain an unorganized embryo comprising only a few hundred cells.

Protoderm (Gr. *protos*, first + *derma*, skin) is a primary meristem tissue that gives rise to epidermis.

Protogyny refers to the condition in which the female reproductive organs (carpels) of a flower mature before the male ones (stamens), thereby ensuring that self-fertilization does not occur.

Protomeristem refers to the primary meristem

Proton pump inhibitors are medicines that affect how acid is produced by the stomach's "proton pump" system, thereby decreasing stomach acid.

Protooncogene is a normal cellular gene that can be changed to an oncogene by mutation or by being incorporated into a retrovirus and then being transcribed at inappropriate times and/or in inappropriate tissues.

Protoplasm refers to the essential, complex living substance of cells, upon which all vital functions of nutrition, secretion, growth and reproduction depend.

Protoplast (Gr. *protoplastos*, formed first) is a bacterial or plant cell for which the relatively rigid wall has been removed either chemically or enzymatically, leaving its cytoplasm enveloped by only a delicate peripheral membrane. Protoplasts are spherical and smaller than the elongate, angular shaped and often vacuolated cells from which they have been released.

Protoplast culture is the isolation and culture of plant protoplasts by mechanical means or by enzymatic digestion of plant tissues or organs, or cultures derived from these. Protoplasts are utilized for selection or hybridization at the cellular level and for a variety of other purposes.

Protoplast fusion refers to the induced or spontaneous coalescence of two or more protoplasts of the same or different species origin. Where fused protoplasts can be regenerated into whole plants, the opportunity exists for the creation of novel genomic combinations.

Prototroph is a nutritionally independent cell. Opposite is auxotroph.

Pro-toxin is a latent, non-active precursor form of a toxin.

Protozoa is any of a large group of single-celled, usually microscopic, organisms such as amoeba.

Protozoan (plural: protozoa) is a microscopic, single-cell organism.

Protruding ends are converted by accomplishing simultaneously by the 3'->5' exonuclease and 5'->3' polymerase activities of T4 DNA Polymerase.

Provenance refers to the geographical and/or genetic origin of an individual.

Provirus is a double stranded DNA copy of the single RNA strand of a retrovirus, which has been integrated into a host genome.

Pseudo-affinity chromatography is a chromatographic technique in which a ligand is immobilized selectively to retain enzymes or other proteins.

Pseudo-autosomal region is a section at one end of the X and Y chromosomes for which there is sufficient homology that there is synapsis between them during meiosis.

Pseudocarp refers to a fruit that incorporates, in addition to the ovary wall, other parts of the flower, such as the receptacle (e.g., strawberry). Synonym: false fruit.

Pseudogene refers to an incomplete or mutated copy of a gene which is not transcribed because it lacks a continuous open reading frame. Those that lack introns are called processed pseudogenes and are most likely cDNA copies synthesized from mRNA by reverse transcriptase

Pseudomonas spp. is a widely distributed Gram-negative bacterial genus. Many of the soil forms produce a pigment that fluoresces under ultraviolet light, hence the descriptive term fluorescent *Pseudomonas*.

P-site is an abbreviation for peptidyl-tRNA site.

Psychoactive agent or **psychotropic agent** is a psychoactive substance.

Psychrophile is a micro-organism that can grow at temperatures below 30°C and as low as 0°C.

p-Type semiconductor is a semiconductor for which the predominant charge carriers responsible for electrical conduction are holes.

Public health helps to protect the persons from injury and disease and for helping them to stay healthy.

Public Health Service Act is a United States federal law enacted in July 1, 1944. The full act is captured under Title 42 of the United States Code (The Public Health and Welfare), Chapter 6A (Public Health Service). It has since been amended many times, some amendments include: (1) Family Planning Services and Population Research Act of 1970 Pub.L. 91–572, which established Title X of the Public Health Service Act, dedicated to providing family planning services for those in need; (2) National Cancer Act of 1971; (3) Health Insurance Portability and Accountability Act of 1996; (4) Muscular Dystrophy Community Assistance Research and Education Amendments of 2001; (5) Hematological Cancer Research Investment and Education Act of 2001; (6) Newborn Screening Saves Lives Act of 2007 Patient Protection and Affordable Care Act of 2010; 7. Pandemic and All-Hazards Preparedness Reauthorization Act of 2013 (H.R. 307; 113th Congress).

Public policy is a set of action guidelines or rules that result from the actions or lack of actions of governmental entities.

PUC refers to a widely used expression plasmid containing a galactosidase gene.

Pulmonary artery is a blood vessel delivering oxygen-poor blood from the right ventricle to the lungs.

Pulmonary autograft valves are new approach for replacement of a diseased aortic valve involves moving the patients own pulmonary valve (the valve on the right side of the heart that leads to the pulmonary artery and the lungs just as the aortic valve leads to the aorta and the body) into the aortic position to replace a stenotic or regurgitant aortic valve.

Pulmonary circulation is the part of the circulation system that delivers blood to and from the lungs for oxygenation.

Pulmonary hypertension refers to abnormally high blood pressure in the arteries of the lungs.

Pulmonary minute volume (pulmonary ventilation) is calculated by: Volume of air respired per minute = (tidal volume) x (breathes/min).

Pulmonary refers to the lungs and respiratory system.

Pulmonary valve is the heart valve between the right ventricle and the pulmonary artery. It controls blood flow from the heart into the lungs.

Pulmonary vein is the blood vessel that carry newly oxygenated blood from the lungs back to the left atrium of the heart.

Pulsatile or pulsating is characterized by a rhythmic beat.

Pulsed ultrasound is a method of administering ultrasound with a pulsed system. Ultrasound is an oscillating sound pressure wave with a frequency greater than the upper limit of the human hearing. Both continuous wave and pulsed systems are used. Low-intensity pulsed ultrasound (LIPUS) is a medical technology, generally using 1.5 MHz frequency pulses, with a pulse width of 200 μs, repeated at 1 kHz, at an intensity of 30 mW/cm^2, 20 minutes/day. Applications of LIPUS include: Promoting bone-fracture healing; Treating orthodontically induced root resorption; Regrow missing teeth; Enhancing mandibular growth in children with hemifacial microsomia; Promoting healing in various soft tissues such as cartilage, inter vertebral disc; and Improving muscle healing after laceration injury.

Pulsed-field gel electrophoresis (PFGE) is a procedure used to separate very large DNA molecules by alternating the direction of electric current in a pulsed manner across a semi-solid gel.

Pump transfers fluids or gases by pressure. It is an equipment to force air or fluid into a cavity, as heart pumps blood.

Punctuated equilibrium refers to the occurrence of speciation events in bursts, separated by long intervals of species stability.

Pure culture refers to an axenic culture.

Pure line is a strain in which all members have descended by self-fertilization or close inbreeding. A pure line is genetically uniform.

Purging is the removal of deleterious recessive alleles from a population through inbreeding which increases homozygosity which in turn increases the ability of selection to act on recessive alleles.

Purine is a double-ring, nitrogen-containing base present in nucleic acids. Adenine (A) and guanine (G) are the two purines normally present in DNA and RNA molecules.

PVA is an abbreviation for Plant Variety Act (USDA).

PVP refers to an: (1) Abbreviation for polyvinylpyrrolidone. (2) Abbreviation for plant variety protection.

PVR is an abbreviation for plant variety rights.

Pyloric sphincter is the muscle between the stomach and the small intestine.

Pyloric stenosis is an enlargement of the muscle between the stomach and the small intestine, blocking the passage of food and liquids forward into the intestines.

Pyloroplasty is an operation that enlarges the opening between the stomach and small intestine so food and liquid can move forward and be digested normally.

Pylorus refers to where the stomach connects to the small intestine.

Pyrethrins are active constituents of pyrethrum (*Tanacetum cinerariifolium*) flowers, used as insecticides.

Pyrimidine is a single-ring, nitrogen-containing base present in nucleic acids. Cytosine (C) and thymine (T) are present in DNA, whereas uracil (U) replaces T in RNA. Thymine is a synonym for 5-methyluracil.

Pyroelectricity is the property of certain crystals, such as tourmaline, of acquiring opposite electrical charges on opposite faces when heated.

Pyrogen refers to bacterial substance that causes fever in mammals.

Pyrophosphate is a phosphate ion dimer; may be released on hydrolysis of ATP.

Q

q denotes the longer of the two chromosome arms, e.g., human 10q is the long arm of human chromosome 10.

Q factor is a rating, applied to coils, capacitors, and resonant circuits, equal to the reactance divided by the resistance. The ratio of energy stored to energy dissipated per cycle in an electrical or mechanical system.

q-beta replicase refers to a viral RNA polymerase secreted by a bacteriophage that infects *E. coli*. It has the property of being able to copy RNA sequences at a rapid rate.

QSAR is an abbreviation for quantitative structure-activity relationship.

QTL is an abbreviation for quantitative trait locus.

Quadrivalent is a chromosome configuration visible in late prophase and metaphase of the first meiotic division, where four chromosomes are linked by chiasmata. It can occur in autotetraploids when four homologous chromosomes pair, or in diploids as a result of heterozygosity for a reciprocal translocation between two non-homologous chromosomes.

Quadruplex refers to the inheritance of alleles in autotetraploids. A genotype *AAAa* will produce gametes *AA*, *Aa* in the ratio 3:1.

Qualitative trait is a trait that shows discontinuous variation, i.e., individuals can be assigned to one of a small number of discrete classes.

Quantitative genetics is the branch of genetics concerned with the inheritance of quantitative traits that show continuous variation, as opposed to qualitative traits. Since many of the critical targets in both plant and animal breeding are of this type, most practical improvement programs involve the application of quantitative genetics.

Quantitative inheritance refers to inheritance of measurable traits (height, weight, color intensity, etc.) that depend on the cumulative action of many genes.

Quantitative structure-activity relationship (QSAR) refers to a computer modeling technique that enables the prediction of the likely activity of a molecule before it is synthesized. QSAR analysis relies on recognizing associations of molecular structures and activity from historical data.

Quantitative trait is a measurable trait that shows continuous variation (e.g., height, weight, color intensity, etc.), i.e., the population cannot be classified into a few discrete classes.

Quantitative trait loci (QTLs) refer to genetic loci that affect phenotypic variation (and potentially fitness), which are identified by a statistically significant association between genetic markers and measurable phenotypes. Quantitative traits are often influenced by multiple loci as well as environmental factors.

Quantization is the concept that energy can occur only in discrete units called quanta.

Quantum (L. *quantum*, how much) is an elemental unit of energy. Quantum energy = hv, where h = Planck's constant = 6.62×10^{-27} erg-second, and v is the frequency of the vibrations or waves with which the energy is associated.

Quantum chemical calculations are molecular property calculations based on the Schrödinger equation, which take into account the interactions between electrons in the molecule.

Quantum computer (quantum supercomputer) is a computation device that makes direct use of quantum-mechanical phenomena, such as superposition and entanglement, to perform operations on data. Quantum computers are different from digital computers based on transistors. Whereas digital computers require data to be encoded into binary digits (bits), each of which is always in one of two definite states (0 or 1), quantum computation uses qubits (quantum bits). A theoretical model is the quantum Turing machine, also known as the universal quantum computer. Quantum computers share theoretical similarities with non-deterministic and probabilistic computers. As of today, quantum computing is still in its infancy but experiments have been carried out in which quantum computational operations were executed on a very small number of qubits. Both practical and theoretical research continues, and many national governments and military funding agencies support quantum computing research to develop quantum computers for both civilian and national security purposes, such as cryptanalysis.

Quantum dots are nanometer sized fragments (the dots) of semiconductor crystalline material, which emits PHOTONS. The wavelength is based on the quantum confinement size of the dot. They are brighter and more persistent than organic chemical INDICATORS. They can be embedded in MICROBEADS for high throughput analytical chemistry. One should not confuse these with microscopic fluorescent bar codes which are micrometer sized. An important strategy for nonisotopic labeling of single molecules is the use of highly luminescent semiconductor nanocrystals, or 'quantum dots,' that can be covalently linked to biological molecules. This class of detectors, which range in size from 1–5 nm, have been exploited for biological labeling by a number of laboratories. Quantum dots offer several advantages over organic dyes, including increased brightness, stability against photobleaching, a broad continuous excitation spectrum, and a narrow, tunable, symmetric emission spectrum. Because quantum dots are nontoxic; and can be made to dissolve in water, efforts are underway to explore their use in labeling single molecules in living cells.

Quantum nanophysics research measure at the edge, and require the fastest signal acquisition, the most comprehensive feature set, the highest frequencies, the fastest scans. Keywords are maximum modulation frequency (50 MHz, 600 MHz) and lowest time constant (780 ns, 30 ns). Applications in this section include: fundamental research in mesoscopic system (small dimensional systems), quantum physics, nano technologies, nano MRI, spintronics, semiconductor material research, graphene, carbon nano tubes, mechanically controllable break junction (MCBJ).

Quantum physics describes fundamental electronic and optical properties of matter at microscopic level and wave and interference phenomena in particular . Quantum electronics, quantum optics and optoelectronics are important areas of application – atomic clocks, lasers, light emitting diodes, optical fibers, tunnel diodes and superconducting systems are important and well-known examples. The trend towards faster and more complicated microprocessors and microelectronics has resulted in electronic components now approaching the domains of quantum physics at research level. In about 20 years time, miniaturization will also be halted in commercial applications. The QNANO (quantum nanophysics) probably provide a target for research and development in semiconductor physics, molecular electronics and bioelectronics in the foreseeable future.

Quantum speciation refers to the rapid formation of new species, primarily by genetic drift.

Quantum speciation is the rapid formation of new species, primarily by genetic drift.

Quarantine (It. *quarantina*, from *quaranta,* forty) refers to keeping a person or living organism in isolation for a period (originally 40 days) after arrival to allow disease symptoms to appear, if there was any disease present.

Quaternary structure refers to a level of protein structure where several individual molecules assemble together and form a functional cluster. A classic example is haemoglobin, a complex of four myoglobin-like units.

Quiescent is a temporary suspension or reduction in the rate of activity or growth, while retaining the potential to resume prior activity.

R

. . . .

R genes is a class of plant genes conferring resistance to a specific strain (or group of strains) of a particular pathogen. Their primary function is to sense the presence of the pathogen and to trigger the defence pathways. *R* genes have been cloned from a number of biological species.

R₁ refers to the first-generation offspring of a recombinant (genetically modified) organism.

Rabies is a viral disease of wild animals that can be transmitted to humans through the bite of an infected animal.

Race refers to a distinguishable group of organisms of a particular species. Criteria for distinctness can be one or a combination of geographic, ecological, physiological, morphological, genetic and karyotypic factors.

Raceme is an inflorescence in which the main axis is elongated but the flowers are borne on pedicels that are about equal in length.

Rachilla (Gr. *rhachis*, a backbone + L. diminutive suffix *-illa*) is a shortened axis of a spikelet.

Rachis (Gr. *rhachis*, a backbone) is the main axis of a spike; axis of fern leaf (frond) from which pinnae arise; in compound leaves, the extension of the petiole corresponding to the midrib of an entire leaf.

Radiation hybrid cell panel (RH) refers to a somatic cell hybrid panel in which the chromosomes from the species of interest have been fragmented by irradiation prior to cell fusion. The resultant small fragments of chromosomes greatly increase the power of physical mapping in the species of interest.

Radicle (L. *radix*, root) refers to that portion of the plant embryo which develops into the primary or seed root.

Radioactive decay (nuclear decay or radioactivity) is the process by which a nucleus of an unstable atom loses energy by emitting ionizing radiation

Radioactive half-life refers to a radioactive decay constant. It is the rate of radioactive decay.

Radioactive isotope (radioisotope) is an unstable isotope that emits ionizing radiation.

Radiochemistry is the chemistry of radioactive materials, where radioactive isotopes of elements are used to study the properties and chemical reactions of non-radioactive isotopes (often within radiochemistry the absence of radioactivity leads to a substance being described as being *inactive* as the isotopes are *stable*). Much of radiochemistry deals with the use of radioactivity to study ordinary chemical reactions. Radiochemistry includes the study of both natural and man-made radioisotopes.

Radioembolization is a radiation therapy that includes patients with: Unrespectable primary liver cancer or metastatic cancer (colorectal cancer, neuroendocrine tumor; liver-dominant tumor burden; and life expectancy of at least 3 months.

Radiofrequency ablation therapy (*Secca* procedure) refers to temperature-controlled radiofrequency energy being delivered to the anal canal, and is marketed as the *SECCA* procedure. This procedure aims to create a controlled scarring and structuring of the anal canal. In theory, it is thought that radiofrequency induced IAS injury may cause collagen deposition and fibrosis (scarring), resulting in the affected area tightening. Specialized surgical instrument called a radiofrequency handpiece is used.

Radiograph is a record produced on a photographic plate, film, or paper by the action of roentgen rays or radium; specifically x-rays.

Radioimmunoassay (RIA) refers to an assay based on the use of a radioactively labeled antibody, where the amount of radiation detected indicates the amount of target substance present in the sample.

Radioimmunotherapy uses an antibody labeled with a radionuclide to deliver cytotoxic radiation to a target cell. In cancer therapy, an antibody with specificity for a tumor-associated antigen is used to deliver a lethal dose of radiation to the tumor cells. The ability for the antibody to specifically bind to a tumor-associated antigen increases the dose delivered to the tumor cells while decreasing the dose to normal tissues. By its nature, RIT requires a tumor cell to express an antigen that is unique to the neoplasm or is not accessible in normal cells.

Radiopharmaceutical is the study and preparation of radiopharmaceuticals, which are radioactive pharmaceuticals. Radiopharmaceuticals are used in the field of nuclear medicine as tracers in the diagnosis and treatment of many diseases. Many radiopharmaceuticals use technetium-99m (Tc-99m), which has many useful properties as a gamma-emitting tracer nuclide. In the book *Technetium*, a total of 31 different radiopharmaceuticals based on Tc-99m are listed for imaging and functional studies of the brain, myocardium, thyroid, lungs, liver, gallbladder, kidneys, skeleton, blood and tumors. The term radioisotope has historically been used to refer to all radiopharmaceuticals, and this usage remains common.

Radiotherapy (Radiation therapy or radiation oncology: abbreviated RT, RTx, or xRT) is the medical use of ionizing radiation, generally as part of cancer treatment to control or kill malignant cells. Radiation therapy may be curative in a number of types of cancer if they are localized to one area of the body. It may also be used as part of adjuvant therapy, to prevent tumor recurrence after surgery to remove a primary malignant tumor (for example, early stages of breast cancer).

Raft culture provides a perfect growing environment for growing vegetables in hydopaunics. RAFT is a proven cell culture system which combines a patented collagen-based process with a range of consumables and reagents to create a realistic extracellular matrix.

RAM (Random Access Memory) is a read-write memory with elements accessible in any order.

Ramet is an individual member of a clone, descended from the ortet.

Random amplified polymorphic DNA (RAPD) is a technique using single, short (usually 10-m) synthetic oligonucleotide primers for PCR. The primer, whose sequence has been chosen at random, initiates replication at its complementary sites on the DNA, producing fragments up to about 2 kb long, which can be separated by electrophoresis and stained with ethidium bromide. A primer can exhibit polymorphism between individuals, and polymorphic fragments can be used as markers.

Random genetic drift refers to one of the basic mechanisms of evolution, along with natural selection, mutation, and migration. In each generation, some individuals may, just by chance, leave behind a few more descendents (and genes) than other individuals. The genes of the next generation will be the genes of the "lucky" individuals, not necessarily the healthier or "better" individuals. That, in a nutshell, is genetic drift. It happens to all populations – there is no avoiding the vagaries of chance. Through sampling error, genetic drift can cause populations to lose genetic variation.

Random mutagenesis is a non-directed change of one or more nucleotide pairs in a DNA molecule.

Random primer method is a protocol for labeling DNA *in vitro*.

Range is a difference between the minimum and maximum values of sensor output in the intended operating range. It defines the overall operating limits of a sensor.

RAPD (Randomly amplified polymorphic DNA) is a method of analysis, where PCR amplification using two copies of an arbitrary oligonucleotide primer is used to create a multilocus fingerprint (i.e., band profile).

Rapid prototype is a machine that can manufacture objects directly (usually, though not necessarily, in plastic) under the control of a computer.

Rate-limiting enzyme refers to the enzyme whose activity controls the output of final product from a multi-enzyme metabolic pathway.

Rational drug design is a systematic method of creating compounds by analyzing their structure, function and stereochemical interactions.

Reactance is a portion of impedance that characterizes

non-dissipative, energy storage effects.

Reactant is a starting substance in a chemical reaction. It appears to the left of the arrow in a chemical reaction.

Reactive oxygen species (ROS) are chemically reactive molecules containing oxygen. Examples include oxygen ions and per-oxides. ROS are formed as a natural byproduct of the normal metabolism of oxygen and have important roles in cell signaling and homeostasis.

Reading frame is a series of trip-lets beginning from a specific nucleotide. Each triplet is rep-resented by a single amino acid in the protein synthesized. The reading frame defines which sets of three nucleotides are read as triplets in the DNA, and hence as codons in the corre-sponding mRNA; this is deter-mined by the initiation codon, AUG. Thus the sequence AUGGCAAAAUUUCCC would read as AUG/GCA/AAA/UUU/CCC/, and not as A/UGC/CAA/AAU/UUC/ CC. Depending on where one begins, each DNA strand contains three different reading frames.

Read-through is a transcription or translation that proceeds beyond the normal stopping point because of the absence of the transcription or translation termination signal of a gene.

Reca is a protein in most bacteria, and that is essential for DNA repair and DNA recombination.

Recalcitrant refers to seeds, that are unable to survive drying and subsequent storage at low temperature.

Receptacle (L. *receptaculum*, a res-ervoir) is an enlarged end of the pedicel or peduncle, to which other flower parts are attached.

Receptor is a trans-membrane protein located in the plasma membrane that can bind with a ligand on the extracellular surface, as a result of which it induces a change in activity on the cytoplasmic surface. More generally, it is a site in a mol-ecule that allows the binding of a ligand.

Receptor binding screening is a biotechnology-based method for drug discovery, which relies on the fact that many drugs act by binding to specific receptors on or in cells. Since receptors *in vivo* bind to hormones or to other cells, and thereby control the cell's behavior, a receptor bound with a drug will likely affect the normal activity of the cell.

Receptor mapping is the technique used to describe the geometric and/or electronic features of a binding site when insufficient structural data for this receptor or enzyme are available. Generally the active site cavity is defined by comparing the superposition of active to that of inactive molecules.

Recessive describes an allele whose effect with respect to a particular trait is not evident in heterozygotes.

Recessive allele is an allelic state of a gene, where homozygosity is required for the expression of the relevant phenotype.

Recessive gene is characterized as having a phenotype expressed only when both copies of the gene are mutated or missing.

Recessive oncogene (recessive-acting oncogene; anti-oncogene) is a single copy of this gene is sufficient to suppress cell proliferation; the loss of both copies of the gene contributes to cancer formation. *See* oncogene.

Reciprocal crosses refer to crosses between the same two strains, but with the sexes reversed; e.g., female A × male B and male A × female B.

Reciprocal monophyly is a genetic lineage is reciprocally monophyletic when all members of the lineage share a more recent common ancestor with each other than with any other lineage on a phylogenetic tree.

Reciprocating shaker is used for agitating culture flasks, with a back and forth action at variable speeds.

Recognition sequence refers to a nucleotide sequence – composed typically of 4, 6, or 8 nucleotides – that is recognized by a restriction endonuclease. Type II enzymes cut (and their corresponding modification enzymes methylate) within or very near the recognition sequence.

Recognition site is a nucleotide sequence – composed typically of 4, 6 or 8 nucleotides – that is recognized by and to which a restriction endonuclease (restriction enzyme) binds. For type II restriction enzymes (those used in gene-cloning experiments), it is also the sequence within which the enzyme specifically cuts (and their corresponding enzymes methylate) the DNA, i.e., for type II enzymes, the recognition site and the target site are the same sequence. Type I enzymes bind to their recognition site and then cleave the DNA at some more or less random position outside that recognition site.

Recombinant is a term used in both classical and molecular genetics: (1) In classical genetics: An organism or cell that is the result of recombination (crossing-over), e.g., Parents: *AB/ab* and *ab/ab*; recombinant offspring: *Ab/ab*; (2) In molecular genetics: A molecule containing DNA from different sources.

Recombinant DNA is the result of combining DNA fragments from different sources.

Recombinant DNA technology is a set of techniques, which enable one to manipulate DNA. One of the main techniques is DNA cloning (because it produces an unlimited number of copies of a particular DNA segment), and the result is sometimes called a DNA clone or gene clone (if the segment is a gene), or simply a clone. An organism manipulated using recombinant DNA techniques is called a genetically modified organism (GMO). Among other things, recombinant DNA technology involves: (1) identifying genes; (2) cloning genes; (3) studying the expression of cloned genes; and (4) producing large quantities of the gene product.

Recombinant human (rh) refers to the *prefix* denoting molecules made through the use of recombinant **DNA** technology.

Recombinant protein is a protein whose amino acid sequence is encoded by a cloned gene.

Recombinant RNA is a term to describe RNA molecules joined *in vitro* by T4 RNA ligase.

Recombinant toxin is a single multifunctional toxic protein encoded by a recombinant gene.

Recombinant vaccine is a vaccine produced from a cloned gene.

Recombinase is a class of enzymes that are able to alter the arrangement of DNA sequences in a site-specific way.

Recombination fraction (recombination frequency) is the proportion of gametes that have arisen from recombination between two loci. It is estimated as the number of recombinant individuals among a set of offspring of a particular mating, divided by the total number of offspring from that mating. Represented by the Greek letter theta. Linkage maps are created from estimates of recombination fraction between all pair-wise combinations of loci.

Recombination frequency is the frequency at which crossing over occurs between two chromosomal loci–the probability that two loci will become unlinked during meiosis.

Recombination is the process that generates a haploid product of meiosis with a genotype differing from both the haploid genotypes that originally combined to form the diploid zygote.

Recombinational hot spot is a chromosomal region, where recombination appears to occur more frequently than expected.

Reconstructed cell is a viable transformed cell resulting from genetic engineering.

Recrystallization is a formation of a new set of strain-free grains within a previously cold-worked material due to an annealing heat treatment.

Rectal manometry is a test that measures the movements and strength of the rectal and anal sphincter muscles.

Rectifier converts bi-directional to one-way current flow.

Rectum refers to the lower end of the large intestine, leading to the anus.

Re-differentiation refers to cell or tissue reversal from one differentiated type to another differentiated type of cell or tissue.

Reducing agent is a substance that acts as an electron donor in a chemical redox reaction.

Reduction is the addition of one or more electrons to an atom, ion, or molecule.

Reduction division is a phase of meiosis in which the maternal

and paternal chromosomes of the bivalent separate.

Reference daily intake or Recommended daily intake (RDI) is the daily intake level of a nutrient that is considered to be sufficient to meet the requirements of 97–98% of healthy individuals in in the United States.

Reflection is a deflection of a light beam at the interface between 2 media.

Reflux esophagitis is an irritation of the lining of the esophagus due to movement of digestive juices backward from the stomach into the esophagus.

Reflux refers to a condition, when the digestive juices, food, and liquids moving backward from the stomach into the esophagus, and possibly into the mouth.

Refraction index (n) is the ratio of the velocity of light in a vacuum to the velocity in some medium.

Refraction is a bending of a light beam when passing from one medium to another, at different velocities of light.

Refractory is a metal or ceramic that does not deteriorate rapidly or does not melt when exposed to extremely high temperatures.

RefSeq (reference sequence) refers to a reference sequence for all naturally occurring molecules from the central dogma (DNA, RNA, Protein).

Refugium (plural: refugia) is an area set aside to provide protection/escape from eco-logical consequences occurring elsewhere.

Regeneration (L. *re*, again + *gen-erate*, to beget) is the growth of new tissues or organs to replace those injured or lost. In tissue culture, regeneration is used to define the development of organs or plantlets from a tissue, callus culture or from a bud. *See* conversion; micro-propagation; organogenesis.

Regenerative medicine is a branch of translational research in Tissue Engineering and Molecular Biology, which deals with the "process of replacing, engineer-ing or regenerating human cells, tissues or organs to restore or establish normal function."

Regulation is a law made by a person or body that has been granted (delegated) law-making authority.

Regulator refers to substance regu-lating growth and development of cells, organs, etc.

Regulatory gene is a gene whose protein controls the activity of other genes or metabolic pathways.

Regulatory sequence is a DNA sequence involved in regulating

the expression of a gene, e.g., a promoter or operator region (in the DNA molecule).

Regurgitation refers to a condition when a valve leaks it is said to be regurgitant or to exhibit regurgitation

Rehabilitation Engineering is a growing specialty area of BME. Rehabilitation engineers enhance the capabilities and improve the quality of life for individuals with physical and cognitive impairments. They are involved in prosthet-ics, the development of home, workplace and transportation modifications and the design of assistive technology that enhance seating and position-ing, mobility, and communica-tion. Rehabilitation engineers are also developing hardware and software computer adap-tations and cognitive aids to assist people with cognitive difficulties.

Reintroduction is the introduction of a species or population into a historical habitat from which it had previously been extirpated.

Rejuvenation refers to reversion from adult to juvenile stage.

Relative fitness is a measure of fit-ness that is the ratio of a given genotype's absolute fitness to the genotype with the greatest absolute fitness. Relative fitness is used to model genetic change by natural selection.

Relative humidity is the ratio of the water vapor content of air to the maximum possible water vapor content of air at the same temperature and air pressure

Relative magnetic permeability (unitless, μ_r) is the ratio of the magnetic permeability of some medium to that of a vacuum: $\mu_r = \mu/\mu_0$, where, μ_0 is the permeability of vacuum (universal constant) = 1.257×10^{-6} H/m.

Relaxed circle (nicked circle) refers to one strand of the DNA becomes nicked, During extraction of plasmid DNA from the bacterial cell. This relaxes the torsional strain needed to maintain supercoiling, producing the familiar form of plasmid.

Relaxed circle plasmid is one five conformations of plasmid DNA and is fully intact with both strands uncut, but has been enzymatically *relaxed* (supercoils removed). This can be modeled by letting a twisted extension cord unwind and relax and then plugging it into itself.

Relaxed plasmid is a plasmid that replicates independently of the main bacterial chromosome and is present in 10–500 copies per cell.

Release factors refers to: (1) Soluble protein that recognizes termination codons in mRNAs and terminate translation in response to these codons; (2) A hormone that is produced by the hypothalamus and stimulates the release of a hormone from the anterior pituitary gland into the bloodstream.

Reliability defines how well a sensor maintains both precision and accuracy over its expected lifetime.

Remediation refers to the cleanup or containment of a hazardous waste disposal site to the satisfaction of the applicable regulatory agency.

Renal angiography (Also called renal arteriography) – a series of x-rays of the renal blood vessels with the injection of a contrast dye into a catheter, which is placed into the blood vessels of the kidney; to detect any signs of blockage or abnormalities affecting the blood supply to the kidneys.

Renal blood flow (RBF) is the volume of blood delivered to the kidneys per unit time. In humans, the kidneys together receive roughly 22% of cardiac output, amounting to 1.1 L/min in a 70-kg adult male. RBF is closely related to renal plasma flow (RPF).

Renal plasma flow is the volume of plasma that reaches the kidneys per unit time. Renal plasma flow is given by the Fick principle.

Renal ultrasound is a non-invasive test in which a transducer is passed over the kidney producing sound waves, which bounce off of the kidney, transmitting a picture of the organ on a video screen. The test is used to determine the size and shape of the kidney, and to detect a mass, kidney stone, cyst, or other obstruction or abnormalities.

Renaturation refers to (1) Of DNA: the reforming of two complementary molecules into a double-stranded structure, following heat or chemical induction of dissociation (denaturation); (2) Of protein, the resumption of three-dimensional conformation, allowing the molecule to function normally. Denaturation of many proteins is irreversible, but denatured DNA molecules will renature readily under appropriate chemical and physical conditions.

Renature is the reannealing (hydrogen bonding) of single-stranded DNA and/or RNA to form a duplex molecule.

Rennin is an enzyme secreted by cells lining the stomach in mammals, and that is responsible for clotting milk.

Repeat unit refers to the sequence of nucleotides that occurs repeatedly, often in a head-to-tail arrangement (tandemly).

Repeatability is the exactness with which a measuring instrument repeats the indications when it measures the same property under the same conditions.

Repetitive DNA refers to DNA sequences that are present in a genome in many copies, some of it originating from retrotransposon activity. A substantial proportion of all eukaryotic genomes is composed of this class of DNA, whose biological function is uncertain. Sometimes referred to as 'junk DNA.'

Replacement or gene replacement is a method of substituting a cloned gene, or part of a gene, which may have been mutated *in vitro*, for the wild-type copy of the gene within the host's chromosome.

Replacement refers to the addition of a cloned corrected copy of a deffective gene.

Replacement therapy refers to the administration of metabolites, co-factors or hormones that are deficient as the result of a genetic disease.

Replica plating refers to duplicating a population of bacterial colonies growing on agar medium in one Petri plate to agar medium in another Petri plate.

Replicase refers to a viral enzyme necessary for the replication of the virus in the host cell.

Replication fork is a Y-shaped structure associated with DNA replication. It represents the point at which the strands of double-stranded DNA are separated so that replication can proceed.

Replication is the synthesis of duplex (double-stranded) DNA by copying from a single-stranded template.

Replicative form (RF) is the molecular configuration of viral nucleic acid that is the template for replication in the host cell.

Replicon is the portion of a DNA molecule which is replicable from a single origin. Plasmids and the chromosomes of bacteria, phages and other viruses usually have a single origin of replication and, in these cases, the entire DNA molecule constitutes a single replicon. Eukaryotic chromosomes have multiple internal origins and thus contain several replicons. The word is often used in the sense of a DNA molecule capable of independent replication.

Replisome refers to the complete replication apparatus present at a replication fork that carries out the semi-conservative replication of DNA.

Repolarize implies to return to a polarized state after a polarizing event.

Reporter gene is a gene that encodes a product that can readily be assayed. Thus reporter genes are used to determinate whether a particular DNA construct has been successfully introduced into a cell, organ or tissue.

Repressible enzyme is an enzyme, whose activity can be diminished by the presence of a regulatory molecule.

Repressible enzyme is an enzyme, whose synthesis is diminished by a regulatory molecule.

Repressible gene is a gene, whose expression can be diminished or extinguished by the presence of a regulatory molecule.

Repression refers to inhibition of transcription by preventing RNA polymerase from binding to the transcription initiation site: a repressed gene is "turned off."

Repressor is a protein, which binds to a specific DNA sequence (the operator) upstream from the transcription initiation site of a gene or operon and prevents RNA polymerase from commencing mRNA synthesis.

Examples of repressors are the C$_1$ protein of bacteriophage and the *lac*1 protein of the *lac* operon.

Reproduction is the production of an organism, cell or organelle like itself (self propagation): (1) Sexual reproduction: the regular alternation (in the life-cycle of haplontic, diplontic and diplohaplontic organisms) of meiosis and fertilization (karyogamy) which provides for the production of offspring; and (2) Asexual or agamic reproduction: the development of a new individual from either a single cell (agamospermy) or from a group of cells (vegetative reproduction) in the absence of any sexual process.

Reproductive cloning is the process of using somatic cell nuclear transfer (SCNT) to produce a normal, full grown organism (e.g., animal) genetically identical to the organism (animal) that donated the somatic cell nucleus. In mammals, this would require implanting the resulting embryo in a uterus, where it would undergo normal development to become a live independent being. The first mammal to be created by reproductive cloning was Dolly the sheep, born at the Roslin Institute in Scotland in 1996.

Reproductive materials are human male or female reproductive cells (sperm or egg), and human embryos and their derivatives.

Repulsion is a double heterozygote in which the dominant (or wildtype) allele at one locus and the recessive (or mutant) allele at a second linked locus occur on the same chromosome (genetic constitution Ab/aB). Synonym is trans configuration. Opposite is coupling, cis configuration.

Rescue effect is a process when immigration into an isolated deme (either genetically or demographically) reduces the probability of the extinction of that deme.

Residual stress is a persisting stress in a material free of external forces or temperature gradients.

Residual volume that remains in the lungs at all times and is about 1200 mL for an adult.

Residues are the components of macromolecules, e.g., amino acids, nucleotides.

Resistance (Ω, Ohm) is a characteristic of a resistor: in a one-ohm resistance a current of one ampere produces a voltage drop of one volt. **Resistance** describes the ability of an organism to withstand a stress, a force or an effect of a disease, or its agent or a toxic substance.

Resistance factor is a plasmid that confers antibiotic resistance to a bacterium.

Resistivity (ρ) is the reciprocal of electrical conductivity, and a measure of a material's resistance to passing electric current.

Resistor is an energy dissipative element consisting of a poor conductor in series with connecting wires.

Resolution is the smallest measurable change in input that will produce a small but noticeable change in the output. In the context of chemical separations. It defines the completeness of separation.

Resonant frequency is the frequency at which a moving member or a circuit has a maximum output for a given input.

Respiration is a gas exchange from air to the blood and from the blood to the body cells.

Response time is the time it takes for the sensor's output to reach its final value. In general, this parameter is a measure of the speed of the sensor and must be compared with the speed of the process.

Rest period refers to an endogenous physiological condition of viable seeds, buds or bulbs that prevents growth even in the presence of otherwise favorable environmental conditions. By some seed physiologists, this is referred to as dormancy.

Restitution nucleus is a single nucleus arising from a failure of nuclear division, either during meiosis, in which a gamete is formed with the unreduced chromosome number; or at mitosis to give a cell with a doubled chromosome number.

Restitution nucleus is a nucleus with unreduced or doubled chromosome number that results from the failure of a meiotic or mitotic division.

Restriction-fragment-length polymorphism (RFLP) refers to differences in nucleotide sequence between alleles at a chromosomal locus result in restriction fragments of varying lengths detected by Southern analysis.

Restriction endonuclease (enzyme) is a class of endonucleases that cleaves DNA after recognizing a specific sequence, such as BamH1 (GGATCC), EcoRI (GAATTC), and HindIII (AAGCTT). **Type I**: Cuts nonspecifically a distance greater than 1000 bp from its recognition sequence and contains both restriction and methylation activities. **Type II**: Cuts at or near a short, and often symmetrical, recognition sequence. A separate enzyme methylates the same recognition sequence. **Type III**: Cuts 24–26 bp downstream from a short, asymmetrical recognition sequence. Requires ATP and contains both restriction and methylation activities.

Restriction enzyme is an enzyme isolated from bacteria, that cleaves DNA at a specific four or six nucleotide sequence. Over 400 such enzymes exist that recognize and cut over 100 different DNA sequences; used in RFLP, AFLP, and RAPD analysis and to construct recombinant DNA (in genetic engineering).

Restriction exonuclease is a class of nucleases that degrades DNA or RNA, starting from an end either 5' or 3'.

Restriction fragment is a shortened DNA molecule generated by the cleavage of a larger molecule by one or more restriction endonucleases.

Restriction Fragment Length Polymorphism (RFLP) refers to a change in the DNA of an organism that changes how a restriction enzyme cuts the DNA into pieces. Two or more organisms can be compared based on the pattern of their DNA fragments when they are run on a gel (by electrophoresis). If the organisms have different spaces between their restriction enzyme sites, they produce different lengths of fragments when the restriction enzyme is used to cut up the DNA.

Restriction map is the linear array of restriction endonuclease sites on a DNA molecule.

Restriction nuclease is a bacterial enzyme that cuts DNA at a specific site.

Restriction site is synonym of recognition site. It refers to the specific nucleotide sequence in DNA that is recognized by a type II restriction endonuclease and within, which it makes a double-stranded cut. Restriction sites usually comprise four or six base pairs that typically are palindromic, e.g., 5'GGCC3' 3'CCGG5'. The two strands may be cut either opposite to one another, to create blunt ends, or in a staggered manner, giving sticky ends, depending on the enzyme involved.

Reticulocyte is a young red blood cell.

Reticulo endothelial system (RES or mononuclear phagocyte system, MPS; or macrophage system) is a part of the immune system that consists of the phagocytic cells located in reticular connective tissue. The cells are primarily monocytes and macrophages, and they accumulate in lymph nodes and the spleen. The Kupffer cells of the liver and tissue histiocytes are also part of the MPS. "Reticuloendothelial system" is an older term for the mononuclear phagocyte system.

Retina is the innermost of the three layers of the eyeball, which is continuous with the optic nerve and contains the light- sensitive rod and cone cells.

Retinitis pigmentosa is a group of hereditary ocular disorders with progressive retinal degeneration. Autosomal dominant, autosomal recessive, and x-linked forms.

Retinoblastoma is a childhood malignant cancer of the retina of the eye. reverse transcriptase viral enzyme used to make cDNA.

Retro-element is any of the integrated retroviruses or the transposable elements that resemble them.

Retro-poson (retro-transposon) is a transposable element that moves via reverse transcription (i.e., from DNA to RNA to DNA) but lacks the long terminal repeat sequences.

Retroposon is a transposable element that moves via reverse

transcription but lacks the long terminal repeat sequences necessary for autonomous transposition. Much of the repetitive DNA that makes up a large proportion of eukaryotic genomes consists of silenced (i.e., inactive) retroposons. Synonym is retro-transposon.

Retrovirus is a class of eukaryotic RNA viruses that can form double-stranded DNA copies of their genomes by using reverse transcription; the double-stranded forms integrate into chromosomes of an infected cell. Many naturally occurring cancers of vertebrate animals are caused by retroviruses. Also, the AIDS virus is a retrovirus.

Retroviral vectors refer to gene transfer systems based on viruses that have RNA as their genetic material.

Reversal transfer refers to transfer of a culture from a callus-supporting medium to a shoot-inducing medium.

Reverse bias is the insulating bias for a p-n junction rectifier; electrons flow into the p side of the junction.

Reverse genetics is an approach to discover the function of a gene by analyzing the phenotypic effects of specific gene sequences obtained by DNA sequencing. This investigative process proceeds in the opposite direction of so-called forward genetic screens of classical genetics.

Reverse mutation rate is a back mutation rate. The rate at which a gene's ability to produce a functional product is restored. This rate is much lower than the forward mutation rate because there are many more ways to remove the function of a gene than restore it. Also it is used to describe mutation at microsatellite loci, where (under the stepwise mutation model, for example) a back mutation yields an allele of length that already exists (i.e., homoplasy) in the population.

Reverse transcriptase (RNA – dependent DNA polymerase) is an enzyme isolated from retrovirus-infected cells that synthesizes a complementary (c)DNA strand from an RNA template.

Reverse transcription refers to the synthesis of DNA from a template of RNA, accomplished by reverse transcriptase.

Reversion refers to the restitution of a mutant gene to the wild-type condition, or at least to a form that gives the wild-type phenotype; more generally, the appearance of a trait expressed by a remote ancestor. Synonym is reverse mutation.

Reversion or reverse mutation is the restitution of a mutant gene to the wild-type condition.

Reynolds number is the most important dimensionless number in biofluid dynamics providing a criterion for dynamic similarity. It is named after Osbourne Reynolds (1842-1912). The Reynolds number is used for determining whether a flow is laminar (for Re<2300) or turbulent (Re>3000).

RF (Radio Frequency) refers to alternating voltages and currents having frequencies between 9 kHz and 3 MHz.

RF is an abbreviation for replicative form.

RFLP (Restriction fragment length polymorphism) is a type of polymorphism detectable in a genome by the size differences in DNA fragments generated by restriction enzyme analysis.

rh is an abbreviation for recombinant human.

Rh disease is a blood group antigen possessed by Rh-positive people; if an Rh-negative person receives a blood transfusion from an Rh-positive person it can result in hemolysis and anemia.

RH mapping (Radiation Hybrid mapping) is a physical mapping

method that estimates linkage and distance relative to radiation-induced chromosome breaks. This is analogous to genetic mapping.

Rhizobacterium is a micro-organism whose natural habitat is near, on or in plant roots.

Rhizobia refers to a bacteria in a symbiotic relationship with leguminous plants that results in nitrogen fixation. See Nitrogen fixation.

Rhizobium (plural: Rhizobia) refer to prokaryotic species, which are able to establish a symbiotic relationship with leguminous plants, as a result of which elemental nitrogen is fixed or converted to ammonia.

Rhizosphere refers to the soil region in the immediate vicinity of growing plant roots.

Ri plasmid is a class of large conjugative plasmids found in the soil bacterium *Agrobacterium rhizogenes*. Ri plasmids are responsible for hairy root disease of certain plants. A segment of the Ri plasmid is found in the genome of tumor tissue from plants with hairy root disease.

RIA is an abbreviation for radioimmunoassay.

Ribbon diagrams are also known as Richardson Diagrams. These are 3-D schematic representations of protein structure and are one of the most common methods of protein depiction used today. The ribbon shows the overall path and organization of the protein backbone in 3D, and serves as a visual framework on which to hang details of the full atomic structure, such as the balls for the copper and zinc atoms at the active site of superoxide dismutase in the image at the right. **Ribbon diagrams** are generated by interpolating a smooth curve through the polypeptide backbone. α-helices are shown as coiled ribbons or thick tubes, β-strands as arrows, and lines or thin tubes for non-repetitive coils or loops. The direction of the polypeptide chain is shown locally by the arrows, and may be indicated overall by a color ramp along the length of the ribbon.

Ribonuclease (RNAse) is any enzyme that catalyzes the hydrolysis of RNA.

Ribonucleic acid (RNA) is an organic acid polymer that is composed of adenosine, guanosine, cytidine and uridine ribonucleotides. The genetic material of some viruses, but more generally is the molecule, derived from DNA by transcription, that either carries information (messenger RNA), provides sub-cellular structure (ribosomal RNA), transports amino acids (transfer RNA), or facilitates the biochemical modification of itself or other RNA molecules.

Ribonucleotide or **ribotide** is a nucleotide containing D-ribose as its pentose component. It is considered a molecular precursor of nucleic acids. Nucleotides are the basic building blocks of DNA and RNA. The monomer itself from ribonucleotides forms the basic building blocks for RNA. However, the reduction of ribonucleotide, by enzyme ribonucleotide reductase (RNR), forms deoxyribonucleotide, which is the essential building block for DNA. There are several differences between DNA deoxyribonucleotides and RNA ribonucleotides. Successive nucleotides are linked together via phosphodiester bonds.

Ribose is a monosaccharide found in all ribonucleosides, ribonucleotides and RNA. Its close analogue, 2-deoxyribose, is similarly found in all deoxyribonucleosides, deoxyribonucleotides and DNA.

Ribosomal binding site is a sequence of nucleotides near the 5' end of a bacterial mRNA molecule that facilitates the binding of the mRNA to the small ribosomal sub-unit. Also it is called the Shine-Delgarno sequence.

Ribosomal DNA refers to the coding locus for ribosomal RNA. This is generally a large and complex locus, typically composed of a large number of repeat units, separated from one another by the intergenic spacer.

Ribosomal RNA (rRNA) refers to the RNA molecules that are essential structural and functional components of ribosomes, where protein synthesis occurs. Different classes of rRNA molecule are identified by their sedimentation (S) values. *E. coli* ribosomes contain one 16S rRNA molecule (1541 nucleotides long) in one (small) ribosomal sub-unit, and a 23S rRNA (2904 nucleotides) and a 5S rRNA (120 nucleotides) in the other (large) sub-unit. These three rRNA molecules are synthesized as part of a large precursor molecule, which also contains the sequences of a number of tRNAs. Special processing enzymes cleave this large precursor to generate the functional molecules. Constitutes about 80% of total cellular RNA.

Ribosome (*ribo*, from RNA + Gr. *soma*, body) is the sub-cellular structure that contains both RNA and protein molecules and mediates the translation of mRNA into protein. Ribosomes comprise large and small sub-units.

Ribosome-binding site refers to the region of an mRNA molecule that binds the ribosome to initiate translation.

Ribosome inactivating protein (RIP) is a class of plant proteins that inhibit normal ribosome function, and are thus highly toxic. **Type 1** RIPs consist of single polypeptide chain proteins; **type 2** (e.g., ricin) consist of two proteins linked by a disulphide bridge, one the toxin and the other a lectin that attaches to recognition sites on a target cell.

Ribozyme is an RNA molecule that can catalyze chemical cleavage of itself or of other RNAs. Synonyms are catalytic RNA, gene shears.

Ribulose biphosphate (RuBP) is a five-carbon sugar that is combined with carbon dioxide to form a six-carbon intermediate in the first stage of the dark reaction of photosynthesis.

Ribulose is a keto-pentose sugar ($C_5H_{11}O_5$) involved in the carbon dioxide fixation pathway of photosynthesis.

Richardson number is the dimensionless number that expresses the ratio of potential to kinetic energy If the Richardson number is much less than unity, buoyancy is unimportant in the flow. If it is much greater than unity, buoyancy is dominant (in the sense that there is insufficient kinetic energy to homogenize the fluids). If the Richardson number is of order unity, then the flow is likely to be buoyancy-driven: the energy of the flow derives from the potential energy in the system originally.

Rights refer to entitlements. Some rights (human rights) belong to everyone by virtue of being human; some rights (legal rights) belong to people by virtue of their belonging to a particular political state.

Rinderpest refers to the cattle plague; a viral infection of cattle, sheep and goats.

RIP is an abbreviation for ribosome-inactivating protein.

Risk refers to a danger that arises unpredictably, such as being allergy to a medicine.

Risk analysis is a process consisting of three components: risk assessment, risk management and risk communication performed to understand the nature of unwanted, negative consequences to human and animal health, or the environment.

Risk assessment is a scientifically based process consisting of the following steps: (i) hazard identification; (ii) hazard characterization; (iii) exposure assessment; and (iv) risk characterization.

Risk communication refers to the interactive exchange of information and opinions throughout the risk analysis process concerning hazards and risks, risk-related factors and risk perceptions, among risk assessors, risk managers, consumers, industry, the academic community and other interested parties, including the explanation of risk assessment findings and the basis of risk management decisions.

Risk management refers to the process of weighing policy alternatives, in consultation with all interested parties, considering risk assessment and other factors relevant for the health protection of consumers and for the promotion of fair trade practices, and, if needed, selecting appropriate prevention and control options.

Ritual purification is a feature of many religions. The aim of these rituals is to remove specifically defined uncleanliness prior to a particular type of activity, and especially prior to the worship of a deity. This ritual uncleanliness is not identical with ordinary physical impurity, such as dirt stains; nevertheless, body fluids are generally considered ritually unclean.

R-loops refer to single-stranded DNA regions in RNA-DNA hybrids formed *in vitro* under conditions where RNA-DNA duplexes are more stable than DNA-DNA duplexes.

rms refers to root mean square.

RNA is an abbreviation for Ribonucleic acid. RNA is an organic acid composed of repeating nucleotide units of adenine, guanine, cytosine and uracil, whose ribose components are linked by phospho-diester bonds. The information-carrying material in some viruses. More generally, a molecule derived from DNA by transcription that may carry information (messenger RNA (mRNA)), provide sub-cellular structure (ribosomal RNA (rRNA)), transport amino acids (transfer RNA (tRNA)) or facilitate the biochemical modification of itself or other RNA molecules. Repeated terms are: antigen RNA; gene splicing; heterogeneous nuclear RNA (hnRNA); mRNA; ribosomal RNA; RNA polymerase; small nuclear RNA; transfer RNA.

RNA editing refers to post-transcriptional processes that alter the information encoded in gene transcripts (RNAs).

RNA polymerase is a polymerase enzyme that catalyzes the synthesis of RNA from a DNA template.

RNase is an abbreviation for ribonuclease. It is a group of enzymes that catalyze the cleavage of nucleotides in RNA.

Robertsonian fission is an event where a metacentric chromosome breaks near the centromere to form two acrocentric chromosomes.

Robertsonian translocation is a special type of translocation where the break occurs near the centromere or telomere and involves the whole chromosomal arm so balanced gametes are usually produced.

Rod refers to light sensitive cells of the retina that are particularly sensitive to dim light and mediate white and black vision.

Roentgen refers to obsolete unit of ionizing radiation. The SI unit is the sievert (Sv).

Rol genes is a family of genes, present on the Ri plasmid of

Agrobacterium rhizogenes, that when transferred to a plant upon infection by the bacterium, induce the formation of roots. Used as a means of root induction on different species and cultivars of micropropagated fruit trees.

Roller pump uses positive displacement with rotating roller head to propel fluid; amount of flow is dependent on degree that tubing is occluded and on number of revolutions per minute; additional roller heads are used for cardiotomy suctions and venting.

ROM (Read only memory) is a memory used for permanent, storage of unalterable data; nonvolatile memory.

Root cap is a mass of reinforced cells covering and protecting the apical meristem of a root.

Root culture refers to the culture of isolated root tips of apical or lateral origin to produce *in vitro* root systems with indeterminate growth habits. Root culture was among the first kinds of plant tissue cultures, and is still largely used in the study of developmental phenomena, and mycorrhizal, symbiotic and plant-parasitic relationships.

Root cutting refers to the cutting made from sections of roots alone.

Root hairs refer to outgrowths from epidermal cell walls of the root, specialized for water and nutrient absorption.

Root nodule is a small round mass of cells that is located on the roots of plants and contains nitrogen-fixing bacteria.

Root tuber refers to a thickened root that stores carbohydrates.

Root zone is the volume of soil or growing medium containing the roots of a plant. In soil science, the depth of the soil profile in which roots are normally found.

Rootstock refers to the trunk or root material to which buds or scions are inserted in grafting.

Rossby number is named after Carl-Gustav Arvid Rossby, and is a dimensionless number used in describing geophysical phenomena in the oceans and atmosphere. It is the ratio of inertial forces in a fluid to the fictitious forces arising from planetary rotation. It is also known as the Kibel number.

Rotary evaporator (or rotavap or rotovap) is a device used in chemical laboratories for the efficient and gentle removal of solvents from samples by evaporation.

Rotary shaker is a rotating apparatus with a platform on which, containers can be shaken, such as Erlenmeyer flasks containing cells in liquid nutrient medium.

Rotational fluid flow can contain streamlines that loop back on themselves. Hence, fluid particles following such streamlines will travel along closed paths. Bounded (and hence nonuniform) viscous fluids exhibit rotational flow, typically within their boundary layers. Since all real fluids are viscous to some amount, all real fluids exhibit a level of rotational flow somewhere in their domain. Regions of rotational flow correspond to the regions of viscous losses in a fluid. Inviscid fluid flows can also be rotational, but these are special nonphysical cases. For an inviscid fluid flow to be rotational, it must be set up that way by initial conditions. The amount of rotation (called the *velocity circulation*) in an inviscid fluid flow is conserved, provided that the fluid is also barotropic and subject only to conservative body forces. This conservation is known as *Kelvin's Theorem* of constant circulation.

Rotavirus is a virus that causes diarrhea. It is the most common cause of infectious diarrhea in the United States, especially in children under 2 years old.

Rouleaux formation refers to agglomeration of RBC in the centerline of flow field

rRNA (ribosomal RNA) refers to the RNA molecules, which are essential structural and functional components of ribosomes, the organelles responsible for protein synthesis. The different rRNA molecules are known by their sedimentation (Svedberg; symbol S) values. *E. coli* ribosomes contain one 16S rRNA molecule (1541 nucleotides long) in the same (small) sub-unit and a 23S rRNA (2904 nucleotides) and a 5S rRNA (120 nucleotides) in the large sub-unit. These three rRNA molecules are synthesized as part of a large precursor molecule which also contains the sequences of a number of tRNAs. Special processing enzymes cleave this large precursor to generate the functional moieties.

Rubinstein-Taybi syndrome is a condition with multiple congenital anomalies including: mental deficiency, broad thumbs, small head, broad nasal bridge and beaked nose.

RuBP is an abbreviation for ribulose biphosphate.

Ruminant is an animal having a rumen – a large digestive sac in which fibrous plant material is fermented by commensal microbes, prior to its digestion in a "true" stomach (the *abomasum*).

Ruminant animals or ruminants are animals having a rumen – a large digestive vat in which fibrous plant material is partially broken down by microbial fermentation, prior to digestion in a "true" stomach (the abomasum). There are also two other stomachs – the reticulum and the omasum.

Runner is a lateral stem that grows horizontally along the ground surface and gives rise to new plants either from axillary or terminal buds. Synonym is stolon. In biology, **stolons** (from Latin stolō "branch") are horizontal connections between organisms. They may be part of the organism, or of its skeleton; typically, animal stolons are external skeletons.

Rust is a generic descriptor for various serious fungal plant pathogens, which infect the leaves and stems of crops. The appearance of spores is reminiscent of metallic rust, although the color varies, according to species, from yellow to reddish-brown.

RVAD (Right Ventricular Assist Device) is a support system for the right ventricle of the heart.

S

. . . .

S phase refers to the phase in the cell cycle during which DNA synthesis occurs.

S_1 mapping is a method for mapping precursor or mature mRNA to particular DNA sequences using the enzyme S1-nuclease.

S_1 nuclease is an enzyme that specifically degrades RNA or single-stranded DNA to 5′ mononucleotides. Purified from the filamentous fungus *Aspergillus oryzae,* S_1 nuclease is used in assessing the extent of a hybridization reaction by removing unpaired regions. It is also used to remove the sticky ends of restriction fragments. In S_1 mapping, the coding region of a gene is detected by performing mRNA-DNA hybridization and removing unpaired DNA with S_1 nuclease.

Saccharification is the hydrolysis of polysaccharides by glucoamylase to maltose and glucose, following liquefaction.

Sacrificial anode is an active metal or alloy that corrodes and protects another metal or alloy to which it is electrically coupled.

Sacrificial layer is a thin film that is later removed to release a microstructure from its substrate.

Safety is a freedom from danger.

SAGE (Serial Analysis of Gene Expression) is a technique for identifying and quantifying transcripts from eukaryotic genomes. This method is based on the isolation and concatenation of short sequence tags (~14 bp) from individual mRNAs into longer DNA molecules that are subsequently sequenced.

Saline resistance is synonym for salt tolerance.

Saliva is a fluid made by glands in the mouth that helps moisten and soften foods we chew, and begins the digestive process.

Salmonella is a genus of rod-shaped, Gram-negative bacteria that are a common cause of food poisoning.

Salt tolerance is the ability of a plant in soil or in culture to withstand a concentration of common salt (sodium chloride), which is damaging or lethal to most other plants. Synonym is saline resistance. An organism with extreme salt tolerance is a halophyte.

Salting out (antisolvent crystallization, precipitation

crystallization, or drowning out) is an effect based on the electrolyte-nonelectrolyte interaction based on the fact that the non-electrolyte could be less soluble at high salt concentrations. It is used as method of separating proteins. The salt concentration is needed for the protein to precipitate out of the solution differs from protein to protein. This process is also used to concentrate dilute solutions of proteins. Dialysis can be used to remove the salt if needed.

Sanger sequence is a "plus and minus" or "primed synthesis" method; DNA is synthesized so it is radioactively labeled and the reaction terminates specifically at the position corresponding to a given base.

Sap refers to the fluid content of the xylem and phloem cells of plants. Fluid content of the vacuole generally referred to as cell sap.

Saprophyte is a vegetable organism that derives its nutriment from decaying organic matter.

SARS (Severe acute respiratory syndrome) is caused by a virus thought to be a combination of the Coronavirus family (a virus that is often a cause of the common cold) and the paramyxovirus family (causes measles and mumps). The syndrome includes fever and coughing or difficulty breathing, and can be fatal.

Satellite DNA is that portion of the DNA in plant and animal cells consisting of highly repetitious sequences (millions of copies) typically in the range from 5 to 500 bases. Thousands of copies occur tandemly (end-on-end) at each of many sites. It can be isolated from the rest of the DNA by density gradient centrifugation.

Satellite RNA (viroids) is a small, self-splicing RNA molecule that accompanies several plant viruses, including tobacco ringspot virus.

Saturated fat has only single bonds in the molecule.

SC is an abbreviation for synaptonemal complex.

SCA is an abbreviation for specific combining ability that is a performance of a parent under consideration, in a specific cross. Features of **Specific Combining Ability** are: It represents deviation from gca; It is due to dominance genetic variance and all the three types of gene interactions; Helps into identification and hence selection of best cross combinations i.e., those with the desired output; When we see that a inbred line combines well in any cross, it is due to specific combining ability; Estimated by full sib mating; and Have relationship with heterosis

Scaffold refers to the central proteinaceous core structure of condensed eukaryotic chromosomes. The scaffold is composed of non-histone chromosomal proteins.

Scale up is the conversion of a process, such as fermentation of a micro-organism, from a small scale to a larger scale.

Scanning electron microscope (SEM) is an electron-beam-based microscope used to examine, in a three dimensional screen image, the surface structure of prepared specimens.

SCAR is an abbreviation for sequence characterized amplified region.

Scarification is the chemical or physical treatment given to some seeds (where the seed coats are very hard or contain germination inhibitors) in order to break or weaken the seed coat sufficiently to permit germination.

SCE is an abbreviation for sister chromatid exchange.

Scientific name is a unique identifier consisting of a genus and a species name (the specific epithet) in Latin, assigned to each recognized and described species of organism.

Scintillator is a material that exhibits scintillation (the property of luminescence) when excited by ionizing radiation.

Scion refers to the twig or bud to be grafted onto another plant, the root stock, in a budding or grafting operation.

Scion-stock interaction is the effect of a rootstock on a scion (and vice versa) in which a scion on one kind of rootstock performs differently than it would on its own roots or on a different rootstock.

Sclerenchyma (Gr. *skleros*, hard + *echyma*, a suffix denoting tissue) is the strengthening tissue in plants, composed of cells with heavily lignified cell walls.

Sclerosing agent is a chemical irritant injected into a vein in sclerotherapy.

SCP is an abbreviation for single-cell protein.

Scrapie is a disease of sheep; a spongiform encephalopathy.

Screen refers to separate by exclusion or collection on the basis of a set of criteria (biochemical, anatomical, physiological, etc.).

Screening is the process of selection for specific purposes, such as disease resistance or improved agronomic qualities in plants, improved performance in animals, specific enzyme properties in micro-organisms, etc.

Scrotum refers to the external sac of skin in males that contains the testes and their accessory organs.

SDS is an abbreviation for sodium dodecyl sulphate.

SDS-PAGE is an abbreviation for sodium dodecyl sulphate poly-acrylamide gel electrophoresis.

Sebaceous glands develop as an outgrowth of the external root sheath of the hair follicle, usually several glands to one follicle. They secrete an oily substance called sebum.

Secondary antibody is the antibody that binds to the primary antibody, in an ELISA or other immunological assay system,. The secondary antibody is often conjugated with an enzyme such as alkaline phosphatase.

Secondary cell wall is the inner-most layer of cell wall, with a highly organized microfibrillar structure, which is formed in certain cells after cell elonga-tion has ceased. It gives rigidity to the cells.

Secondary growth is a type of growth characterized by an increase in thickness of stem and root and resulting from formation of secondary

vascular tissues by the vascular cambium.

Secondary immune response is the rapid immune response that occurs during the second (and subsequent) encounters of the immune system of a mammal with a specific antigen.

Secondary messenger is a chemi-cal compound within a cell that is responsible for initiating the response to a signal from a chemical messenger (such as a hormone) that cannot enter the target cell itself.

Secondary metabolism is the production by living organisms of substances not essential for primary metabolic functions or physiology. Their role is associated with interaction with the environment, for example for defence, as elicitors or as attractants.

Secondary metabolite is a com-pound that is not necessary for growth or maintenance of cellu-lar functions but is synthesized, generally, for the protection of a cell or micro-organism, dur-ing the stationary phase of the growth cycle.

Secondary oocyte is an oocyte in which the first meiotic division is completed. The second mei-otic division usually stops short of completion unless fertiliza-tion occurs.

Secondary phloem refers to the phloem tissue formed by the **vascular cambium** during secondary growth in a vascular plant.

Secondary plant products are metabolic products not having a known functional or structural use in plant cells. They have been extracted from plant tissue cultures for pharmaceutical and food processing purposes (e.g., essential oils, food additives, flavors).

Secondary structure is a localized three dimensional conformations adopted by macromolecules, in particular nucleic acids and polypeptides. Examples are alpha-helix regions and beta-pleated sheets in proteins, and hairpin loops in nucleic acids.

Secondary thickening refers to the deposition of secondary cell wall materials which result in an increase in thickness in stems and roots.

Secondary vascular tissue is a vascular tissue (xylem and phloem) formed by the vascular cambium during secondary growth in a vascular plant.

Secondary vasodilatation refers to dilation following exposure to cold to sustain viable tissues.

Secondary xylem is a xylem tissue formed by the vascular cambium during secondary growth in a vascular plant.

Secretion is the passage of a molecule from the inside of a cell through a membrane into the periplasmic space, or the extracellular medium.

SED (Single Electron Devices) are nanoscale devices that control the movement of individual electrons, may one day make it possible for integrated circuits to have as many as 10 billion electronic devices in a square centimeter, a density 1000 times greater than that believed feasible for conventional integrated circuits. These devices consist of two electrodes (typically 30 nm wide) separated by a 1 nm-deep insulating layer through which single electrons can tunnel. These devices have many potential applications, from building more sensitive measurement devices to understanding fundamental problems in physics. Two-junction devices share a middle electrode. These devices are called "single-electron transistors," because, like conventional transistors, their current can be controlled by modifying the surface charge on the middle electrode, making it an ideal element for an integrated circuit. A circuit made of single-electron devices operates at a temperature of 4 K or below to reduce thermal effects, which disturb the movements of single electrons in the solid.

Seed refers to the matured ovule without accessory parts.

Seed storage proteins are proteins that are accumulated in large amounts in seeds not because of their enzymatic or structural properties but simply as a convenient source of amino acid for use when the seed germinates. They are of interest to biotechnologists: (1) **As a source of protein**: A substantial amount of the world's food protein comes from plant storage protein. Any improvement of the nutritional content of those proteins could correspondingly improve human diet; (2) **As expression systems**: Storage proteins are produced in very large amounts relative to other proteins, and are stored in stable, compact bodies in the plant seed. Several workers are seeking to make the plants produce other proteins in similarly large amounts and in as convenient a form, by splicing the gene for a desired protein into the middle of a plant storage protein gene.

Segment polarity gene is a gene that functions to define the anterior and posterior components of body segments in *Drosophila*.

Segmental duplication is a region of genomic DNA ranging from 1 to 400 kb that may be found at more than one site in the genome. Segmental duplications often share >90% sequence identity.

Segment polarity gene is a gene that functions to define the anterior and posterior components of body segments in *Drosophila*.

Segregant is a hybrid resulting from the crossing of two genetically unlike individuals.

Segregation is the separation of allele pairs from one another and their resulting assortment into different cells at meiosis. For chromosomes, the separation and re-assortment of the two homologues in anaphase of the first meiotic division.

Selectable refers to having a gene product that, when present, enables the identification and preferential propagation of a particular genotype.

Selectable marker is a gene whose expression allows the identification of: (1) A specific trait or gene in an organism; (2) Cells that have been transformed or transfected with a vector containing the marker gene.

Selection is the process of determining the relative share allotted individuals of different genotypes in the propagation of a population; the selective effect of a gene can be defined by the probability that carriers of the gene will reproduce.

Selection coefficient refers to the proportion by which the fitness of a genotype is less than the fitness of a standard genotype, which is usually the genotype with the highest fitness. In general, relative fitness = $[1 - s]$.

Selection culture is a selection based on differences in environmental conditions or in culture medium composition, such that preferred variant cells or cell lines (presumptive or putative mutants) are favored over other variants or the wild type.

Selection differential refers to the difference between the mean of the individuals selected to be parents and the mean of the overall population. It represents the average superiority of the selected parents.

Selection pressure is the intensity of selection acting on a population of organisms or cells in culture. Its effectiveness is measured in terms of differential survival and reproduction, and consequently in change in the frequency of alleles in a population.

Selection response is the difference between the mean of the individuals selected to be parents and the mean of their offspring. Predicted response = heritability (narrow-sense) × selection differential.

Selection unit is the minimum number of organisms or cells effective in the screening process.

Selective agent is an environmental or chemical agent characterized by its lethal or sub-lethal stress on growing plants, or portion thereof in culture. A selective agent is mainly used when selection of resistant or tolerant individuals is the research aim.

Selective breeding is a process in which new or improved strains of plants or animals are developed, mainly through controlled mating or crossing and selection of progeny for desired traits.

Selective internal radiation therapy (SIRT) is a form of radiation therapy used in interventional radiology to treat cancer. It is generally for selected patients with un-resectable cancers, those that cannot be treated surgically, especially hepatic cell carcinoma or metastasis to the liver.

Selective sweep is the rapid increase in frequency by natural selection of an initially rare allele that also fixes (or nearly fixes) alleles at closely linked loci thus reduces the genetic variation in a region of a chromosome.

Selectivity is the ability of a sensor to measure only one metric or, in the case of a chemical sensor, to measure only a single chemical species.

Self fertilization is the process by which pollen of a given plant fertilizes the ovules of the same plant. Plants fertilized in this way are said to have been selfed. An analogous process occurs in some animals, such as nematodes and molluscs.

Self incompatibility is the inability of the pollen to fertilize ovules (female gametes) of the same plant.

Self pollination refers to when pollen of one plant is transferred to the female part of the same plant or another plant with the same genetic makeup.

Self renewal is the ability of a stem cell to divide and form more stem cells with identical properties to the parent cell, thereby allowing the population to be replenished indefinitely.

Self replicating elements refer to extrachromosomal DNA elements that have origins of replication for the initiation of their own DNA synthesis.

Self replication is an effective route to truly low cost manufacturing. Self replicating systems can seriously mislead about the properties and characteristics of artificial self replicating systems designed for manufacturing purposes. Artificial systems able to make a wide range of non- biological products (like diamond) under programmatic control are likely to be more brittle and less adaptable in their response to changes in their environment than biological systems. At the same time, they should be simpler and easier to design. The complexity of such systems need not be excessive by present engineering standards.

Self sterility is synonym of self-incompatibility.

SEM is an abbreviation for scanning electron microscope.

Semen is a fluid composed of sperm suspended in various glandular secretions that is ejaculated from the penis during orgasm.

Semen sexing is synonym of sperm sexing.

Semi conductor is a material with an electrical conductivity between that of an insulator and that of a conductor.

Semi conservative replication refers to when, during DNA duplication, each strand of a parent DNA molecule is a template for the synthesis of its new complementary strand. Thus, one half of a preexisting DNA molecule is conserved during each round of replication.

Semi continuous culture refers to cells in an actively dividing state which are maintained in culture by periodically draining off the medium and replenishing it with fresh medium.

Semi empirical methods are molecular orbital calculations using various degrees of approximation and using only valence electrons.

Semi empirical quantum mechanical methods use parameters derived from experimental data to simplify computations. The simplification may occur at various levels: simplification of the Hamiltonian (e.g., as in the Extended Hückel method), approximate evaluation of certain molecular integrals, simplification of the wave function.

Semi permeable membrane is a cell or plasma membrane that is partially permeable; certain ions or molecules (water,

solvents) can pass through it but others cannot (such as certain solutes).

Semi solid is gelled but not firmly so; small amounts of a gelling agent are used to obtain a semi-solid medium; called also semi-liquid.

Semi sterility refers to the condition of partial fertility. Often associated with chromosomal aberrations or the result of mutagenesis.

Semilunar valves are valves between the ventricles of the heart and the arteries that carry blood away from the heart; aortic and pulmonary valves.

Seminal vesicles are glandular sacs that secrete a component of seminal fluid.

Senescence is the last stage in the post-embryonic development of multicellular organisms, during which loss of functions and degradation of biological components occur. It is a physiological ageing process in which cells and tissues deteriorate and finally die.

Sense RNA refers to the RNA transcript of the coding strand DNA (often represented as the ± strand). Opposite is antisense RNA. When both sense and antisense transcripts of a gene are present simultaneously, gene silencing is resulted.

Sentinel lymph node is the hypothetical first lymph node or group of nodes draining a cancer. In case of established cancerous dissemination it is postulated that the sentinel lymph node/s is/are the target organs primarily reached by metastasizing cancer cells from the tumor. Thus, sentinel lymph nodes can be totally void of cancer, due to the fact that they were detected prior to dissemination. The concept of the sentinel lymph node is important because of the advent of the sentinel lymph node biopsy technique, also known as a sentinel node procedure. This technique is used in the staging of certain types of cancer to see if they have spread to any lymph nodes, since lymph node metastasis is one of the most important prognostic signs. It can also guide the surgeon to the appropriate therapy.

Sensitivity is the amount of change in a sensor's output in response to a change at a sensor's input over the sensor's entire range. It provides an indication of a sensor's ability to detect changes. For some sensors, the sensitivity is defined as the input parameter change required to produce a standardized output change.

Sensitivity testing is a method used in population viability analyses where the effects of parameters on the persistence of populations are determined by testing a range of possible values for each parameter.

Sepsis is the destruction of tissue by pathogenic micro-organisms or their toxins, especially through infection of a wound.

Septate (L. *septum*, fence) refers to one that is divided by cross walls into cells or compartments.

Septum (L. *septum*, fence) is any dividing wall or partition; frequently a cross wall in a fungal or algal filament.

Sequence is the linear order of nucleotides along a DNA or RNA molecule, and the process of obtaining this. Genome sequencing aims to generate the linear order of all nucleotides present in the nuclear DNA of an organism.

Sequence characterized amplified region (SCAR) refers to a molecular marker obtained by the conversion to a sequence-tagged site of a single random amplified polymorphic DNA product.

Sequence divergence is the percent difference in the nucleotide sequence between related nucleic acid sequences, or in the amino acid sequence in a comparison between related proteins.

Sequence hypothesis is a Francis Crick's seminal concept that genetic information exists as a linear DNA code; DNA and protein sequence are collinear.

Sequence tagged site (STS) refers to a short unique DNA sequence (200–500 bp long) that can be amplified by PCR and is thus tagged to the site on the chromosome from which it was amplified.

Sequence tandem repeat is an abbreviated as STR.

Sequencing is determining the order of bases in a length of DNA. Provides information on where genes start and stop and where mutations or changes have occurred. It also allows you to translate the sequence of bases within a gene into what amino acids it codes for, and therefore what protein is produced.

Sequencing of DNA molecules is the process of finding the order of nucleotides (guanine, adenine, cytosine and thymine) that make up a DNA or RNA fragment.

Sequential Bonferroni correction is a method, similar to the Bonferroni correction, that is used to reduce the probability

of a Type I statistical error when conducting multiple simultaneous tests.

Serial divisions refer to splitting at about monthly intervals of excised shoot-tip material growing on culture medium, in order to induce additional plantlets.

Serial float culture is a technique of floating anthers on liquid medium developed by Sunderland. Anther dehiscence, pollen release and development occur at intervals of several days, and in different nutrient media.

Serology (adj.: serological) is the study of serum reactions between an antigen and its antibody. Serology is mainly used to identify and distinguish between antigens, such as those specific to micro-organisms or viruses. Serology is also employed as an indicator technique to assay plants suspected of being virus-infected.

Serum albumin is a globular protein obtained from blood and body fluids. Bovine and human serum albumins are abbreviated BSA and HSA respectively.

Serum is a blood plasma that has had its clotting factor removed.

Sewage treatment is a biotechnological process to deal with the huge amounts of human and animal waste that such societies produce. Sewage treatment methods vary widely, but all have a biological basis to break down the organic material in sewage and convert it into something that can be safely discharged into the environment (usually rivers or seas).

Sex chromosome is a differentiated chromosome which is responsible for the determination of sex of the individual. For all mammals, a small number of flowering plants and many insects, female individuals carry a pair of X chromosomes, and males carry one X and one Y. For birds, reptiles and most amphibians, male individuals carry a pair of W chromosomes, and females carry one W and one Z. In some insects there is only one sex chromosome, X, and sex is determined by the number of these present.

Sex determination is the mechanism in a given species by which sex is determined; in many species sex is determined at fertilization by the nature of the sperm that fertilizes the egg.

Sexduction ia the incorporation of bacterial genes into F factors and their subsequent transfer, by conjugation, to a recipient cell.

Sex factor is a bacterial episome (e.g., the F plasmid in *E. coli*) that enables the cell to be a donor of genetic material. The sex factor may be propagated in the cytoplasm, or it may be integrated into the bacterial chromosome.

Sex hormones are steroid hormones that control sexual development.

Sex influenced dominance is the tendency for the type of gene action to vary between the sexes within a species. Thus the presence of horns in some breeds of sheep appears to be dominant in males and recessive in females.

Sex limited is an expression of a trait in only one sex; e.g., milk production in mammals; egg production in chickens.

Sex linkage is the location of a gene on a sex chromosome, typically on the X chromosome.

Sex linked locus is a locus that is located on a sex chromosome.

Sex mosaic is synonym of gynandromorph.

Sexed embryos Embryos separated according to sex.

Sexual reproduction is the process where two cells (gametes) fuse to form one hybrid, fertilized cell.

Sexual selection is the selection due to differential mating success either through competition for mates or mate choice.

Signal transduction refers to the biochemical events that conduct the signal of a hormone or growth factor from the cell exterior, through the cell membrane, and into the cytoplasm. This involves a number of molecules, including receptors, proteins, and messengers.

Shadow effect is a case usually caused by low marker polymorphism in mark recapture studies in which a novel capture is labeled as a recapture due to identical genotypes at the loci studied.

Shake culture is an agitated suspension in culture providing adequate aeration for cells in the liquid medium. Usually an Erlenmeyer flask containing the culture is attached to a horizontal or platform shaker, or agitated with a magnetic stirrer.

Shaker is a platform, with set or variable speed control, used to agitate vessels containing liquid cell cultures. Also described as a platform shaker.

Shawnn cells are supporting cells found in nervous tissue outside the central nervous system.

Shear refers to: (1) The sliding of one layer across another, with deformation and fracturing in the direction parallel to the movement. This term usually refers to the forces that cells are subjected to in a bioreactor or a mechanical device used for cell breakage; (2) To fragment DNA molecules into smaller pieces. DNA, as a very long and fairly stiff molecule, is very susceptible to hydrodynamic shear forces. Forcing a DNA solution through a hypodermic needle will fragment it into small pieces. The size of the fragments obtained is inversely proportional to the diameter of the needle's bore. The actual sites at which the shear force breaks a DNA molecule are approximately random. Therefore DNA fragments may be generated by random shear and then cloned (by either tailing their ends or using linkers) so as to create a complete gene library of an organism. This method is little used now, having been replaced by the use of partial digests with four-basepair cutters, such as *Sau*3A, as a means of generating random DNA fragments.

Shine Dalgarno sequence refers to a conserved sequence of prokaryotic mRNAs that is complementary to a sequence near the 5' terminus of the 16S ribosomal RNA and is involved in the initiation of translation.

Shoot is a young branch that grows out from the main stock of a tree, or the young main portion of a plant growing above ground.

Shoot differentiation is the development of growing points, leaf primordia and finally shoots from a shoot tip, axial bud, or even a callus surface.

Shoot tip graft or micrograft is a shoot tip or meristem tip that is grafted onto a prepared seedling or micropropagated rootstock in culture. Meristem tip grafting is mainly used for *in vitro* virus elimination with *Citrus* spp. and other plants.

Shoot tip is the terminal bud (0.1 – 1.0 mm) of a plant, which consists of the apical meristem (0.05 – 0.1 mm) and the immediately surrounding leaf primordia and developing leaves, and adjacent stem tissue. Synonym is shoot apex.

Short day plant is a plant which will not flower until triggered to do so by exposure to one or a number of dark periods equal to or longer than its critical period.

Short interspersed nuclear element (SINE) refers to families of short (150–300 bp), moderately repetitive DNA elements of eukaryotic genomes. They appear to be DNA copies of certain tRNA molecules, created presumably by the unintended action of reverse transcriptase during retroviral infection.

Short template is a DNA strand that is synthesized during the polymerase chain reaction and has a primer sequence at one end and a sequence complementary to the second primer at the other end.

Shot gun genome sequencing is a strategy for sequencing a whole genome, in which the genomic DNA is initially fragmented into pieces small enough to be sequenced. Specialized computer software is then used to piece together the individual sequences to create long contiguous tracts of sequenced DNA.

Shuttle vector (bifunctional vector) is a plasmid capable of replicating in two different host organisms because it carries two different origins of replication and can therefore be used to 'shuttle' genes from one to the other. For example, the YEp, pJDB219, is a shuttle vector able to replicate in *E. coli* from its pMB9 origin and in *Saccaromyces cerevisiae* from its 2 μm-plasmid origin.

SI units (International System of Units) is based on the metric system and units derived from the metric system.

Sib mating refers to the deliberate crossing of siblings. Generally done where self-incompatibility prevents the production of self-fertilized progeny.

Sickle cell anemia is an hereditary, chronic form of hemolytic anemia characterized by break-down of the red blood cells; red blood cells undergo a reversible alteration in shape when the oxygen tension of the plasma falls slightly and a sickle-like shape forms.

Siderophore is a low molecular weight entity that binds very tightly to iron. Siderophores are synthesized by a variety of soil micro-organisms to ensure that the organism is able to obtain sufficient amounts of iron from the environment.

Sieve cell is a long and slender sieve element with relatively unspecialized sieve areas and with tapering end walls that lack sieve plates.

Sieve element refers to the phloem cell concerned with longitudinal conduction of food materials.

Sieve plate refers to the perforated wall area in a sieve tube element, through which strands connecting sieve tube protoplasts can pass.

Sieve tube is a tube within the phloem tissue of a plant, and composed of joined sieve elements.

Sievert (Sv) is a SI unit of ionizing radiation.

Sigma factor is the sub-unit of prokaryotic RNA polymerases that is responsible for the initiation of transcription at specific initiation sequences.

Sigmoid colon refers to the lower part of the large intestine that empties into the rectum.

Sigmoidoscopy is a test that uses a thin, flexible tube with a camera lens at the end to look at the inside of the rectum and lower large intestine for abnormalities.

Signal sequence is a stretch of 15–30 amino acid residues at the N terminus of a protein, which is thought to enable the protein to be secreted (pass through a cell membrane). The signal sequence is removed as the protein is secreted. Synonyms are signal peptide, leader peptide.

Signal transduction refers to the biochemical events that conduct the signal of a hormone or growth factor from the cell exterior, through the cell membrane, and into the cytoplasm. This involves a number of molecules, including receptors, ligands and messengers.

Signals are internal and external factors that control changes in cell structure and function. They can be chemical or physical in nature.

Signal-to-noise ratio is a specifically produced response compared to the response level when no specific stimulus (activity) is present.

Silencer is a DNA sequence that helps to reduce or shut off the expression of a nearby gene.

Silencing is the loss of gene expression either through an alteration in the DNA sequence of a structural gene, or its regulatory region; or because of interactions between its transcript and other mRNAs present in the cell.

Silent mutations are DNA mutations that do not significantly alter the phenotype of the organism in which they occur. Silent mutations can occur in non-coding regions (outside of genes within introns), or they may occur within exons. When they occur within exons they either do not result in a change to the amino acid sequence of a protein (i.e., a synonymous substitution), or result in the insertion of an alternative amino acid with similar properties to that of the original amino acid, and in either case there is no significant change in phenotype.

Simple sequence repeat is abbreviated as SSR. SSR is also known as microsatellites or short tandem repeats (STRs). SSRs are repeating sequences of 2–5 base pairs of DNA. It is a type of Variable Number Tandem Repeat (VNTR). Microsatellites are typically co-dominant. They are used as molecular markers in STR analysis, for kinship, population and other studies. They can also be used for studies of gene duplication or deletion, marker assisted selection, and fingerprinting.

Simplicity is the ease with which an assay can be implemented.

Simulated annealing is a procedure used in molecular dynamics simulations, in which the system is allowed to equilibrate at high temperatures, and then cooled down slowly to remove kinetic energy and to permit trajectories to settle into local minimum energy conformations.

SINEs (Short interspersed nuclear elements) are families of short (150 to 300 bp), moderately repetitive elements of eukaryotes, occurring about 100,000 times in a genome. SINES appear to be DNA copies of certain tRNA molecules, created presumably by the unintended action of reverse transcriptase during retroviral infection.

Single cell line is a culture initiated from a single cell, usually from suspension cultures of single cells or small aggregates plated on solidified medium. The latter may incorporate a selective agent, from which tolerant or resistant individual cell lines or cell clones can be selected.

Single cell protein (SCP) is a protein produced by micro-organisms. The dried mass of a pure sample of a protein-rich-microorganism, which may be used either as feed (for animals) or as a food (for humans).

Single chain antigen is abbreviated as SCA.

Single copy is a gene or DNA sequence which occurs only once per (haploid) genome. Most structural genes, those encoding functional proteins, are single-copy genes.

Single crystal is a crystalline solid for which the periodic and repeated atomic pattern extends throughout its entirety without interruption.

Single domain antibody (sdAb, called Nanobody by Ablynx, the developer) is an antibody fragment consisting of a single monomeric variable antibody domain. Like a whole antibody, it is able to bind selectively to a specific antigen. With a molecular weight of only 12–15 kDa, single-domain antibodies are much smaller than common antibodies (150–160 kDa) which are composed of two heavy protein chains and two light chains, and even smaller than Fab fragments (~50 kDa, one light chain and half a heavy chain) and single-chain variable fragments (~25 kDa, two variable domains, one from a light and one from a heavy chain).

Single node culture is a culture of separate lateral buds with each carrying a piece of stem tissue.

Single nucleotide polymorphism (SNP) is a polymorphism at a particular base site in a coding sequence, e.g., at base 306 in a particular gene, one individual could be heterozygous for A and G: the maternal allele could have an A at this site, while the paternal allele has a G at this site. This type of polymorphism is extensive throughout the genome, and has the great advantage of being detectable without the need for gel electrophoresis, which opens the way for large-scale automation of genotyping.

Single photon emission tomography (SPECT, OR SPET) is a nuclear medicine tomographic imaging technique using gamma rays. It is very similar to conventional nuclear medicine planar imaging using a gamma camera. However, it is able to provide true 3-D information. This information is typically presented as cross-sectional slices through the patient, but can be freely reformatted or manipulated as required. The technique requires delivery of a gamma-emitting radioisotope (a radionuclide) into the patient, normally through injection into the bloodstream.

Single primer amplification reaction (SPAR) is a PCR-based genotyping technique in which genomic template is amplified with a single primer.

Single strand conformational polymorphism (SSCP) is a technique for detection of mutations in a defined DNA sequence. Single-stranded polynucleotides are electrophoretically separated on non-denaturing gels. Intrachain base pairing results in a limited number of conformers stabilized by intrachain loops, and mutated DNA shows on electrophoresis an altered assortment of such conformers.

Single strand DNA (ssDNA) is a DNA molecules separated from their complementary strand, either by its absence or following denaturation.

Single strand DNA binding protein is a protein that coats single-stranded DNA, preventing renaturation and so maintaining the DNA in an extended state.

Single strand nucleic acid refers to nucleic acid molecules consisting of only one polynucleotide chain. The genomes of many viruses are single-stranded DNA molecules, as are most biologically effective RNAs. Many RNA molecules do include double-stranded regions formed by the intra-strand base-pairing of self-complementary sequences, and these determine the 3-dimensional shape (conformation) that they adopt *in vivo*.

Single stranded describes nucleic acid molecules consisting of only one polynucleotide chain. The genomes of certain phages, e.g., MI3, are single-stranded DNA molecules; rRNA, mRNA and tRNA are all single-stranded nucleic acids, but they all contain double-stranded regions formed by the intra-strand base-pairing of self-complementary sequences.

Single-walled carbon nanotubes (SWCNT) have the ability to conduct electricity. This conduction can be ballistic or diffusive.

Sinoatrial node is a mass of specialized cardiac muscle in which the impulse triggering the heartbeat originates; the peacemaker of the heart.

Sintering is particle coalescence of a powdered aggregate by diffusion that is accomplished by firing at an elevated temperature.

Sister chromatid exchange (SCE) is a reciprocal interchanges of the two chromatid arms within a single chromosome.

Site directed mutagenesis is the introduction of base changes – mutations – into a piece of DNA at a specific site, using recombinant DNA methods.

Site specific describes any process or enzyme which acts at a defined sequence within a DNA or RNA molecule. Type II restriction enzymes are site-specific endonucleases and the recombination systems encoded by some transposons are site-specific, such as is the integration of phage into the *E. coli* chromosome.

Site specific mutagenesis is a technique to change one or more specific nucleotides within a cloned gene in order to create an altered form of a protein with one or more specific amino acid changes.

Six base cutter is a type II restriction endonuclease that binds (and subsequently cleaves) DNA at sites that contain a sequence of six nucleotide pairs that is uniquely recognized by that enzyme. Because any sequence of six bases occurs less frequently by chance than any sequence of four bases, six-base cutters cleave less frequently than do four-base cutters. Thus, six-base cutters create larger fragments than four-base cutters.

Skin (integument, cutis) forms the external covering of the body. The skin and its derivatives constitute the integumentary system. It consists of two main layers: The epidermis and the dermis.

Skull fracture is braking of the bony skull caused by an accident.

Slip is a plastic deformation resulting from dislocation motion; also, the shear displacement of two adjacent planes of atoms.

Slip casting is a forming technique used to shape ceramic materials. A slip or suspension of solid particles in water is poured into a porous mold. A solid layer forms on the inside wall as water is absorbed by the mold, leaving a shell (or a solid piece) in the shape of the mold.

Slipped strand mispairing refers to denaturation of the new strand from the template during replication, followed by renaturation in a different spot ("slipping"). This can lead to insertions or deletions.

Small interfering RNA (siRNA), sometimes known as short interfering RNA or silencing RNA, is a class of double-stranded RNA molecules, 20-25 base pairs in length. The siRNA plays many roles, but it is most notable in the RNA interference (RNAi) pathway, where it interferes with the expression of specific genes with complementary nucleotide sequences. siRNA functions by causing mRNA to be broken down after transcription, resulting in no translation. siRNA also acts in RNAi-related pathways, e.g., as an antiviral mechanism or in shaping the chromatin structure of a genome.

Small intestine is a portion of the digestive tract that extends from the stomach to the large intestine. Most of digestion occurs here as nutrients are absorbed from food.

Small nuclear ribonucleoprotein (snRNP) is a complex of small nuclear RNA and nuclear protein, heavily involved in the post-transcriptional processing of mRNA, especially the removal of introns. snRNPs are a major component of spliceosomes.

Small nuclear RNA (snRNA) is a short RNA transcripts of 100–300 bp that associate with proteins to form small nuclear ribonucleoprotein particles (snRNPs); most snRNAs are components of the spliceosomes that excise introns from pre-mRNAs in RNA processing.

Smart sensor is a sensor in which the electronics that process the output from the sensor, and forms the modifier, are partially or fully integrated on a single chip.

Smooth muscle refers to a muscle that performs automatic tasks, such as constricting blood vessels.

SNP (Single Nucleotide Polymorphism) is a single base difference found when comparing the same DNA sequence from two different individuals.

snRNA is an abbreviation for small nuclear RNA.

snRNP is an abbreviation for small nuclear ribonucleoprotein.

Social hierarchy is an arrangement within a group of animals, such as rabbits, where some individuals are dominant over others.

Sodium is a mineral essential to life found in nearly all plant and animal tissue. Table salt (sodium chloride) is nearly half sodium.

Sodium channel blocking agent is any of a class of antiarrhythmic agents that prevent ectopic beats by acting on partially inactivated sodium channels to inhibit abnormal depolarizations.

Sodium dodecyl sulphate (SDS) is a detergent used to solubilize protein and DNA from biological materials. Specific use in sodium dodecyl sulphate polyacrylamide gel electrophoresis.

Sodium dodecyl sulphate polyacrylamide gel electrophoresis (SDS-PAGE) is a widely employed electrophoretic method for the separation of proteins from biological samples. The sodium dodecyl sulphate gives a uniform charge density to the surface of proteins or nucleic acids, so that their rate of migration through the gel is determined largely by their molecular weight.

Soil amelioration is the improvement of poor soils, usually using bacteria or fungi. This contrasts with bioremediation, which is the cleaning up of toxins, usually in soils. Amelioration includes breaking down organic matter; forming humus; by solubilizing them, making minerals in the soil available to plants; fixing nitrogen; and sometimes an element of bioremediation as well.

Soilless culture refers to tissue culture and hydroponics. Growing plants in nutrient solution without soil.

Solid media is a nutrient media that has been solidified, such as by addition of agar.

Solid medium refers to nutrient medium solidified by the addition of a gelling agent, commonly agar.

Solubility is the property of a solid, liquid, or gaseous chemical substance called solute to dissolve in a solid, liquid, or gaseous solvent to form a homogeneous solution of the solute in the solvent.

Solvent is the component of a solution that dissolves a solute.

Solvent-activated system employs a semi-permeable membrane (reservoir) containing a small, laser-drilled hole(s). Within the membrane there is a high concentration of an osmotic agent, either the drug itself or a salt, which causes water to enter through the membrane. The drug is then forced out through the hole because of the increased pressure. Drug release proceeds at a constant rate in solvent-activated systems.

Solvent diffusion is the migration of solvent molecules into or out of a polymer as driven by the concentration gradient.

Solvent drag (bulk transport) is a phenomenon primarily in renal physiology, but it also occurs in gastrointestinal physiology. It is when solutes in the ultra-filtrate are transported back from the renal tubule by the flow of water rather than specifically by ion pumps or other membrane transport proteins. It generally occurs in the paracellular, rather than transcellular, pathway across the tubular cells.

Solvent evaporation is a gentle of removal of salts by evaporation.

Solvent extraction (Liquid–liquid extraction or partitioning) is a method to separate compounds based on their relative solubilities in two different immiscible liquids, usually water and an organic solvent. It is an extraction of a substance from one liquid into another liquid phase. Liquid–liquid extraction is a basic technique in chemical laboratories, where it is performed using a variety of apparatus, from separatory funnels

to countercurrent distribution equipment. This type of process is commonly performed after a chemical reaction as part of the work-up. The term *partitioning* is commonly used to refer to the underlying chemical and physical processes involved in *liquid–liquid extraction* but may be fully synonymous. The term *solvent extraction* can also refer to the separation of a substance from a mixture by preferentially dissolving that substance in a suitable solvent. In that case, a soluble compound is separated from an insoluble compound or a complex matrix. Solvent extraction is used in nuclear reprocessing, ore processing, the production of fine organic compounds, the processing of perfumes, the production of vegetable oils and biodiesel, and other industries.

Somaclonal variation refers to epigenetic or genetic changes induced during the callus phase of plant cells cultured *in vitro*. Sometimes visible as changed phenotype in plants regenerated from culture.

Somatic refers to vegetative or non-sexual stages of a life-cycle.

Somatic (adult) stem cells are relatively rare undifferentiated cells found in many organs and differentiated tissues with a limited capacity for both self renewal (in the laboratory) and differentiation. Such cells vary in their differentiation capacity, but it is usually limited to cell types in the organ of origin.

Somatic cell are cells not involved in sexual reproduction, i.e., not germ cells.

Somatic cell embryogenesis are embryos that are produced either from somatic cells of explants (direct embryogenesis) or by induction on callus formed by explants (indirect embryogenesis).

Somatic cell gene therapy refers to the repair or replacement of a deffective gene within somatic tissue.

Somatic cell hybrid is a hybrid cell line derived from two different species; contains a complete chromosomal complement of one species and a partial chromosomal complement of the other; human/hamster hybrids grow and divide, losing human chromosomes with each generation until they finally stabilize, the hybrid cell line established is then utilized to detect the presence of genes on the remaining human chromosome.

Somatic cell hybrid panel is a panel of cells created by cell fusion, typically involving a reference species (e.g., hamster) and the species of interest (e.g., sheep) with each member of the panel containing a different mixture of chromosomes from the two species. By relating the presence or absence of cloned fragments (via *in situ* hybridization) or PCR products to the presence or absence of particular chromosomes from the species of interest, such panels can be used for physical mapping.

Somatic cell is any cell in the body except the germ cells (egg and sperm).

Somatic Cell Nuclear Transfer (SCNT) is sometimes known as therapeutic cloning. A process by which a nucleus from a single cell (for example a skin cell) is transferred into an unfertilized egg, from which the nucleus (the genetic contents) have been removed. The resulting reconstructed embryo is then allowed to develop to the blastocyst stage. Embryonic stem cells derived from this blastocyst are genetically identical to the donor of the original nucleus.

Somatic cell variant is a somatic cell with unique characters not present in the other cells, such as might be selected for in a screening trial that following a mutation event.

Somatic cells are any cells in a body other than a germ cell.

Somatic embryo or somatic embryoid is an organized embryonic structure morphologically similar to a zygotic embryo but initiated from somatic (non-zygotic) cells. Under *in vitro* conditions, somatic embryos go through developmental processes similar to embryos of zygotic origin.

Somatic hybridization refers to naturally occurring or induced fusion of somatic protoplasts or cells of two genetically different parents. The difference may be as wide as interspecific. Wide synthetic hybrids formed in this way (i.e., not via gametic fusion) are known as cybrids. Not all cybrids contain the full genetic information (nuclear and non-nuclear) of both parents.

Somatic hypermutation is a high frequency of mutation that occurs in the gene segments encoding the variable regions of antibodies during the differentiation of B lymphocytes into antibody producing plasma cells.

Somatic mutation is a mutation occurring in any cell that is not destined to become a germ cell; if the mutant cell continues to divide, the individual will come to contain a patch of tissue of genotype different from the cells of the rest of the body.

Somatic reduction refers to halving of the chromosomal number of somatic cells; a possible method of producing "haploids" from somatic cells and calli by artificial means.

Somatostatin growth is also known as hormone-inhibiting hormone (GHIH) or somatotropin release-inhibiting factor, SRIF or somatotropin release-inhibiting hormone) is a peptide hormone that regulates the endocrine system and affects neurotransmission and cell proliferation via interaction with G protein-coupled somatostatin receptors and inhibition of the release of numerous secondary hormones. Somatostatin regulates insulin and glucagon. Somatostatin has two active forms produced by alternative cleavage of a single preproprotein: one of 14 amino acids, the other of 28 amino acids.

Somatotropin is a growth hormone that is produced by the anterior pituitary (the front part of the pituitary gland). Somatotropin acts by stimulating the release of another hormone called somatomedin by the liver, thereby causing growth to occur. Somatotropin is given to children with pituitary dwarfism (short stature due to underfunction of the anterior pituitary) to help them grow. Also known as somatropin, growth hormone.

Sonication is disruption of cells or DNA molecules by high frequency sound waves.

SOS response is the synthesis of a whole set of DNA repair, recombination and replication proteins in bacteria containing severely damaged DNA (e.g., following exposure to UV light).

Sound barrier is the apparent physical boundary stopping large objects from becoming supersonic.

Source DNA is the DNA from an organism that contains a target gene; this DNA is used as starting material in a cloning experiment.

Source organism is a bacterium, plant or animal from which DNA is purified and used in a cloning experiment.

Southern blot is a cellulose or nylon membrane to which DNA fragments previously separated by gel electrophoresis have been transferred by capillary action. Named after Ed Southern.

Southern blotting is a technique for transferring electrophoretically resolved DNA segments from an agarose gel to a nitrocellulose filter paper sheet via capillary action; the DNA segment of interest is probed with a radioactive, complementary nucleic acid, and its position is determined by autoradiography.

Southern hybridization (also Southern blotting) refers to a procedure in which DNA restriction fragments are transferred from an agarose gel to a nitrocellulose filter, where the denatured DNA is then hybridized to a radioactive probe (blotting).

Spacer sequence is a DNA sequence separating neighboring genes; spacer sequences are not usually transcribed.

Spallation refers to the release of micro-particles of plastic from the inner walls of tubing due to compression by roller-pumps.

Span is the difference between the highest and lowest scale values of an instrument.

SPAR is an abbreviation for single primer amplification reaction. SPAR is a polymerase chain reaction (PCR) technique using core motifs of microsatellite

DNA, which can detect inter- and intra-specific natural genetic variations in plants.

Sparger is a device that introduces into a bioreactor air in the form of separate fine streams.

Spasm is a movement of a muscle that causes cramping and pain.

Spatial autocorrelation statistics is a set of statistical parameters aimed to depict the spatial (geographical) pattern of genetic diversity in a population.

Specialized refers to one that is anatomically or physiologically adapted for particular functions or habitats.

Speciation is the development of one or more species from an existing species.

Species (L. *species*, appearance, form, kind) is a class of potentially interbreeding individuals that are reproductively isolated from other such groups having many characteristics in common. A somewhat arbitrary and sometimes blurred classification; but still quite useful in many situations.

Species concepts refer to the ideas of what constitutes a species, such as reproductive isolation (BSC), or monophyly of a lineage (PSC).

Species scale is the spatial scale encompassing an entire species' distribution.

Species-specific pertains to individuals of only one species. For example, a pesticide that is species-specific affects only one species.

Specific combining ability (SCA) is a component of genetic variance calculable where a number of genotypes are intercrossed in all possible combinations. The SCA measures the deviation of the performance of a particular cross from the average general combining ability of its two parents.

Specific heat is the amount of heat, measured in calories, required to raise the temperature of one gram of a substance by one Celsius degree.

Specific modulus (or specific stiffness) is the ratio of elastic modulus to specific gravity for a material.

Specific strength is the ratio of tensile strength to specific gravity for a material.

Specificity is the ability of a probe to react precisely with a specific target molecule.

Spectroscopy is the study of the interaction between matter and radiated energy. Historically, spectroscopy originated through the study of visible light dispersed according to its wavelength, by a prism.

Speed of sound increases proportionally to the absolute

temperature. Since temperature varies within the troposphere, the speed of sound will vary with altitude also. At sea level, at 15 degrees, it is 1224 km/hr (about 340 m/sec) and at 9100 meters altitude it is 1090.8 km/hr. The Earth's troposphere ends near 11000 meters, and since the temperature in the stratosphere is fairly constant, the speed of sound is also approximately a constant there – about 1060 km/hr. Strictly speaking, the speed of sound is to do with compressible flow.

Spent medium refers to the medium that is discarded because it has been depleted of nutrients, dehydrated or accumulated toxic metabolic products.

Sperm is an abbreviation for spermatozoon. A sperm (plural **spermatozoa**; from Ancient Greek: σπέρμα "seed" and Ancient Greek: ζῷον "living being") is a motile sperm cell, or moving form of the haploid cell that is the male gamete. A spermatozoon joins an ovum to form a zygote. A zygote is a single cell, with a complete set of chromosomes, that normally develops into an embryo. Sperm cells contribute approximately half of the nuclear genetic information to the diploid offspring (excluding, in most cases, mitochondrial DNA). In mammals, the sex of the offspring is determined by the sperm cell: a spermatozoon bearing a Y chromosome will lead to a male (XY) offspring, while one bearing an X chromosome will lead to a female (XX) offspring. Sperm cells were first observed by Anton van Leeuwenhoek in 1677.

Sperm competition refers to the competition between different spermatozoa to reach and fertilize the egg cell of a single female.

Sperm disorders refer to the condition with the production and maturation of sperm; the single most common cause of male infertility. Sperm may be immature, abnormally shaped, unable to move properly, or, normal sperm may be produced in abnormally low numbers (oligospermia).

Sperm sexing The separation of mammalian sperm into those bearing an X chromosome and those bearing a Y chromosome, in order to be able to produce, via artificial insemination or *in vitro* fertilization, animals of a specified sex. Methods for achieving this include the inactivation of X-bearing or Y-bearing sperm by antibodies recognizing sex-specific sperm surface peptides, and fluorescence-activated cell sorting.

Spermatid is a immature spermatozoon. It is one of the four cells formed at the end of the second meiotic division in spermatogenesis.

Spermatocyte refers to the premeiotic parental cell of the spermatids; the primary spermatocyte before the initiation of the first meiotic division; the secondary spermatocyte after completion of the first meiotic division, but before the initiation of the second division. Synonym is sperm mother cell.

Spermatogenesis refers to the series of cell divisions in the testis by which maturation of the gametes (sperm) of the male takes place.

Spermatogonium (plural: spermatogonia) is a primordial male germ cell. These can either divide by mitosis to produce daughter cells, or enter a growth phase and differentiate into a primary spermatocyte.

Spermatozoon is the mature, mobile reproductive cell of male animals, produced by the testis.

Spheroplast (Alternative spelling for sphaeroplast) is a microbial or plant cell from which most of the cell wall has been removed, usually by enzymatic treatment. Strictly, in a spheroplast, some of the cell wall remains, while in a protoplast the cell wall has been completely removed. In practice, the two words are often used interchangeably.

Sphincter is a circular muscle that opens and closes at an entrance to an organ. Examples include the lower esophageal sphincter and the anal sphincter. Sphincter muscle keeps urine from leaking by closing tightly like a rubber band around the opening of the bladder.

Sphygmomanometer is an instrument for measuring blood pressure, especially arterial blood pressure.

Spike (L. *spica*, an ear of grain) is an inflorescence in which the main axis is elongated and the flowers are sessile.

Spikelet refers to the unit of inflorescence in grasses, made up of a small group of florets.

Spina bifida is a congenital condition that results from altered fetal development of the spinal cord, part of the neural plate fails to join together and bone and muscle are unable to grow over this open section.

Spinal cord refers to the body's major nerve tract in vertebrates. In humans, it is about 18 in. (45 cm) long.

Spinal nerves are nerves that emerge from the spinal cord.

Spindle refers to the spindle-shaped intracellular structure in which the chromosomes move.

Spine is a hard, sharp structure on the surface of a plant; usually a modified leaf.

Spinning is a fiber forming process: A multitude of fibers are spun as molten material is forced through many small orifices.

Spirochaete is a non-rigid, corkscrew-shaped bacterium that moves by means of muscular flexions of the cell.

Spirogram is a record of the amounts of air being moved in and out of the lungs.

Spirometer is an instrument for measuring the air entering and leaving the lungs.

Spleen is an organ found on the left side of the abdomen, next to the stomach. Makes white blood cells that help fight infection and filters and cleanses the blood. Plays a role in immunity.

Spliceosome is a complex of small nuclear ribonucleoproteins and other proteins that assemble on an immature mRNA and catalyze the excision of an intron.

Splicing is the process that eliminates intervening intron sequences and covalently joins exon sequences of RNA, during the maturation of eukaryotic mRNA. In recombinant DNA technology, the term refers to the latter of the two processes just described, namely joining fragments of DNA together.

Splicing junction refers to the DNA sequence immediately surrounding the boundary between an exon and an intron. There is a degree of sequence conservation in these regions, allowing the identification of introns in newly sequenced genes.

Split gene refers the pattern of interruption in the coding sequence. The encoding DNA of many structural genes is made up of exons and introns, in eukaryotes,.

Spontaneous mutation is a mutation occurring in the absence of any known mutagen.

Sporangium (plural: sporangia) is a reproductive structure in plants that produces spores. A megasporangium produces megaspores, which give rise to the female gametophyte; in seed plants it is represented by the ovule. A microsporangium produces microspores, which give rise to the male gametophyte; it is represented in seed plants by the pollen sac.

Spore refers to: (1) A reproductive cell that develops into an individual without union with other cells; some spores such as meiospores occur at a critical stage in the sexual cycle, but others are asexual in nature; (2) A small, protected reproductive form of a micro-organism, often synthesized when nutrient levels are low.

Spore mother cell is a synonym of sporocyte.

Sporocyte is a diploid cell that gives rise to four haploid spores by meiosis.

Sporophyll is a leaf that bears spore producing structures (sporangia).

Sporophyte is the diploid generation in the life cycle of a plant, and that produces haploid spores by meiosis.

Sport is an individual plant, or portion thereof, showing a recognizably different phenotype from the parent, presumably as a result of spontaneous mutation.

Spririllum is a rigid, spiral-shaped bacterium.

Squeeze-film damping defines effect of ambient fluid and spacing on the vertical movement of a structural member with respect to a substrate.

SSAHA is a hashing algorithm developed for rapid searching of large amounts of genome sequence. This program is similar to BLAT but does not use splice information to align mRNA sequences, nor can it perform translated searches.

ssDNA is an abbreviation for single-stranded DNA. ssDNA is DNA molecule consisting of only a single strand contrary to the typical two strands of nucleotides in helical form.

SSLP (Simple sequence length polymorphism) are used as genetic markers with Polymerase Chain Reaction (PCR). An SSLP is a type of polymorphism: a difference in DNA sequence amongst individuals. SSLPs are repeated sequences over varying base lengths in intergenic regions of deoxyribonucleic acid (DNA). Variance in the length of SSLPs can be used to understand genetic variance between two individuals in a certain species. Common examples of these in mammalian genomes include runs of dinucleotide or trinucleotide repeats (CACACACACA-CACACACA).

SSR is an abbreviation for simple sequence repeat. SSR or **short tandem repeats** (STRs), are repeating sequences of 2–5 base pairs of DNA. It is a type of Variable Number Tandem Repeat (VNTR). Microsatellites are typically co-dominant. They are used as molecular markers in STR analysis, for kinship, population and other studies. They can also be used for studies of gene duplication or deletion, marker assisted selection, and fingerprinting.

Stability is the ability of a sensor to retain specified characteristics after being subjected to designated environmental or electrical test conditions.

Stabilizer is a polymer additive that counteracts the deteriorative process. It is a chemical which tends to inhibit the reaction between two or more other chemicals. It can be thought of as the antonym to a catalyst. The term can also refer to a chemical that inhibits separation of suspensions, emulsions, and foams. Stabilizers are usually added to polymers during production. The trend is towards fluid systems, pellets, and increased use of masterbatches. There are monofunctional, bifunctional, and polyfunctional stabilizers.

Stabilizing selection is the selection for a phenotype with a more intermediate state.

Stable polymorphism is a polymorphism that is maintained at a locus through natural selection.

Stacked genes refers to the insertion of two or more genes into the genome of an organism. An example is a plant carrying a *Bt* transgene giving insect resistance, and a *bar* transgene giving resistance to a specific herbicide.

Stages of culture (I-IV) refers to micropropagation, which is the practice of rapidly multiplying stock plant material to produce a large number of progeny plants, using modern plant tissue culture methods. Micropropagation is used to multiply novel plants, such as those that have been genetically modified or bred through conventional plant breeding methods. It is also used to provide a sufficient number of plantlets for planting from a stock plant which does not produce seeds, or does not respond well to vegetative reproduction. Four stages of micropropagation are: Establishment, multiplication, pretransplant and transfer from culture.

Staggered cuts refer to symmetrically cleaved phospho-diester bonds that lie on both strands of duplex DNA but are not opposite one another.

Stamen (L. *stamen*, the standing-up things) is a flower structure made up of an anther (pollen-bearing portion) and a stalk or filament. The stamen is the male part of the flower.

Standard atmosphere is a unit of pressure and = 760 mm Hg at m.s.l.

Standard deviation is a statistical measure of variability in a population of individuals or in a set of data.

Standard error is a statistical measure of variation in a population of means, used to indicate how well sample estimates represent population parameters.

Standard hydrogen electrode (SHE) is a platinum conductor in contact with a one M H^+ ions and bathed by hydrogen gas at one atmosphere.

Starch refers to the major plant carbohydrate storage substance, particularly but not exclusively found in seeds, and used both as food and feed source and for various industrial processes. A large water-insoluble heterogenous group of polysaccharides, consisting of various proportions of the two glucose polymers, amylose and amylopectin. Starch is broken down into simple metabolisable sugars *in vivo* by the action of amylases.

Start codon or initiator codon refers to the set of three nucleotides in an mRNA molecule with which the ribosome starts the process of translation. The start codon sets the reading frame for translation. The most commonly used start codon is AUG, which is decoded as methionine in eukaryotes and as *N*-formylmethionine in prokaryotes. AUG appears to be the only start codon used by eukaryotes, while in bacteria, GUG (valine) may sometimes be employed. *See* initiation codon; initiator.

Starter culture refers to the microorganisms that are deliberately added to foods to alter flavor, color, texture, smell, or taste.

Stationary culture is a culture maintained in the growth chamber with no agitation movement. The antonym is shake culture.

Stationary phase is the plateau of the growth curve after log growth, during which cell number remains constant. New cells are produced at the same rate as older cells die.

Statistic is an estimate based on a sample or samples of a population, providing an indication of the true population parameter.

Steady state is the condition under which the number of cells removed with the outflow is exactly balanced by the number of newly synthesized cells.

Steady-state diffusion is a diffusion condition for which there is no net accumulation or depletion of diffusing species. The diffusion flux is independent of time.

Steatorrhea refers to the loose, greasy bowel movements caused by an inability of the body to absorb fat.

Stele (Gr. *stele*, a post) is the central cylinder, inside the cortex, of roots and stems of vascular plants. *See* meristele.

Stem (*stemn*) is the main body of the above-ground portion of a tree, shrub, herb or other plant; the ascending axis, whether above or below ground, of a plant.

Stem cell is an undifferentiated somatic cell that is capable of either division to give rise to daughter stem cells, or differentiating into any specialized cell type given the appropriate signals. Cultured stem cells are critical to the concept of therapeutic cloning.

Stem cell differentiation is the process by which a stem cell can become a specific cell type. Stem cell differentiation begins when they are exposed to certain biochemical cues – whether physiological or experimental. Biochemical cues in different parts of the body stimulate stem cells to grow into the specific cells needed in that location. All stem cells have the capacity to differentiate, but to different degrees: Totipotent stem cells can become any cell in the human body; Pluripotent stem cells can become almost any cell in the human body, but they cannot become placental tissue needed for development in the human uterus; and Multipotent stem cells can become only a certain type of cell, such as blood cells.

Stem cell homing is the migration of stem cells through the blood or tissue to an ultimate destination where it differentiates and replaces or builds tissue. Stem cell homing is triggered by interactions between the cell surface adhesion molecules (such as selectins, integrins and ICAMs) and the cell's surrounding environment.

Stem cell niche is the microenvironment in which a stem cells is situated. During development, the niche may contain various factors and elements that alter gene expression within the stem cell, causing the cell to differentiate and proliferate into various tissues of the fetus. In developed tissue, the niche may help maintain stem cells in a quiescent state, until injury or disease signals them to self-renew and differentiate to replace the damaged tissue. Niche elements may include interactions with other cells, adhesion molecules, growth factors, cytokines and parameters such as pH, ionic strength and gas composition. Scientists study niche characteristics in order to replicate them *in vitro*, to control and direct the differentiation of stem cells in the laboratory.

Stenosis is the narrowing or constriction of an opening, such as a blood vessel or heart valve.

Step response is the response of a system to an instantaneous jump in the input signal.

Stepping stone model of migration is a model of migration in which the probability of migration between nearby or adjacent populations is higher than the probability of migration between distant populations.

Stepwise mutation model (SMM) is a model of mutation in which the microsatellite allele length has an equal probability of either increasing or decreasing (usually by a single repeat unit, as in the strict one-step SMM).

Steric hindrance occurs when the large size of groups within a molecule prevents chemical reactions that are observed in

related molecules with smaller groups.

Sterile refers to: (1) Medium or object free of viable micro-organisms; (2) Incapable of producing viable gametes.

Sterile room refers to a space for the carrying out of activities that require sterile conditions. It can usually be achieved more economically with a laminar air-flow cabinet.

Sterility refers to complete or partial failure of an individual to produce functional gametes or viable zygotes under a given set of environmental conditions. It is an inability to produce offspring.

Sterilization is the act of sterilizing.

Sterilize refers to: (1) The process of elimination of micro-organisms, such as by chemicals, heat, irradiation or filtration; (2) The operation of making an animal incapable of producing offspring.

Sternotomy is the surgical operation of cutting through the sternum.

Steward bottle is a flask developed by Steward for the growth of cells and tissues in a liquid medium, in which they can be periodically submerged during rotation.

Stewardship is the preservation of public good by ensuring that the social and the ethical issues related to biotechnology are addressed, and that the federal government has an effective regulatory regime and the science capacity to protect human and animal health and the environment.

Sticky end refers to the properties of the end of a molecule of DNA or a recombinant DNA molecule. The concept is important in molecular biology, especially in cloning or when subcloning inserts DNA into vector DNA. The sticky ends or cohesive ends form base pairs. Any two complementary cohesive ends can anneal, even those from two different organisms. This bondage is temporary however, and DNA ligase will eventually form a covalent bond between the sugar-phosphate residue of adjacent nucleotides to join the two molecules together.

Stiction is an adhering of thin micromachined layers to a substrate.

Stigma (L. *stigma*, a prick, a spot) is a receptive portion of the style, to which pollen adheres.

Stirred-tank fermenter is a growth vessel in which cells or micro-organisms are mixed by mechanically-driven impellers.

Stochastic is the presence of a random variable in determining the outcome of an event.

Stock is a term generally used in fisheries management that refers to a population that is demographically

Stock plant is the source plant from which cuttings or explants are made. Stock plants are usually maintained carefully in an optimum state for (sometimes prolonged) explant use. Preferably they are certified-pathogen-free plants.

Stock solution refers to the pre-prepared solution of commonly used reagents.

Stoichiometry is the state of having exactly the ratio of cations to anions specified by the chemical formula, for ionic compounds,. Stoichiometric quantities refers to quantities of reactants mixed in exactly the correct amounts so that all are used up at the same time.

Stokesian (or non-Newtonian) **fluid** is a viscous fluid whose shear stresses are a non-linear function of the fluid strain rate.

Stolon (L. *stolo*, a shoot) is a lateral stem that grows horizontally along the ground surface. The runners of white clover, strawberry and bermuda grass are examples of stolons.

Stoma (Gr. *stoma*, mouth; plural: stomata) refers to: (1) Any of various small openings or pores in an animal body, especially an opening resembling a mouth in various invertebrates; (2) Botany: A minute pore in the epidermis of the leaf or stem of a plant, forming a slit of variable width between two specialized cells (guard-cells), which allows movement of gases, including water vapor, to and from the intercellular spaces. Also, the whole pore with its associated guard-cells.

Stomach refers to muscular region of the digestive tract extending from the esophagus to the small intestine.

Stomatal complex includes the stoma, together with its guard cells and, when present, any related subsidiary cells.

Stomatal index is equal to (number of stomata per mm^2 × 100)/ (number of stomata per mm^2 + number of epidermal cells per mm^2). This value is useful in comparing leaves of different sizes. Relative humidity and light intensity during leaf development affect the value of stomata index.

Stool is a waste product that remain after food is digested, including fiber, bacteria, mucus, undigested foods, and cells from the inside of the intestine. This is passed through the rectum as a bowel movement.

Stop codon or termination codon is a set of three nulecotides for which there is no corresponding tRNA molecule to insert an amino acid into the polypeptide chain. Protein synthesis is hence terminated and the completed polypeptide released from the ribosome. Three stop codons are found: UAA (ochre), UAG (amber) and UGA (opal). Mutations which generate any of these three codons in a position which normally contains a codon specifying an amino acid are known as *nonsense mutations*. Stop codons can also be called nonsense codons.

STR is an abbreviation for sequence tandem repeat. The STR occur in DNA when a pattern of one or more nucleotides is repeated and the repetitions are directly adjacent to each other. An example is: ATTCG ATTCG ATTCG, in which the sequence ATTCG is repeated three times.

Strain (ε) is the change in gauge length of a specimen, in the direction of an applied stress, divided by its original gauge length. **Strain** is a group of individuals derived by descent from a single individual within a species.

Strain gauge is an element (wire or foil) that measures a strain based on electrical resistance changes of the gauge that result from a change in length or dimension strain of the wire or foil.

Strain isolation is an isolation of any bacterium, animal or plant from the outside world.

Stratification is the treatment of moist seeds at low temperature (+2° ± 4°C) to break dormancy.

Stratum basale is adjacent to the basal lamina. It is also called the stratum germinativum because it contains move toward the surface to replace those that have sloughed off.

Stratum corneum is the exposed layer of both thick skin and thin skin; it is made up of dead cells.

Streamlines are curves whose tangent at any point is in the direction of the velocity vector at that point. By analogy, in general, a streamline pattern is like a single frame in a moving picture – each frame will be different. A pathline, on the other hand, is like a time-exposure picture of a particular fluid element. Pathlines and streamlines will, in general, be different. However for steady flow, the pathlines and streamlines will coincide, and the streamline pattern will remain unaltered from one 'frame' to the next.

Streptavidin is a microbial protein with a high affinity for the B complex vitamin biotin. The specific interaction of these two molecules has been exploited in labeling technology and in applications where a specific molecule needs to be captured or purified.

Stress concentration is the concentration or amplification of an applied stress at the tip of a notch or small crack.

Stress corrosion is a form of failure resulting from the combined action of a tensile stress and a corrosion environment, occurring at lower stress levels than required when the corrosion environment is absent.

Stress incontinence is the most common type of incontinence that involves the leakage of urine during exercise, coughing, sneezing, laughing, lifting heavy objects, or other body movements that put pressure on the bladder.

Stress protein (Heat shock proteins, HSP) are a group of proteins induced by heat shock. The most prominent members of this group are a class of functionally related proteins involved in the folding and unfolding of other proteins. Their expression is increased when cells are exposed to elevated temperatures or other stress. This increase in expression is transcriptionally regulated. The dramatic upregulation of the heat shock proteins is a key part of the heat shock response and is induced primarily by heat shock factor (HSF).

Stress refers to non-optimal conditions for growth. Stresses may be imposed by biotic (pathogens, pests) or abiotic (environment, such as heat, drought etc.) factors.

Stress ulcer is an ulcer in the esophagus, stomach, or upper small intestine caused by excess acid produced as a result of surgery, major burns, head injury, or other trauma.

Stricture is an abnormal narrowing of a part of the organ.

Stringency refers to reaction conditions – notably temperature, salt, and pH – that dictate the annealing of single-stranded DNA/DNA, DNA/RNA, and RNA/RNA hybrids. At high stringency, duplexes form only between strands with perfect one-to-one complementarity; lower stringency allows annealing between strands with some degree of mismatch between bases.

Stringent plasmid refers to a plasmid that only replicates along with the main bacterial chromosome and is present as a single copy, or at most several copies, per cell.

Stroma is a tissue that forms the framework of an organ.

Stromal cells are connective tissue cells found in virtually every organ. In bone marrow, stromal cells support blood formation.

Strouhal number is a dimensioless quantity describing oscillating flow mechanisms. Often, it is given as $Sr = f*D/V$, where Sr is the Strouhal number, f is the frequency of vortex shedding, D is the hydraulic diameter of the object in the fluid flow and V is the velocity of the fluid. The Strouhal number is a function of the Reynolds number Re. It is assumed to be equal to 0.2 in the region $200<Re<200,000$.

Structural gene is a DNA sequence that forms the blueprint for the synthesis of a polypeptide.

Structure analysis is the integration of gene identification and promoter recognition programs will be very important point for a complete gene structure analysis.

Structure functionalism refers to the scientific tradition that stresses the relationship between a physical structure and its function, e.g., the related disciplines of anatomy and physiology.

STS (Sequence tag site) is a short DNA sequence (200–500 bp) that has a single occurrence in the genome and whose location and base sequence are known. STS are produced throughout a genome. Oligonucleotide primers are generated such that this sequence can be amplified using PCR to produce a discrete band when analyzed by electrophoresis. STS markers can be polymorphic or monomorphic. They are critical to integrating non-sequence based maps (such as genetic or RH) with sequence based maps.

Style (Gr. *stylos*, a column) is a slender column of tissue that arises from the top of the ovary and through which the pollen tube grows.

Sub-clone is a procedure in which a large cloned DNA molecule is divided into smaller fragments, each one of which is then separately cloned.

Subcloning is the process of transferring a cloned DNA fragment from one vector to another.

Sub-culture interval is the time between consecutive sub-cultures of cells.

Sub-culture number is the number of times cells, etc., have been sub-cultured, i.e., transplanted by inoculation from one culture vessel to another.

Sub-culture refers to division and transfer of a portion or inoculum of a culture to fresh medium. Sometimes used to denote the adding of fresh liquid to a suspension culture.

Subculturing are transferring cultured cells, with or without dilution, from one culture vessel to another.

Subcutaneous (SC) **tissue** (Latin *subcutaneous*, meaning beneath the skin) is also called the hypodermis, hypoderm, subcutis, or superficial fascia. It is the lowermost layer of the integumentary system in vertebrates.

Subgenomic promoter is a promoter added to a virus for a specific heterologous gene, resulting in the formation of mRNA for that gene alone.

Sublingual immunotherapy involves putting drops or tablets of allergen extracts under the tongue of a patient with allergic reactions and then swallowing the extracts. It allows the organism to become tolerant to the allergen by absorbing the allergen through the stomach lining.

Subpopulations are groups within a population delineated by reduced levels of gene flow with other groups.

Subspecies is a taxonomically defined subdivision within a species that is physically or genetically distinct, and often geographically separated.

Subspecies refers to population(s) of organisms sharing certain characteristics that are not present in other populations of the same species.

Substantial derivative or total derivative is physically the time rate of change due to the movement of the fluid element from one location to another in the flow field where the flow properties are spatially different.

Substitution is a point mutation in which one base pair in the DNA sequence is replaced by another.

Sub-strain is derived from a strain by isolating a single cell or groups of cells having properties or markers not shared by all cells of the strain.

Substrate refers to: (1) A compound that is altered by an enzyme; (2) Food source for growing cells or micro-organisms; (3) Material on which a sedentary organism lives and grows.

Subunit vaccine refers to one or more immunogenic proteins either purified from the disease-causing organism or produced from a cloned gene. A vaccine composed of a purified antigenic determinant that is separated from the virulent organism.

Sucker is a shoot that arises from an underground root or stem and grows at the expense of the parent plant.

Suckering is a type of vegetative propagation where lateral buds grow out to produce an individual that is a clone of the parent.

Sucrose density gradient centrifugation is a procedure used to fractionate mRNAs or DNA fragments on the basis of size.

Sum rule is a statistical rule that states that the probability of n_i mutually exclusive, independent events occurring is equal to the sum of the probabilities of each n event.

Superbug refers to Jargon for the bacterial strain of *Pseudomonas* developed by Chakrabarty, who combined hydrocarbon-degrading genes carried on different plasmids into one organism. Although this genetically engineered micro-organism is neither "super" nor a "bug."

Supercoil refers to the conformation of a double-stranded DNA molecule placed under torsional stress as a result of interactions with proteins. The stress is accommodated by a twist imposed on the duplex. A left-handed supercoil favors unwinding of the double helix; a right-handed supercoil favors tighter winding.

Supercoiled plasmid refers to the predominant *in vivo* form of plasmid, in which the plasmid is coiled around histone-like proteins. Supporting proteins are stripped away during extraction from the bacterial cell, causing the plasmid molecule to supercoil around itself *in vitro*.

Superconductivity is a phenomenon characterized by the disappearance of the electrical resistivity at temperatures approaching 0 K.

Supercontig (scaffold) is a supercontig is formed when an association can be made between two contigs that have no sequence overlap. This commonly occurs using information obtained from paired plasmid ends. For example, both ends of a BAC clone are sequenced. It can be inferred that these two sequences are approximately 150–200 Kb apart (based on the average size of a BAC). If the sequence from one end is found in a particular sequence contig, and the sequence from the other end is found in a different sequence contig, the two sequence contigs are said to be linked. In general, it is useful to have end sequences from more than one clone to provide evidence for linkage.

Supergene are allelic combinations found at closely linked loci that affect related traits and are inherited together. An example of a supergene is the major histocompatibility complex (MHC), which in humans contains more than 200 genes adjacently located over several megabases of sequence on chromosome 6.

Superior vena cava refers to the main vein feeding back to the heart from systemic circulation above the heart.

Supernatant is the soluble liquid fraction of a sample after centrifugation or precipitation of insoluble solids.

Supernumerary chromosome is a chromosome, often present in varying numbers, that is not needed for normal development, lacks functional genes, and does not segregate during meiosis. These small chromosomes, which are also called B chromosomes, are present in addition to the normal complement of functional chromosomes in an organism.

Superparamagnetic iron oxide-based colloids (SPIOC: median diameter > 50 nm) are compounds that consist of nonstoichiometric microcrystalline magnetite cores, which are coated with dextrans (in ferumoxide) or siloxanes (in ferumoxsil). After injection, they accumulate in the reticuloendothelial system (RES) of the liver (Kupffer cells) and the spleen. At low doses circulating iron decreases the T1 time of blood, and at higher doses predominates the T2 effect. SPIO agents are much more effective in MR relaxation than paramagnetic agents.

Supersonic flow is a flow with speed above the speed of sound, 1,225 km/h at sea level, is said to be supersonic.

Supportive breeding is the practice of removing a subset of individuals from a wild population for captive breeding and releasing the captive-born offspring back into their native habitat to intermix with wild-born individuals and increase population size or persistence.

Suppressor refers to mutations in suppressor genes are able to overcome (suppress) the effects of mutations in other, unlinked, genes. A common, and very useful, kind of suppressor mutation occurs within the gene encoding a tRNA molecule and results in a change in the tRNA's anticodon. Such a mutant tRNA can reverse the effect of chain-terminating mutations, such as amber or ochre, in protein-encoding genes.

Suppressor mutation is a mutation that partially or completely cancels the phenotypic effect of another mutation.

Suppressor-sensitive mutant is an organism that can grow when a second genetic factor – a suppressor – is present, but not in the absence of this factor.

Supramolecular electronics is the experimental field of supramolecular chemistry that bridges the gap between molecular electronics and bulk plastics in the construction of electronic circuitry at the nanoscale. In supramolecular electronics, assemblies of pi-conjugated systems on the 5 to 100 nanometer length scale are prepared by molecular self-assembly with the aim to fit these structures between electrodes.

Surface Enhanced Raman Spectroscopy

(surface-enhanced Raman scattering, SERS) is a surface-sensitive technique that enhances Raman scattering by molecules adsorbed on rough metal surfaces or by nanostructures such as plasmonic-magnetic silica nanotubes. The enhancement factor can be as much as 10^{10} to 10^{11}, which means the technique may detect single molecules.

Surface markers are proteins on the outside surface of a cell that are unique to certain cell types and that can be visualized using antibodies or other detection methods.

Surface micromaching technology makes thin micromechanical devices on the surface of a silicon wafer. Large numbers of devices can be inexpensively made, and this technology integrates well with electronics. On the surface of a silicon wafer, thin layers of structural and sacrificial material are deposited and patterned. At the end of the processing the sacrificial material is removed, and completely assembled micro mechanical devices remain. This technique was first demonstrated in 1967 by Nathanson, Newell, Wickstrom, and Davis using gold with a sacrificial photoresist layer.

Surface plasmon is a collective motion of electrons in the surface of a metal conductor, excited by the impact of light of appropriate wavelength at a particular angle.

Surface Plasmon Resonance (SPR) is the collective oscillation of electrons in a solid or liquid stimulated by incident light. The resonance condition is established when the frequency of light photons matches the natural frequency of surface electrons oscillating against the restoring force of positive nuclei. SPR in nanometer-sized structures is called localized surface plasmon resonance.

Surface tension is a force within the surface layer of a liquid that causes the layer to behave as an elastic sheet.

Surface-active agent is a substance that exerts a change on the surface properties of a liquid, especially one that reduces its surface tension, as a detergent.

Surfactant is a surface-active agent or wetting agent, such as: Tween 20™ or Tween 80™, Teepol™, Lissapol F™, Alconox™, etc. Surfactants act by lowering the surface tension and are common addenda to solutions used to surface sterilize materials prior to aseptic excision of explants.

Surrogate is a person or animal that functions as a substitute for another. In the case of a surrogate mother, a woman or female animal carries an embryo and ultimately gives birth to a baby that was formed from the egg of another female.

Surveillance is a systematic collection, analysis, interpretation and dissemination of data (generated by the laboratory and private and public domain literature) related to the biotechnology field to assist in the planning and implementation of research, evaluation and management of risks and public health interventions and programs (if needed).

Susceptible refers to inability to withstand injury due to biotic or abiotic stress. *Opposite*: resistance, tolerance.

Suspension culture is a type of culture in which (single) cells and/or clumps of cells grow and multiply while suspended in a liquid medium.

Sustainable development is an approach to development that meets the needs of the present without compromising the ability of future generations to meet their own needs. It seeks to ensure that current development does not alter the environment's ability to recover from any damage sustained, and also makes use of renewable resources.

Sustainable intensification of animal production systems refers to the manipulation of inputs to, and outputs from, livestock production systems aimed at increasing production and/or productivity and/or changing product quality, while maintaining the long-term integrity of the systems and their surrounding environment, so as to meet the needs of both present and future human generations.

SWOT analysis is a business tool for strategic planning.

Symbiont is an organism living in symbiosis with another, dissimilar organism.

Symbiosis (Gr. *syn*, with + *bios*, life) refers to the close association of two different kinds of living organisms where there is benefit to both or where both receive an advantage from the association. An example is the association of the mycelium of mycorrhizal fungi with roots of seed plants.

Symbiotic association is an intimate partnership between two organisms, in which the mutual advantages normally outweigh the disadvantages.

Sympatric refers to populations or species that occupy the same geographic area.

Sympatric speciation is the formation of new species by populations that inhabit the same or overlapping geographic regions.

Symplast is the system of protoplasts in plants, that are interconnected by plasmodesmata.

Sympodial is a type of plant development in which the terminal bud of the stem stops growing due either to its abortion, or to its differentiation into a floral meristem. Frequently, the uppermost lateral bud then takes over the further axial growth of the stem.

Synapomorphy is a shared derived trait between evolutionary lineages. A homology that evolved in an ancestor common to all species on one branch of a phylogeny, but not common to species on other branches.

Synapsis refers to the pairing of homologous chromosome pairs during prophase of the first meiotic division, when crossing over occurs.

Synaptonemal complex (SC) is a ribbon-like proteinaceous structure formed between paired homologous chromosomes at the end of the first meiotic prophase. The SC binds the chromatids along their length, and facilitates crossing over.

Synchronous culture is a culture in which the majority of the cells are dividing at the same time or are at a specific phase of the cell cycle.

Syncitiotrophoblast are multinucleate cells that cover placental villi. These are currently thought to form by the fusion of another trophoblast cell the cytotrophoblasts, within the trophoblast layer (shell) of the implanting conceptus. In early development, these cells mediate implantation of the conceptus into the uterine wall and secrete the hormone (human Chorionic Gonadotrophin, hCG) responsible for feedback maintenance of the corpus luteum (in maternal ovary) and therefore maintaining early pregnancy.

Syncope refers to fainting; temporary loss of consciousness.

Syncytium is a group of cells in which cytoplasmic continuity is maintained; the effect is of a multinucleate cell.

Syndrome is a recognizable pattern or group of multiple signs, symptoms or malformations that characterize a particular condition; syndromes are thought to arise from a common origin and result from more than one developmental error during fetal growth.

Synergid is one of the two haploid nuclei at the micropylar end of the embryo sac of higher plants. The third nucleus is the egg.

Synergism is an interaction between two organisms (e.g., Rhizobium and legumes) in which the growth of one is helped by the other. Opposite is antagonism.

Syngamy is synonym of fertilization (also known as conception, fecundation and syngamy). It is the fusion of gametes to initiate the development of a new individual organism.

Synkaryon is the initial hybrid nucleus of the zygote, formed by the fusion of the gametic nuclei upon fertilization. A hybrid nucleus formed by the fusion of two different somatic cells during somatic cell hybridization is called a heterokaryon.

Synteny refers to the occurrence of two or more loci on the same chromosome, without regard to the distance between them.

Syringohydromyelia is also called syrinx. It is a fluid collection in the spinal cord.

System Physiology (or Systems physiology) is a term used to describe that aspect of BME in which engineering strategies, techniques and tools are used to gain a comprehensive and integrated understanding of the function of living organisms ranging from bacteria to humans.

Systematic error is an error that always occurs in the same direction.

Systemic circulation refers to the circulation of blood throughout the entire body

Systemic relates to a process that affects the body generally; in this instance, the way in which blood is supplied through the aorta to all body organs except the lungs.

Systole refers to the contraction, or period of contraction, of the heart, especially that of the ventricles. It coincides with the interval between the first and second heart sound, during which blood is forced into the aorta and the pulmonary trunk.

Systolic pressure is the upper limit of pressure to which blood pressure rises with the contraction of the ventricles.

T

· ·

T is an abbreviation for thymine, which is one of the four nucleobases in the nucleic acid of DNA that are represented by the letters G–C–A–T. The others are adenine, guanine, and cytosine. Thymine is also known as 5-methyluracil, a pyrimidine nucleobase.

T cells (T lymphocyte) refers to lymphocyte that pass through the thymus gland during maturation. Different kinds of T cells play important roles in the immune response, being primarily responsible for the T cell-mediated response or cellular immune response.

T cell receptor is an antigen-binding protein that is located on the surfaces of killer T cells and mediates the cellular immune response of mammals. The genes that encode T cell antigens are assembled from gene segments by somatic recombination processes that occur during T lymphocyte differentiation.

T lymphocyte refers to T-cells.

T_0, T_1 and T_2 refer to the successive generations of plants following a transformation event. The parent transformed plant is T_0, its immediate progeny is T_1, and the progeny of the T_1 are T_2 plants etc. Of particular interest is the stability of transgene expression from T_0 to T_2, and beyond.

T4 DNA ligase is an enzyme, present in bacteria infected with bacteriophage T4, which catalyzes the joining (ligation) of, and repairs nicks in, duplex DNA molecules. Ligation activity requires that one DNA molecule has a 5'-phosphate group, and that the other has a free 3'-hydroxyl group.

TAB (Tape Automated Bonding) is a semiconductor packaging technique that uses a tiny lead-frame to connect circuitry on the surface of the chip to a substrate instead of wire bonds.

Tailing refers to the *in vitro* addition, to the 3'-hydroxyl ends of a double-stranded DNA molecule, of multiple copies of a single nucleotide by the enzyme terminal transferase. Synonym is homopolymeric tailing.

Tandem array is a series of copies of a gene arranged in tandem along a chromosome. Nucleolar organizers, for example, can contain up to 250 copies of a single ribosomal RNA (rRNA) gene in tandem. Genes for histone proteins also occur in tandem arrays. Such arrays ensure that large amounts of the gene product are synthesized by the cell.

Tandem repeat refers to two (or more) contiguous identical DNA sequences. The orientation can be either head-to-tail, or head-to-head. Synonyms are tandem array, sequence tandem repeat.

Tank bioreactor is a vessel in which fermentation takes place. A tank bioreactor is a vessel in which a micro-organism is grown in a large volume of liquid. This contrasts with fiber or membrane bioreactors and immobilized cell reactors. The large majority of bioreactors used in biotechnology are tank bioreactors, and most tank bioreactors are stirred-tank bioreactors, because stirring helps to distribute effectively gas and nutrients to the growing organism.

Tap root refers to the root system in which the primary root has a much larger diameter than any lateral roots (e.g., carrot). Opposite is fibrous root.

Taq is the bacterium *Thermus aquaticus* from which a heat stable DNA polymerase used in PCR was isolated.

Taq **polymerase** is a heat-stable DNA polymerase isolated from the thermophilic bacterium *Thermus aquaticus,* widely used in PCR.

Target refers to the molecule or nucleic acid sequence assayed in a sample, in diagnostic tests. In mutagenesis, the gene sequence that needs to be altered to generate the desired change in phenotype.

Target site duplication is a sequence of DNA that is duplicated when a transposable element inserts; usually found at each end the insertion.

Targeted drug delivery is a method of delivering the activated form of a drug molecule to the site in the body where it is needed, rather than allowing it reach the target by uncontrolled diffusion.

Targeted therapy or molecularly targeted therapy is a type of medication that blocks the growth of cancer cells by interfering with specific targeted molecules needed for carcinogenesis and tumor growth, rather than by simply interfering with all rapidly dividing cells (e.g., with traditional chemotherapy). Radiotherapy is not considered a 'targeted therapy' despite its often being aimed at the tumors. Targeted cancer therapies are expected to be more effective than current treatments and less harmful to normal cells. There are targeted therapies for breast cancer, multiple myeloma, lymphoma, prostate cancer, melanoma and other cancers.

Targeting vector refers to a cloning vector that carries a DNA sequence capable of participating in a crossing-over event at a specified chromosomal location in the host cell.

TATA box is a widely conserved adenine- and thymine-rich DNA sequence found 25-30 bp upstream of the transcription initiation point of many eukaryotic genes. The TATA box is implicated in the promotion of gene transcription as it acts as a binding site for RNA polymerase. Analogous to the Pribnow box in prokaryotic promoters. Synonym is Hogness box.

Tautomeric shift is the transfer of a hydrogen atom from one position in an organic molecule to another position.

Tautomerism is a type of isomerism in which the two isomers arising from a tautomeric shift are in equilibrium.

Taxoprexin is an investigational drug (from Protarga Inc.) made by linking paclitaxel to docosahexaenoic acid (DHA).

Tay-Sachs disease is a lethal hereditary disease. The progressive accumulation of a substance called ganglioside in the brain causes paralysis, mental deterioration and blindness. Death usually occurs before the age of four.

T-cell-mediated (cellular) immune response refers to the synthesis of antigen-specific T cell receptors and the development of killer T cells in response to an encounter of immune system cells with an unrecognized immunogenic molecule.

T-DNA refers to the **DNA** segment of the Ti plasmid, present in pathogenic *Agrobacterium tumefaciens*, that is transferred to plant cells and inserted into the plant's DNA as part of the infection process.

Telemeter is the unique structure found at the end of eukaryotic chromosomes containing specialized sequences of DNA that assures the completion of a cycle of DNA replication.

Telomerase is an enzyme that adds telomeric sequences to the ends of eukaryotic chromosomes.

Telomere is the end of a chromosome, associated with a characteristic DNA sequence that is replicated in a special way. A telomere counteracts the tendency of the chromosome to shorten with each round of replication.

Telophase (Gr. *telos*, end + phase) is the last stage in each mitotic or meiotic division, in which the chromosomes are assembled at the poles of the division spindle.

TEM is an abbreviation for transmission electron microscope.

Temperate phage is a phage (virus) that invades but does not normally destroy (lyse) the host bacterial cell. Under specific circumstances, the lytic cycle is induced, resulting in the release of infective phage particles.

Temperature is the degree of hotness or coldness of a body or environment. A measure of the average kinetic energy of the particles in a sample of matter, expressed in terms of units or degrees designated on a standard scale.

Temperature-sensitive protein is a protein that is functional at one temperature but loses function at another (usually higher) temperature.

Template is a strand of DNA or RNA (mRNA) that specifies the base sequence of a newly synthesized strand of DNA or RNA, the two strands being complementary.

Template strand refers to the sequence of DNA that is copied during the synthesis of mRNA. The opposite strand (that is, the strand with a base sequence directly corresponding to the mRNA sequence) is called the coding strand or the mRNA-like strand because the sequence corresponds to the codons that are translated into protein.

Tensile strength is the maximum engineering tensile stress, sustainable without fracture; also called "ultimate (tensile) strength."

Teratogens is any agent that raises the incidence of congenital malformations.

Teratoma is a multi-layered benign tumor that grows from pluripotent cells injected into mice with a dysfunctional immune system. Scientists test whether they have established a human embryonic stem cell (hESC) line by injecting putative stem cells into such mice and verifying that the resulting teratomas contain cells derived from all three embryonic germ layers.

Term finalization refers to the repelling movement of the centromeres of bivalents in the diplotene stage of the meiotic prophase, that tends to move the visible chiasmata toward the ends of the bivalents.

Terminal bud is a branch tip, an undeveloped shoot containing rudimentary floral buds or leaves, enclosed within protective bud scales.

Terminal transferase is an enzyme that adds nucleotides to the 3' terminus of DNA molecules.

Terminalization refers to the repelling movement of the centromeres of bivalents in the diplotene stage of the meiotic prophase, that appears to move visible chiasmata toward the ends of the bivalents.

Termination codon (stop codon) is a nucleotide triplet within messenger RNA that signals a termination of translation.

Termination signal is a nucleotide sequence that specifies RNA chain termination.

Terminator refers to: (1) A DNA sequence just downstream of the coding segment of a gene, which is recognized by RNA polymerase as a signal to stop synthesizing mRNA; (2) A term used in GMO technology for a transgenic method which genetically sterilizes the progeny of the planted seed, thereby preventing the use of farm-saved seed.

Terminator codon is a UAA, UAG, or UGA trinucleotide in messenger ribonucleic acid (mRNA) that specifies termination of synthesis of the polypeptide (protein) product of the gene. Also known as stop codon.

Terminator gene is a specific variety-level genetic use restriction technology. It is a patented technique.

Terminator region is a DNA sequence that signals the end of transcription.

Tertiary structure refers to the three-dimensional conformation taken up by complete macromolecules as a result of intramolecular interactions, such as hydrogen-bonding. *See*: primary structure, secondary structure, quaternary structure.

Tesla [T] is a unit of magnetic induction: one T = one weber/ m^2 = 104 gauss.

Test cross refers to backcross to the recessive parental type, or a cross between genetically unknown individuals and a fully recessive tester to determine whether an individual in question is heterozygous or homozygous for a certain allele. It is also used as a test for linkage, i.e., to estimate recombination fraction.

Test tube is a tube in which cells, tissues, etc., can be cultured.

Testis (plural: testes) refers to a male sex organ where spermatozoa mature and are stored.

Testosterone is a male hormone, synthesized in the testis of mammals; used to induce sex reversal in fish.

Test-tube fertilization (*in vitro* fertilization) involves fertilizing an egg outside the body, in a laboratory dish, and then implanting it in a woman's uterus.

Tetracycline is an antibiotic that interferes with protein synthesis in prokaryotes. A gene encoding resistance to tetracycline has been widely used as a marker to distinguish between transformed and non-transformed cells in the production of transgenic plants.

Tetrad refers to: (1) The four cells arising from the second meiotic division in plants (pollen tetrads) or fungi (ascospores); (2) The quadruple group of chromatids that is formed by the association of duplicated homologous chromosomes during synapsis in meiosis I.

Tetraploid complementation assay is an assay that can be used to test a stem cell's potency. Scientists studying mouse chimeras (mixing cells of two different animals) noted that fusing two 8-cell embryos produces cells with 4 sets of chromosomes (tetraploid cells) that are biased toward developing into extra-embryonic tissues such as the placenta. The tetraploid cells do not generate the embryo itself; the embryo proper develops from injected diploid stem cells. This tendency has been exploited to test the potency of a stem cell. Scientists begin with a tetraploid embryo. Next, they inject the stem cells to be tested. If the injected cells are pluripotent, then an embryo develops. If no embryo develops, or if the resultant embryo cannot survive until birth, the scientists conclude that the cells were not truly pluripotent.

Tetrasomic (Noun: tetrasome) pertains to a nucleus or an organism with four members of one of its chromosomes, whereas the remainder of its chromosome complement is diploid. Chromosome formula is $2n + 2$.

Tetratype is a tetrad of spores that contains four different types; e.g., *AB, aB, Ab* and *ab*.

TFT (Thin Film Transistor).

TGGE is an abbreviation for thermal gel gradient

electrophoresis. TGGE and Denaturing Gradient Gel Electrophoresis (DGGE) are forms of electrophoresis which use either a temperature or chemical gradient to denature the sample as it moves across an acrylamide gel. TGGE and DGGE can be applied to nucleic acids such as DNA and RNA, and (less commonly) proteins. TGGE relies on temperature dependent changes in structure to separate nucleic acids. DGGE was the original technique, and TGGE a refinement of it.

Thalassaemia is a hereditary anaemia resulting from reduced production of either alpha or beta haemoglobin. Depending on the type, the condition can be fatal before or just after birth, or can result in varying levels of anaemia and development difficulties.

Thallus (Gr. *thallos*, a sprout) is a plant body without true roots, stems, or leaves.

Therapeutic agent is any chemical substance formulated or compounded as single active ingredient or in combination of other pharmacologically active substance, it may be in a separate but packed in a single unit pack as combination product intended for internal, or external or for use in the medical diagnosis, cure, treatment, or prevention of disease.

Therapeutic cloning is a term used to refer to somatic cell nuclear transfer (SCNT). Embryonic stem cells derived from therapeutic cloning (or SCNT) can then be instructed to form particular cell types, for example heart muscle. If the stem cells are placed back into the individual who gave the DNA for the somatic cell nuclear transfer, these cells are genetically identical and will not be rejected by the donors immune system.

Thermal conductivity (κ) is the proportionality constant between the heat flux and the temperature gradient, for steady-state heat flow.

Thermal expansion coefficient, linear (α) is the fractional change in length divided by the change in temperature.

Thermal fatigue is a type of fatigue failure that introduces the cyclic stresses by fluctuating thermal stresses.

Thermal gel gradient electrophoresis (TGGE) is method for separating DNA fragments according to their mobility under increasingly denaturing conditions imposed by heat.

Thermal refers to heat.

Thermal shock is the fracture of a brittle material resulting from stresses introduced by a rapid temperature change.

Thermal stress is a residual stress introduced within a body resulting from a change in temperature.

Thermistor is a temperature-measuring device, that contains a resistor or semiconductor whose resistance varies with temperature.

Thermocouple is a temperature-measuring device, which contains a pair of end-joined dissimilar conductors in which an electromotive force is developed by thermoelectric effects when the joined ends and the free ends of the conductors are a different temperature.

Thermolabile refers to one not resistant to heat, often in the context of a molecule which is unstable upon heating. Opposite is thermostable.

Thermophile is an organism, which grows at a higher temperature than most other organisms grow. In general, a wide range of bacteria, fungi and simple plants and animals can grow at temperature up to 50°C; thermophiles are considered to be organisms which can grow at above 50°C. They can be classified according to their optimal growth temperature, into: simple thermophiles (50–65°C), thermophiles (65–85°C), and extreme thermophiles (>85°C).

Thermoplastic polymer is a substance that when molded to a certain shape under appropriate conditions can later be remelted.

Thermosensitivity is a loss of biological activity of a molecule at high temperature.

Thermoset polymer is a substance that when molded to a certain shape under pressure and high temperatures. This type of polymer cannot be softened again or dissolved.

Thermostability refers to retention of activity at high temperature.

Thermostable is a molecule which retains its biological activity at some specified higher temperature. Opposite is thermolabile.

Thermotherapy is a technique that is used for virus or mycoplasma elimination. Plants are exposed to elevated temperatures as a treatment. Thermotherapy is used alone or in combination with meristem culture or meristem tip culture.

Theranostics (Rx/Dx) refers to the development of molecular diagnostic tests and targeted therapeutics in an interdependent, collaborative manner with the goals of individualizing treatment by targeting therapy to an individual's' specific disease subtype and genetic profile. This strategy will enable optimization of drug efficacy and safety and will assist in streamlining the drug development process.

Thick skin refers to the hairless skin where the epidermis is much thicker.

Thin skin contains hair except in certain locations; the epidermis is thinner.

Thinning refers to: (1) Removal of older stems to promote new growth; (2) Removal of excess fruits to improve the size and quality of the remaining fruits; (3) Removal of seedlings spaced too closely for optimum growth.

Threshold is the point at which environmental (or genetic) changes produce large phenotypic changes in an organism (or population). For example, there could be a threshold effect of inbreeding on fitness such that after a certain level of inbreeding is reached, individual fitness declines increasingly rapidly.

Threshold character is a phenotypic character that contains a few discrete states that are controlled by many genes underlying continuous variation, which affects a character phenotypically only when a certain physiological threshold is exceeded.

Thrombophilias (protein C or S deficiency, factor V Leiden, sickle cell disease, antiphospholipid antibody) can generate an increased fibrin/fibrinoid deposition in the maternal or intervillous space; this can trap and kill villi.

Thrombosis is a blood clot that forms inside the blood vessel or cavity of the heart.

Thrombus is a blood clot obstructing a blood vessel or a cavity of the heart.

Thymidine kinase (tk) is an enzyme that allows a cell to utilize an alternate metabolic pathway for incorporating thymidine into DNA. It is used as a selectable marker to identify transfected eukaryotic cells.

Thymidine refers to the deoxyribonucleoside resulting from the combination of the base thymine (T) and the sugar 2-deoxy-D-ribose.

Thymidine triphosphate is abbreviated as TTP (dTTP is strictly correct but rarely used). The TTP is one of the four nucleoside triphosphates that are used in the *in vivo* synthesis of DNA.

Thymidylic acid (TMP or dTMP) is synonym for thymidine 5'-monophosphate: a deoxy-ribonucleotide containing the nucleoside thymidine.

Thymine is a pyrimidine base found in DNA. The other three organic bases – adenine, cytosine, and guanine – are found in both RNA and DNA; in RNA, thymine is replaced by uracil.

Thyroid gland is an endocrine gland that lies anterior to the trachea and releases hormones that regulate the rate of metabolism.

Thyroid hormones refer to the hormones, including thyroxin, secreted by the thyroid gland; stimulate rate of metabolism.

Ti plasmid is a tumor-inducing plasmid. A large plasmid present in pathogenic *Agrobacterium tumefaciens*, responsible for the induction of tumors in plant with crown gall disease. Engineered forms of this plasmid are central to the production of transgenics in many crop species.

Tidal volume is a volume of gas inspired or expired during each quiet respiration cycle.

Tiling (Targeting induced local lesions in genomes) is a reverse genetics technique that permits the directed identification of mutations in genes of interest.

Time constant is the time it takes for the output change to reach 63% of its final value.

Tissue is a group of cells of similar structure which sometimes performs a special function.

Tissue culture is the separation of cells from each other and their growth in a container of liquid nutrients.

Tissue engineering (regenerative medicine) is an emerging multidisciplinary field involving biology, medicine, and engineering that is likely to revolutionize the ways we improve the health and quality of life for millions of people worldwide by restoring, maintaining, or enhancing tissue and organ function. In addition to having a therapeutic application, where the tissue is either grown in a patient or outside the patient and transplanted, tissue engineering can have diagnostic applications where the tissue is made *in vitro* and used for testing drug metabolism and uptake, toxicity, and pathogenicity. The foundation of tissue engineering/regenerative medicine for either therapeutic or diagnostic applications is the ability to exploit living cells in a variety of ways. Tissue engineering research includes (1) Biomaterials: including novel biomaterials that are designed to direct the organization, growth, and differentiation of cells in the process of forming functional tissue by providing both physical and chemical cues; (2) Cells: including enabling methodologies for the proliferation and differentiation of cells, acquiring the appropriate source of cells such as autologous cells, allogeneic cells, xenogeneic cells, stem cells, genetically engineered cells, and immunological manipulation; (3) Biomolecules: including angiogenic factors, growth factors, differentiation factors and bone morphogenic proteins; (4) Engineering Design Aspects: including 2-d cell expansion, 3-d tissue growth, bioreactors, vascularization, cell and tissue storage and shipping (biological packaging); (5) Biomechanical Aspects of Design: including properties of native tissues, identification of minimum properties required of engineered tissues, mechanical signals regulating engineered tissues, and efficacy and safety of engineered tissues; (6) Informatics to support tissue engineering: gene and protein sequencing, gene expression analysis, protein expression and interaction analysis, quantitative cellular image analysis, quantitative tissue analysis, *in silico* tissue and cell modeling, digital tissue manufacturing, automated quality assurance systems, data mining tools, and clinical informatics interfaces; (7) Stem cell research involves stem cells, whether from embryonic, fetal, or adult sources, human and non-human. It should include research in which stem cells are isolated, derived or cultured for purposes such as developing cell or tissue therapies, studying cellular differentiation, research to understand the factors necessary to direct cell specialization to specific pathways, and other developmental studies. It should not include transgenic studies, gene knock-out studies nor the generation of chimeric animals.

Titre refers to: (1) The concentration of infectious virus particles present in a suspension; (2) A measure of antibody concentration, given by the highest dilution of the sample that results either in a useable immunoassay, or in the formation of visible precipitate when challenged by the appropriate antigen.

tk is an abbreviation for thymidine kinase.

TMP is an abbreviation for the deoxyribonucleotide thymidine 5'-monophosphate.

TMRCA is time since the most recent common ancestor.

Tolerance refers to an incomplete resistance to a given biotic or abiotic stress. Tolerant genotypes are less inhibited by the stress, but are not immune.

Tonoplast (Gr. *tonos*, stretching tension + *plastos*, moulded, formed) is the cytoplasmic membrane bordering the vacuole, with a role in regulating the pressure exerted by the cell sap.

Top–down nanotechnology refers to when engineers taking existing devices, such as transistors, and making them smaller. Top–down or mechanical nanotechnology will have the greatest impact on our everyday lives in the near future.

Topo-isomerase is an enzyme that regulates the overwinding or underwinding of DNA. The winding problem of DNA arises due to the intertwined nature of its double-helical structure. For example, during DNA replication, DNA becomes overwound ahead of a replication fork. If left unabated, this tension would eventually halt DNA replication. (A similar event happens during transcription.) In order to help overcome these types of topological problems caused by the double helix, topoisomerases bind to either single-stranded or double-stranded DNA and cut the phosphate backbone of the DNA. This intermediate break allows the DNA to be untangled or unwound, and, at the end of these processes, the DNA backbone is resealed again. Since the overall chemical composition and connectivity of the DNA do not change, the tangled and untangled DNAs are chemical isomers, differing only in their global topology, thus their name. Topoisomerases are isomerase enzymes that act on the topology of DNA.

Torr is an obsolete unit of pressure equal to that exerted by a column of mercury 1 mm high at 0°C and standard gravity (1 mm Hg); named after Evangelista Torricelli (1608–1647), the inventor of the mercury barometer. 1 Torr = 1/760 atm = 133.322 Pa.

Total lung capacity is a total volume of air that the lungs can hold. TLC = VC + RV.

Total parenteral nutrition (Also see parenteral nutrition).

Totipotency refers to having the potentiality of forming all the types of cells in the body. The property of somatic cells to be induced to undergo regeneration. The diploid zygote formed at fertilization is a single cell which is capable of division and differentiation to give rise to the total range of cell types found in the adult organism.

Totipotent is the state of a cell that is capable of giving rise to all types of differentiated cells found in an organism, as well as the supporting extra-embryonic structures of the placenta. A single totipotent cell could, by division in utero, reproduce the whole organism.

Totipotent cell (totipotent nucleus) is an undifferentiated cell (or nucleus), such as a blastomere, that, when isolated or suitably transplanted, can develop into a complete embryo.

Totipotent stem cells are bone marrow cells that (when signaled) mature into both red blood cells and white blood cells. Receptors on the surface of totipotent stem cells "grasp" passing blood cell growth factors (for example, Interleukin-7, Stem Cell Growth Factor), bringing them inside these stem cells and thus causing the maturation and differentiation into red and white blood cells. These receptors are called FLK-Z receptors.

Toughness is a measure of the amount of energy absorbed by a material as it fractures, indicated by the total area under the material's tensile stress-strain curve.

Toxic refers to poisonous.

Toxic Substances Control Act (TSCA) is a United States law, passed by the United States Congress in October 11 of 1976, that regulates the introduction of new or already existing chemicals. It grandfathered most existing chemicals, in contrast to the Registration, Evaluation and Authorization of Chemicals (REACh) legislation of the European Union. The TSCA specifically regulates polychlorinated biphenyl (PCB) products. Contrary to what the name implies, TSCA does not separate chemicals into categories of toxic and non-toxic. Rather it prohibits the manufacture or importation of chemicals that are not on the TSCA Inventory (or subject to one of many exemptions). Chemicals that are listed on the TSCA Inventory are referred to as "existing chemicals."

Chemicals not listed are referred to as new chemicals. Major amendments are: P.L. 99-519 (1986); P.L. 100–551 (1988); P.L. 101-637 (1990); P.L. 102–550 (1992). Since May 22, 2013, Senate Bill 1009 has been pending in Congress to reform TSCA, entitled the Chemical Safety Improvement Act. This would be the first major overhaul in many years.

Toxicity refers to negative effect of a compound, as shown by altered morphology or physiology. It is meaningful only when the effect itself is also described, such as changes in the rate of cell growth, cell death, etc.

Toxicogenomics is a fusion of genomics and toxicology disciplines intended to identify, classify and manage the latent (inherent susceptibility), incipient and overt adverse (toxic) effects on genome structure and expression levels (RNA, protein, cell/tissue/organ type) as a consequence of an organism's exposure to environmental substances (contaminants such as chemicals, drugs and micro/multicellular organisms and/or components) and stressors (for example, quality of air, climate, soil, solar radiation and water).

Toxin (L. *toxicum*, poison) is a compound produced by an organism and poisonous to plants or animals.

TPF (Tiling path file) is a simple file that simply lists the order of clones along a chromosome. These files are often used in genome assemblies in an effort to convey mapping information to the assembly program.

Tracer is a substance (typically a radioactive isotope or a fluorescent dye) that can be detected by physical means, and which is used to analyse the progress of a chemical reaction or a biological process.

Trachea is the main trunk of the system of tubes by which air passes to and from the lungs.

Tracheid (Gr. *tracheia*, windpipe) is an elongated, tapering xylem cell, with lignified pitted walls, and adapted for conduction and support. Found in conifers, ferns and related plants.

Tracheoesophageal fistula is caused by improper development of the baby's trachea (windpipe) and esophagus during pregnancy. The esophagus does not connect to the stomach, and there is also an abnormal connection between the esophagus and the trachea. food cannot pass through to the stomach, and may pass into the trachea and then into the lungs, causing breathing problems.

TRAFFIC is a wildlife trade monitoring network sponsored by the WWF and IUCN.

Trait is one of the many characteristics that define an organism. The phenotype is a description of one or more traits. Synonym is character.

Trans acting describes substances that are diffusible and that can affect spatially separated entities within cells.

Trans **acting factor** is any of the multiple ancillary DNA-binding proteins that interact with the *cis*-regulatory DNA sequences to control gene expression.

Trans and *cis* are from Latin, in which *cis* means "on the same side" and *trans* means "on the other side" or "across."

Trans capsidation is the partial or full coating of the nucleic acid of one virus with a coat protein of a differing virus.

Trans **configuration** is an arrangement in which at least one mutant gene and one wild-type gene of a pair of pseudoalleles are present on each chromosome of a homologous pair. Also called trans arrangement, trans position. Related term is cis-configuration.

Trans **heterozygote** is a double heterozygote that contains two mutations arranged in the *trans* configuration.

Trans **test** is a complementation test with two or more interacting genes placed in cis and in trans relationships to each other. A double mutant genome is used in the cis test made from the two single mutant genomes used in the trans test by recombination. If the wild type phenotype is restored by both cis and trans arrangements it is concluded that the two mutations are in different genes and hence that the phenotype is determined by more than one gene. If the trans test is negative and the cis positive this means that the two mutations are in the same gene. If both tests are negative then at least one of the mutations must be dominant. Thus the double test provides a means of fine mapping of genes.

Transcript is a RNA molecule that has been synthesized from a specific DNA template. In eukaryotes, the primary transcript produced by RNA polymerase must often be processed or modified in order to form the mature, functional mRNA, rRNA or tRNA.

Transcription factor is a protein that regulates the transcription of genes.

Transcription is a process in the cell where the DNA is used as a template to make the messenger RNA.

Transcription unit is a segment of DNA that contains signals for the initiation and termination of transcription, and is transcribed into one RNA molecule.

Transcription vector is a cloning vector that allows the foreign gene or DNA sequence to be transcribed *in vitro.*

Transcriptional anti-terminator is a protein that prevents RNA polymerase from terminating transcription at specific transcription termination sequences.

Transcriptional roadblock is a DNA-binding protein which affects the rate at which RNA polymerases transcribe genes. The protein/DNA complex interferes with the passage of the elongation complex. In some cases these obstacles are readily bypassed, but in others a significant level of pausing or termination occurs, and this can then act as a control point for gene expression.

Transcutaneous electrical nerve stimulation (TENS) is a method of providing pain relief using electrical signals which are sent to the nerve endings.

Transdermal is a route of administration wherein active ingredients are delivered across the skin for systemic distribution.

Transdermal drug delivery is the transport of drug through the skin.

Transdermal patch is a medicated adhesive patch that is placed on the skin to deliver a specific dose of medication through the skin.

Transdifferentiation is the ability of a particular cell of one tissue, organ or system, including stem or progenitor cells, to differentiate into a cell type characteristic of another tissue, organ, or system; e.g., blood stem cells changing to liver cells.

Transducer is a device that changes energy from one type to another.

Transducing phage is a transducing phage is a phage that can package host DNA into its progeny, instead of only the phage DNA.

Transduction (t) is the transfer of DNA sequences from one bacterium to another via lysogenic infection by a bacteriophage (transducing phage). Genetic recombination in bacteria mediated by bacteriophage. Abortive t: Bacterial DNA is injected by a phage into a bacterium, but unable to replicate.

Transduction mode (direct or indirect) defines how the sensor acquires the desired information from the material. In general, it is an indication of the ability of the sensor signal to provide information regarding a material property or state of interest.

Transfection is the transfer of DNA to an eukaryotic cell.

Transfer is the process of moving cultured tissue or cells to a fresh medium.

Transfer RNA (tRNA) is an adaptor molecule composed of RNA, typically 73 to 94 nucleotides in length, that serves as the physical link between the nucleotide sequence of nucleic acids (DNA and RNA) and the amino acid sequence of proteins. It does this by carrying an amino acid to the protein synthetic machinery of a cell (ribosome) as directed by a three-nucleotide sequence (codon) in a messenger RNA (mRNA). As such, tRNAs are a necessary component of protein translation, the biological synthesis of new proteins according to the genetic code.

Transferase is a class of enzymes that catalyzes the transfer of a

group of atoms from one molecule to another.

Transferrins are iron-binding blood plasma glycoproteins that control the level of free iron in biological fluids. Human transferrin is encoded by the *TF* gene. Transferrin glycoproteins bind iron very tightly, but reversibly.

Transformant refers to a cell in prokaryotes that has been genetically altered through the uptake of foreign DNA. In higher eukaryotes, a cultured cell that has acquired a malignant phenotype.

Transformation refers to the natural or induced uptake and expression of a foreign DNA sequence in prokaryotes – typically a recombinant plasmid in experimental systems. In higher eukaryotes, the conversion of cultured cells to a malignant phenotype–typically through infection by a tumor virus or transfection with an oncogene.

Transformation efficiency or frequency is the fraction of a cell population that takes up and integrates the introduced transgene; expressed as the number of transformed cells recovered divided by the total number of cells in a population.

Transformation efficiency is the number of cells that take up foreign DNA as a function of the amount of added DNA; expressed as transformants per microgram of added DNA.

Transformer is a device using magnetically linked inductors to change AC voltage level.

Transforming oncogene is a gene that upon transfection converts a previously immortalized cell to the malignant phenotype.

Transgene is a gene from one genome that has been incorporated into the genome of another organism. Often refers to a gene that has been introduced into a multicellular organism.

Transgenesis is the introduction of exogenous DNA into a cell. Typically, this term refers to the introduction of a gene into an embryo or other eukaryotic cell. In general, this DNA will insert into the genome at random, although specific loci can be targeted. The size of the DNA molecule introduce can be small (a few basepairs) to quite large (over 100 Kb).

Transgenic is an organism in which a foreign DNA gene (a transgene) is incorporated into its genome early in development. The transgene is present in both somatic and germ cells, is expressed in one or more tissues, and is inherited by offspring in a Mendelian fashion.

Transgenic animal refers to a genetically engineered animal or offspring of genetically engineered animals. The transgenic animal usually contains material from at lease one unrelated organism, such as from a virus, plant, or other animal.

Transgenic organism refers to one into which a cloned genetic material has been experimentally transferred, a subset of these foreign gene express themselves in their offspring. Turner syndrome a chromosomal condition in females (usually 45, XO) due to monosomy of the X-chromosome; characterized by short stature, failure to develop secondary sex characteristics, and infertility.

Transgenic plant is a genetically engineered plant or offspring of genetically engineered plants. The transgenic plant usually contains material from at least one unrelated organisms, such as from a virus, animal, or other plant.

Transgenics is the insertion or splicing of specific genetic sequences from one species into the functioning genome of an unrelated species to transfer desired properties for human purposes. This may be viewed as a more precise form of hybridization or plant/animal breeding, with the added consideration that genetic material from species significantly different from one another is involved (for example, the insertion of genetic material from an animal into a plant or vice versa). Another possibility is the transfer of genetically controlled properties between different animal species, such as the breeding of goats whose milk yields spider silk for possible development of new structural materials.

Transgressive segregation are hybridization events that produce progeny that express phenotypic values outside the range of either parental phenotypic value. These differences are usually due to the disruption of polygenic traits.

Transgressive variation refers to the appearance in the F_2 (or later) generation of individuals showing more extreme development of a trait than either of the original parents.

Transient refers to that is of short duration.

Transient expression refers to short-term activity of a transgene following its introduction into target tissue. Transient expression usually implies non-integration of the transgene into the host genome.

Transient response is the response of the sensor to a step change in the measurand.

Transistor is semiconductor device used for amplification and switching.

Transition is the more common single nucleotide mutation (or polymorphism) that results from a point mutation in which a purine is substituted with a purine or a pyrimidine is substituted with a pyrimidine.

Transition stage is the integration period of juvenile and reproductive stages of growth.

Transition-state intermediate is an unstable and high-energy configuration assumed by reactants on the way to making products. Enzymes are thought to bind and stabilize the transition state, thus lowering the energy of activation needed to drive the reaction to completion.

Translation is the process of converting the genetic information of an mRNA on ribosomes into a polypeptide. Transfer RNA molecules carry the appropriate amino acids to the ribosome, where they are joined by peptide bonds.

Translational initiation signal (Translational start codon)

concerns both the ATG initiation codon and the sequences flanking the initiation codon are required to direct the position of initiation. In eukaryotes, this is important because not all of the mRNA codes – for the polypeptide – are untranslated regions on mRNA (UTRs). Translating the non-coding regions would produce non-functional polypeptides. Prokaryotes are a whole different bag altogether, and don't have precise discrete start and stop codons per se (Shine-Dalgarno sequences, attenuators and rho-dependant termination etc. instead), as well as producing polycistronic mRNA (which is not really relevant). Both eukaryotes and prokaryotes produce multiple polypeptide chains from one strand of mRNA though, but this doesn't have much to with why stop and start codons are important, and I feel C is more relevant.

Translational medicine is the area of focus or effort to transition basic research discoveries into clinical applications that benefit patients.

Translational stop signal is defined in the genetic code as UAA, UAG and UGA, although the mechanism of their decoding via protein factors is clearly different from that of the other codons. There are strong biases in the upstream and downstream nucleotides surrounding stop codons. Experimental tests have shown that termination-signal strength is strongly influenced by the identity of the nucleotide immediately downstream of the codon (+4), with a correlation between the strength of this four-base signal and its occurrence at termination sites. The +4 nucleotide and other biases downstream of the stop codon may reflect sites of contact between the release factor and the mRNA, whereas upstream biases may be due to coding restrictions, with the release factor perhaps recognizing the final tRNA and the last two amino acids of the polypeptide undergoing synthesis.

Translocation refers to: (1) The movement of individuals from one population (or location) to another that is usually intended to achieve either genetic or demographic rescue of an isolated population; (2) A rearrangement occurring when a piece of one chromosome is broken off and joined to another chromosome.

Transmission electron microscope (TEM) is a microscope that produces an image by using electron beams to transmit (pass through) the specimen, making examination of internal features at high magnifications possible.

Transmission refers to system for carrying electric power at voltages above 100,000 volts.

Transonic is a flow with speed at velocities just below and above the speed of sound is said to be transonic.

Transplant refers to: (1) noun: A plant grown in a cold frame, greenhouse, tissue culture or indoors for later planting outdoors. (2) To dig up and move a plant to another location.

Transplantation biology is the science that studies the transplantation of organs and cells. Transplantation biologists investigate scientific questions to understand why foreign tissues and organs are rejected, the way transplanted organs function in the recipient, how this function can be maintained or improved, and how the organ to be transplanted should be handled to obtain optimal results.

Transplantation is the implanting of cells, tissues, or organs which have been retrieved from a living or deceased donor into a recipient.

Transposable element (TE, transposon or **retrotransposon**) is a DNA sequence that can change its position within the genome, sometimes creating or reversing mutations and altering the cell's genome size. Transposition often results in duplication of the TE. Barbara McClintock's discovery of these jumping genes earned her a Nobel prize in 1983.

Transposable genetic element is a DNA element that can move from one location in the genome to another.

Transposase is an enzyme encoded by a transposon gene and that catalyzes the movement of a DNA sequence to a different site in a DNA molecule, by catalyzing the excision of the transposon from one site and its insertion into a new chromosomal site.

Transposition is the process whereby a transposon or insertion sequence inserts itself into a new site on the same or another DNA molecule. The exact mechanism is not fully understood and different transposons may transpose by different mechanisms. Transposition in bacteria does not require extensive DNA homology between the transposon and the target DNA.

Transposon is a synonym of transposable genetic element. It is a sequence of DNA that can move to new positions within the genome of a single cell. Transpoon is also called jumping gene, but it is not correct to call them 'genes.' Transposons were first found by Barbara McClintock while working on maize. She received a Nobel Prize for her work in 1983.

Transposon tagging is a method of gene isolation that exploits the disruption of normal gene expression that is the result of an insertion of a transposon within, or close to the target. Since the sequence of the transposon is known, this can be used as a DNA probe to define the DNA fragment containing the target gene. Large-scale experiments to generate populations of gene mutations are colloquially referred to as gene machines.

Transrectal ultrasound of the prostate is a test using sound wave echoes to create an image of an organ or gland to visually inspect for abnormal conditions like gland enlargement, nodules, penetration of tumor through capsule of the gland and/or invasion of seminal vesicles. It may also be used for guidance of needle biopsies of the prostate gland and guiding the nitrogen probes in cryosurgery.

Transurethral hyperthermia is an investigative procedure that uses heat, usually provided by microwaves, to shrink the prostate.

Transurethral incision of the prostate (TUIP) is a procedure that widens the urethra by making some small cuts in the bladder neck, where the urethra joins the bladder, and in the prostate gland itself.

Transurethral laser incision of the prostate (TULIP) refers to the use of laser through the urethra that melts the tissue.

Transurethral resection of the prostate (TURP) is a surgical procedure by which portions of the prostate gland are removed through the penis.

Transurethral surgery refers to an operation in which no external incision is needed. For prostate transurethral surgery, the surgeon reaches the prostate by inserting an instrument through the urethra.

Transverse colon is a part of the intestine that lies horizontally in the abdomen, running straight across the abdomen from right to left.

Transversion is the substitution in DNA or RNA of one purine by a pyrimidine or *vice versa*. It can only be reversed by a spontaneous reversion. Because this type of mutation changes the chemical structure dramatically, the consequences of this change tend to be more drastic than those of transitions. Transversions can be caused by ionizing radiation and alkylating agents.

Tribology is the science and technology of two interacting surfaces in relative motion and of related subjects and practices. The popular equivalent is friction, wear, and lubrication in surfaces sliding against each other, as in bearing and gears.

Tribrid protein is a fusion protein that has three segments, each encoded by parts of different genes.

Trichome (Gr. *trichoma*, a growth of hair) is a short filament of cells.

Tricuspid aortic valve is the normal aortic valve has three cusps or leaflets, and is therefore called tricuspid (as opposed to a bicuspid aortic valve). The valve connects the right atrium to the right ventricle.

Tri-hybrid refers to the hybrid offspring of a cross between parents carrying contrasting alleles at three loci.

Tri-nucleotide repeats are tandem repeats of three nucleotides that are present in many genes. In several cases, these trinucleotide repeats have undergone expansions in copy number, and that has resulted in inherited diseases.

Tripartite mating is a process in which conjugation is used to transfer a plasmid vector to a target cell when the plasmid vector is not self-mobilizable. When (1) cells that have a plasmid with conjugative and mobilizing functions are mixed with (2) cells that carry the plasmid vector and (3) target cells, mobilizing plasmids enter the cells with the plasmid vector and mobilize the plasmid vector to enter into the target cells. Following tripartite mating, the target cells with the plasmid vector are separated from the other cell types in the mixture by various selection procedures.

Triplet code is a code in which a given amino acid is specified by a set of three nucleotides.

Triplet is a sequence of three nucleotides of DNA which specifies an amino acid. The elucidation of the genetic code involved the binding of charged tRNA species to chemically synthesized ribonucleotide triplets.

Triploid is a cell or organism containing three times the haploid number of chromosomes.

Trisomy is the presence of an extra chromosome, in addition to the normal pair. In humans, this would result in a total of 47 chromosomes. An example of trisomy is trisomy 21, which is also known as Down syndrome.

Triticale refers to the hybrid man-made species formed by the crossing of tetraploid or hexaploid wheat with diploid rye.

tRNA (transfer RNA) is a class of small RNA molecules that transfer amino acids to the ribosome during protein synthesis. Transfer RNA molecules are folded into a 'clover-leaf' secondary structure by intrastrand base pairing. The anticodon loop contains a nucleotide-triplet complementary to a specific codon within the mRNA molecule. Each tRNA is 'charged' with the correct amino acid molecule, via its 3' adenosine moiety, by an enzyme called aminoacyl-tRNA synthetase.

Trophoblast is a process when the tissue of the developing embryo responsible for implantation and formation of the placenta. In contrast to embryonic stem cells, the trophoblast does not come from the inner cell mass, but from cells surrounding it.

Tropism (Gr. *trope*, a turning) is an involuntary plant response to a stimulus, in which a bending, turning or growth occurs, such as phototropism, geotropism or hydrotropism. The response may be positive (towards) or negative (away from) to the stimulus.

True-to-type refers to a plant or propagation source, this term denotes correct cultivar identification and lack of variation in productivity or performance. Verification is determined visually by an expert or through biochemical, serological or other means.

Trypsin inhibitor refers to substances inactivating trypsin, typically found in seed tissue of certain plants, where they are thought to have evolved as antifeedant agents against insect predators.

Trypsin refers to a proteolytic enzyme that hydrolyzes peptide bonds on the carboxyl side of the amino acids arginine and lysine.

TSCA is an abbreviation for the Toxic Substances Control Act.

TTP is an abbreviation for thymidine 5'-triphosphate. TTP is required for DNA synthesis since it is a direct precursor molecule.

Tuber are food-storing modified roots in plants like potato.

Tubulin refers to the major protein component of the microtubules of eukaryotic cells. It is one of several members of a small family of globular proteins. The tubulin superfamily includes five distinct families: the alpha-, beta-, gamma-, delta-, and epsilon-tubulins; and a sixth family (zeta-tubulin) which is present only in kinetoplastid protozoa. The most common members of the tubulin family are α-tubulin and β-tubulin, the proteins that make up microtubules. Each has a molecular weight of approximately 55,000 Daltons. Microtubules are assembled from dimers of α- and β-tubulin. These subunits are slightly acidic with an isoelectric point between 5.2 and 5.8. Tubulin was long thought to be specific to eukaryotes. Recently, however, the prokaryotic cell division protein FtsZ was shown to be related to tubulin.

Tumble tube is a glass tube mainly used *in vitro* to agitate and consequently aerate suspension cultures. The tube, which is commonly attached to a slowly revolving platform, is closed at both ends, with a side-neck opening.

Tumor is an abnormal benign or malignant mass of tissue that is not inflammatory, arises without obvious cause from cells of preexistent tissue, and possesses no physiological function.

Tumor suppressor gene refers to genes that normally function to restrain the growth of tumors; the best understood case is for hereditary retinoblastoma.

Tumor virus is a virus capable of transforming a cell to a malignant phenotype.

Tumor-inducing plasmid (Ti plasmid) are double stranded circular DNA present in *Agrobacterium tumefaciens*. Tumor-suppressor gene is a gene that regulates cell growth. If such a gene becomes dysfunctional, and potentiating damage occurs to the cell, then uncontrolled growth and a cancer may result.

Tunica refers to the outer one- to four-cell layer region of the apical meristem, where cell division is anticlinal, i.e., perpendicular to the surface.

Tunica vaginalis is a thin pouch that holds the testes within the scrotum.

Turbidostat is an open continuous culture in which a pre-selected biomass density is uniformly maintained by automatic removal of excess cells. The fresh medium flows in response to an increase in the turbidity (usually cell density) of the culture.

Turbulent is a flow field that cannot be described with streamlines in the absolute sense. However, time-averaged streamlines can be defined to describe the average behavior of the flow. In turbulent flow, the inertia stresses dominate over the viscous stresses, leading to small-scale chaotic behavior in the fluid motion.

Turgid (L. *turgidus*, swollen, inflated) refers to one that is swollen, distended; a cell that is firm due to water uptake.

Turgor potential (pressure potential) refers to that component of the water potential due to the hydrostatic pressure; equal to the turgor pressure. An important component in turgid cells and in the xylem.

Turgor pressure (L. *turgor*, a swelling) is the pressure within the cell resulting from the absorption of water into the vacuole and the imbibition of water by the protoplasm.

Turion refers to an underground bud or shoot from which an aerial stem arises.

Turn-on-voltage is an applied voltage required to produce conduction in a diode.

Twin refers to one of two individuals originating from the same zygote.

Type I statistical error is the probability of rejecting a true null hypothesis. Usually chosen, by convention, to be 0.05 or 0.01.

Type II statistical error is the probability of accepting a false null hypothesis.

U

U is an abbreviation for uracil. Uracil (juərəsıl) is one of the four nucleobases in the nucleic acid of RNA that are represented by the letters A, G, C and U. The others are adenine (A), cytosine (C), and guanine (G). In RNA, uracil binds to adenine via two hydrogen bonds. In DNA, the uracil nucleobase is replaced by thymine. Uracil could be considered a demethylated form of thymine. Uracil is a common and naturally occurring pyrimidine derivative. It is a planar, unsaturated compound that has the ability to absorb light.

U.S. Department of Agriculture is the U.S. agency responsible for regulation of biotechnology products in plants and animals. The major laws under which the agency has regulatory powers include the Federal Plant Pest Act (PPA), the Federal Seed Act, and the Plant Variety Act (PVA). In addition, the Science and Education (S&E) division has nonregulatory oversight of research activities that the agency funds.

Ubiquitin is a small protein, present in all eukaryotic cells, which plays an important role in tagging proteins destined for proteolytic cleavage (because they are damaged or no longer needed).

Ulcer is a sore in the lining of the digestive tract.

Ulcerative colitis is a disease that causes irritation and ulcers in the lining of the large intestine and rectum. Also known as Inflammatory Bowel Disease.

ULSI (Ultra large scale integration) is a chip with over 1,000,000 components.

Ultrasensitivity describes an output response that is more sensitive to stimulus change than the hyperbolic Michaelis-Menten response. Ultrasensitivity is one of the Biochemical switches in the cell cycle and has been implicated in a number of important cellular events, including exiting G2 cell cycle arrests in *Xenopus laevis* oocytes, a stage to which the cell or organism would not want to return.

Ultrasonication (Sonication) is the act of applying sound energy to agitate particles in a sample, for various purposes. Ultrasonic frequencies (>20 kHz) are usually used, leading to the process also being known as ultra-sonication. In the laboratory, it is usually applied using an *ultrasonic bath* or an *ultrasonic probe*, colloquially known as a *sonicator*. In a paper machine, an ultrasonic foil can distribute cellulose fibers more uniformly and strengthen the paper.

Ultrasound (Also called sonography) is a diagnostic imaging technique which uses high-frequency sound waves and a computer to create images of blood vessels, tissues, and organs. Ultrasounds are used to view internal organs as they function, and to assess blood flow through various vessels.

Ultrasound imaging is a technique in which high frequency sound waves are used to provide an image (sonogram) of an internal

Ultraviolet is the portion of the electromagnetic spectrum associated with chemical changes, located adjacent to the violet portion of the visible light spectrum.

Ultraviolet light or ultraviolet radiation (UV) is the portion of the electromagnetic spectrum with wavelengths from about 100 to 400 nm; between ionizing radiation (X-rays) and visible light. UV is absorbed by DNA and is highly mutagenic to unicellular organisms and to the epidermal cells of multicellular organisms. UV light is used in tissue culture for its mutagenic and bactericidal properties.

Umbilical cord stem cells are hematopoietic stem cells that are present in the blood of the umbilical cord during and shortly after delivery. These stem cells are in the blood at the time of delivery, because they move from the liver, where blood-formation takes place during fetal life, to the bone marrow, where blood is made

after birth. Umbilical cord stem cells are similar to stem cells that reside in bone marrow, and can be used for the treatment of leukemia and other diseases of the blood. Efforts are now being undertaken to collect these cells and store them in freezers for later use. However, one problem is that there may not be enough umbilical cord stem cells in any one sample to transplant into an adult.

UMP is an abbreviation for the (ribo)nucleotide uridine 5'-monophosphate (also known as 5'-uridylic acid) is a nucleotide that is used as a monomer in RNA. It is an ester of phosphoric acid with the nucleoside uridine. UMP consists of the phosphate group, the pentose sugar ribose, and the nucleobase uracil; hence, it is a ribonucleoside monophosphate. Another common shorthand for the molecule is uridylate – the deprotonated form of the molecule, which is predominant in aqueous solution. As a substituent it takes the form of the prefix uridylyl-. The deoxy form is abbreviated dUMP.

Undefined is a medium or substance added to medium in which not all of the constituents or their concentrations are chemically defined, such as media containing coconut milk, malt extract, casein hydrolysate, fish emulsion or other complex compounds. *cf* organic complex.

Understock refers to the host plant for a grafted scion, a branch or shoot from another plant; an understock may be a fully grown tree or a stump with a living root system.

Undifferentiated cells are cells which have not been committed to become part of a specialized tissue.

Unencapsidated is a virus not enclosed by a coat protein or capsid.

Unequal crossing over refers to crossing over between repeated DNA sequences that have paired out of register, creating duplicated and deficient products.

UNESCO refers to United National Educational, Scientific, and Cultural Organization.

Unicellular refers to tissues, organs or organisms consisting of a single cell.

Uniform Flow follows when any time derivative vanishes.

Uniparental inheritance refers to the inheritance of genes exclusively from one parent, e.g., chloroplast DNA is inherited either maternally (many angiosperms) or paternally (most gymnosperms).

Unipotent stem cells are stem cells that self-renew as well as give rise to a single mature cell type; e.g., spermatogenic stem cells.

Unisexual refers to higher organisms (animals or plants) possessing either male or female reproductive organs, but not both.

Unit cell is the basic structural unit of a crystal structure, defined in terms of atom (or ion) positions within a parallelepiped volume.

Units are standards for measurement of physical quantities that need clear definitions to be useful. We have S.I. system of units, English system of units, c.g.s. system of units, m.k.s. system of units.

Univalent is an unpaired chromosome at the first division of meiosis.

Universal constructor is a machine that can replicate itself and – in addition – make other industrial products.

Universal donor cells are cells that, after introduction into a recipient, will not induce an immune response that leads to their rejection.

Universality refers to the genetic code, the triplet codons are translated to the same amino acid, with minor exceptions, in virtually all species.

Unorganized growth refers to *in vitro* formation of tissues with few differentiated cell types and no recognizable structure; many call are unorganized.

UNOS is an abbreviation for United Network for Organ Sharing.

Unspecialized refers to having no specific function.

Upper GI series is a test that looks at the organs of the upper part of the digestive system: the esophagus, stomach, and duodenum (upper small intestine). A liquid that shows up well on x-rays called barium is swallowed. X-rays are then taken to evaluate the digestive organs.

Upstream refers to: (1) **In molecular biology**, the stretch of DNA base pairs that lie in the 5′ direction from the site of initiation of transcription. Usually the first transcribed base is designated +1 and the upstream nucleotides are marked with minus signs, e.g., −1, −10. Also, to the 5′ side of a particular gene or sequence of nucleotides; (2) In **chemical engineering**, those phases of a manufacturing process that precede the biotransformation step. Refers to the preparation of raw materials for a fermentation process. Also called upstream processing.

Upstream processing is a specific bioprocess that uses complete living cells or their components (e.g., bacteria, enzymes, chloroplasts) to obtain desired products. The upstream stage of the production process involves searching for and extracting raw materials. The upstream part of the production process does not do anything with the material itself, such as processing the material. This part of the process simply finds and extracts the raw material.

Uracil is a pyrimidine base found in RNA but not in DNA. In DNA, uracil is replaced by thymine.

Urea breath test is a test that measures the amount of urease in the breath, which is an enzyme that the bacteria Helicobacter pylori makes. This helps diagnose H. pylori infection, which can help determine the cause for ulcers in the digestive tract.

Urea is the nitrogen part of urine produced from the breakdown of protein.

Ureter is one of the paired tubular structures that conducts urine from the kidney to the bladder.

Ureterocele is the portion of the ureter closest to the bladder becomes enlarged because the ureter opening is very tiny and obstructs urine outflow; urine backs up in the ureter tube.

Ureteroscope is an optical device which is inserted into the urethra and passed up through the bladder to the ureter; to inspect the opening of the ureters.

Urethra is the tube that conducts urine from the bladder to the outside of the body.

Urethritis is an infection limited to the urethra.

Urge incontinence refers to the inability to hold urine long enough to reach a restroom. It is often found in people who have conditions such as diabetes, stroke, dementia, Parkinson's disease, and multiple sclerosis, but may be an indication of other diseases or conditions that would also warrant medical attention.

Uridine refers to the (ribo)nucleoside resulting from the combination of the base uracil (U) and the sugar D-ribose.

Uridine triphosphate (uridine 5'-triphosphate, UTP) is required for RNA synthesis since it is a direct precursor molecule.

Uridylic acid is synonym for uridine 5'-monophosphate (abbreviation: UMP), a (ribo) nucleotide containing the base uracil.

Urinalysis refers to laboratory examination of urine for various cells and chemicals, such as red blood cells, white blood cells, infection, or excessive protein.

Urinary bladder is an organ that receives urine from the ureters and temporarily stores it.

Urinary incontinence is the loss of bladder control.

Urinary system is a part of body system that consists of kidneys, urinary bladder, and associated ducts.

Urinary tract infection (UTI) is an infection that occurs in the urinary tract, often caused by bacteria such as *Escherichia coli*. A UTI often causes frequent urination, pain and burning when urinating, and blood in the urine.

Urine flow test measures how quickly the urine is flowing. A reduced flow may suggest benign prostatic hyperplasia (BPH).

URL (Universal Resource Locator) addresses a World Wide Web site.

Urogenital refers to the urinary and reproductive systems.

Urology is the branch of medicine concerned with the urinary tract in both genders, and with the genital tract or reproductive system in the male.

USDA is an abbreviation for The U.S. Department of Agriculture.

Usenet are interlinked bulletin boards available via Internet and commercial on-line services.

Uterine wall is a wall of the uterus.

Uterus (Also called the womb) is a hollow, pear-shaped organ located in a woman's lower abdomen, between the bladder and the rectum, that sheds its lining each month during menstruation and in which a fertilized egg (ovum) becomes implanted and the fetus develops.

Utilization of farm animal genetic resources refers to the use and development of animal genetic resources for the production of food and agriculture.

UTP is an abbreviation for uridine triphosphate.

UV (Ultraviolet) is a characterization of short-wavelength light for exposing photoresist in making semiconductor devices.

V

V region refers to the variable region in antibodies.

v/v (volume per volume basis) is the percent of the volume of a constituent in 100 units of volume, e.g., (ml/100 ml) × 100.

Vaccination is used as preventive immunization. Vaccination is the administration of antigenic material (a vaccine) to stimulate an individual's immune system to develop adaptive immunity to a pathogen. Vaccines can prevent or ameliorate morbidity from infection. The effectiveness of vaccination has been widely studied and verified; for example, the influenza vaccine, the HPV vaccine, and the chicken pox vaccine. Vaccination is the most effective method of preventing infectious diseases.

Vaccine is a preparation that contains an agent or its components, administered to stimulate an immune response that will protect a person from illness due to that agent. A therapeutic (treatment) vaccine is given after disease has started and is intended to reduce or arrest the progress of the disease. A preventive (prophylactic) vaccine is intended to prevent disease from starting. Agents used in vaccines may be whole-killed (inactive), live-attenuated (weakened) or artificially manufactured. It can be created using the recombinant DNA process.

Vaccinia refers to the cowpox virus used to vaccinate against small-pox and, experimentally, as a carrier of genes for antigenic determinants cloned from other disease organisms.

VACTERL is a syndrome involving birth defects affecting several organs. V stands for vertebral defects (spinal cord), A stands for anal deformities, C stands for cardiac problems, TE stands for tracheoesopha-geal fistula, R stands for renal abnormalities (urinary system and kidneys), and L stands for limb deformities (arms and legs).

Vacuole (L. diminutive of *vacuus*, empty) is a cavity in a plant cell, bounded by a membrane; in which various plant products and by-products are stored.

Vacuum is created through use of a vacuum pump or aspira-tor pump, to facilitate specific biological preparations, such as inclusions or disinfection of material for *in vitro* culture, etc.

VAD (Ventricular Assist Device) is a medical-technological device supporting the heart and circulation, commonly known as "artificial heart."

Valence band is the electron energy band that contains the valence electrons in solid materials.

Valence electrons are the electrons in the outermost occupied electron shell, that participate in interatomic bonding.

Value-added traits are modified crops produced with traits such as improved taste, nutritional value, or utility to provide value for the consumer.

van der Waals bond is a second-ary, permanent or induced, interatomic bond between adja-cent molecular dipoles.

van der Waals forces are he attrac-tive or repulsive forces between molecular entities (or between groups within the same molecu-lar entity) other than those due to bond formation or to the electrostatic interaction of ions or of ionic groups with one another or with neutral mole-cules. ... The term is sometimes used loosely for the totality of nonspecific attractive or repul-sive forces.

Vapor pressure is a pressure at which the vapor of that sub-stance is in equilibrium with its liquid or solid forms.

Variable domain refers to regions of antibody molecules that have different amino acid sequences in different antibody molecules.

These regions are responsible for the antigen-binding specificity of the antibody.

Variable expressivity refers to variation in the phenotype caused by different alleles of the same gene and/or by the action of other genes and/or by the action of non-genetic factors. *cf* expressivity.

Variable number tandem repeat (VNTR) is a DNA sequence, present as tandem repeats, for which the number of copies varies greatly between unrelated genotypes.

Variable surface glycoprotein (VSG) is one of a battery of antigenic determinants expressed by a micro-organism to elude immune detection.

Variance is a statistical term representing a measure of the dispersion of data from the overall mean. Used to quantify the variability of a population.

Variance effective number is the size of the ideal population that experiences changes in allele frequency at the same rate as the observed population.

Variant is an organism that is genetically different from the wild type organism. a.k.a. mutant.

Variation refers to differences between individuals within a population or among populations.

Variegated are plants having both green and albino tissues. This difference in color may result from viral infection, nutritional deficiency, or may be under genetic or physiological control.

Variegation refers to the occurrence, within a single tissue, organ or organism, of mosaicism. Usually referring to plants showing either both green and albino coloration in a leaf, or flecks of contrasting color in a flower. The origin of variegation can be through viral infection, nutritional deficiency, or genetic instability caused by transposon activity.

Variety is a naturally occurring subdivision of a species, with distinct morphological characters and given a Latin name according to the rules of the International Code of Nomenclature.

Vascular (L. *vasculum*, a small vessel) refers to any plant tissue or region consisting of or giving rise to conducting tissue, e.g., bundle, cambium, ray.

Vascular bundle is a strand of tissue containing primary xylem and primary phloem (and procambium if present) and frequently enclosed by a bundle sheath of parenchyma or fibers.

Vascular cambium is a cambium giving rise to secondary phloem and secondary xylem.

Vascular headache is an abnormal stretching of the arterial walls in the cranium as a result of vessel – wall disease.

Vascular plant refers to plant species possessing organized vascular tissues.

Vascular refers to the blood vessel.

Vascular system refers to: (1) A specialized network of vessels for the circulation of fluids throughout the body tissue of an animal. (2) The system of vascular tissue in plants.

Vascular tissue is the tissue that conducts water and nutrients throughout the plant body in higher plants.

Vasoconstrictions is narrowing of the blood vessels.

Vasodilatation is a dilatation of the blood vessels.

Vasodilator is an agent that widens blood vessels.

Vector (L. *vehere*, to carry) refers to: (1) An organism, usually an insect, that carries and transmits disease-causing organisms. (2) A plasmid or phage that is used to deliver selected foreign DNA for cloning and in gene transfer.

Vegetative propagation (vegetative propagation, vegetative multiplication, vegetative cloning) is a form of asexual reproduction in plants. It is a process by which new organisms arise without production of seeds or spores. It can occur naturally or be induced by horticulturists.

Vehicle is the host organism used for the replication or expression of a cloned gene or other sequence The term is little used and is often confused with vector. *See* vector.

Vein is any one of a series of blood vessels of the vascular system that carries blood from various parts of the body back to the heart; returns oxygen-depleted blood to the heart.

Velocity density gradient centrifugation is a procedure used to separate macromolecules based on their rate of movement through a density gradient.

Velogenetics is the combined use of marker-assisted selection and embryo technologies such as OPU, IVM, and IVF, in order to increase the rate of genetic improvement in animal populations.

Venous reservoir collects venous return and stores excess volume; may be incorporated into the oxygenator system.

Ventilation is the movement of air (gases) in and out of the lungs.

Ventricle is one of the two pumping chambers of the heart; right ventricle receives oxygen-poor blood from the right atrium and pumps it to the lungs through the pulmonary artery; left ventricle receives oxygen-rich blood from the left atrium and pumps it to the body through the aorta.

Ventricular fibrillation is a condition, in which the ventricles contract in a rapid, unsynchronized fashion. When fibrillation occurs, the ventricles cannot pump blood throughout the body.

Venturi is a system for speeding the flow of the fluid, by constricting it in a cone-shaped tube.

Venule is a small vein; especially one of the minute veins connecting the capillary bed with the larger systemic veins.

Vermiculite refers to the material made from expanded mica used as a rooting medium and as a soil additive.

Vernalization refers to the chilling juvenile plants for a minimum period in order to induce flowering. Some plants require vernalization to flower, but others have no such requirement.

Vertebral Column (or spinal column or spine or backbone) is a flexible column extending the length of the torso.

Vesico ureteral reflux (VUR) refers to the abnormal flow of urine from the bladder back into the ureters; often as a result of a urinary tract infection or birth defect.

Vessel (L. *vasculum*, a small vessel) refers to: (1) A series of xylem elements whose function is to conduct water and nutrients in plants. (2) A container, such as a Petri dish or test tube, used for tissue culture.

Vessel element is a type of cell occurring within the xylem of flowering plants. Many are water-conducting vessels.

Viability refers to the capability to live and develop normally.

Viability test refers to an assay of the number or percent of living cells or plants in a population following a specific treatment. Often used to describe quality of seed following long-term storage.

Viable refers to capable of normal completion of life cycle.

Vibration refers to a shaking massage technique; a fine tremulous movement made by the hand or fingers placed firmly against a body part and causing that part to vibrate. Often used for a soothing effect; may be stimulating when more energy is applied.

Vibrio refers to comma-shaped bacterium. It is a genus of

Gram-negative bacteria possessing a curved rod shape (comma shape), several species of which can cause foodborne infection, usually associated with eating undercooked seafood.

Villi is a plural of villus, which is a thin projection from a surface. The term is used to describe the individual functional units together of the fetal placenta.

Villi, primary (primary chorionic villi) is a term describing the earliest stage of embryonic placenta development. In humans, the conceptus during week two this first stage of chorionic villi development consists of only the trophoblastic shell cells (syncitiotrophoblasts and cytotrophoblasts) forming finger-like extensions into maternal decidua. Initially these finger-like projections cover the entire surface of chorionic sac and later become restricted to the placental surface. Placental villi stages are primary villi, secondary villi and tertiary villi.

Villi, secondary (secondary chorionic villi) is a term describing the second stage of embryonic placenta development. In humans, the conceptus during week 3 onward this stage of chorionic villi development consists of the trophoblastic shell cells (syncitiotrophoblasts and cytotrophoblasts) filled with extraembryonic mesoderm forming finger-like extensions into maternal decidua. Initially these finger-like projections cover the entire surface of chorionic sac and later become restricted to the placental surface.

Villi, tertiary (tertiary chorionic villi) is a term describing the final stage of embryonic placenta development. In humans, the conceptus after week 3 the chorionic secondary villi now develop placental blood vessels within the core extraembryonic mesoderm. The villi form finger-like extensions that are either anchoring chorionic villi attached to the maternal decidua or floating chorionic villi in maternal lacunae. The villi stages are ongoing as the placenta continues to grow through both the embryonic and fetal development.

Villitis, chronic can occur following placental infection leading to maternal inflammation of the villous stroma, often with associated intervillositis. The inflammation can lead to disruption of blood flow and necrotic cell death.

Vir **genes** is a set of genes on a Ti plasmid that prepare the T-DNA segment for transfer into a plant cell.

Viral coat protein is a protein present in the layer surrounding the nucleic acid core of a virus. It envelopes viral genetic material is known as a capsid. A capsid is composed of protein subunits called capsomeres. Capsids can have several shapes: polyhedral, rod or complex. Capsids function to protect the viral genetic material from damage.

Viral oncogene is a gene in the viral genome that contributes to malignancies in vertebrate hosts.

Viral pathogen refers to a disease – causing virus.

Viral vaccines are vaccines consisting of live viruses rather than dead ones or separated parts of viruses. However, as the virus itself cannot be used, because that would simply give the patient the disease, the virus is genetically engineered so that it elicits the immune response to the viral pathogen without causing the disease itself.

Virion is an infectious virus particle. A plant pathogen that consists of a naked RNA molecule of approximately 250–350 nucleotides, whose extensive base pairing results in a nearly correct double helix.

Viroid is a plant pathogen that consists of a naked RNA molecule of approximately 250–350 nucleotides, whose extensive base pairing results in a nearly correct double helix.

Virtual screening is the selection of compounds by evaluating their desirability in a computational model. Also termed *in silico* screening.

Virulence (*L. virulentia,* a stench) is the degree of ability of an organism to cause disease. The relative infectiousness of a bacterium or virus, or its ability to overcome the resistance of the host metabolism.

Virulent phage is a phage (virus) that destroys the host (bacterium).

Viruliferous refers to a vector (usually insect) organism that carries virions and spreads the virus from host to host by mechanical means.

Virus (*L. virus,* a poisonous or slimy liquid) is an infectious particle composed of a protein capsule and a nucleic acid core (DNA or RNA), which is dependent on a host organism for replication. The DNA or a double-stranded DNA copy of an RNA virus genome is integrated into the host chromosome during lysogenic infection or replicated during the cystic cycle. *See* coat protein; DNA; genome; host; nucleic acid; prion; RNA; tumor virus; viroid.

Virus-free is a plant, animal, cell, tissue or meristem which exhibits no viral symptoms or contains no identifiable virus-particles.

Virus-tested is the description of a organism or a cell stock certi-fied as being free of certain specified viruses following recognized procedures of virus diagnosis.

Viscoelastic property refers to the property of a material to show sensitivity to rate of loading.

Viscoelasticity is a deformation exhibiting the mechanical char-acteristics of viscous flow and elastic deformation.

Viscosity (η) is the ratio of the magnitude of an applied shear stress to the velocity gradient. It is a measure of a noncrystalline material's resistance to perma-nent deformation.

Visible light is the part of the electromagnetic spectrum with wavelengths between 380 nm and 750 nm and perceived by the human eye.

Vital capacity is the maximum vol-ume that can be exhaled after taking the deepest breathe. VC = TV + IRV+ ERV.

Vitamin B complex is a large group of water soluble vitamins that function as co-enzymes, including thiamine (B_1); ribo-flavin or vitamin G (B_2); niacin or nicotinic acid (B_3); panto-thenic acid (B_5); pyridoxine (B_6); cyanocobalamin (B_{12}); biotin or vitamin H; folic acid or vitamin M (Bc); inositol; choline; and others.

Vitamins (L. *vita*, life + amine) are naturally occurring organic substances required by living organisms in relatively small amounts to maintain normal health, and which are added to tissue culture media to enhance growth, usually acting as enzyme co-factors.

Vitrification is the formation of a liquid phase that becomes a glass-bonding matrix upon cooling, during firing of a ceramic body.

Vitrified refers to a cultured tissue having leaves and sometimes stems with a glassy, transpar-ent or wet and often swollen appearance. The process of vitrification is a general term for a variety of physiological disorders that lead to shoot tip and leaf necrosis. Synonym is water soaked.

Vivipary (viviparous (adj.) refers to: (1) A form of reproduc-tion in animals in which the developing embryo obtains its nourishment directly from the mother via a placenta or by other means; (2) A form of asexual reproduction in certain plants, in which the flower develops into a bud-like structure that forms a new plant when detached from the parent; (3) The development of young plants in the inflorescence of the parent plant.

VLSI (Very Large Scale Integration) is a chip with 100,000 to 1,000,000 components.

V_{max} refers to the maximal rate of an enzyme-catalyzed reaction. V_{max} is the product of E_o (the total amount of enzyme) and K_{cat} (the catalytic rate constant).

VNTR is an abbreviation for variable number tandem repeat. VNTR is a location in a genome where a short nucleotide sequence is organized as a tandem repeat. These can be found on many chromosomes, and often show variations in length between individuals. Each variant acts as an inherited allele, allowing them to be used for personal or parental identification. Their analysis is useful in genetics and biology research, forensics, and DNA fingerprinting.

Volatilization is the conversion of a solid or liquid into a gas or vapor.

Volt refers to the electromotive force that must be applied to produce a movement of electrons.

Voltage [V] is a potential difference between two points: Energy to move a one C charge through a one V potential difference is one J.

Voltage sensitive permeability is the quality of some cell membranes that makes them permeable to different ions based on the electric charge of the ions. Nerve and muscle cell membranes allow negatively charged ions into the cell while actively transporting some positively charged ions outside the cell membrane.

Volume contraction is a decrease in body fluid volume, with or without a concomitant loss of osmolytes.

Volvulus is a twisting of the stomach or large intestine that leads to blockage of the digestive tract.

Vomiting is the release of contents in the stomach through the mouth; also known as throwing-up.

Von Hippel-Lindau syndrome is an autosomal dominant condition characterized by the

anomalous growth and pro-
liferation of blood vessels on
the retina of the eye and the
cerebellum of the brain; cysts
and cancers in the kidneys, pancreas, and adrenal glands.

Vorticity is not equal to zero at every
point in a rotational flow, implying that the fluid elements have
a finite angular velocity. This is
equal to zero at every point in
a irrotational flow, where any
motion is pure translation.

VRML (Virtual Reality Modeling
Language) is an open language
under development. XML
(Extensible Markup Language)
is emerging as the most likely
alternative to or fix for VRML.

VSG is an abbreviation for variable surface glycoprotein or
variant surface glycoproteins,
which are type of proteins

coating the surface of some
infectious microorganisms
(e.g., *Trypanosoma brucei*)
and helping them to evade the
host's immune system by mean
of antigenic variation. The VSG
protein is a key molecule for
immune escape and parasitic
success. Combinatorial processes increase the diversity of
variable surface glycoproteins.
The parasite is expressing a
series of antigenically distinct
VSGs from an estimated 1000
VSG genes. The genes are
located in subtelomeric region
and are often activated by the
duplicative transposition of a
silent basic copy gene into an
unlinked telomerically located
expression site, producing an
active expression-linked copy
of that gene.

W

w/v (Weight per volume) is the
weight of a constituent in 100
cm³ of solution, expressed as a
percentage.

Wafer is semiconductor disk out
of which integrated circuits are
made.

Wahlund principle is the deficit
of heterozygotes in subdivided

populations, compared to
expected Hardy-Weinberg
proportions, due to subdivision
into small panmictic (random
mating) demes within the large
population.

Walking is related to: chromosome
walking and primer walking.
Primer walking is a sequencing

method of choice for sequencing DNA fragments between 1.3 and 7 kilobases. Such fragments are too long to be sequenced in a single sequence read using the chain termination method. This method works by dividing the long sequence into several consecutive short ones. The term "primer walking" is used where the main aim is to sequence the genome. The term "chromosome walking" is used instead when we know the sequence but don't have a clone of a gene. For example the gene for a disease may be located near a specific marker such as an RFLP on the sequence.

Wall pressure is the pressure that a cell wall exerts against the turgor of the cell contents. Wall pressure is equal and opposite to the turgor potential.

Wash-out refers to the loss of the slower growing micro-organism when two organisms are being grown together.

Water intoxication is a process of consuming too much water too quickly.

Water potential refers to the difference between the activity of water molecules in pure distilled water at atmospheric pressure and 30°C (standard conditions), and the activity of water molecules in any other system. The activity of these water molecules may be greater (positive) or less (negative) than the activity of the water molecules under standard conditions.

Water soaked refers to one that is soaked or drenched with or in water.

Water stress occurs when plants are unable to absorb enough water to replace that lost by transpiration. Short-term water stress leads to turgor loss (wilting). Prolonged stress leads to cessation of growth, and eventually plant death.

Watt (W) is a unit of power. One watt is the power that gives rise to an energy of one joule/second.

Wave drag refers to a sudden and very powerful drag that appears on aircrafts flying at high-subsonic speeds.

Wave length is the distance from one peak to the next; energy from electromagnetic radiation is inversely proportional to its wave length.

Waveform is the shape of an electrical current as displayed on an oscilloscope.

Wax refers to esters of alcohol higher than glycerol, which are insoluble in water and difficult to hydrolyze; wax forms protective waterproof layers on leaves, stems, fruits, animal fur and integuments of insects.

Weber is a unit of magnetic flux. One weber is a magnetic flux that, linking a circuit of one turn, would produce in it an electromotive force of one volt if it were reduced to zero at a uniform rate in one second.

Weed is simply any plant growing where it is not wanted. In agriculture, used for a plant which has good colonizing capability in a disturbed environment, and can usually compete with a cultivated species therein. Weeds are typically considered as unwanted, economically useless or pest species.

Weediness is the ability of a plant to colonize a disturbed habitat and compete with cultivated species.

Weight percent (wt%) is a concentration specification on the basis of weight (or mass) of a particular element relative to the total weight (or mass).

Western blot is a technique in which protein is transferred from an electrophoretic gel to a cellulose or nylon support membrane following electrophoresis. Related terms are Northern blot; Southern blot.

Western blotting analysis is a technique used to identify a specific protein; the probe is a radioactively labeled antibody raised against the protein in question.

Wet weight (fresh weight or curb weight) is the weight with all fluids topped off. Dry weight is the weight with no fluids.

Wetting agent is a substance that improves surface contact by reducing the surface tension of a liquid: e.g., Triton X-10TM added to disinfecting solutions promotes the disinfestation process. *See* detergent; surfactant.

WGSS (Whole genome shotgun sequencing) is a sequencing method by which an entire genome is cut into chunks of discrete sizes (usually 2,10, 50 and 150 Kb) and cloned into an appropriate vector. The ends of these clones are sequenced. The two ends from the same clone are referred to as mate pairs.

Whisker is a very thin, single crystal of high perfection which has an extremely large length-to-diameter ratio. Whiskers are used as the reinforcing phase in some composites.

White matter is a nervous tissue in the brain and spinal cord that contains myelinated axons. Compare with gray matter.

Wild type is an organism as found in nature; the dominant allele usually found in nature, and from which mutations produce other dominants or recessives alleles.

Wilt refers to drooping of stems and foliage due to loss of cell turgor. May be caused by water **stress** or by disease.

Winter's formula evaluates respiratory compensation when analyzing acid-base disorders and a metabolic acidosis are present. It is described by the equation: $P_{CO2} = [1.5 \times HCO^{-3}] + [8 \pm 2]$, where HCO_{3-} is given in units of mEq/L and P_{CO2} will be in units of mmHg. Winter's formula gives an expected value for the patient's P_{CO2}; the patient's actual (measured) P_{CO2} is then compared to this. If the two values correspond, respiratory compensation is considered to be adequate. If the measured P_{CO2} is higher than the calculated value, there is also a primary respiratory acidosis. If the measured P_{CO2} is lower than the calculated value, there is also a primary respiratory alkalosis.

Wobble hypothesis refers to an explanation of how one tRNA may recognize more than one codon. The first two bases of the mRNA codon and anticodon pair properly, but the third base in the anticodon has some flexibility that permits it to pair with either the expected base or an alternative.

World Wide Web (WWW) is a graphical hypertext system linking many Internet computers.

WrightFisher model refers to random mating population model with complete random union of gametes (including the possibility of selfing).

WSSD (Whole genome shotgun sequence detection) is a computational method for the comparison of whole genome shotgun sequence (WGS) to a reference genome, commonly used for the detection of segmental duplications. A method complementary to Whole Genome Alignment Comparison (WGAC).

X

X is the basic number of chromosomes in a polyploid series, monoploid = x; diploid = 2x; triploid = 3x; etc.

Xanthophyll (Gr. *xanthos*, yellowish brown + *phyllon*, leaf) is a yellow chloroplast pigment.

X-chromosome is one of the two sex-determining chromosomes (allosomes) in many animal species, including mammals (the other is the Y chromosome), and is found in both males and females. It is a part of the XY sex-determination system and X0 sex-determination system. The X chromosome was named for its unique properties by early researchers, which resulted in the naming of its counterpart Y chromosome, for the next letter in the alphabet, after it was discovered later.

Xenia is the immediate effect of pollen on some characters of the endosperm.

Xenobiotic is a chemical compound that is not produced by, and often cannot be degraded by, living organisms.

Xenogeneic organs refer to genetically engineered (for example, "humanized") organs that have been grown within an animal of another species. Xenogeneic literally means "strange genes."

Xenogeneic refers to organs, genetically engineered ("humanized") to decrease the chance of rejection, that have been grown in an animal of another species for potential transplant to humans.

Xenograft valves are artificial valves made from animal tissue. Most often the valves are made from pig aortic valves. More recently, some valves have been made from cow tissues.

Xenografts is a type of tissue graft in which the donor and recipient are of different species. Also called heterographs.

Xenosis describes the transfer of infections by transplantation of xenogeneic tissues or organs. It potentially poses unique epidemiological hazards due to the efficiency of transmission of pathogens, particularly viruses, with viable, cellular grafts. It is a term coined from the word "xenozoonoses."

Xenotransplantation is the transplantation of living cells, tissues and organs from one species to another. The term is usually used to describe animal-to-human transplants. An example is the transplant of a kidney from a pig to a human. The principal reason for medical and scientific inquiry in this area is to find alternatives to human organs and tissue transplants.

Xerophyte (Gr. *xeros*, dry + *phyton*, a plant) is a plant very resistant to drought or that lives in very dry places.

Xerophyte is a plant very resistant to drought, typically adapted to extremely dry environments.

X-inactivation is the repression of one of the two X-chromosomes in the somatic cells of females as a method of dosage compensation; at an early embryonic stage in the normal female, one of the two X-chromosomes undergoes inactivation, apparently at random, from this point on all descendent cells will have the same X-chromosome inactivated as the cell from which they arose, thus a female is a mosaic composed of two types of cells, one which expresses only the paternal X-chromosome, and another which expresses only the maternal X-chromosome.

X-linked disease is a genetic disease caused by an allele at a locus on the X-chromosome. In X-linked recessive conditions, a normal female "carrier" passes on the allele on her X chromosome to an affected son.

X-linked is the presence of a gene on the X chromosome.

X-ray crystallography refers to the deduction of crystal structure from analysis of the diffraction pattern of X-rays passing through a pure crystal of a substance.

X-ray is a diagnostic test which uses invisible electromagnetic energy beams to produce images of internal tissues, bones, and organs onto film.

Xylem (Gr. *xylon*, wood) is a complex tissue specialized for efficient conduction of water and mineral nutrients in solution. Xylem may also function as a supporting tissue, particularly secondary xylem.

XYY syndrome is a genetic condition in males with extra Y chromosome (in 1 in 1000 male births). Symptoms: tall stature (over 6'), may including sterility, developmental delay, learning problems.

Y

YAC is an abbreviation for yeast artificial chromosome. The YACs are genetically engineered chromosomes derived from the DNA of the yeast, *Saccharomyces cerevisiae*, which is then ligated into a bacterial plasmid. By inserting large fragments of DNA, from 100–1000 kb, the inserted sequences can be cloned and physically mapped using a process called chromosome walking. This is the process that was initially used for the Human Genome Project, however due to stability issues, YACs were abandoned for the use of Bacterial artificial chromosomes (BAC).

Y-chromosome is one of two sex chromosomes (allosomes) in mammals, including humans, and many other animals. The other is the X chromosome. Y is the sex-determining chromosome in many species, since it is the presence or absence of Y that determines male or female sex. In mammals, the Y chromosome contains the gene SRY, which triggers testis development. The DNA in the human Y chromosome is composed of about 59 million base pairs. The Y chromosome is passed only from father to son, so analysis of Y chromosome DNA may thus be used in genealogical research. With a 30% difference between humans and chimpanzees, the Y chromosome is one of the fastest evolving parts of the human genome. To date, over 200 Y-linked genes have been identified. All Y-linked genes are expressed and (apart from duplicated genes) hemizygous (present on only one chromosome) except in the cases of aneuploidy such as Klinefelter's Syndrome (47,XXY) or XXYY syndrome.

Yeast artificial chromosome (YAC) refers to a vector which can be propagated in budding yeast (*Saccharamyces pombe*), consisting of the minimal elements required for a chromosome to replicate, and allowing for the cloning of large DNA fragments (hundreds of kilobase pairs).

Yeast cloning vectors refer to the yeasts, and especially *Saccaromyces cerevisiae*, are favorite organisms in which to clone and express DNA. They are eukaryotes, and so can splice out introns, the non-coding sequences in the middle of many eukaryotic genes.

Yeast episomal vector (YEp) refers to a cloning plasmid vector for the yeast *Saccharomyces cerevisiae* maintained as an extrachromosomal nuclear DNA molecule.

Yeast extract is a mixture of substances from yeast. *cf* organic complex; undefined.

Yeast is a unicellular ascomycete fungus, commonly found as a contaminant in plant tissue culture.

Yield strength is the stress required to produce a very slight yet specified amount of plastic strain; a strain offset of 0.002 is commonly used.

Yielding is the onset of plastic deformation.

Yoctomole is one 10 to the negative 24th power. This is even smaller than the inverse of Avogadro's constant. It is a unit of measure in mass spectrometry. The record for detection is 800 yoctomoles, or 480 molecules of a substance. Zeptomole is one 10^{-21}th part of a mole, or about 600 molecules of a substance.

Z

....

Z-DNA is a form of DNA, in which the double helix is wound in a left-hand, instead of a right-hand, manner. DNA adopts the Z conformation when purines and pyrimidines alternate on each strand, e.g., 5'CGCGCGCG 3' or 3'GCGCGCGC5.' Synonym is zig-zag DNA.

Z-DNA or zig-zag DNA is a form of DNA duplex in which the double helix is wound in a left-hand, instead of a right-hand, manner. DNA adopts the Z configuration when purines and pyrimidines alternate on a single strand, e.g., 5'CGCGCGCG3' or 5'CACACACACA3', 3'GCGCGCGC5' 3'GTGTGTGTGT5'

Z-DNA refers to a region of DNA that is "flipped" into a left handed helix, characterized by alternating purines and pyrimidines, and which may be the target of a DNA-binding protein.

Zener diode is semiconductor diode that has a well-defined turn-on voltage for conduction in the reverse direction.

Zeptomole is one 10^{-21}th part of a mole, or about 600 molecules of a substance

Zero offset is the output of a sensor at zero input for a specified supply voltage or current.

Zeta potential is a scientific term for electrokinetic potential in colloidal systems. In the colloidal chemistry literature, it is usually denoted using the Greek letter zeta (ζ), hence *ζ-potential*. From a theoretical viewpoint, the zeta potential is the electric potential in the interfacial double layer (DL) at the location of the slipping plane versus a point in the bulk fluid away from the interface. In other words, zeta potential is the potential difference between the dispersion medium and the stationary layer of fluid attached to the dispersed particle. A value of 25 mV (positive or negative) can be taken as the arbitrary value that separates low-charged surfaces from highly charged surfaces. The significance of zeta potential is that its value can be related to the stability of colloidal dispersions (e.g., a multi vitamin syrup). The zeta potential indicates the degree of repulsion between adjacent, similarly charged particles (the vitamins) in dispersion.

Zig-zag DNA is one of the many possible double helical structures of DNA. It is a left-handed double helical structure in which the double helix winds to the left in a zig-zag pattern (instead of to the right, like the more common B-DNA form). Z-DNA is thought to be one of three biologically active double helical structures along with A- and B-DNA.

Zinc finger is a DNA-binding protein motif, characterized by two closely spaced cysteine and two histidine residues that serve as ligands for a single Zn^{2+} ion. When bound, the structure takes on a conformation in which amino acid side chains protrude in a way that allows interaction with the DNA major groove.

Zone of elongation is the section of the young root or shoot just behind the apical meristem, in which the cells are enlarging and elongating rapidly.

Zone refining is a metallurgical process for obtaining a highly pure metal that depends on continuously melting the impure material and recrystallizing the pure metal.

Zoo blot is the hybridization of cloned DNA from one species to DNA from other organisms to determine the extent to which the cloned DNA is evolutionarily conserved.

Zoo FISH (Fluorescent *in situ* hybridization) is a zoo-FISH of DNA from one species on metaphase chromosomes of another species. Typically, the hybridization is done separately for DNA libraries representing each chromosome. The result is a fascinating picture of the regions of chromosomal homology between species.

Zoonosis is a disease that is communicable from animals to humans.

Zoospore is a spore that possesses flagella and is therefore motile.

Zygonema (adj.: zygotene) is the stage in meiosis during which synapsis occurs; coming after the leptotene stage and before the pachytene stage in the meiotic prophase.

Zygospore (Gr. *zygon*, a yoke + spore) is a thick-walled resistant spore developing from a zygote resulting from the fusion of isogametes.

Zygote (Gr. *zygon*, a yoke) is a diploid cell formed by the fusion of two haploid gametes during fertilization in eukaryotic organisms with sexual reproduction. It is the first cell of the new individual.

Zymogen is inactive enzyme precursor that after secretion is chemically altered to the active form of the enzyme.

3.2 SUMMARY

This chapter includes glossary (or a definition) of technical terms that are commonly used in the specialized areas of Bioengineering and Biotechnology.

KEYWORDS

This list will include all the technical terms in bold letters at the beginning of each glossary in this chapter.

REFERENCES

1. www.acronyms.thefreedictionary.com/ANGR.
2. Allendorf, F. W., & Luikart, G. (2007). *Conservation and the Genetics of Populations*. Blackwell Publishing. 642 pp.
3. Bellomo, M., (2006). The Stem Cell Divide: The Facts, the Fiction, and the Fear Driving the Greatest Scientific, Political, and Religious Debate of Our Time. New York: AMACOM.
4. Berger, E., & Roth, J., (1997). The Golgi Apparatus. Basel, Switzerland. Birkhäuser.
5. Bevington, L. K., Ray, G. B., Gary, P. S., John, F. K., & Christopher, H. C., (2002). Basic Questions on Genetics, Stem Cell Research and Cloning: Are These Technologies Okay to Use? Grand Rapids: Kregel.
6. Black, L., (2006). The Stem Cell Debate: The Ethics and Science behind the Research. Berkeley Heights, NJ: Enslow.

7. Brock, T. D., (1961). Milestones in Microbiology. Science Tech Publishers. Madison, Wisconsin. pp. 273.

8. Brock, T. D., (1990). The Emergence of Bacterial Genetics. Cold Spring Harbor Laboratory Press. Cold Spring Harbor, New York. pp. 346.

9. Brown, N., (2006). Blood Ties: Banking the Stem Cell Promise. *Technology Analysis & Strategic Management, 18*(3/4), 313–327.

10. Bunch, B., & Hellemans, A., (1993). The Timetables of Technology. Simon & Schuster. New York, New York. pp. 490.

11. Cameron, N. M. de S., (1987). Embryos and Ethics: The Warnock Report in Debate. Edinburgh: Rutherford House.

12. Capps, B. J., & Alastair, V. C., (2010). Contested Cells: Global Perspectives on the Stem Cell Debate. London: Imperial College Press.

13. Carrier, E., & Gracy, L., (2004). 100 Questions & Answers about Bone Marrow and Stem Cell Transplantation. Sudbury, MA: Jones and Bartlett.

14. www.chemwiki.ucdavis.edu/.../Chemiluminescence

15. Cohen, C. B., (2007). Renewing the Stuff of Life: Stem Cells, Ethics, and Public Policy. New York: Oxford University Press.

16. Collins English Dictionary – Complete & Unabridged, 10th Edition 2009. William Collins Sons & Co. Ltd.

17. Collins English Dictionary – Complete and Unabridged, 2003. HarperCollins Publishers.

18. Committee on Guidelines for Human Embryonic Stem Cell Research and National Research Council, 2005. Guidelines for Human Embryonic Stem Cell Research. Washington, D.C.: National Academies Press.

19. DeGette, D., (2008). Sex, Science, and Stem Cells: Inside the Right Wing Assault on Reason. Guilford, CT: Lyons.

20. www.dictionary.com 21st Century Lexicon 2003–2014, www.Dictionary.com, LLC.

21. www.dictionary.com Unabridged (Random House Dictionary), Random House, Inc. 2014.

22. Dorland's Medical Dictionary for Health Consumers. 2007 by Saunders, an imprint of Elsevier, Inc.

23. FAO, Agricultural biotechnology for developing countries – Results of an electronic forum, 2001.

24. FAO, Agricultural biotechnology in the developing world, 1995.

25. FAO, Glossary of biotechnology and genetic engineering, 1999.

26. FAO, The role of universities in national agricultural research systems, 1993.

27. Feneque, J., (2000). Nanotechnology: a new challenge for veterinary medicine. *The Pet Tribune. 6*(5), 16.

28. Fox, Cynthia, 2007. Cell of Cells: The Global Race to Capture and Control the Stem Cell. New York: W.W. Norton.

29. Friedman, L. S., & Hal, M., (2009). Is Stem Cell Research Necessary? Reference Point Press.

30. Futuyma, D. J., (2013). Evolution. 3rd ed. Sinauer Associates, Sunderland, Massachusetts.

31. Gale Encyclopedia of Medicine 2008. The Gale Group, Inc.

32. Gillis, M., (2009). Ethics, Stem Cells, and Women: A Feminist Perspective. Saarbrücken, Germany: LAP Lambert Academic Publishing.

33. Gottweis, H., & Brian, S., (2009). The Global Politics of Human Embryonic Stem Cell Science. New York: Palgrave Macmillan.

34. Green, R. M., (2001). The Human Embryo Research Debates: Bioethics in the Vortex of Controversy. New York: Oxford University Press.

35. Gruen, L., Laura, G., & Peter, S., (2007). Stem Cell Research: The Ethical Issues. New York: Wiley-Blackwell.

36. Haerens, M., (2009). Embryonic and Adult Stem Cells. San Diego: Greenhaven Press.

37. Hall, S. S., (2001). Adult stem cells. *Technology Review, 104*(9), 42.

38. Hellemans, A., & Bunch, B., 1988. The Timetables of Science. Simon & Schuster. New York, New York. Pp. 660.

39. Herold, E., & George, D., (2007). Stem Cell Wars: Inside Stories from the Frontlines. New York: Palgrave Macmillan.

40. Holland, S., Karen, L., & Laurie, Z., (2001). The Human Embryonic Stem Cell Debate: Science, Ethics, and Public Policy . Cambridge, MA: MIT Press.

41. http://academic.research.microsoft.com/Keyword/51187/Dna-Amplification-Fingerprinting

42. http://agbiosafety.unl.edu/education/backcross.htm

43. http://archive.today/DzU2

44. http://bioethics.georgetown.edu/nbac/stemcell.pdf National Bioethics Advisory Commission. Report and Recommendations of the National Bioethics Advisory Commission. Vol. 1 of Ethical Issues in Human Stem Cell Research. Rockville, MD: NBAC, 1999.

45. http://bioethics.georgetown.edu/nbac/stemcell2.pdf National Bioethics Advisory Commission. Commissioned Papers. Vol. 2 of Ethical Issues in Human Stem Cell Research. Rockville, MD: NBAC, 2000.

46. http://bioethics.georgetown.edu/nbac/stemcell3.pdf National Bioethics Advisory Commission. Religious Perspectives. Vol. 3 of Ethical Issues in Human Stem Cell Research. Rockville, MD: NBAC, 2000.

47. http://bioethics.georgetown.edu/pdbe/reports/cloningreport/pcbe_cloning. The President's Council on Bioethics. Human Cloning and Human Dignity: The Report of the President's Council on Bioethics. New York: Public Affairs, 2002.

48. http://biology.about.com/od/virology/ss/viruses_2.htm

49. http://biotechterms.org/

50. http://bmes.org/cmbesig Cellular and Molecular Bioengineering Special Interest Group

51. http://bmes.org/faseb. *Federation of American Societies for Experimental Biology.*

52. http://bse.unl.edu/history

53. http://cbhd.org/content/biotechnology-bibliography

54. http://cbhd.org/themes/bioethics/images/menu-leaf.gif

55. http://dictionary.reference.com/browse/nanophysics

56. http://dictionary.reference.com/browse/phyto-

57. http://dictionary.reference.com/browse/reverse+mutation

58. http://dictionary.reference.com/browse/Water-soaked

59. http://en.docsity.com/answers/192445/define-artificial-inembryonation-and-its-significance

60. http://en.mimi.hu/biology/cos_site.html

61. http://en.termwiki.com/EN:parasexual_hybridization

62. http://en.wikipedia.org/wiki/

63. http://en.wikipedia.org/wiki/Repressor

64. http://en.wiktionary.org/wiki/cytotype

65. http://encyclopedia2.thefreedictionary.com/bifunctional+vector

66. http://evolution.berkeley.edu/evosite/evo101/IIIDGeneticdrift.shtml

67. http://genesolutions.com/page8.html

68. http://groups.molbiosci.northwestern.edu/holmgren/Glossary/Definitions/Def-N/nicked_plasmid%20.html

69. http://gutenberg.llnl.gov/~colvin/ Michael Colvin. What is ab initio quantum chemistry? Lawrence Livermore National Lab, USA.

70. http://home.earthlink.net/~trimmerw/mems/SM_surface.html

71. http://link.springer.com/proto-col/10.1385/1-59259-117-5:69
72. http://mass-spec.lsu.edu/msterms/index.php/Nanoelectrospray
73. http://no.cyclopaedia.net/wiki/Endog-amy
74. http://onlinebooks.library.upenn.edu/webbin/book/browse?type=lcs ubc&key=Biotechnology%20–%20 Bibliography&c=x
75. http://onlinelibrary.wiley.com/doi/10.1046/j.1365–2958.1996.6391352.x/abstract
76. http://pubs.rsc.org/en/con-tent/articlelanding/2013/cs/c2cs35427f#!divAbstract
77. http://seq.mc.vanderbilt.edu/DNA/html/ola.html
78. http://stemcell.childrenshospital.org /?gclid=CJv497qJ2rMCFeuPPAodp H0Ahw. Stem Cell Research, Bos-ton Children's Hospital. Stem Cell Research Boston Children's Hospital.
79. http://stemcellresearch1311.weebly.com/annotated-bibliography.html
80. http://stemcells.nih.gov/info/basics/basics1.asp Stem Cell Basics: Intro-duction [Stem Cell Information]. NIH Stem Cell Information Home Page.
81. http://svtc.org/our-work/nano/timeline/
82. http://theagricos.com/plant-breeding/combining-ability/specific-conbining-ability/
83. http://thesciencedictionary.org/turgor-potential/
84. http://timelines.ws/subjects/Technol-ogy.HTML
85. http://wiki.answers.com/Q/What_ are_the_steps_in_hanging_drop_ preparation?#slide=1
86. http://www.aaas.org/spp/sfrl/projects/stem/report.pdf Chapman, Audrey R., Mark S. Frankel, and Michele S. Garfinkel. "Stem Cell Research and Applications: Monitoring the Fron-tiers of Biomedical Research." AAAS / ICS, November 1999.
87. http://www.academon.com/term-paper/stem-cell-research-a-bibliog-raphy-125772/ Stem Cell Research: A Bibliography (2008, December 01) Retrieved April 19, 2014.
88. http://www.accessexcellence.org/RC/AB/BC/1977-Present.php
89. http://www.accessexcellence.org/RC/AB/BC/1977-Present.php
90. http://www.allacronyms.com/EGS/External+Guide+Sequence+Oligoz yme/883979
91. http://www.answers.com/topic/amino-acyl-site
92. http://www.answers.com/topic/lethal-mutation
93. http://www.antibodybeyond.com/applications/eia.htm
94. http://www.antibodybeyond.com/applications/eia.htm
95. http://www.azonano.com/article.aspx?ArticleID=1344
96. http://www.bio.davidson.edu/people/sosarafova/Assets/Bio307/liwoeste/Cellular%20Immune%20Response.html
97. http://www.bio.org/articles/glossary-agricultural-biotechnology-terms
98. http://www.biologydir.com/a-molecu-lar-biology-glossary-info-5637.html
99. http://www.biologydir.com/glossary-of-biotechnology-terms-info-30908.html
100. http://www.biology-online.org/dic-tionary/Cis_trans_test
101. http://www.biology-online.org/dic-tionary/Embryo_technology
102. http://www.biology-online.org/dic-tionary/Hot_spot
103. http://www.biology-online.org/dic-tionary/Single_stranded_dna
104. http://www.biotech.kth.se/glycosci-ence/Main_Page
105. http://www.biotechinstitute.org/go.cfm?do=Page.View&pid=22
106. http://www.biotechnologyforums.com/thread-1757.html
107. http://www.bme.fiu.edu/about/history/
108. http://www.cdc.gov/ncidod/dvrd/bse/
109. http://www.definitions.net/definition/biomedical+technology

110. http://www.dummies.com/how-to/
content/nanotechnology-timeline-and-
predictions.html

111. http://www.ehow.com/
about_6364029_history-bioengineer-
ing.html

112. http://www.ehow.com/
about_6364029_history-bioengineer-
ing.html#ixzz31PBIg8vV

113. http://www.ehow.com/
facts_5723915_agarose-gels-vs_-
acrylamide-gels.html

114. http://www.els.net/WileyCDA/ElsAr-
ticle/refId-a0002657.html

115. http://www.encyclopedia.com/
doc/1O6-genetracking.html

116. http://www.encyclopedia.com/topic/
tandem_array.aspx

117. http://www.everythingbio.
com/?s=product%20of%20meiosis

118. http://www.expertglossary.com/food-
biotechnology/definition/artificial-
inembryonation

119. http://www.expertglossary.com/tech-
nology

120. http://www.experts123.com/q/what-
is-a-transducing-phage.html

121. http://www.fao.org/biotech/biotech-
glossary/en/

122. http://www.fao.org/docrep/004/
y2775e/y2775e00.htm

123. http://www.fao.org/docrep/004/
y2775e/y2775e00.htm

124. http://www.fao.org/glossary/

125. http://www.fao.org/glossary/
spec-term-n.asp?id_glo=3252&id_
lang=TERMS_E

126. http://www.fao.org/glossary/
spec-term-n.asp?id_glo=4906&id_
lang=TERMS_E

127. http://www.filebox.vt.edu/cals/cses/
chagedor/glossary.html

128. http://www.genscript.com/prod-
uct_003/molecular_biology_glossary/
id/11205/category/glossary/relaxed_
circle.html

129. http://www.gmo-safety.eu/glos-
sary/667.gene-construct.html

130. http://www.gravertech.com/PDF/
Product_Sheets/LPF/Glossary_of_
Biotech_Terms.pdf

131. http://www.greatachievements.
org/?id=3824

132. http://www.greatachievements.
org/?id=3837

133. http://www.greathairtransplants.com/
micrograft.html

134. http://www.grida.no/graphicslib/
detail/biotechnology-and-modern-
biotechnology-defined_b9d8

135. http://www.hawaii.edu/microbiology/
MO/blunting.htm

136. http://www.historyworld.net/
timesearch/default.asp?conid=static_
timeline&timelineid=406&page=1&k
eywords=Engineering%20timeline

137. http://www.koko.gov.my/CocoaBio-
Tech/Glossaryd.html

138. http://www.larapedia.com/glossary_
of_technology_terms/bioengineer-
ing_meaning_and_definition.html

139. http://www.lexic.us/definition-of/
embryo_technology

140. http://www.mathgoodies.com/lessons/
vol6/addition_rules.html

141. http://www.medicalglossary.org/bio-
logical_sciences_biotechnology_defi-
nitions.html

142. http://www.medicalnewstoday.com/
articles/262798.php

143. http://www.medscape.com/viewarti-
cle/566133_2

144. http://www.medscape.com/viewarti-
cle/769128_9

145. http://www.medterms.com/script/
main/art.asp?articlekey=16690

146. http://www.medterms.com/script/
main/art.asp?articlekey=9803

147. http://www.merriam-webster.com/
dictionary/gyrase

148. http://www.merriam-webster.com/
medical/homeodomain

149. http://www.microbelibrary.org/
library/laboratory-test/3139-exam-
ination-for-motility-by-hanging-
drop-technique

150. http://www.microtissues.com/hanging_drop_method_or_technique.htm
151. http://www.microtransponder.com/
152. http://www.mondofacto.com/facts/dictionary?beta-DNA
153. http://www.nano.gov/timeline
154. http://www.nano.gov/timeline
155. http://www.nap.edu/books/0309076374/html/ National Academies. Committee on Science, Engineering, and Public Policy (COSEPUP) and Board on Life Sciences (BLS). Scientific and Medical Aspects of Human Reproductive Cloning. Committee on Science, Engineering and Public Policy; National Academy of Sciences; National Academy of Engineering; Institute of Medicine. Washington, DC: National Academy, 2002.
156. http://www.nature.com/nmat/journal/v8/n6/fig_tab/nmat2441_F1.html
157. http://www.ncbi.nlm.nih.gov/pmc/articles/PMC449985/
158. http://www.ncbi.nlm.nih.gov/projects/genome/glossary.shtml#CHIP National Center for Biotechnology Information, US National Library of Medicine.
159. http://www.newworldencyclopedia.org/entry/Asparagine
160. http://www.nisenet.org/catalog/exhibits/nano_museum_labels_graphic_signs_nanodays_2013
161. http://www.princeton.edu/~achaney/tmve/wiki100k/docs/Stop_codon.html
162. http://www.protocol-online.org/biology-forums/posts/19419.html
163. http://www.psc.edu/index.php/biomedical-applications-group
164. http://www.sci.sdsu.edu/~smaloy/Glossary/C.html
165. http://www.sci.sdsu.edu/~smaloy/MicrobialGenetics/topics/chroms-genes-prots/temp-strand.html
166. http://www.sci.sdsu.edu/~smaloy/MicrobialGenetics/topics/in-vitro-genetics/SDM.html
167. http://www.scidev.net/global/link/fao-glossary-of-biotechnology-for-food-and-agricul.html
168. http://www.sciencedirect.com/science/article/pii/S0300908497867168
169. http://www.sciencemag.org/content/249/4970/783.short
170. http://www.science-of-aging.com/timelines/cell-history-timeline-detail.php
171. http://www.springer.com/life+sciences/ecology/journal/12686
172. http://www.springer.com/medicine/radiology/book/978-1-62703-136-3
173. http://www.stemcellresearch.org. Do No Harm: The Coalition of Americans for Research Ethics.
174. http://www.studymode.com/essays/Annotated-Bibliography-Of-Stem-Cell-Research-45367642.html
175. http://www.thefreedictionary.com/lyophilize
176. http://www.the-scientist.com/?articles.view/articleNo/13180/title/Cell-free-Transcription-and-Translation/
177. http://www.timetoast.com/timelines/tissue-engineering-and-regenerative-medicine
178. http://www.usda.gov/wps/portal/usda/usdahome?navid=BIOTECH_GLOSS&navtype=RT&parentnav=BIOTECH
179. http://www.usgs.gov/ecosystems/genetics_genomics/glossary.html#abc
180. http://www.web-books.com/MoBio/Free/Ch4C1.htm
181. http://www.wisegeek.com/what-is-molecular-computing.htm
182. http://www.wisegeek.org/what-is-gene-sequencing.htm#didyouknowout
183. http://www.zhinst.com/applications/quantumnano
184. https://answers.yahoo.com/question/index;_ylt=A0LEVw585F9T-aXUAN-LJXNyoA;_ylu=X3oDMTEzcDBoZW VpBHNlYwNzcgRwb3MDMDMwRjb2-xvA2JmMQR2dGlkA1-ZJUDA3N18x?qid=20090408223124AAOZ0je.
185. https://engineering.purdue.edu/BME/AboutUs/History
186. https://www.bcm.edu/research/centers/imaging-research/?PMID=5056

187. https://www.google.com/patents/
US20100099110
188. https://www.raft3dcellculture.com/
system_overview.php
189. Hug, K. & Göran, H., eds. Translational Stem Cell Research: Issues
Beyond the Debate on the Moral Status of the Human Embryo. New York:
Springer, 2011.
190. Humber, J. M., & Robert, F. A.,
(2004). Stem Cell Research. Totowa,
NJ: Humana.
191. IAASTD, The International Assessment of Agricultural Science and
Technology for Development.
192. Jochemsen, H., Elisa, G., Asher, M., &
Ron, H., (2005). Human Stem Cells:
Source of Hope and of Controversy.
Chicago: The Bioethics Press.
193. Jones, G., & Mary, B., (2005). Stem
Cell Research and Cloning: Contemporary Challenges to Our Humanity.
Hindmarsh, SA: ATF Press.
194. Juengst, E., & Michael, F., (2000).
The Ethics of Embryonic Stem Cells:
Now and Forever, Cells Without End.
JAMA 284, 3180–3184.
195. Kass, L. R., (2002). Life, Liberty, and
the Defense of Dignity: The Challenge
for Bioethics. San Francisco: Encounter.
196. Kelly, E. B., (2006). Stem Cells. New
York: Greenwood.
197. Kilner, J., (2002). Basic Questions
on Genetics, Stem Cell Research and
Cloning: Are These Technologies
Okay to Use? Grand Rapids: Kregel.
198. Korobkin, R., & Stephen, R. M.,
(2009). Stem Cell Century: Law and
Policy for a Breakthrough Technology. New Haven, CT: Yale University
Press.
199. Lanza, R., Roger, P., Douglas, M.,
John, G., Donnall, T. E., James, A.
T., & Brigid, H., (2005). Essentials of
Stem Cell Biology. Burlington, MA:
Academic.
200. Leeb, C. C., Jurga, M. M., McGu-
ckin, C. C., Forraz, N. N., Thallinger,
C. C., Moriggl, R. R., & Kenner,

L. (2011). New perspectives in stem
cell research: beyond embryonic stem
cells. Cell Proliferation, 449.
201. Maienschein, J., (2005). Whose View
of Life? Embryos, Cloning, and Stem
Cells. Cambridge, MA: Harvard University Press.
202. Marzilli, A., (2007). Stem Cell
Research and Cloning. New York:
Chelsea House.
203. McGraw-Hill Dictionary of Scientific
& Technical Terms, 6E, McGraw-Hill
Co. Inc.
204. Monroe, K. R., Ronald, M., & Jerome,
T., (2007). Fundamentals of the Stem
Cell Debate: The Scientific, Religious,
Ethical, and Political Issues. Berkeley:
University of California Press.
205. Mosby's Medical Dictionary, 8th edi-
tion, 2009, Elsevier.
206. Mulkay, M., (1997). The Embryo
Research Debate: Science and the
Politics of Reproduction. Cambridge:
Cambridge University Press.
207. Murphy, A., & Judy, P., (1993). A Fur-
ther Look at Biotechnology. Princeton,
NJ: The Woodrow Wilson National
Fellowship Foundation, Woodrow
Wilson Foundation Biology Institute.
208. Nill, K., (2002). Glossary of Biotech-
nology Terms, Third Edition, CRC
Press.
209. Ostnor, L., (2008). Stem Cells, Human
Embryos and Ethics: Interdisciplinary
Perspectives. New York: Springer.
210. Pacela's Bioengineering Education
Directory. Quest Publishing Co., 1990.
211. Panno, J., (2006). Stem Cell Research:
Medical Applications and Ethical
Controversy. New York: Checkmark.
212. Peters, Pamela, 1993. From Biotech-
nology: A Guide To Genetic Engi-
neering. Wm. C. Brown Publishers,
Inc.
213. Peters, T., Karen, L., & Gaymon, B.,
(2008). Sacred Cells? Why Christians
Should Support Stem Cell Research.
Lanham, MD: Rowman and Littlefield
Publishers, Inc.

214. Peters, T., (2010). Sacred Cells? Why Christians Should Support Stem Cell Research. Rowman & Littlefield.

215. Peters, T., (2007). The Stem Cell Debate. Minneapolis: Fortress.

216. Peterson, C., (1995). *Nanotechnology: From Concept to R&D Goal.* HotWired.

217. Peterson, C., (1995). Nanotechnology: evolution of the concept. In: *Prospects in Nanotechnology: Toward Molecular Manufacturing* by ed. Markus Krummenacker and James Lewis. Wiley.

218. Peterson, C., (2000). *Molecular Nanotechnology: the Next Industrial Revolution.* IEEE Computer, January.

219. Peterson, C., (2004). *Nanotechnology: from Feynman to the Grand Challenge of Molecular Manufacturing.* IEEE Technology and Society, Winter 2004.

220. Plomer, A., & Paul, T., (2009). Emryonic Stem Cell Patents: European Patent Law and Ethics. New York: Oxford University Press.

221. Potten, C. S., Robert, B. C., James, W., & Andrew, G. R., (2006). Tissue Stem Cells. New York: Informa Healthcare.

222. Prentice, D. A., (2008). Stem Cells and Cloning. 2nd ed. San Francisco: Benjamin Cummings.

223. Quiqley, M., Sarah, C., & John, H., (2012). Stem Cells: New Frontiers in Science and Ethics. Singapore: World Scientific Publishing Company.

224. Regenerative medicine glossary. Regenerative Medicine 4 (4 Suppl): 81–88. July 2009.

225. Regenerative Medicine, 2008, 3(1), 1–5.

226. Regis, E., (1995). *Nano: The Emerging Science of Nanotechnology.* Little Brown.

227. Robertson, J. A., (2002). Science and Society: Human Embryonic Stem Cell Research: Ethical and Legal Issues. Nature Reviews Genetics, 2, 74–78.

228. Ruse, Michael, and Christopher A. Pynes, 2006. The Stem Cell Controversy: Debating the Issues. Amherst, NY: Prometheus.

229. Saunders Comprehensive Veterinary Dictionary, 3 ed. 2007 Elsevier, Inc.

230. Segen's Medical Dictionary. 2012 Farlex, Inc.

231. Shostak, S., (2002). Becoming Immortal: Combining Cloning and Stem-cell Therapy. Albany: State University of New York Press.

232. Snow, Nancy E., 2005. Stem Cell Research: New Frontiers in Science and Ethics. Notre Dame, IN: University of Notre Dame Press.

233. The American Heritage Medical Dictionary. 2007, Houghton Mifflin Company.

234. Waters, B., & Ronald, C.-T., (2003). God and the Embryo: Religious Voices on Stem Cells and Cloning. Washington, DC: Georgetown University Press.

235. Wertz, D. C., (2002). Embryo and Stem Cell Research in the USA: A Political History. *Trends in Molecular Medicine 8,* 143–146.

236. WordNet 3.0, Farlex clipart collection. 2003–2012, Princeton University, Farlex Inc.

237. www.bmes.org

238. www.cnm.es

239. www.hc-sc.gc.ca

240. www.nap.edu/books/0309076307/html/ Committee on the Biological and Biomedical Applications of Stem Cell Research, Commission on Life Sciences, National Research Council, Board on Neuroscience and Behavioral Health, Institute of Medicine. Stem Cells and the Future of Regenerative Medicine. Washington, DC: National Academies, 2002.

241. www.ogtr.gov.au

242. Zaid, A., et al., (2001). Glossary of Biotechnology for Food and Agriculture – A Revised and Augmented Edition of the Glossary of Biotechnology and Genetic Engineering. FAO research and technology paper 9, FAO, Viale delle Terme di Caracalla, 00100 Rome, Italy.

INDEX

2 μm plasmid, 112
22q deletion syndrome, 112
2D PAGE, 112
3'-end, 112
3'-extension, 112
3'-hydroxyl end, 112
5'-end, 112
5' end, 112
5'-extension, 112
3'-extension, 112

A

ab initio, 113, 201, 257, 350
  calculations, 113
  gene prediction, 113
  quantum chemistry, 113
Abdominal
  aorta, 113
  cavity, 388, 395
  muscles, 255
Abdominal
  ultrasound, 113
  x-ray, 113
Abiotic, 497, 503, 517
Ablation, 113, 195, 437
Abnormal depolarizations, 480
Abomasum, 459
Abruptio placenta, 114
Absces, 114
Abscisic acid, 114, 131
Abscissa, 114, 381
Absorb, 114, 242, 290–292, 304, 306, 318,
  339, 363, 373, 386, 403, 492, 532, 547
Absorption, 114, 117, 139, 146, 225, 229,
  250, 293, 327, 330, 398, 400, 403, 458,
  531
Absorption, distribution, metabolism, and
  excretion (ADME), 117, 398
Abzyme, 114, 168
Acaricide, 114
ACC synthase, 114
Acceptor
  control, 114

junction site, 114
Accession number, 114
Accessory bud, 114
Acclimatization, 114
Accommodation, 114
Acellular, 114
Acentric chromosome, 114
Acetyl co-enzyme A (acetyl CoA), 115
Achondroplasia, 115
Achromasia, 120
Achromatosis, 120
Achromia, 120
Acidity and alkalinity, 397
Acoustic-spectrum, 115
Acquired, 115, 120, 160, 220, 261, 294,
  418, 522
Acquired immunodeficiency, 120
Acridine dyes, 115
Acridinium esters, 315
Acrocentric, 115, 457
Acrocentric chromosomes, 457
Acropetal, 115
Acrosome, 115
Acrylamide gel, 115, 411, 512
Act, 21, 58, 60, 76, 78, 80, 97, 113, 115,
  122, 125, 226, 232, 242, 248, 255, 274,
  276, 285, 294, 327, 345, 352, 360, 370,
  377, 381, 407, 412, 414, 430, 432, 440,
  441, 493, 502, 517, 519, 521, 529, 532
Actin, 115
Action potential, 115
Activated
  carbon, 115
  charcoal, 115, 118
Active
  collection, 115
  electrode, 115
  immunization, 294
  pharmaceutical ingredient, 116
  transport, 116, 238, 240, 392
Acute, 74, 78, 80, 83, 84, 116, 119, 277,
  461
  transfection, 116

Acyl carrier protein, 115, 116
Adaptation, 114, 116, 303
    traits, 116
Adaptive
    Neural Systems Laboratory, 38
    radiation, 116
Adaptor, 116, 522
Addendum, 116
Addition rule establishes, 116
Additive
    allelic effects, 116, 117
    gene, 117, 280, 365
    genetic variance, 117, 365
Adenilate cyclase, 117
Adenine, 68, 113, 117, 143, 186, 200, 241,
    339, 370, 375, 411, 456, 470, 506, 515,
    531
Adenosine
    diphosphate, 118
    triphosphate, 136
Adenylate cyclase, 270
Adenylic acid, 117, 200
Adhesion, 117, 239, 403, 492, 493
Adipocyte, 171
Adjuvant, 117, 438
Admission requirements, 43, 44
Admixture, 118, 403
Adoptive immunization, 118, 295
Adrenal
    cortex, 118, 424
    gland, 118, 231, 546
Adrenergic
    blocking agent, 118
    neuron blocking agent, 118
Adriamycin, 218
Adsorbent, 118, 283
Adsorption, 118
Adult cloning, 118
Adult stem cells, 118, 124, 164
Advanced Materials Engineering Research
    Institute, 38
Advanced maternal age, 118
Adventitious, 118, 235, 319, 337, 382
Aerate, 118, 530
Aerobe, 118
Aerobic, 63, 118, 119, 141, 150, 241
    bacteria, 119
    respiration, 119
Aerogels, 360

Aerosol research, 343
Afferent, 119
Affinity chromatography, 119
Affinity tag, 119, 245
Aflatoxin, 119, 357
Agamospermy, 119, 448
Agarose, 119, 223, 293, 372, 484
    gel, 223, 372, 484
Agent Orange, 119
Aggregate, 119, 172, 184, 229, 246, 271,
    339, 340, 346, 478
Agonist, 119, 128
Agricultural
    applications, 3, 364
    production, 189, 296
Agrobacterium, 120, 145, 195, 453, 457,
    508, 515, 530
    rhizogenes, 120, 453, 457
    tumefaciens, 120, 145, 195, 508, 515,
    530
Agrobiodiversity, 120
Agronomic qualities, 462
Airlift fermenter, 120
Albinism, 120
Albino tissues, 120, 539
Albuminoid, 356
Albuminuria, 418
Alcoholism, 120
Aleurone, 120
Algal biomass, 121
Alginate, 121, 225, 293
Alien species, 121, 308
Alimentary canal, 114, 226
Alkali sodium hydroxide, 176
Alkaline solutions, 324
Alkaloids, 268
Alkylating agents, 528
Allantois, 121, 405
Allele, 121, 134, 135, 139, 142, 167, 180,
    183, 186, 211, 214, 216, 245, 254, 257,
    261, 262, 276, 278, 284, 286, 293, 319,
    320, 352, 356, 367, 369, 375, 378, 414,
    421, 441, 448, 452, 465, 467, 476, 493,
    510, 539, 545, 548, 551
    frequency, 121, 257, 539
Allelic
    diversity, 121, 262
    exclusion, 121
    richness, 121

series, 121
Allelomorphic, 121
Allelopathy, 122
Allergen, 122, 498, 499
Allergy, 122, 306, 455
Allogenic cells, 93, 122, 516
Allometric, 122
Allometry, 28, 62
Allopatric
   speciation, 122
   species, 122
Allopolyploid, 122
Allosome, 122
Allosteric
   control, 122
   enzyme, 122
   site, 123
   transition, 123
Allotetraploid, 123
Allotype, 123
Allozygote, 123
Allozygous, 123
Allozyme, 122
Allozyme, 123
Alpha
   fetoprotein, 123
   lactalbumin, 123
   particles, 123
Alternative mRNA splicing, 123
Alu repetitive sequence, 123
Alveoli, 123, 161, 301
Alveolus, 123, 124
Amber, 351, 371, 495, 501
Ambient temperature, 124
Amelioration, 480
Amine group, 201
Amino acid, 67, 82, 119, 124, 135, 139,
   144, 167, 170, 173, 177, 182, 183, 189,
   190, 199, 203, 214, 227, 229, 233, 261,
   284, 297, 301, 310, 338, 339, 347, 354,
   380, 387, 390, 394, 395, 404, 413, 414,
   426, 427, 440, 442, 448, 454, 456, 465,
   470, 474, 475, 478, 483, 495, 522, 524,
   525, 528, 529, 534, 538, 554
Aminoacyl
   site (A-site), 124
   tRNA synthetase, 124
Aminoglycoside, 313
Amitosis, 124

Amniocentesis, 124
Amniocyte, 124
Amnion, 124, 176, 177
Amniotic, 119, 123, 124, 264, 378, 406,
   412
   fluid, 123, 124, 264, 378, 406, 412
   membrane, 406
   sac, 124
   stem cells, 124
Amorph, 125
Amphidiploid, 125
Amphimixis, 125
Amphiphilic, 125
Ampicillin, 125, 388, 393
   resistance gene, 393
Amplified fragment length polymorphism,
   119, 125
Amplify, 125, 133, 236
Amplitude, 125, 253, 283, 326
Amylase, 125, 206
Amylolytic, 126
Amylopectin, 126, 491
Amylose, 126, 491
Anabolic pathway, 126
Anabolism, 126, 154
Anaemia, 126, 512
Anaerobic
   digestion, 126
   respiration, 126
Anagenesis, 126
Anal deformities, 538
Analgesia, 126, 372
Analgesic, 126, 227
Analogous, 126, 150, 287, 351, 360, 400,
   453, 467, 508
Analysis of
   molecular variation, 125
   variance, 128
Analyte concentration, 310, 417
Anaphase, 126, 184, 212, 339, 465
Anastomosing, 126
Anastomosis, 127
Anchor gene, 127
Androgen, 127
Androgenesis, 129
Anemia, 127, 452
Anesthesia, 127
Aneuploid, 127, 278, 348
Aneuploidy, 127, 178, 353, 409, 552

Aneurysm, 127
Angelman syndrome, 127
Anger camera, 127, 254
Angina, 127, 163
Angiogenesis, 127
Angiogenin, 127
Angiography, 127, 299
Angioplasty, 89
Angioplasty, 26, 127
Angiosperm, 127
Angiotensins, 128
Animal
    cell immobilization, 128
    genetic resources databank, 128
    genome (gene) bank, 128
    locomotion, 28
    model, 38, 128, 379
    tissue, 480, 550
Anneal, 128, 248, 494
Annealing, 128, 193, 195, 214, 443, 497
Annotation, 115, 128, 187
Annual, 26, 91, 93, 94, 99, 104, 128
Anode, 128
Anonymous DNA marker, 128
Anorectal atresia, 128
Antagonist, 128
Anterior choroidal artery, 128
Anterior pituitary, 248, 424, 445, 483, 484
Anther, 129, 203, 411, 470, 490
Anther culture, 129
Anthocyanin, 129
Anthropocentrism, 129
Antiacids, 129
Antiarrhythmic agents, 417, 480
Antiauxin, 129
Antibiosis, 129
Antibiotic, 67, 74, 77, 125, 129, 133, 137,
    154, 172, 176, 193, 291, 313, 367, 408,
    449, 511
    resistance, 74, 77, 133, 193, 367, 408,
    449
        gene, 193
        marker gene, 133, 367
Antibody, 82, 86, 92, 94, 113, 117, 119,
    121, 123, 129, 140, 157, 158, 168, 169,
    181, 185, 190, 200, 222, 230, 231, 240,
    242, 247, 278, 288, 289, 291, 293–295,
    313, 315, 316, 324, 370, 390, 407, 419,

438, 463, 470, 476, 483, 514, 516, 538,
    539, 548
    binding site, 129, 390
    class, 129, 293
    molecules, 123, 190, 538
    response, 324, 419
    structure, 129, 200
Anticholinergics, 130
Anticlinal, 130, 530
Anticoagulant, 64, 130
Anticoding strand, 130
Anticodon, 130, 501, 529, 549
Antidiarrheals, 130
Anti-digoxin monoclonal antibody, 140
Antiemetics, 130
Antifungal, 137
Antigen, 80, 119, 122, 129, 140, 169, 171,
    181, 185, 200, 222, 224, 230, 231, 242,
    262, 276, 278, 284, 285, 291, 294, 295,
    315, 324, 336, 381, 390, 411, 419, 420,
    426, 438, 452, 457, 463, 470, 476, 508,
    517, 539
Antigenbinding protein, 506
Antigenic
    determinant, 129, 130, 415, 499, 538,
    539
    material, 537
    switching, 130
Antihaemophilic globulin, 120, 130
Anti-idiotype antibodies, 130
Antimicrobial agent, 67, 130
Antinflamatory steroid, 290
Antinutrient, 130
Anti-oncogene, 441
Anti-oncogene, 130
Antioxidant, 130, 135, 415
    solution, 130
Antiparallel orientation, 130
Antisense
    DNA, 131
    gene, 131
    RNA, 131
    therapy, 131
Antiseptic, 131
Antiserum, 131
Antisolvent crystallization, 460
Antispasmodics, 131
Anti-terminator, 131
Antitoxin, 61, 62, 150

Antitranspirant, 131
Antiviral mechanism, 479
Antixenosis, 131
Anus, 126, 128, 131, 137, 159, 164, 255, 443
Aorta, 69, 113, 131, 192, 219, 430, 505, 541
Aortic and pulmonary valves, 468
Aortic valve, 131, 145, 430, 528, 550
Apert syndrome, 131
Apex, 103, 115, 132, 143, 184, 304, 473
Apical
    cell, 132
    dominance, 132, 139
    meristem, 132, 172, 192, 298, 320, 337, 420, 457, 473, 530, 554
Apoenzyme, 132, 284
Apomixes, 132
Apoptosis, 89, 132, 387
Appendectomy, 132
Appendicitis, 132
Appendix, 2, 6, 132
Applications, 3, 4, 6, 8, 13, 16, 19, 38, 39, 77, 90, 101, 147, 151, 166, 279, 332, 340, 342, 343, 351, 361, 365, 373, 394, 435, 464, 496, 525
Applicator, 132
Approximate Bayesian computation, 132
Aptamers, 132
Aquaculture, 132, 332
Arabidopsis thaliana, 113, 133, 348
Arbitrarily primed polymerase chain reaction, 132, 133, 213
Arbitrary primer, 133, 328, 355
Archaea, 133
Arginine, 124, 166, 309, 529
Arrhenius equation, 133
Arrhythmia, 66
Arrhythmia, 133
Arterioles, 133
Arteriosclerosis, 136
Artery, 126, 127, 133, 134, 158, 395
Arthritis, 106, 134, 252, 384
Artificial
    chromosom, 82, 552
    devices, 152
    heart, 18, 61, 66, 69–71, 73–76, 81–83, 94, 105, 538
    inembryonation, 134, 224, 300

insemination, 68, 134, 486
    seed, 134
    selection, 134, 262
    valves, 150, 334, 550
Ascending colon, 134
Ascertainment bias, 134
Ascites, 135
Ascomycete fungus, 553
Ascomycetes, 135
Ascospore, 135
Ascus, 135
Aseptic, 129, 135, 140, 233, 344, 381, 388, 503
Asexual, 119, 132, 135, 162, 170, 181, 188, 245, 256, 377, 425, 448, 489, 540, 545
    propagation, 135, 181
    reproduction, 135, 162, 170, 181, 245, 256, 377, 540, 545
Asian Technology Information Program, 135
Asparagine, 124, 135
Asparamide, 135
Aspartic acid, 124, 309
Aspirator pump, 538
Assay, 135, 207, 222, 223, 230, 241, 283, 294, 306, 346, 419, 437, 463, 470, 475, 511, 541
Assignment test, 136
Assisted
    human reproduction, 136
    reproductive technologies, 136
Associative overdominance, 136
Assortative mating, 136
Astrocyt, 136
Asymmetric hybrid, 136
Asynapsis, 136
Atherosclerosis, 136
Atomic
    diffusion, 208
    displacement, 408
    force microscope, 4–6, 84, 362
    mass unit, 136
    nucleus, 136, 373
    number, 253, 311
    number, 136
    weight, 136
Atonic colon, 136
Atria, 137, 162
Atrioventicular valve, 137

Atrium, 137, 248, 528, 541
Attenuated vaccine, 137, 372
Attenuation, 61, 137
Attenuator, 137
Aureofacin, 137
Authentic protein, 137
Autocatalysis, 137
Autoclave, 137
Autogamy, 137
Autogenous control, 138
Auto-immune disease, 138
Auto-immunity, 138
Autologous cells, 138, 516
Autolysis, 138, 367
Autonomous, 133, 138, 371, 407, 451
Autopolyploid, 138
Autoradiography, 138, 158, 316, 422, 484
Autosomal dominant, 138, 220, 331, 336, 357, 368, 451, 545
Autosome, 138, 139
Autotetraploid, 138
Autotroph, 139, 282
Autotrophic, 139
Autotrophy, 139
Autozygosity, 139
Autozygous, 139
Auxin, 129, 139
Auxotroph, 139, 429
Availability, 114, 139
Avian influenza, 139
Avidin, 140
Avidity, 140
Avirulence gene, 140
Avr gene, 140
Axenic culture, 140, 431
Axial pump, 140
Axilla, 140
Axillary bud, 114, 140, 318
Axon, 143, 369
Azotobacter spp, 370

B

B cell, 140, 157, 158, 289, 336
B chromosome, 140
B lymphocyte, 121, 140, 181, 379, 407, 483
BAC end sequence, 141
Bacillus, 141, 161, 195, 203, 390
    thuringensis, 141

Back flow, 137
Back mutation, 141, 452
Bacteria, 57, 58, 61–65, 67–69, 74, 77, 79, 82, 83, 119, 120, 141, 144, 146, 154–156, 165, 181, 199, 203, 206, 228, 230, 256, 269, 271, 277, 294, 301, 303, 306, 309, 316, 327, 337, 343, 348, 359, 370, 371, 373, 380, 392, 394, 397, 401, 407, 408, 414, 424, 440, 447, 450, 453, 458, 460, 480, 484, 491, 495, 505, 506, 513, 521, 526, 535, 536, 542
Bacterial
    artificial chromosome, 140, 141, 552
    cell, 68, 69, 141, 185, 297, 327, 340, 369, 414, 445, 500, 509
    chromosomes, 389
    acterial enzyme, 273, 373, 450
    fertility, 141
    genes, 369, 471
    ribonuclease, 142
    species, 161, 218, 317, 365
    toxin, 141
Bactericidal properties, 533
Bacteriocide, 141
Bacteriocin, 141
Bacteriology, 301
Bacteriophage, 66, 141, 193, 304, 317, 327, 328, 354, 397, 433, 448, 506, 521
    lambda DNA, 193, 304
Bacteriostat, 141
Bacterium, 59, 61, 65, 70, 79, 80, 91, 120, 141, 161, 182, 189, 192, 225, 226, 232, 237, 240, 248, 297, 300, 316, 327, 340, 348, 392, 395, 412, 449, 453, 457, 484, 488, 489, 496, 507, 521, 522, 541, 543
Baculovirus, 145, 193
    expression vector, 142, 145
Balanced
    lethal system, 142
    polymorphism, 142
Balancing selection, 142
Balloon urethroplasty, 142
Band gap energy, 142
Bar, 142, 392, 434, 490
Barium, 142, 228, 242, 326, 535
    enema, 142, 228
    titanate, 242
Barnase, 142
Barotropic, 458

Barotropic, 142
Barr bodies, 142
Barstar protein, 142
Basal, 142, 143, 315, 385, 389, 496
  lamina, 143, 389, 496
Base, 59, 73, 82, 91, 114–118, 120, 123,
  134, 142, 143, 159, 162, 178, 182, 186,
  193, 195, 198, 201, 203, 205, 213, 217,
  221, 234, 250, 255, 259, 261, 262, 281,
  303, 304, 314, 315, 325, 335, 347, 375,
  377, 388, 393, 420, 425, 432, 450, 461,
  475–479, 489, 494, 498, 499, 509, 514,
  515, 525, 529, 535, 536, 543, 549, 552
  analogue, 143
  collection, 115, 143
  pair, 82, 91, 116, 118, 143, 159, 178,
    186, 203, 217, 234, 255, 281, 304, 314,
    347, 393, 420, 425, 450, 475, 477, 479,
    489, 494, 499, 529, 535, 543, 552
  sequence, 143, 182, 262, 303, 498, 509
  substitution, 143
Basic fibroblast growth factor, 143
Basilar plate, 406
Basipetal, 143
Basophil, 143
Batch culture, 143
Batch fermentation, 143
Battery-powered external pacemaker, 69
Bayesian inference, 143
Becker muscular dystrophy, 144, 219
Bench scale process, 144
Benign
  prostatic hyperplasia, 144, 536
  tumors, 144, 197
Bernoulli Equation, 144
Beta
  error, 144
  particles, 145
Bicuspid aortic valve, 145, 528
Biennial, 145
Bifunctional, 473, 490
  vector, 145, 473
Bilateral symmetry, 420
Bilayer lipid membrane, 145
Bile, 145, 176, 185, 253, 280, 324
Bile ducts, 145, 176
Biliary atresia, 145
Bilirubin, 145, 291, 312
Binary vector system, 145

Binding, 122, 123, 130, 140, 145, 168, 200,
  216, 230, 242, 252, 285, 295, 313–315,
  377, 390, 425, 440, 441, 447, 454, 455,
  477, 508, 521, 522, 528, 539, 553, 554
Binomial
  expansion, 145
  nomenclature, 146
  proportion, 146
  sampling, 264
Bioaccumulation, 146, 150
Bioassay, 146
Bioaugmentation, 146
Bioavailability, 146
Biocatalysis, 146, 156
Biochemical
  cues, 492
  environment, 267
  modification, 454, 456
  pathway, 63, 242
  reaction, 230, 241, 246, 259
Biochemistry, 2, 39, 164, 171, 189, 235,
  342, 343, 349, 355, 411
Biochip technologies, 153
Biocompatibility, 413
Biocompatible prostheses, 19, 147, 151
Biocomplexity, 146
Biocontainment, 146
Biocontrol, 146
Bioconversion, 146
Biodegradability, 413
Biodegradable polymers, 411
Biodegradation, 147, 203, 413
Biodegrade, 147
Biodesulphurization, 147
Biodetectors, 362
Biodiesel, 481
Biodiversity, 120, 147, 149
Bioelectricity, 148, 152
Bioenergetics, 147
Bioenergy, 147
Bioengineering, 1, 2, 13, 24, 25, 34, 39, 55,
  70, 107, 111, 112, 151, 209, 360
Bioenrichment, 148
Bioethics, 148
Biofilms, 148, 239
Biofiltration, 245
Biofluid dynamics, 452
Biofouling, 148
Biofuel, 148

Biogas, 148
Biogenesis, 149
Biogeography, 149
Biohazard, 149
Bioheat transfer, 27, 148, 149, 152
Bio-imaging, 43
Bioinformatics, 19, 148, 149, 152, 201
Bioinstrumentation, 19, 148, 149, 152
Bioleaching, 149
Biolistics, 149, 257, 344
Biological
    action, 157
    affinity, 119
    communities, 221
    diversity, 154, 238
    engineering, 1, 4, 55, 75, 107, 111, 112
    macromolecule, 411
    material, 150, 246, 247, 324, 386, 398,
    480
    membranes, 16, 145, 400
    organisms, 2, 146
    oxygen demand, 158
    processes, 126, 154, 155, 295, 358, 398
    reaction, 153, 203
    systems, 3, 16, 19, 28, 29, 46, 77, 153,
    155, 295, 307, 338, 364, 467
    valves, 150
Biologics/biopharmaceuticals, 3, 19, 20, 22,
    150, 276, 364
Biology, 3, 4, 13, 16, 18, 20, 24, 26, 46, 51,
    58, 87, 88, 133, 146, 147, 151, 152, 154,
    218, 221, 261, 265, 295, 349, 358, 363,
    424, 459, 515, 545
Bioluminescence, 150, 326
Bioluminescent compounds, 315
Biomagnification, 150
Biomarker detection, 51
Biomarkers, 360
Biomass, 106, 147, 148, 150, 152, 531
    concentration, 150
    density, 531
    transport, 148, 152
Biomaterial compatibility, 148
Biomaterials, 13, 19, 27, 29, 30, 32, 33, 41,
    43, 48, 76, 147, 148, 150, 152, 153, 383,
    516
Biome, 30, 150
Biomechanical Aspects of Design, 516

Biomechanics, 19, 28–31, 34, 77, 79, 93,
    94, 97, 106, 147, 148, 150, 152
Biomedical
    Engineering Department at Florida
    International University, 13, 36
    engineering education, 2, 36
    Engineering Partnership Program, 43,
    96
    engineers, 1–3, 18, 19, 21, 22, 24, 26,
    34, 36, 37, 39, 40, 43–45, 77, 78, 93–95,
    97, 100, 101, 104, 107, 147, 148, 152
    imaging, 19, 21, 400
    industry, 36, 40, 44
    research, 40, 44, 81, 105, 151
BioMEMS, 39, 152
Biometric authentication, 152
Biometrics, 152, 153
Biomimetic materials, 153
Biomimetics, 28, 153
Biomolecular systems, 351
Biomolecule, 279
Bionanoscience, 3, 365
Bionanotechnology, 1, 34, 55, 107, 148,
    151, 152, 155, 364
Biopesticide, 153
Biopharmaceuticals, 20, 153, 398
    products, 159
Biopharming, 153, 351
Biopiracy, 153
Biopolymer, 153, 412
Bioprocess, 153, 535
Biopsy, 153, 469
Biopsy technique, 469
Bioreactor, 153, 162, 191, 242, 244, 295,
    337, 384, 400, 472, 485
Biorecovery, 153
Biorefineries, 2
Bioremediation, 5, 18, 153, 156, 480
Biosafety, 154, 167
Biosafety protocol, 154, 167
Biosensing, 154
Biosensor, 91, 154, 230, 295
Biosilk, 154
Biosorbents, 154
Biosphere, 154, 220
Biosynthesis, 72, 154, 199, 229, 301, 349
Biosynthetic antibody binding sites, 140,
    154
Biot number, 154

Biotech service, 196
Biotechnological
   process, 218, 470
   techniques, 5, 348
Biotechnologists, 154, 465
Biotechnology, 1–3, 16, 25, 32, 34, 55, 65, 78, 81, 88, 90, 92, 94, 107, 111, 114, 115, 148, 151, 152, 154, 155, 218, 248, 269, 309, 342, 358, 364, 366, 377, 404, 411, 440, 494, 503, 507, 532, 555
   industry, 90, 309, 404
Bioterrorism, 155
Biotherapeutic strategy, 155
Biotic, 150, 155, 497, 503, 517
   factor, 155
   stress, 150, 155
Biotin, 140, 155, 156, 496, 544
Biotope, 156
Biotoxin, 156
Biotransformation, 146, 156, 218, 535
Biotreatment, 156
Biotribology, 29
Bioweaponry, 8
Bird flu, 139
Birth defect, 243, 538, 541
Bivalent, 156, 443
Bladder, 100, 104, 121, 156, 198, 229, 306, 307, 314, 368, 385, 396, 426, 487, 496, 527, 535–537, 541
Bladder instillation, 156
Blast cell, 157
Blastocyst, 157, 202, 221, 226, 255, 287, 301, 419, 482
Blastocyst implantation, 202
Blastoderm, 157
Blastomere, 157, 518
Blastula, 157, 255
Bleach, 85, 157
Bleeding, 83, 126, 157, 194, 279, 405
Blindness, 508
Blocking agent, 157
Blood
   clot, 130, 157, 202, 240, 278, 279, 514
   components, 158
   filter, 158
   granulocytes, 379
   plasma, 171, 278, 327, 470, 522
   pressure, 133, 158, 207, 291, 292, 312, 334, 407, 416, 431, 487, 505

   products, 158
   serum, 267
   vessel, 127, 158, 164, 166, 168, 191, 219, 228, 268, 278, 279, 291, 316, 389, 405, 406, 430, 431, 445, 479, 493, 514, 532, 540, 546
Blot, 158
Blunt end, 116, 158, 450
B-lymphocytes, 158
Body
   fluid volume, 234, 545
   temperature, 159, 277, 291, 292, 331
   water, 159
   marrow, 18, 140, 143, 158, 159, 171, 264, 337, 357, 408, 497, 518, 533
      stromal cell, 159, 337
Bonferroni correction, 159, 470
Bony skull, 478
Bootstrap analysis, 159
Boring platform, 159
Boron nitride, 365
Bottleneck, 159
Bottomup methods, 360
Bound water, 159, 250
Boundary layer, 159
Bovine, 67, 159, 161, 329, 421
   growth hormone, 159
   serum albumin, 161
   somatotrophin, 159, 161
Bowel, 131, 159, 190, 207, 306, 492, 495, 532
   contents, 131
   movement, 159, 190, 207, 492, 495
Brachial, 159
Bract, 160
Brain, 19, 39, 57, 84, 99, 151, 157, 160, 171–173, 194, 208, 222, 252, 267, 271, 292, 307, 312, 316, 322, 329, 330, 336, 368, 399, 438, 508, 546, 548
   activity, 252
   injury, 208
   mapping, 39
Branch length, 160
Branemark Osseointegration Center, 85
Brassicas, 268
Breast
   cancer, 39, 85–87, 89, 90, 92, 97, 160, 280, 427, 438, 508
   milk protein, 316

Breed, 14, 160, 194, 196, 211, 216, 226, 238, 385
Breeding, 132, 142, 160, 181, 189, 194, 195, 224, 226, 233, 238, 243–245, 249, 257, 258, 260, 289, 296, 332, 348, 365, 371, 384, 385, 415, 433, 490, 501, 524
    program, 224, 296
    value, 160, 233, 365, 371
Brewing, 154, 160, 330
Bridge, 37, 83, 89, 160, 161, 212, 240, 242, 455, 459
Brinell, 161
Broad sense heritability, 161, 280
Broad-host-range plasmid, 161
Bronchiole, 161
Bronchiolitis, 161
Bronchoscopy, 161
Bronchus, 161
Brood stock, 161
Brothersister mating, 297
Browning, 130, 161, 399, 415
Brucella, 161
Brucellosis, 161
Bt crops, 161
Bt toxins, 161
Bubble
    column fermenter, 162
    oxygenator, 71, 83, 162
Bud, 132, 140, 162, 270, 318, 347, 358, 409, 444, 462, 473, 510, 531, 545
Budding, 162, 462, 552
Buffer, 162, 198
Bulk
    degradation, 162
    micromachining, 162
Bulked segregant analysis, 162
Buoyant density, 162
Burns, 96, 497
Byte, 162

C

Cable electrode, 163
Calcium
    channel blocking agent, 163
    hypochlorite, 157, 292
Calf scours, 163
Calibration, 163, 322, 343
Calicivirus, 163

Callipyge, 163
Callus culture, 67, 163, 444
Calorie, 163
Calyx, 164
Cambial zone, 164
Cambium, 164, 305, 422, 464, 539, 540
Cancer, 10, 11, 19, 84, 85, 87, 89, 90, 92, 96, 98, 100, 105, 119, 153, 164, 166, 171, 174, 200, 209, 218, 327, 335, 339, 356, 357, 359, 360, 363, 367, 379, 386, 387, 401, 402, 430, 438, 441, 451, 466, 469, 507, 508, 530
    cells, 96, 200, 339, 363, 401, 469, 507
    chemotherapy, 218, 387
    stem cells, 164
    technology, 19
    therapy, 438
Cancerous cells, 425
Candida, 164
Candidate gene, 164, 252
Canine, 58, 73, 164
Cannula, 164
Canola, 165
Cap site, 165
Capacitance, 165, 241
Capacitor, 165, 175, 208, 387
Capillaries, 57, 158, 165, 268, 428
    veins, 158
    capillary electrophoresis, 165, 223
Capsid, 165, 225, 534, 543
Capsomeres, 543
Capstone project, 44
Capsule, 159, 165, 226, 268, 422, 527, 543
Carbapenems, 144
Carbohydrate, 165, 171, 172, 278, 334, 379, 491
Carbon
    nanofoam, 6, 165
    nanotube actuators, 6
    nanotubes, 5, 6, 57, 88, 90, 92, 95, 165, 365, 478
Carbowax, 166
Carboxyl group, 166, 309, 394
Carboxylase enzymes, 155
Carboxypeptidase, 166
Carcinogen, 166
Carcinoma, 166, 466

Cardiac, 65, 67, 69, 70, 75–77, 84, 87, 106, 137, 162, 163, 166, 207, 299, 302, 306, 345, 445, 478, 538
  arrest, 166
  cycle, 166, 207
  muscle, 162, 302, 478
  output, 76, 166, 299, 445
  problems, 538
Cardiology, 19, 32, 166
Cardiomyocytes, 171
Cardiomyopathy, 249
Cardioplegia, 166
Cardiopulmonary bypass, 68, 194
Cardiotomy, 166, 306, 457
  suction, 166, 457
Cardiovascular, 3, 19, 28, 29, 39, 92, 97, 166, 247, 331
  disease, 166
  mechanobiology, 39
  system, 3, 19, 166, 247, 331
  technology, 19
Carotenoid, 166
Carpel, 167
  tunnel syndrome, 167
Carrier, 117, 167, 270, 276, 348, 538, 551
  DNA, 167
  gas, 167
  mediated transport, 167
  molecule, 167
Cartagena protocol, 154, 167
Casein hydrolysate, 167, 382, 534
Catabolic pathway, 167, 168
Catabolism, 167
Catabolite
  activator protein, 165, 167
  regulator protein, 167, 195
  repression, 168
Catalase, 168
Catalysis, 137, 168
Catalytic
  antibody, 114, 168
  RNA (ribozyme), 168
  site, 168
Catheter, 71, 157, 168, 445
Cathode, 168
Cation, 168, 174, 447
Cauliflower mosaic virus 35S, 164, 168
CD molecules, 168
Cecum, 132, 169, 184, 318

Cell
  culture, 18, 68, 169, 201, 276, 281, 291, 407, 438, 472
  cycle, 169, 282, 306, 356, 386, 460, 504, 532
  differentiation, 169, 199, 492
  division, 60, 121, 124, 130, 132, 139, 157, 169, 170, 172, 181, 199, 202, 205, 215, 231, 314, 315, 335, 337, 389, 391, 395, 423, 424, 487, 530
  elongation, 463
  free, 170
    transcription, 170
    translation, 170
  fusion, 170, 281, 437, 482
  generation time, 170, 260
  hybridization, 75, 170, 505
  hybridization, 170
  line, 70, 169, 170, 196, 266, 294, 303, 322, 346, 387, 420, 466, 476
  membrane, 59, 116, 157, 163, 169, 170, 205, 223, 226, 269, 270, 308, 323, 392, 394, 417, 422, 472, 474, 545
  nucleus, 59, 313, 373, 419
  number, 170, 233, 272, 316, 322, 491
  phosphoglycerides, 302
  plate, 170
  proliferation, 4, 216, 379, 441
  sap, 170, 461, 517
  selection, 170
  strain, 170
  suspension, 105, 119, 170, 191, 316
  therapies, 169
  type, 124, 168–171, 202, 206, 211, 263, 278, 286, 287, 294, 326, 351, 355, 412, 482, 492, 502, 512, 518, 521, 528, 534, 535
  wall, 121, 125, 130, 170–172, 175, 228, 244, 269, 271, 321–323, 328, 345, 393, 395, 407, 408, 420, 462–464, 487, 547
Cellomics, 171
Cellular
  function, 18, 383, 463
  immune response, 171, 506
  oncogene (proto-oncogene), 171
  processes, 167
  responses, 270
  tumor antigen p53, 386

Cellulose, 172, 184, 244, 278, 322, 370, 372, 414, 484, 532, 548
  fibers, 532
  nitrate, 172, 370
Cellulosome, 172
Census population size, 172
Center of origin, 172
CentiMorgan (cM), 182, 323, 331
Central
  dogma, 172, 444
  mother cell, 172
  nervous system, 13, 151, 173, 267, 368, 372, 472
Centrifugal
  forces, 172
  pump, 79, 172
Centrifugation, 172, 204, 387, 461, 500
Centrifuge, 60, 172
Centriole, 172, 425
Centromere, 114, 115, 172, 210, 310, 315, 338, 389, 395, 425, 457
Centrosome, 172
Cephalic, 172
Cephalosporin, 172
Cephamycins, 144
Cephem-type antibiotic, 172
Ceramic, 4, 57, 72, 173, 210, 245, 267, 271, 387, 443, 478, 544
  engineering, 4
  glazes, 57
Cereal species, 383
Cerebellum, 173, 174, 546
Cerebral
  cortex, 173, 331
  hemispheres, 192
Cerebrospinal fluid, 173, 336
Cerebrum, 173, 322, 376
Certified Clinical Engineer, 26
Cervix, 173
Chain terminator, 173
Chakrabarty decision, 173
Chaperone, 173, 349
Chaperonins, 173
Character, 145, 173, 367, 398, 514, 520
Characterization, 17, 173, 247, 275, 331, 344, 456, 537
Charcoal, 115, 174
Charcot-Marie tooth disease, 174
Chelate, 174

Chelating agent, 174
Chelator, 174
Chemical
  bond, 93, 194, 212, 308, 321, 394
  compound, 125, 154, 205, 302, 357, 423, 463, 550
  engineering, 18, 19, 87, 209, 218, 535
  messenger, 302, 369, 402, 463
  semisynthesis, 218
  transducers, 342
  vapor deposition, 6, 102
  weapons, 8
Chemiluminescence, 174, 315
Chemistry, 2, 4, 6, 10, 12, 13, 20, 41, 42, 88, 113, 146, 187, 199, 201, 254, 275, 283, 342, 346, 349, 350, 359, 360, 362, 363, 382, 434, 437, 553
Chemoembolization, 174
Chemostat, 174, 403
Chemotaxis, 174
Chemotherapy, 89, 174, 215, 507
  medication, 215
Chiari malformation, 174
Chiasmata, 175, 433, 510
Chicken pox vaccine, 537
Chimera, 175, 336, 395
Chimeraplasty, 175
Chimeric DNA, 175
Chimeric
  gene, 175, 252
  proteins, 175
  selectable marker gene, 175
Chip, 7, 39, 74, 102, 175, 264, 303, 333, 336, 342, 344, 348, 353, 387, 479, 506, 532, 545
Chi-square test, 175
Chitin, 175
Chitinase, 175
Chitosan, 176, 309
Chitosanase, 176
Chloramphenicol, 176
Chlorella spp, 121
Chlorenchyma, 176
Chlorophyll, 120, 176, 408
Chloroplast DNA, 176
Chloroplast transit peptide, 176, 196
Chloroplastids, 408
Chloroplasts, 153, 176, 194, 199, 239, 271, 382, 388, 408, 424, 535

Cholangiography, 176
Cholecystectomy, 176
Cholesterol, 176, 199, 368
Cholinergic blocking agent, 176
Chondrocyte, 171
Chorioamnionitis, 176
Chorion, 176, 177, 202
    frondosum, 176
    laeve, 177
Chorionic cavity, 177
Chorionic
    membrane, 406
    plate, 334, 406
    sac, 542
    somatomammotropin, 177
    villi development, 542
    villus sampling, 177
Chroma, 177
Chromatid, 115, 177, 275, 425, 478
Chromatin, 61, 175, 177, 284, 375, 479
    fibers, 177
    immunoprecipitation, 175
Chromatograms, 138
Chromatographic technique, 339, 429
Chromatography, 67, 119, 177, 254, 255,
    283, 294, 343, 411
Chromocenter, 177
Chromogenic substrate, 177
Chromomeres, 177
Chromonema, 177
Chromophores, 316
Chromoplast plastid, 177
Chromosomal
    aberration, 178, 468
    arms, 395
    genes, 348
    integration site, 178
    material, 234
    polymorphism, 178
    rearrangement, 178
    region, 190, 443
    segment, 125, 374
    site, 249, 526
Chromosome, 62, 63, 67, 71, 87, 112,
    114, 115, 120, 121, 124, 125, 127, 136,
    138–142, 156, 172, 177–179, 181, 187,
    190, 193, 194, 197, 203, 207, 210, 212,
    217, 219, 227, 231, 232, 234, 235, 239,
    240, 245, 247, 250, 257–259, 261–263,

274, 275, 281, 282, 288, 291, 292, 304,
    308, 310, 315, 323, 325, 327, 331, 335,
    338, 348, 353, 356, 373, 374, 376, 377,
    386, 390, 402, 409, 414, 418, 425, 432,
    433, 445, 446, 448, 449, 453, 467, 470,
    471, 478, 482, 497, 500, 504–506, 509,
    511, 519, 520, 523, 525, 529, 534, 543,
    546, 547, 550–552, 554
    aberration, 178
    banding, 178
    jumping, 178
    landing, 178
    mutation, 178, 356
    theory of inheritance, 179
    walking, 178, 179, 546, 547, 552
Chronic, 72, 73, 75, 76, 86, 120, 179, 195,
    232, 306, 474
Chronological
    order, 55
    events, 107
        2030 A.D., 107
        7000 B.C., 107
Chrysanthemum, 56
Chyme, 179
Chymosin, 179
Cilium, 143, 179, 315
Circadian, 92, 179
Circuit, 41, 100, 165, 168, 175, 179, 222,
    223, 255, 279, 299, 300, 303, 329, 332,
    383, 421, 449, 464, 548
Circularization, 179
Circulatory system, 114, 166, 179
Cirrhosis, 135, 145, 179
Cis
    acting protein, 179
    configuration, 179, 193, 448
    heterozygote, 179
Cistron, 179
Citric acid, 130
Citrus spp, 473
Civil liberties, 8
Civil organizations, 8
Clade, 179, 390
Cladistics, 180
Cladogenesis, 180
Cladogram, 180
Class switching, 180
Cleave, 158, 180, 209, 374, 441, 454, 459,
    478

Cleaved amplified polymorphic sequence, 165, 180, 393
Cleft lip/palate, 180
Cline, 180
Clinical
    engineering, 21, 34, 148, 152
    trial, 78, 94, 96, 180
Clitoris, 180
Cloaca, 121, 405
Clonal
    propagation, 181
    selection, 181
Clone, 79, 141, 169, 181, 185, 201, 218, 238, 241, 245, 256, 267, 333, 347, 383, 439, 442, 499, 500, 547, 548, 552
Cloning vector, 125, 140–142, 181, 235, 238, 248, 249, 303, 347, 393, 397, 508, 521, 552
Closed continuous culture, 181
Clostridia bacteria, 62
Clostridium difficile, 181
Clotting factor, 157, 279, 470
Cluster of differentiation, 181
Coalescent, 182
Coanda effect, 182
Coat protein (capsid), 182
Coaxial cable, 182
Coccus, 141
Coconut milk, 182, 382, 534
Cocoon, 182
Co-culture, 182, 242
Codex Alimentarius Commission, 182
Coding, 169, 175, 182, 183, 236, 238, 247, 252, 258, 265, 284, 305, 316, 372, 425, 454, 460, 468, 475, 476, 488, 509, 510, 525, 552
    sequence, 169, 175, 182, 236, 238, 252, 425, 476, 488, 552
    strand, 182, 372, 468, 509
Co-dominance, 183
Co-dominant alleles, 183
Codon, 73, 130, 183, 203, 301, 347, 371, 491, 495, 522, 524, 525, 529, 549
Coefficient, 183, 198, 419
Coelocentesis, 183
Coenzyme (cofactor), 183
Co-evolution, 183
Co-factor, 183, 446
Co-fermentation, 183

Coherent light, 318
Cohesion, 183
Cohesive ends, 116, 183, 494
Coincidence, 184
Co-integrate, 184
Co-integrate vector system, 184
Colchicine, 184
Colchicum autumnale, 184
Cold-induced vasodilatation, 184
Coleoptile, 184
Coleorhiza, 184
Colic, 184
Co-linearity, 184
Colitis, 184
Collagen tissue, 184
Collecting duct, 184
Collenchyma, 184
Collision frequency, 133
Colloid, 7, 184, 255, 339, 343
Colloidal crystal, 7
Colon, 184, 195, 232, 318, 387
Colony, 158, 184, 296, 428
Colony hybridization, 184, 296
Colorectal cancer, 437
Combinatorial
    library, 184
    processes, 546
Commensal, 232, 356, 459
Commensalism, 185
Common bile duct, 185
Commonwealth Scientific and Industrial Research Organization, 196
Companion cell, 185, 399
Comparative
    gene mapping, 185
    mapping, 185
    positional candidate gene, 185
Compartmental model, 185
Competence, 185, 297
Competent, 185
Complement proteins, 185, 288
Complementarity, 169, 185, 186, 497
Complementarity determining regions, 185
Complementary
    DNA, 185
    entity, 186
    genes, 186
    homopolymeric tailing, 186, 200, 206
    nucleotides, 186, 289

RNA, 186
Complementation test, 186, 520
Complete digest, 186, 391
Component, 46, 120, 123, 138, 177, 186, 202–205, 228, 235, 268, 273, 283, 300, 323, 331, 333, 346, 370, 377, 400, 417, 454, 468, 479, 480, 485, 522, 530, 531
Composite transposon, 187
Compound chromosome, 187
Compressible fluid flow, 187, 204
Computational
    gene recognition, 187
    modeling, 43
Computed tomography scan, 187
Concatemer, 187
Concordance, 187
Condensation, 380, 394, 400
Condenser electrode, 188
Conditional lethal mutation, 188
Conditioning, 188
Conduction, 119, 142, 188, 277, 376, 430, 474, 478, 519, 531, 551, 553
Conductor, 188, 378, 414, 449, 468, 491, 502
Confocal laser scanning microscopy, 188
Confocal microscopy, 188
Conformation, 173, 188, 216, 277, 371, 446, 477, 500, 510, 553, 554
Conformational analysis, 188
Congenital
    condition, 180, 487
    malformations, 509
Congestion, 188
Conidium, 188
Conjugate, 188, 221
Conjugation, 68, 69, 189, 240, 471, 528
Conjugative functions, 189
Connective tissue, 143, 158, 184, 205, 219, 222, 244, 284, 292, 317, 331, 337, 368, 388, 451, 497
Consanguinity, 189
Consejo Superior de Investigaciones Científicas, 365
Consensual heat vasodilatation, 189
Consensus sequence, 163, 189, 419
Conservation
    breeding, 189
    collection, 189
    laws, 189

Conserved sequence, 163, 189, 354, 472
Conspecific, 189, 203
Constant domains, 190
Constipation, 190, 226
Constitutive
    enzyme, 190
    gene, 190
    promoter, 190
    synthesis, 190
Construct, 2, 82, 186, 190, 244, 359, 364, 450
Contaminant, 119, 190, 214, 553
Contig, 120, 186, 190, 357, 358, 500
Contiguous
    genes, 190
    map, 190
Continuous
    culture, 190, 380
    distribution model, 190
    variation, 153, 191, 211, 433, 514
Contrast
    agent, 191, 340, 343
    bath, 191
Control, 7, 31, 42, 71, 77, 78, 82, 90, 102, 114, 120, 130, 138, 146, 149, 153, 164, 174, 180, 191, 215, 229, 230, 245, 246, 248, 266, 279, 288, 294, 295, 304, 318, 330, 334, 342, 346, 364, 373, 377, 380, 381, 392, 398, 416, 417, 438–440, 456, 464, 467, 471, 472, 475, 493, 520–522, 529, 536, 539
Controlled environment, 180, 191, 197, 272, 344
Controlling element, 191
Convention, 154, 155, 168, 179, 191, 388, 531
    biological diversity, 154
    International Trade in Endangered Species, 179, 191
Converged, 191
Conversion, 146, 147, 156, 192, 227, 230, 257, 345, 346, 370, 374, 444, 462, 469, 522, 545
Coordinate repression, 192
Co-polymers, 192, 412
Copy DNA, 192
Copy number, 115, 192, 217, 528
Copy number variation, 115
Co-repressor, 192

Cornelia de Lange syndrome, 192
Coronary arteries, 127, 192
Coronavirus, 192, 461
Corpus, 192, 424, 504
Corpus collosum, 192
Corpus luteum, 424, 504
Correlation, 192, 220, 282, 525
Corrosion, 192, 253, 404, 496
Cortex, 192, 492
Cortical, 192, 227
Corticosteroids, 192
Cos ends, 193
Cos site, 193
Co-segregation, 193
Cosmid, 193
Co-suppression, 193
Cosurfactant, 341
Cot curve, 193
Co-transfection, 193
Co-transformation, 193
Cotyledon, 193, 352
Coulomb, 193
Coulombic interatomic bond, 308
Countergradient variation, 193
Coupling, 179, 193, 398, 448
Coupling agent, 193
Covalent bond, 194, 195, 290, 494
Covalently closed circle, 194
Covariance, 194
CP4 5-enolpyruvyl-shikimate, 194
CpG islands, 194
Cranial nerves, 194
Craniocerebral trauma, 194
Creep, 194
Cri-du-chat syndrome, 194
Critical breed, 194
Crop rotation, 56
Cross, 5, 14, 24, 58, 70, 90, 113, 122, 175,
    184, 187, 190, 195, 209, 240, 271, 272,
    305–307, 323, 352, 461, 462, 469, 477,
    485, 510, 528
  hybridization, 195
  pollination, 195
  pollination efficiency, 195
Crossing over, 64, 195, 388, 442, 504, 534
Crown, 120, 192, 195, 377, 381, 515
Crown gall, 195, 381
Crustacean shells, 176
Cry proteins, 195

Cryoablation, 195
Cryobiological preservation, 196, 251
Cryodesiccation, 327
Cryogenic, 196, 224
Cryogenic preservation, 224
Cryoprotectant, 196
Cryptic, 196
Crystal, 99, 188, 196, 231, 240, 271, 311,
    319, 326, 359, 534, 551
  structure, 188, 240, 311, 534, 551
Crystallization, 314, 427, 461
Crystallographic
  directions, 311
  orientations, 271
Cultigen, 196
Cultivar, 196, 233, 264, 529
Culture, 6, 39, 75, 101, 120, 129, 139, 140,
    161, 170, 174, 181, 182, 190, 191, 193,
    196, 197, 210, 211, 219, 224, 235, 237,
    242, 244, 267, 272, 275, 281, 294, 296,
    298, 299, 301, 302, 324, 330, 334, 337,
    340, 346, 370, 381, 383, 387, 388, 391,
    403, 407, 408, 410, 420, 425, 428, 429,
    441, 451, 457, 460, 466, 468, 470, 472,
    473, 476, 481, 490, 491, 498, 503, 504,
    513, 531, 538
  alteration, 196
  medium, 120, 190, 197, 244, 330, 334,
    381, 466, 470
  room, 197, 298
Curing, 197
Current, 17, 32, 69, 79, 94, 99, 115, 125,
    165, 179, 188, 197, 208, 210, 217, 222,
    223, 236, 243, 251, 279, 283, 295, 299,
    302, 310, 320, 326, 329, 343, 344, 348,
    354, 359, 364, 378, 401, 402, 406, 417,
    431, 443, 448, 449, 464, 503, 508, 547,
    553
Cut, 65, 77, 129, 158, 161, 163, 175, 180,
    193, 197, 216, 238, 302, 368, 393, 441,
    450, 517, 548
Cuticle, 197
Cutting, 44, 187, 197, 212, 235, 260, 319,
    457, 493
Cyanine dye, 197, 299
Cyanosis, 197
Cyberknife, 197
Cybrid, 197
Cyclic dimers, 413

Cyclic oligosaccharides, 198
Cyclodextrin (cycloamyloses), 198
Cyclohexane, 357
Cycloheximide, 198
Cyclotron, 198
Cy-dyes, 198
Cysteine, 124, 198, 427, 554
Cystic fibrosis, 85, 87, 89, 198, 260
Cytidine, 182, 196, 198, 204, 453
    monophosphate, 182, 198
Cytidylic acid, 182, 198
Cytochrome, 13, 199
Cytogenetics, 199
Cytokinin, 139, 199, 315
Cytology, 199
Cytolysis, 199
Cytometers, 343
Cytoplasm, 170, 197, 199, 200, 239, 243,
    269, 307, 327, 347, 374, 375, 402, 408,
    410, 424, 428, 471, 472, 474
Cytoplasmic, 136, 176, 197, 199, 227, 239,
    315, 333, 374, 390, 395, 407, 408, 419,
    440, 504, 517
    division, 199
    genomes, 390
    hybrid, 197
    inheritance, 239
    membrane, 517
    structure, 315
Cytoplast, 197
Cytosine, 68, 143, 163, 182, 186, 198, 200,
    339, 370, 375, 432, 456, 470, 506, 515,
    531
Cytosol, 200, 306
Cytotoxic T cell, 200
Cytotoxicity, 200, 221
Cytotrophoblast, 200
Cytotype, 200

D

D loop, 200
DA–dT tailing, 186
Dalton, 201, 314
Dansyl chloride, 316
Darwinian cloning, 201
Daughterless carp, 201
De novo, 201, 350, 382, 421
    transcriptome, 201

De-activation, 169
Deamination, 201
Death phase, 201
Decaying organic matter, 461
Deceleration phase, 202, 322
Decidua, 202, 404
    parietalis, 202
Decidual
    cell, 202
    stroma, 306
Decidualization, 202
    process, 202
Dedifferentiation, 202
Deep vein thrombosis, 202
Defective virus, 202
Deficiency, 86, 89, 156, 178, 202, 240, 274,
    404, 514
Defined, 22, 77, 87, 94, 102, 129, 160, 170,
    172–175, 190, 202, 207, 209, 211, 214,
    219, 243, 251, 276, 287, 299, 331, 347,
    356, 365, 370, 376, 392, 395, 414, 415,
    422, 428, 441, 456, 466, 469, 477, 478,
    499, 525, 531, 533, 534, 553
Degeneracy, 203
Degeneration, 174, 203, 329, 451
    joint disease, 384
Degradation, 203, 222, 328, 396, 413, 468
Degradative pathway, 167
Degrees of freedom, 203
Dehalogenation, 203
Dehiscence, 203, 470
Dehydration, 203, 307, 327
Dehydrogenase, 203
Dehydrogenation, 203, 246
De-ionized water, 203
Deleterious mutations, 356
Deletion, 112, 178, 194, 203, 211, 250, 348,
    409, 418, 475, 490
Deliberate release, 203
Delta endotoxins, 195, 203
Deme, 203, 448
Dementia, 536
Demineralize, 203
Demographic stochasticity, 203
Denaturation, 194, 204, 446, 477, 479
Denature, 204, 512
Denatured DNA, 204
Denaturing gradient gel electrophoresis,
    204, 207, 223

Dendrimer, 204
Dendrite, 204, 271
Dendritic cells, 278, 379
Dendrogram, 204
Denitrification, 204
Density, 73, 74, 115, 142, 162, 172, 174,
    187, 204, 232, 245, 289, 293, 298, 329,
    330, 346, 428, 461, 464, 480, 531, 540
Density gradient centrifugation, 204
Deoxy form, 533
Deoxyadenosine, 201, 204, 205
    5'-triphosphate, 201
Deoxycytidine, 186, 198, 201
    5'-triphosphate, 201
Deoxygenated blood, 405
Deoxyguanosine, 186, 204, 205, 207
    5'-triphosphate, 207
Deoxyguanosine monophosphate, 205
Deoxyguanylic acid, 273
Deoxyhemoglobin, 279
Deoxyribonuclease, 205, 215, 230, 373
Deoxyribonucleic acid, 145, 204–206, 213,
    374, 397, 489
Deoxyribonucleoside, 198, 204, 369, 375,
    514
Deoxyribonucleotide, 205, 273, 375, 454,
    515, 517
Deoxyribose (2-deoxyribose), 205
Deoxyribosylthymine, 205
Depolarization, 162, 205
Depurination, 205
Derepression, 205
Derivative, 204, 205, 244, 273, 370, 499,
    534
Derived character, 205
Dermis, 205, 292, 478
Descending colon, 206
Desiccant, 206
Desiccate, 206
Desiccation, 206
Design, 4, 18, 20–23, 29, 37, 40, 41, 46, 72,
    73, 76, 92, 102, 147, 148, 151, 152, 163,
    164, 187, 206, 215, 287, 342, 345, 350,
    383, 422, 444, 467, 516
Desoxyribonucleic acid, 206, 213
Dessicator, 206
Desulphurization, 206
Detergent, 206, 480, 502, 548
Determinate growth, 206

Determination, 13, 113, 206, 224, 286, 471,
    550
Determined, 67, 91, 206, 212, 224, 255,
    268, 271, 323, 337, 345, 346, 380, 421,
    427, 440, 469, 471, 480, 484, 486, 520,
    529
Deterministic, 206, 434
Development, 206
Developmental biology, 206
Deviation, 160, 206, 461, 485
Dextran, 206
Dextrin, 206
Dextrose, 206
DG–dC tailing, 206
Diabetes, 82, 207, 536
Diagnostic
    applications, 515
    bioimaging and sensor systems, 47
    imaging technique, 113, 532
    procedure, 330
    techniques, 5
Diakinesis, 207, 210
Dialysis, 20, 27, 32, 72, 74, 79, 207, 461
Dialyzer, 67–70, 72
Diamondoids, 7
Diarrhea, 130, 181, 246, 267, 458
Diastolic pressure, 334
Diathermy, 207, 222, 329, 387
Diazotroph, 207
Dicentric chromosome, 207
Dichogamy, 207
Dicotyledon, 208
Di-deoxynucleotide, 208
Dideoxynucleotide triphosphate, 201
Dielectric, 208, 242, 339, 403, 410
    displacement, 208
    material, 208, 242, 410
    strength, 208, 403
Differential
    centrifugation, 208
    display, 208
    scanning calorimetry, 219
Diffuse axonal injury, 208
Diffusion, 70, 208, 213, 239, 240, 244, 252,
    335, 337, 376, 383, 384, 392, 410, 478,
    492, 507
    coefficient, 208
    system, 208
Digest, 145, 180, 209, 253, 290, 316, 381

Digestion, 123, 125, 126, 145, 179, 198, 209, 229, 243, 255, 274, 291, 298, 306, 315, 322, 324, 328, 459, 479
Digestive
  process, 460
  system, 254, 393, 535
  tract, 114, 131, 195, 209, 227, 255, 318, 354, 376, 398, 479, 495, 532, 535, 545
Digital, 15, 42, 94, 100, 209, 219, 306, 434, 516
  rectal exam, 209
Dihaploid, 209
Dihybrid, 209
Dilate, 209
Dimensionless number, 154, 209, 376, 418, 452, 455, 458
Dimer, 209, 211, 285, 432
Dimethyl sulphoxide, 209, 213
Dimorphism, 209
Dinucleotide, 209, 489
Diode, 98, 175, 210, 320, 401, 531, 553
Dioecious, 210
Dip, 6, 94, 210
  pen nanolithography, 6, 210
Diploblastic, 420
Diplochromosome, 210
Diploid, 121, 123, 127, 132, 210, 275, 281, 287, 335, 353, 376, 409, 443, 486, 489, 511, 518, 529, 549, 555
  rye, 529
Diplonema, 210, 387
Diplontic and diplohaplontic organisms, 448
Diplophase, 210
Diplotene stage, 510
Diptera, 177
Direct
  current, 128, 210
  embryogenesis, 210, 299, 482
  organogenesis, 210, 299
  reprogramming, 211
Directed
  amplification of minisatellite DNA, 211
  differentiation, 211
  mutagenesis, 211, 296, 378, 379, 427
Directional
  cloning, 211
  selection, 211
Disaccharide, 211, 316

Disarm, 211
Disassortative mating, 211
Discontinuous variation, 191, 211, 433
Discordant, 211
Discrete
  generations, 211
  microelectronic devices, 341
Disease, 211
  indexing, 212
  organisms, 538
  resistance, 212, 462
Disinfestation, 212, 548
Disjunction, 212, 371
Disomy, 212
Dispense, 212
Dispersal, 212
Disrupter gene, 212
Dissecting microscope, 212
Dissolve, 149, 168, 170, 177, 208, 212, 234, 279, 290, 323, 324, 349, 369, 382, 435, 480, 513
Distention, 212
Distilling process, 212
Distinct population segment, 212, 218
Di-sulphide bond, 212
Diterpenoid, 249
Ditype, 213
Diurnal, 213
Divergent evolution, 213
Diverticulitis, 213
Diverticulum, 213, 334
Dizygotic twins, 213
DNA
  carrier, 214
  construct, 214, 447
  delivery system, 214
  diagnostics, 214, 346
  fingerprint, 5, 82, 90, 94, 133, 162, 214, 545
  helicase (gyrase), 214
  hybridization, 214, 460
  ligase, 214, 321, 378, 494
  micro-array, 214
  polymerase, 71, 169, 185, 214, 215, 315, 369, 420, 421, 425, 452, 507
  polymorphism, 213, 214
  primase, 215, 421
  probe, 115, 195, 215, 281, 372, 374, 527
  repair, 215, 235, 347, 401, 440, 484

replication, 197, 215, 320, 378, 420, 447, 508, 517
transfer procedures, 167
Dobzhansky Muller incompatibilities, 215
Docetaxel, 215
Dock, 215, 216, 351
Docking, 215, 216, 351
  programs, 216
  studies, 216
Docosahexaenoic acid, 508
Doctoral academic program, 43
Dolly, 92, 216, 448
Domain, 187, 216, 231, 243, 252, 270, 285, 458, 476, 503
Domestic animal diversity, 201, 216
Dominance, 117, 216, 265, 461
  genetic variation, 216
Dominant, 121, 150, 186, 193, 216, 416, 437, 448, 455, 471, 475, 480, 490, 520, 548
  alleles, 216
  gene, 216
  marker selection, 216
  oncogene, 216
  selectable marker gene, 216
Donor, 66, 70, 86, 122, 136, 138, 216, 227, 240, 260, 283, 321, 342, 354, 373, 443, 471, 482, 526, 550
  insemination, 136
  junction site, 216
  plant, 216, 354
  immune system, 512
Doping, 217
Doppler spectroscopy, 217
Dorito, 362
Dormancy, 114, 217, 449, 496
Dosage compensation, 217, 551
Double
  crossing over, 217
  fertilization, 217
  helix, 118, 143, 185, 214, 217, 278, 347, 371, 378, 500, 517, 543, 553, 554
  layer, 554
  recessive, 217
  stranded complementary DNA, 217
Doubling time, 217
Douche, 217
Down
  promoter mutation, 217

syndrome, 217, 529
  regulate, 218
Downstream processing, 218
Doxil, 218
Doxorubicin, 218
D-ribose sugar molecule, 273
Drift, 218, 262, 322, 348, 369
Drosophila, 200, 218, 285, 293, 386, 388, 410, 465
Drosophila melanogaster, 218, 285, 293
Drug
  delivery, 3, 18, 20, 27, 28, 80, 88, 218
  development, 154, 404, 418, 514
Dry weight, 219, 272, 548
Dual culture, 219
Duchenne muscular dystrophy, 144, 355
Ductility, 219
Ductus arteriosus, 67, 219
Ductus venosus, 219
Duodenal ulcer, 219
Duodenum, 179, 219, 394, 535
Duplex DNA, 213, 219
Duplication, 125, 156, 178, 210, 219, 227, 468, 475, 490, 526
Dut ung method, 379
Dwarf technology, 364
Dwarfism, 89, 115, 219
Dynamic
  characteristics, 219
  error, 220
  light scattering, 220
  range, 220
Dysphagia, 220
Dyspnea, 220
Dystonia, 220

E

E. coli, 69, 70, 82, 91, 92, 167, 220, 232, 236, 249, 273, 304, 317, 321, 369, 393, 433, 454, 459, 471, 473, 478
EC number, 230
Eccrine sweat glands, 220
Ecdysone, 220
Echocardiogram, 220
Eclosion, 220
Ecocentrism, 220
Ecological diversity, 147, 220
Ecology, 221

Economic trait locus, 221, 234
Economy, 308
Ecosystem, 120, 147, 221, 244, 296, 308, 366, 371, 376
Ecosystem diversity, 120, 147
Ecotone, 221
Ecotype, 221
Ectoderm, 221, 409, 420
Ectopic, 221, 480
   beats, 480
Edema, 221
Edible vaccine, 221
Effective
   number of alleles, 221
   population size, 159, 221, 283
Effector
   cells, 221
   molecule, 122, 123, 192, 221
Elastic deformation, 222, 544
Elastin, 222
Elastomer, 222
Electrical
   breakdown, 222
   conductivity, 449, 467
   connection, 272
   field, 222, 255
   potential, 115, 222, 403
   resistance, 378, 496
   signals, 223, 521
Electro-active compounds, 368
Electrocardiogram, 220, 222
Electrocardiograph, 63, 222
Electrochemical sensor, 12, 222
Electrochemistry, 39
Electrode, 71, 80, 115, 128, 132, 163, 168, 230, 243, 298, 310, 354, 368, 386, 387, 398, 417, 464
Electrodialysis, 203
Electroencephalogram, 221, 222
Electroencephalograph, 222
Electrokinetic, 553
Electroluminescence, 222
Electrolyte, 222, 223, 246, 253, 307, 461
Electrolyte-nonelectrolyte interaction, 461
Electromagnetic
   energy, 300, 551
   field, 163
   forces, 172
   radiation, 62, 247, 254, 373, 402, 547

   spectrum, 300, 309, 339, 366, 533, 544
Electromotive force, 223, 279, 299, 329, 513, 545, 548
Electromyography, 223
Electron, 5, 6, 21, 67, 69, 78, 90, 92, 144, 167, 174, 192, 194, 223, 227, 242, 249, 271, 275, 290, 315, 321, 342, 348, 383, 385, 400, 416, 443, 462, 464, 465, 526, 538
Electron level, 223
   microscope, 5, 67, 192, 223, 227, 271, 342
   microscopy, 21, 69, 383
   transport, 167, 385
Electronegative, 223
Electronic
   components, 4, 404, 435
   devices, 12, 39, 381, 464
   nanoelectronics, 5
Electrooptic devices, 341
Electrophoresis, 115, 123, 165, 204, 214, 223, 255, 377, 439, 450, 477, 498, 512, 548
Electrophoretic
   gel, 138, 548
   mobility, 281, 411
Electrophysiology, 28, 69
Electroporation, 223
Electrospray ionization, 359
Electrostatic
   attraction, 308
   interaction, 538
Electrovalent bond, 308
Elite tree, 224
Ellipsoid, 6
Elongation factors, 224
Embrua, 224
Embruon, 224
Embruos, 224
Embryo, 64, 124, 127, 132, 134, 157, 184, 185, 192, 206, 217, 224, 225, 228, 230, 233, 247, 255, 266, 269, 279, 283, 287, 289, 292, 296, 299, 301, 312, 320, 331, 344, 352, 373, 385, 391, 405, 406, 409, 410, 412, 417–419, 421, 423, 428, 437, 448, 482, 483, 486, 503, 505, 511, 518, 523, 529, 540, 545
   cloning, 224
   culture, 224, 344

multiplication and transfer, 224, 225
rescue, 224
sac, 217, 224, 228, 373, 385, 505
sexing, 224
splitting, 224
storage, 224
technology, 224
transfer, 225, 233
Embryoid bodies, 225
Embryoids, 210
Embryological stage, 255
Embryonic germ
    cells, 90
    layers, 509
Embryonic
    placenta development, 542
    signals, 202
    stage, 243, 247, 551
    stem cell, 90, 93, 95, 225, 232, 285–287,
    299, 315, 529
    structure, 483, 518
    tissue, 206, 511
Embryum, 224
Emulsifying agent, 225
Emulsion, 225, 534
Encapsidation, 225
Encapsulating agents, 225
Encapsulation, 218, 225
    techniques, 218
Encode, 168, 226, 310, 374, 380, 506
Encopresis, 226
Endangered
    breed, 194, 226
    species, 191, 226
Endemic, 226
Endinspiratory, 303
End-labeling, 226
Endocrine
    gland, 118, 226, 312, 388, 515
    interference, 226
    system, 226, 483
Endocytosis, 226, 397, 403
Endoder, 226
Endoderm, 121, 226, 409, 420
Endogamy, 227, 297
Endogenote, 227, 236
Endogenous, 144, 227, 233, 236, 410, 449
Endometrium, 283, 424
Endomitosis, 227

Endonuclease, 123, 227, 235
Endophyte, 227
Endoplasmic reticulum, 227, 232, 382
Endopolyploidy, 227
Endoprotease, 227
Endoreduplication, 227
Endorphins, 227
Endoscope, 227, 232, 346
Endoscopy, 227
Endosome, 228
Endosperm, 120, 182, 228, 268, 410, 550
    mother cell, 228
Endothelial cells, 228, 288
Endothelium, 228, 288
Endotoxin, 141, 228
Endovascular trophoblast, 228
Endovenous laser ablation, 228
End-product inhibition, 228
Enema, 228, 326
Energy
    applications, 5
    function, 228
    production, 5, 347
    source, 139, 347, 376, 383
Enhanced permeability and retention, 228
Enhancer element, 228
Enkephalin, 144, 229
Enkephalinergic neurons, 229
Enolpyruvyl – shikimate-3-phosphate
    synthase, 229
Enteral administration, 229
Enteral route, 229
Enterotoxin, 229
Enthalpy, 229, 250
Enucleated ovum, 229
Enuresis, 229
Environment, 2, 3, 7, 8, 23, 60, 114–116,
    126, 147, 154, 155, 191, 202, 220, 221,
    229, 235–237, 246, 264, 276, 277, 279,
    290, 296, 297, 301, 316–318, 323, 325,
    339–341, 345, 346, 352, 353, 365, 377,
    398, 399, 418, 419, 438, 456, 463, 467,
    471, 474, 492, 494, 496, 497, 503, 509,
    548
Enzymatic, 39, 167, 201, 202, 229, 235,
    323, 328, 348, 374, 385, 427, 429, 465,
    487
    biofuel cells, 39
    degradation, 427

digestion, 323, 427, 429
methylation, 348
Enzyme, 64, 67, 75, 79, 89, 114, 115, 117,
    120, 122–124, 132, 137, 144, 150, 154,
    156, 168, 175, 176, 179, 183, 190, 194,
    198, 203, 205, 214–216, 220, 222–230,
    236, 237, 241, 245, 246, 248, 250, 259,
    268, 269, 271, 274, 284, 297, 299, 312–
    316, 319, 321, 338, 339, 367, 393–400,
    404, 411, 412, 426, 427, 439, 441, 446,
    447, 449–454, 457, 460, 463, 478, 499,
    506, 508, 510, 514, 517, 526, 529, 535,
    544, 545, 555
    bioreactor, 230
    co-factors, 544
    electrode, 222, 230
    immunoassay, 222, 230
    kinetics, 230
    properties, 463
    stabilization using antibodies, 230
Eosinophils, 278
Epicotyl, 230
Epidemiology, 230, 261
Epidermal
    cell, 273, 274, 335, 458, 495, 533
    growth factor, 221, 280
    layers, 338
    tissues, 409
Epidermis, 192, 205, 230, 274, 314, 388,
    389, 428, 478, 495, 514
Epididymis, 230
Epigenesis, 231
Epigenetic, 231, 259, 481
Epiglottis, 171, 231
Epinasty, 231
Epinephrine, 231
Epiphyte, 231
Episome, 231, 471
Epistasis, 117, 231
Epistatic genetic variation, 231
Epitaxial, 231
Epithelial tissue, 166
Epitheliod decidual cells, 202
Epithelium, 143, 177, 231, 317
Epitope, 231, 390
Epizootic, 231
Epogen, 85
Equational division, 232
Equatorial plate, 170, 232

Equilibrium, 232, 250, 253, 262, 263, 276,
    279, 332, 384, 508, 538
    density gradient centrifugation, 232
Equimolar, 232
Erectile structure, 180
Ergonomy, 28
Erlenmeyer flask, 212, 232, 458, 472
Ertapenem, 144
Erythrocytes, 232, 274, 278, 315
Erythropoietin, 232
ES cells, 232
Escherichia coli, 59, 79, 140, 220, 231,
    232, 236, 237, 282, 283, 317, 319, 536
Escherichia coli exonuclease, 236
E-site, 232
Esophageal
    atresia, 232
    manometry, 232
    pH monitoring, 232
    stricture, 232
Esophagogastroduodenoscopy, 232
Esophagus, 137, 142, 209, 232, 255, 277,
    326, 370, 394, 398, 443, 495, 497, 519,
    535
Essential
    amino acid, 198, 233
    derivation of varieties, 221, 233
    element, 233, 329, 343
    nutrient, 233
    requirement, 233
Established culture, 233
Estimated breeding value, 220, 233
Estrogen, 233, 377
    replacement therapy, 233
Ethephon, 234
Ethical, legal and social implications, 224
Ethics, 234, 56
Ethidium bromide, 197, 234, 304, 439
Ethnic groups, 227
Ethylene, 221, 234, 389
Ethylenediamine tetraacetic acid, 234
Etiolation, 234
Euchromatin, 234, 281
Eugenics, 61, 234
Eukaryote, 234
Euler equations, 234
Eumetazoans, 420
Euploid, 234, 348
Eutrophication, 234

Euvolemia, 234
Evangelista Torricelli, 517
Evolution, 55, 96, 116, 147, 183, 203, 235, 286, 369, 439
Evolutionary significant unit, 235
Ex situ conservation, 235
Ex vitro, 235, 248
Ex vivo, 235
Exact tests, 235
Excinuclease, 235
Excision, 235, 293, 304, 488, 503, 526
    repair, 235
Excitation wavelength, 225, 236
Excrete, 236, 314, 338
Excretion, 236
Exhalation, 236
Exocrine gland, 236
Exocyclic amino group, 134
Exodeoxyribonuclease III, 236
Exogamy, 236
Exogenote, 227, 236
Exogenous, 236, 264, 274, 523
Exogenous DNA, 236
Exo-III, 236
Exon, 114, 216, 236, 237, 488
    amplification, 236
    parsing, 237
    prediction, 237
Exonuclease, 236, 237, 315, 369, 429
Exonuclease III, 237
Exopolysaccharide, 237
Exotoxin, 237
Expectation maximization, 224
    algorithm, 237
Expected progeny difference, 230, 233, 237
Expiration, 236, 237
Explant, 210, 216, 233, 237, 302, 311, 337, 346, 420, 494
Explant donor, 237
Explantation, 237
Explosion method, 237
Export, 8, 238
Express, 136, 190, 204, 238, 276, 351, 417, 438, 523, 524, 552
Expressed sequence tag, 233, 238
Expression, 13, 77, 78, 80, 82, 83, 85, 117, 138, 146, 154, 163, 179, 192, 193, 201, 205, 214, 218, 238, 257, 260, 266, 299, 301, 316, 324, 331, 348, 354, 367, 380,

397, 412, 414, 425, 427, 430, 441–444, 447, 460, 465, 471, 475, 479, 497, 506, 516, 519, 522, 524, 540, 546
    library, 238
    system, 80, 193, 238, 465
    vector, 82, 85, 238
Expressivity, 238, 539
Ex-situ conservation, 238
Extant, 238
Extension, 112, 116, 121, 238, 310, 359, 363, 402, 405, 420, 436, 445
External
    guide sequence, 222, 238
    respiration, 238
    root sheath, 238, 463
Extinct breed, 238
Extinction, 137, 198, 226, 238, 265, 366, 448
Extirpation, 238, 415
Extracellular
    fluid, 239, 288, 403
    growth factors, 197
    matrix, 102, 104, 239, 438
    medium, 464
Extrachromosomal, 69, 199, 231, 239, 327, 333, 467, 552
    inheritance, 199, 239
Extrachromosomes, 239
Extracoelomic fluid, 183
Extracorporeal, 66, 67, 73, 74, 77, 78, 81, 239
    membrane oxygenation, 78
    shock wave lithotripsy, 239
Extraembryonic
    membrane, 121, 176
    mesoderm, 176, 177, 542
Extranuclear genes, 239
Extrasolar planets, 217
Extrinsic, 201, 239
Exude, 239
Ex-vitro, 239

F

Face-centered cubic, 240
Facilitated diffusion, 240, 392
Factor VIII deficiency, 240
Factorial mating, 240
Facultative anaerobe, 241

False fruit, 241, 429
False negative, 144, 241
False positive, 241
Farad, 241
Faraday, 60, 241
Farm animal genetic resources, 128, 189, 241
Fascicle, 241
Fatal Familial Insomnia and kuru, 421
Fats, 145, 242, 253, 324, 353, 391
Fecal fat test, 242
Fecal occult blood test, 242
Fecundity, 242
Fed-batch fermentation, 242
Federal Insecticide, Fungicide, and Rodenticide Act, 242, 244
Federal Plant Pest Act, 242, 532
Federal Seed Act, 242, 532
Feedback inhibition, 242
Feeder layer, 242
Female pronucleus, 289
Female sex hormones, 233, 377
Femtoengineering, 242
Feral, 241, 242
Fermentation, 56, 57, 59, 60, 62, 120, 147, 153, 190, 218, 242, 254, 384, 462, 507, 535
Fermenter, 242
Fermi energy, 242
Ferroelectric material, 242, 292, 403
Ferromagnetism, 243
Fertile, 243, 261, 265
Fertility, 136, 231, 240, 243, 264, 294, 468
  factor, 231, 240, 243
Fertilization, 58, 59, 119, 122, 135, 137, 181, 189, 195, 213, 224, 243, 267, 273, 296–298, 312, 344, 373, 391, 411, 415, 428, 431, 448, 463, 471, 505, 518, 555
Fertilized
  egg (ovum) , 157, 266, 288, 353, 419, 423, 424, 537
  ova, 425
  ovum, 221, 425
Fertilizer, 243
Fetal
  alcohol syndrome, 243
  circulation, 368
  contributions, 404
  development, 487, 542

tissue, 243, 283
Fetus, 118, 123, 124, 177, 219, 243, 279, 405, 406, 492, 537
Feulgen staining, 243
Fiber, 15, 75, 76, 99, 102, 106, 154, 192, 243, 335, 337, 342, 423, 488, 495, 507
  forming process, 488
  optic, 243
Fibril, 244
Fibrillation, 244, 541
Fibrin matrix, 130
Fibrinogen, 244, 407
Fibroblast, 145, 171, 202
Fibrosis, 437
Fibrous
  root, 244, 507
  tissue, 397
Field effect transistor, 223
Field gene bank, 244, 256
Filial generation, 240, 244
Filler, 244
Filtration, 245, 360, 392, 493
Firing, 147, 173, 245, 478, 544
Fisher Wright model, 245
Fission, 58, 245
Fistula, 75, 245
Fitness, 136, 215, 245, 251, 263, 264, 282, 289, 318, 324, 356, 369, 384, 433, 444, 466, 514
  rebound, 245
Fixation index, 245
Flagellum, 143, 245, 315
Flaming, 245
Flanking region, 246
Flat-band potential, 246
Flavin adenine dinucleotide, 241, 246
Flavonoid pigments, 129
Flip-flop, 246
Flocculant, 246
Floccule, 246
Florida International University, 2, 34, 46, 47, 97, 107
Flow cytometry, 246, 247
Flower, 113, 164, 195, 208, 246, 267, 385, 388, 394, 396, 404, 428, 429, 440, 473, 490, 491, 539, 541, 545
Fluctuating asymmetry, 246
Fluctuation test, 67
Fludeoxyglucose, 246

Fluid
  balance, 246
  bonding, 247
  collection, 505
  retention, 418
Fluorescence, 76, 84, 198, 243, 247, 422, 487
  activated cell sorter, 247
  immunoassay, 243, 247
  in situ hybridization, 247
Fluorescent
  dye, 247, 519
  lamp, 271, 272
Fluorophores, 315
Fluoroscopy, 21, 247
Foetus, 221, 243, 247, 267, 269, 287, 332, 392
Fog, 6, 248
Folate, 248, 426
Fold-back, 248
Folded genome, 248
Follicle, 192, 248, 251, 463
  stimulating hormone, 248, 251
Food
  biotechnology, 248
  poisoning, 460
  processing, 5, 464
  enzyme, 248
Food and Agriculture Organization, 68, 241
Food and Drug Administration, 22, 242, 248, 412
Food, Drug, and Cosmetic Act, 248
Foodborne infection, 542
Foot-candle, 248, 327
Foramen ovale, 248
Force field, 248, 350
Forced
  cloning, 248
  expiratory flow, 249
  midexpiratory flow, 249
  vital capacity, 249
Foreign
  antigen, 130, 170
  DNA, 178, 181, 185, 236, 248, 303, 522, 523, 540
  substance, 77, 129, 244, 294, 329, 344
  tissues, 526
Forensics, 5, 249, 411, 545
Formulation, 143, 218, 249, 334, 426

Forskolin, 249
Fortify, 249
Forward bias, 249
Forward mutation, 249, 452
Fouling, 249
Founder
  animal, 249
  effect, 249
  principle, 249
Four-base cutter, 250, 478
Fourier transform infrared spectroscopy, 250
Fractionation, 250
Fragile sites, 250
Fragile-X syndrome, 250
Fragment, 114, 179, 215, 227, 236, 250, 259, 260, 359, 369, 380, 404, 413, 449, 450, 470, 472, 476, 498, 527
Frameshift mutation, 115, 250
Francis Crick's seminal concept, 470
Free energy, 250
Free water, 159, 250
Freeze preservation, 196, 251
Frequency, 113, 133, 184, 207, 211, 217, 242, 245, 247, 251, 253, 254, 258, 262, 263, 281–284, 348, 352, 373, 401, 414, 431, 434, 435, 442, 449, 452, 466, 467, 483, 484, 497, 502, 522, 533
  correspondence analysis, 242
  distribution, 251, 348
  modulation, 247
Fresh weight, 251, 548
Friable, 119, 251
Friedrich Miescher, 374
Froude number, 251
Fullerene, 5, 6, 99, 251, 365
Functional
  food, 251
  gene cloning, 164, 251
  genomics, 251, 265
  incontinence, 252
  residual capacity, 252
  components, 359
  molecular pathways, 350
fundamental
  components, 359
  molecular pathways, 350
Fungicide, 72, 252, 396
Fungus, 60, 252, 340, 356, 392, 395, 460

Fusarium spp, 252
Fusion
    biopharmaceuticals, 252
    gene, 252
    protein, 175, 252, 295, 332, 528
    toxin, 252
Fusogenic agent, 252

G

G cap, 165, 253
G protein, 253, 270, 272, 483
    linked receptors, 270
Gadolinium, 253
Gait analysis, 148
Galactomannan, 253
Galactose, 253, 316
Gall bladder, 253
Galvanic corrosion, 253
Gamete, 253, 254, 266, 380, 385, 409, 449,
    486
    embryo storage, 253
    forming tissues, 266
    disequilibrium, 253, 284
        fusion, 483
        tissue, 253
Gametoclone, 253
Gametogenesis, 253
Gametophyte, 224, 253, 329, 335, 342, 344,
    488
Gametophytic incompatibility, 254
Gamma camera, 127, 254, 477
    emission, 254
    globulins, 267
    particles, 391
    radiation (gamma rays), 254
    rays, 309, 381, 417, 477
    system, 254
Ganglion, 254
Ganglionic
    blocking agent, 254
    synapses, 254
Ganglioside, 508
Gas, 2, 66, 81, 118, 124, 158, 162, 167,
    191, 249, 250, 252, 254, 279, 290, 293,
    298, 303, 314, 318, 324, 326, 336, 340,
    357, 370, 386, 449, 491, 493, 507, 515,
    545
    chromatography, 167, 254

transfer, 254
Gastrin, 171, 254
Gastritis, 254
Gastroenteritis, 254
Gastroenterologist, 254
Gastroenterology, 254
Gastroesophageal reflux disease, 232, 254,
    274, 370
Gastrointestinal tract, 191, 255, 409
Gastroparesis, 255
Gastrostomy, 255
Gastrula, 226, 255, 266
Gastrulation, 255
Gate, 243, 255, 354, 358, 372, 381
Gauss, 255, 510
Gel, 21, 76, 115, 119, 125, 176, 199, 204,
    222, 223, 255, 290, 310, 397, 406, 411,
    431, 450, 476, 480, 484, 511, 512
    electrophoresis, 119, 125, 223, 255, 310,
        476, 484
    filtration, 255
    matrix, 397
    permeation chromatography, 255
Gelatin, 61, 256
Gelatinization, 256
Gelatinous protein mixture, 333
Gelrite, 256
Gene, 256
    addition, 256
    amplification, 256
    bank, 181, 244, 256, 391
    behavior, 265
    cloning, 80, 117, 256
    conservation, 257
    construct, 257, 303
    drop, 257
    encoding, 216, 312, 501, 511
    expression, 89, 137, 142, 217, 221, 231,
        251, 257, 260, 268, 284, 292, 340, 367,
        380, 416, 475, 492, 516, 520, 521, 527
    flow, 213, 257, 264, 311, 317, 339, 345,
        499
    frequency, 257, 262
    genealogies, 257
    gun (biolistics), 257
    identification, 257, 497
    imprinting, 257
    insertion, 257
    interaction, 257, 461

knock-in, 258
knockout, 258
library, 125, 256, 258, 472
linkage, 63, 258
machine, 258, 527
mapping, 75, 77, 90, 258
modification, 258
parsing, 258
pool, 257, 258, 336
prediction, 113, 258, 259
probe, 259
replacement, 259, 446
segments, 483, 506
sequence, 83, 91, 213, 286, 305, 451, 507
shear technology, 259
shears, 259, 455
silencing, 193, 259, 468
splicing, 260, 321, 457
stacking, 260
structure analysis, 497
targeting, 260
technology, 155, 260, 269, 377
testing, 260
therapy, 83, 86, 90, 91, 175, 235, 239, 260, 266, 287, 482
tracking, 260
transfer systems, 451
translocation, 260
trapping, 260
Genealogy, 93, 398
Genera, 244, 260, 266, 305, 370
Generally Regarded as Safe (GRAS), 260, 271
Generate, 17, 83, 129, 133, 162, 201, 207, 260, 273, 328, 334, 345, 355, 385, 403, 444, 454, 459, 469, 495, 507, 511, 514, 527
Generative nucleus, 261
Genet, 261
Genetic algorithms, 351
code, 74, 93, 183, 203, 380, 522, 525, 528, 534
composition, 382
condition, 291, 292, 419, 551
defect, 90, 199
differentiation, 190, 311
disease, 131, 260, 261, 279, 409, 446, 551

disorder, 239, 297
distance, 262
divergence, 160, 261
diversity, 121, 147, 249, 262, 264, 296, 485
drift, 159, 213, 221, 264, 283, 435, 436, 439
effects, 193
engineering, 3, 5, 67, 79, 119, 148, 151, 152, 171, 256, 258, 259, 265, 348, 351, 364, 443, 450
epidemiology, 264
fingerprinting, 249
information, 68, 71, 172, 179, 205, 214, 238, 257–259, 263, 280, 427, 470, 483, 524
instability, 539
linkage, 172, 193
map, 63, 195, 258, 453
mapping, 453
marker, 70, 275, 323, 350, 433, 489
material, 68, 69, 155, 235, 241, 244, 256, 257, 261, 263–266, 311, 324, 347, 348, 351, 451, 453, 471, 523, 524, 543
polymorphism, 263
replica, 181
resources, 128, 153, 173, 189, 257, 296, 306, 331, 537
stocks, 153, 406
techniques, 258
transformation, 237, 294, 367
use restriction technology, 273, 510
variance, 461, 485, 489
variation, 117, 126, 143, 147, 159, 196, 216, 221, 231, 249, 260, 262, 280, 282, 366, 371, 416, 439, 467, 485
Genetically engineered organism, 265, 266
modified food, 256, 269
modified organism, 154, 203, 265, 351, 377, 418, 442
Genetics, 63, 64, 119, 133, 160, 199, 218, 264, 265, 288, 336, 340, 349, 350, 360, 406, 411, 415, 433, 442, 452, 545
Genome, 82, 85, 88–93, 95, 96, 123, 128, 133, 157, 175, 178, 190, 192, 201, 213, 214, 219, 227, 236, 251, 254, 256, 258, 259, 262–265, 273, 278, 285, 287, 303, 304, 310, 317, 322, 325, 327, 331, 354–356, 386, 387, 416, 423, 429, 446,

452, 453, 465, 469, 473, 475, 476, 479,
489, 490, 498, 519, 520, 523, 524, 526,
527, 543–545, 547–549, 552
Genomic combinations, 429
   extinction, 261
   library, 265
   sequence, 169, 251, 287, 306
Genomics, 95, 149, 171, 265, 295, 399, 519
Genotype, 64, 181, 186, 214, 245, 265, 266,
275, 276, 285, 291, 292, 311, 318, 336,
337, 394, 398, 422, 424, 433, 443, 444,
465, 466, 483
Genotyping, 252, 476, 477
Genus, 120, 133, 146, 161, 260, 266, 307,
429, 460, 462, 541
Geotropism, 266, 529
Germ, 11, 61, 221, 225, 249, 266, 267, 288,
336, 337, 356, 379, 409, 410, 420, 482,
483, 487, 523
   cell, 225, 266, 267, 336, 356, 379, 482,
483, 487, 523
   layers, 266, 337, 409, 420
   line, 249, 266, 410
Germicide, 266
Germinal epithelium, 266
Germination, 114, 220, 266, 267, 293, 388,
410, 462
Germinativum, 496
Germline, 266
Germplasm, 196, 235, 241, 266
Gestation, 267
Giardia lamblia, 267
Gibberellins, 233, , 267, 402
   ethylene, 402
Gland, 226, 236, 248, 267, 357, 368, 388,
424, 426, 445, 506, 527
Glandular
   hair, 274
   sacs, 468
Glass, 56, 94, 214, 267, 296, 332, 340, 358,
396, 530, 544
   transition temperature, 267
Glaucoma, 249
Glaucous, 267
Glaucus, 267
Gliadin and glutenin, 268
Glial cells, 267
Globular protein, 470, 530
Globulins, 267, 294

Glomerular kidney disease, 267
Glomerulonephritis, 267
Glomerulosclerosis, 267
Glomerulus, 159, 268
Glossaries, 55, 112
Glucagon, 171, 483
Glucoamylase, 460
Glucocorticoid, 268
Gluconeogenesis, 155
Glucose, 125, 126, 165, 168, 171, 206, 207,
233, 246, 268, 311, 316, 353, 384, 396,
413, 460, 491
   invertase, 268
   isomerase, 268
   polymers, 491
Glucosinolates, 268
Glucuronidase, 144, 268, 273
Glutamate, 426, 427
Glutamate carboxypeptidase II, 426
Glutamic acid, 124, 309
Glutamine, 124, 427
Gluten, 268
Glyco nanotechnology, 268
Glycoalkaloids, 268
Glycoform, 268
Glycolic acid, 413
Glycolysis, 268
Glycoprotein, 130, 140, 268, 269, 539
Glycoprotein remodeling, 268
Glycoscience, 269
Glyphosate, 269
   oxidase, 269
   oxidoreductase, 269
Golden rice, 269
Goldman equation, 269
Golgi apparatus, 62, 269
Gonad ridge, 269
Good laboratory practice, 269
   manufacturing practice, 269, 270
   pasture syndrome, 270
G-protein coupled receptor, 270
G-proteins, 270
Gr. *chroma*, 177, 178
Gr. *protoplastos*, 428
Gr. *tracheia*, 519
Gr. *trichoma*, 528
Graafian follicles, 248
Graft, 96, 162, 216, 270, 271, 298, 342,
395, 473, 550

chimera, 270
hybrid, 270
inoculation test, 270
union, 271
Grafting, 271, 270, 458, 462, 473
operation, 462
Grain boundary, 271
growth, 271
size, 271, 361
Gram staining, 271
technique, 271
Graphene, 6, 12, 72, 270, 435
Graphical visualization techniques, 187, 350
Gratuitous inducer, 271
Gravity, 64, 251, 266, 362, 384, 485, 517
Gray matter, 173, 271, 548
Gregor Mendel's theory, 336
Grey goo, 6, 7
Gro-luxä, 271
Gro-lux™, 272
Ground, 5, 59, 231, 272, 375, 390, 459, 473, 492, 494
Growth, 272
cabinet, 272, 298
curve, 272, 325, 491
deficiency, 127, 192
factor, 4, 127, 145, 199, 221, 272, 302, 405, 472, 474, 493, 516, 518
hormone, 89, 159, 177, 267, 272, 483, 484
phase curve, 272
rate, 122, 174, 272
regulator, 234, 272, 274, 314
retardant, 272
ring, 272
substance, 272
Guanine, 68, 143, 186, 200, 253, 273, 370, 375, 432, 456, 470, 506, 515, 531
Guanosine, 165, 205, 255, 269, 272, 273, 453
monophosphate, 273
Guanylic acid, 269, 273
Guard cell, 273, 495
Guide RNA, 273
Guide sequence, 238, 273
Gus gene, 273
Gymnosperm, 273
Gynandromorph, 273, 471

Gynodioecy, 273
Gynogenesis, 273
Gyrase, 273

H

H prefix, 274
H. pylori, 230, 277, 535
H2 blockers, 274
Habituation, 274
Haematopoietic stem cells, 274
Haemoglobin, 126, 274, 276, 279, 436, 512
Haemolymph, 274, 279
Haemophilia, 240, 274
Haemostasis, 199
Hair, 38, 57, 120, 155, 166, 238, 274, 314, 331, 342, 388, 463, 514, 528
follicle, 238, 274, 388, 463
pin loop, 274
Hairy root culture, 275
Half life, 275
Hall effect, 275
Halophyte, 275, 460
Halothane, 275
Hamiltonian, 468
Hanging droplet technique, 275, 341
Haploid, 127, 129, 209, 273, 275, 278, 289, 335, 409, 425, 442, 443, 476, 486, 489, 505, 529, 555
cell, 275, 486
gamete nuclei, 425
number, 409, 529
plants, 129
spores, 489
Haplotype, 253, 366
Haplozygous, 276
Hapten, 276
Haptoglobin, 276
Hardening off, 276
Hardness, 276
Hardy Weinberg proportion, 203
Harvesting, 132, 276
Hazardous, 23, 156, 445
Head injury, 497
Health care, 2, 37, 151, 180
Heart assist device, 277
attack, 84, 158, 277
beat, 166, 277
block, 75, 277

burn, 277
valve, 21, 39, 72, 431, 493
Heat capacity, 277
cramps, 277
exchanger, 72, 277
exhaustion, 277
flux, 512
hyper pyrexia, 277
shock
factor, 497
protein, 277, 287, 496, 497
strain, 277
stroke, 277
therapy, 277
transfer, 154, 188, 291, 376
Helical form, 489
structure, 73, 517, 554
Helicobacter pylori, 277, 394, 535
Helium atom, 276
Helix, 71, 87, 205, 278, 371, 464, 553
Helminth, 278
Helper cells, 278
plasmid, 278
virus, 278
Hematocrit, 278
Hematomas, 278
Hematopoietic cell, 171, 278
Hematopoietic stem cells, 278, 533
Hematuria, 278
Heme, 278, 279
Hemicellulose, 184, , 278, 322
Hemifacial microsomia, 431
Hemizygous, 276, 278, 552
Hemodialysis, 28, 64, 65, 67, 72–75
Hemodilution, 278
Hemodynamics, 42, 278
Hemoglobin, 386, 278, 400
Hemolymph, 279
Hemolysis, 68, 279, 452
Hemolytic anemia, 474
Hemophilia, 87, 279
Hemostasis, 279
Hemotrophic nutrition, 279, 283
HEPA filter, 279, 283
Heparin, 64, 67, 73, 279
Hepatic, 279, 280, 299, 466
function, 299
Hepatocyte, 171, 280
Herbicide, 86, 119, 280, 392, 490

resistance, 280, 392
Herceptin, 88, 92, 280
Hereditary disorder, 280, 399
process, 336
Heredity, 56, 63, 256, 265, 280, 336
Heritability, 280, 365, 466
Hermaphrodite, 280, 306
Hermaphroditic species, 137
Hernia, 280
Herniation, 306
Hertz, 281
Heteroallele, 281
Heterochromatic regions, 177
Heterochromatin, 281
Heterochromosomes, 340
Heterocyclic aromatic ring, 200
Heteroduplex, 281, 378
analysis, 281
Heterogametic sex, 281
Heterogeneity, 193, 262, 281
Heterogeneous
nuclear RNA, 281
selection, 142
Heterographs, 550
Heterokaryon, 281, 285, 505
Heterologous probe, 281
protein, 281
Heteroplasmy, 281, 286
Heteroploid, 281
Heteropyknosis, 282
Heterosis, 282289, 462
Heterosporous plants, 329, 342, 344
Heterotroph, 139, 282
Heterozygosity, 221, 282, 286, 297, 433
Heterozygote, 116, 179, 216, 282, 335, 448, 520
Heterozygous, 123, 142, 183, 196, 209, 263, 282, 298, 311, 318, 476, 510
advantage, 142, 263, 282
disadvantage, 282
Hetrotrophic, 282
Hfr strain, 240
see fertility factor or F factor
Hidden Markov model, 282
High efficiency particulate air filter, 279, 283
electron mobility transistor, 279
throughput genome sequence, 287
voltage current, 283

Hill Robertson effect, 283
Histamine, 274
Histidine, 554
Histiocytes, 333, 451
Histiotrophic nutrition, 279, 283
Histocompatibility, 283, 284, 500
    complex, 284
Histocompatible, 283
Histoglobulin, 284
Histological preparations, 138
Histology, 284
    technicians, 284
    technologists, 284
Histone, 284, 371, 462, 500, 507
Histotechnicians, 284
Hitchhiking, 262, 284
Hofbauer cells, 284
Hogness box, 284, 508
Hollow fiber, 72, 284
Holoenzyme, 132, 284
Holometabolous, 284, 293
    insects, 293
Homeobox, 284, 285
Homeostasis, 246, 279, 440
Homeotic genes, 285
    mutation, 285
Homoallele, 285
Homodimer, 285
Homoduplex DNA, 285
Homoeologous, 285
Homogametic sex, 285
Homogeneous assay, 285
Homogenotization, 285
Homogenous, 193
Homokaryon, 281, 285
Homologous, 63, 121, 136, 156, 175, 181,
    195, 212, 227, 236, 259, 275, 282, 285,
    286, 304, 336, 371, 383, 388, 389, 433,
    504, 511, 520
    chromosomes, 121, 136, 156, 175, 195,
        212, 275, 282, 286, 371, 388, 433, 504,
        511
    recombination, 259, 285, 286, 304
Homology, 236, 257, 281, 286, 351, 429,
    504, 526, 555
Homomultimer, 286, 307
Homoplasmy, 281, 286, 452
Homopolymer, 200, 286
Homopolymeric, 506

    tailing, 286
Homozygosity, 139, 245, 432, 286441
Homozygote, 286
Homozygous, 123, 186, 209, 217, 282, 286,
    297, 311, 320, 352, 378, 510
    recessive, 186
Hopkins artificial kidney, 65
Hormone, 119, 127, 139, 144, 163, 177,
    188, 207, 220, 231–233, 248, 268, 286,
    312, 326, 388–390, 399, 424, 445, 463,
    472, 474, 483, 484, 504, 511
Host, 61, 62, 68, 125, 130, 175, 178, 181,
    193, 197, 202, 214, 216, 231, 238, 256,
    260–262, 270, 286, 300, 304, 305, 317,
    318, 325, 327, 328, 349, 351, 354, 365,
    392, 396, 412, 429, 446, 447, 473, 508,
    509, 521, 524, 534, 540, 543, 544, 546
    genomics, 286
    specific toxin, 286
Hot spot, 287
House keeping genes, 287
Hückel method, 468
Human artificial chromosome, 274, 287
    chromosomes, 88, 92, 93, 288, 313, 482
    clone, 287
    embryonic stem cell, 93, 287, 419, 509
    embryos, 88, 448
    genome project, 86, 90, 95, 97, 141,
        288, 552
    growth hormone, 80, 159, 274, 282, 288
    health, 8, 151, 154, 156, 288, 308
    immunodeficiency virus, 82, 284
    leukocyte antigen system, 288
    reproduction, 136
    serum albumin, 288, 470
    umbilical vein endothelial cells, 288
Huntington disease, 288
Hybrid, 24, 65, 71, 77, 125, 128, 136, 170,
    195, 240, 252, 261, 265, 281, 282, 288,
    289, 305, 306, 307, 400, 425, 437, 452,
    465, 471, 482, 505, 528, 529
    arrested translation, 288
    cell, 170, 288, 289, 482
    dysgenesis, 288
    offspring, 528
    released translation, 289
    seed, 289
    selection, 289
    sink, 289

vigor, 282, 289
zone, 289
Hybridization, 65, 115, 118, 121, 123, 215, 247, 265, 286, 289, 296, 372, 390, 403, 422, 429, 460, 524, 554
Hybridoma, 289
Hydatiform mole, 289
Hydrate, 289
Hydraulics, 289
Hydrocephalus, 290
Hydrochloric acid, 7, 290
Hydrocollator, 290
Hydrocortisone, 290
Hydrodynamics, 290
Hydrogel, 102, 290
Hydrogen bond, 128, 186, 204, 205, 289, 290, 371, 446, 531
   breath test, 290
Hydrogymnastics, 290
Hydrolysis, 114, 137, 144, 167, 201, 206, 240, 242, 268, 290, 393, 394, 400, 423, 426, 427, 453, 460
Hydrolytic cleavage, 205
Hydrolyzes
   esters, 400
   peptide bonds, 529
Hydronephrosis, 290
Hydrophilic, 76, 124, 125, 145, 189, 290, 340
   molecule, 290
   parts, 145
   state, 189
Hydrophobic, 124, 145, 290, 291, 319, 340
   interaction, 290
   molecules, 290, 291
   parts, 145
Hydroponics, 291, 480
Hydrostatic pressure, 291, 531
Hydrotherapy, 291
Hydrotropism, 529
Hydroxydaunorubicin, 218
Hydroxyl group, 112, 204, 205, 208, 246, 399, 421, 506
Hygromycin, 291
Hygroscopic liquid, 209
Hyperactive, 291
Hyperbilirubinemia, 291
Hyperkalemia, 291
Hyperploid, 291, 292

Hypersensitive site, 291
Hypertension, 163, 291, 418
Hypertext transfer protocol, 287
Hyperthermia, 291
Hypertonic, 291, 292
Hypervariable region, 291
   segment, 291
Hyperventilation, 292
Hypochlorite, 292
Hypocotyl, 292
Hypodermis, 292, 498
Hypogastric arteries, 405
Hypomorph, 292
Hypoplastic, 292
Hypoploid, 291, 292
Hypotension, 292
Hypothalamic peptides, 292
Hypothalamus, 160, 292, 445
Hypothermia, 72, 292
Hypothesis, 59, 67, 144, 164, 292, 322, 366, 417, 421, 427, 531
Hypotonic, 291, 292
   solution, 292
Hypoxanthine, 201
Hysteresis, 292

I

I/E region, 293
Ichthyosis, 293
Identical by descent, 123, 139, 293
   twin, 69, , 293, 353
Idiotype, 293
IEF gel, 310
IgD, 129
IgE, 129, 293
IgG, 129, 240, 242, 293
IgM, 129, 293
Ileum, 195, 293
Illuminance, 326
Illuminate, 293
Imaginal disc, 293
Imbibition, 114, 293, 531
Immediate early gene, 293
Immobilized cells, 293
Immortalization, 294
Immortalizing oncogene, 294
Immune reactions or allergy, 295

response, 77, 122, 129, 130, 138, 158, 170, 171, 215, 288, 294, 295, 305, 327, 330, 352, 355, 381, 420, 463, 506, 508, 534, 537, 543
system, 168, 270
Immunity, 150, 267, 288, 294, 488, 537
Immunization, 262, 294, 295, 301, 419
Immuno, 119, 230, 294
affinity, 119
Immunogen, 130, 294
Immunogenic molecule, 508
proteins, 499
Immunogenicity, 294
Immunoglobulin, 113, 129, 294, 312
Immunology, 301
Impedance, 295, 310, 417, 439
Impeller, 295
Imprinting, 295, 360
In silico, 295, 365, 387, 516, 543
modeling, 295
proteomics, 295
In situ, 102, 236, 245, 256, 257, 295, 296, 482, 554
colony, 296
conservation, 296
hybridization, 236, 245, 482, 554
plaque hybridization, 296
In vitro embryo technology, 134
fertilization, 136, 296, 313, 486, 511
maturation, 296, 313
mutagenesis, 183
In vivo, 3, 64, 80, 102, 125, 131, 170, 173, 225, 235, 297, 307, 345, 440, 446, 477, 491, 493, 500, 514
gene therapy, 297
Inactivated agent, 297
Inbred line, 297
Inbreeding, 183, 245, 297, 431, 432, 514
coefficient, 297
depression, 245, 297
effect, 297
effective number, 297
Inclusion body, 297
Incompatibility group, 298
Incomplete digest, 298, 391
dominance, 298
penetrance, 298
Incompressible fluid, 144, 293, 298
Incubation, 329, 298

Incubator, 298
Indehiscent, 298
Independent assortment, 179, 298, 336
Indeterminate growth, 298, 457
Indigestion, 298
Indirect embryogenesis, 210, 299, 482
organogenesis, 210, 299
Indocyanine green, 299
Indole-acetic acid, 139
Induced pluripotent stem cells, 100, 105, 299
Inducer, 299, 309
Inducible, 299
enzyme, 299
gene, 299
promoter, 299
Inductance, 163, 279, 299, 329
Induction, 204, 255, 299, 300, 329, 395, 446, 457, 482, 510, 515
electrode, 299
media, 300
Inductor, 300, 382
Inembryonation, 134, 224, 300
Inert, 3, 72, 119, 244, 300, 318
Infection, 60, 72, 114, 161, 164, 171, 176, 181, 195, 254, 261, 293, 294, 298, 300, 301, 321, 324, 336, 367, 395, 416, 419, 457, 469, 488, 508, 522, 535–537
process, 508
Infectious agent, 300, 355, 395, 421, 427
diseases, 537
Inferior vena cava, 300
Infertile, 300
Infiltrate, 300
Inflammation, 80, 132, 144, 161, 167, 192, 254, 280, 300, 368, 371, 397, 419, 542
Inflammatory arthritis, 300
bowel disease, 195, 300
Inflorescence, 388, 394, 436, 487, 545
Influenza vaccine, 537
virus, 139
Informed consent, 300
Infrared, 250, 300, 320, 339, 363, 366, 402
wavelengths, 402
Inguinal hernia, 301
Inhalation, 301
Inheritance, 56, 60, 64, 87, 138, 192, 193, 199, 239, 260, 263, 265, 301, 332, 333, 336, 355, 396, 412, 428, 433, 534

Inherited, 63, 87, 90, 115, 160, 163, 188, 199, 200, 217, 257, 258, 260, 261, 263, 265, 274, 275, 282, 285, 286, 288, 301, 357, 368, 500, 523, 528, 534, 545
  diseases, 528
Inhibitor, 84, 122, 142, 198, 234, 272, 301, 387, 426
Initial, 122, 132, 220, 240, 266, 283, 293, 301, 309, 316, 374, 420, 421, 425, 458, 505
  public offering, 309
Initiation, 139, 217, 218, 301, 319, 382, 421, 440, 447, 467, 472, 474, 487, 491, 508, 521, 524, 525, 535
  codon, 301, 319, 440, 491, 525
  factor, 301
Inner cell mass, 221, 226, 255, 287, 301, 529
Inoculate, 301
Inoculation cabinet, 302
Inoculum, 301, 302, 346, 498
Inorganic, 4, 137, 139, 147, 183, 203, 233, 282, 302, 309, 346, 358, 363, 370, 376, 401
  chemistry, 302
  compound, 302
  elements, 139, 282
Inositol, 302, 357, 544
  lipid, 302
Inotropic agent, 302
Input
  traits, 302
  output, 293
Insecticide, 56, 302
Insert, 149, 181, 187, 211, 245, 303, 304, 351, 398, 416, 495, 523
Insertion element, 303
  mutation, 303
  point, 240, 303
  sequence, 303, 309, 526, 354
  site, 303
Inspiration, 207, 249, 301, 303, 326
Instability, 303
Instrumentation, 26, 27, 43, 48, 97, 152, 180
Insulator, 165, 222, 223, 246, 303, 320, 468
Insulin, 5, 27, 67, 79, 81,–83, 152, 171, 207, 211, 303, 351, 388, 483
  and glucagons, 388

Integrated circuit, 71, 210, 293, 303, 341, 354, 415, 464, 546
Integrating vector, 304
Integration, 82, 293, 303, 304, 325, 416, 423, 478, 497, 524, 532, 545
Integument, 304, 478
Integumentary system, 304, 478, 498
Intellectual property, 304, 309
Intensifying screen, 304
Interaction, 46, 123, 128, 145, 157, 181, 185, 188, 231, 251, 287, 290, 298, 304, 317, 392, 463, 483, 485, 496, 505, 516, 554
Interatomic bond, 538
Intercalary growth, 304
Intercalating agent, 304
Intercellular fluid, 184
  space, 159, 227, 250, 305, 495
Interfascicular cambium, 305
Interference, 225, 226, 277, 305, 426, 435, 479
Interferon, 80, 81, 86, 305
  gamma, 81
Interferons, 199, 305, 327
Intergeneric, 305
  cross, 305
Interheimispheric fissure, 133
Interleukin, 85, 305, 518
Internal guide sequence, 293, 305
  transcribed spacer, 305
International Code of Nomenclature, 539
  Society for Pharmaceutical Engineering, 305
  Treaty on Plant Genetic Resources for Food and Agriculture, 305
  Undertaking on Plant Genetic Resources, 305
  Union for Conservation of Nature, 312
Internet, 306, 537, 549
Internode, 306
Interphase, 169, 227, 281, 306, 375, 415, 425
Intersex, 280, 306
Inter-simple sequence repeat, 306, 312
Interspecific, 125, 224, 306, 483
Interspecific cross, 224, 306
Interstitial cystitis, 306
  diffusion, 306
  site, 306

Interstitial trophoblast, 306
Intervertebral discs, 383
Intervillositis, 542
Intestinal bacteria, 156
Intestine, 130, 213, 228, 229, 278, 280,
    306, 308, 316, 367, 384, 424, 495, 527
Intolerance, 290, 306
Intracardiac vent, 306
Intracellular, 163, 197, 248, 306, 338, 422,
    424, 488
    fluid, 306
Intracytoplasmic sperm injection, 293, 307
Intrageneric, 307
    cross, 307
Intraoperative magnetic resonance imaging,
    307
Intraspecific cross, 307
Intrathoracic, 307
Intravenous pyelogram, 307
    therapy, 307
Intravital microscopy, 307
Introduction, 2, 14, 29, 30, 42, 55, 56, 63,
    74, 81, 85, 186, 223, 226, 294, 307, 308,
    392, 418, 425, 444, 478, 518, 523, 524,
    534
Introgression, 307, 332
Intron, 114, 216, 273, 308, 488
    sequences, 273, 488
    splicing, 308
Intussusception, 308
Invariant, 308
Invasive procedure, 308
    species, 308
Invasiveness, 308
Inversion, 178, 308, 389, 395
Inverted repeat, 248, 308
Inviscid, 308, 309, 395, 458
    fluids, 309
Invivo, 308
Ion channel, 308
    exchange, 203
    gel, 308
    selective electrode, 309
    sensitive field effect transistor, 310
Ionic bond, 308
    bonding, 308
    gelation, 309
    groups, 538
    interactions, 309

Ionizing radiation, 309, 437, 438, 457, 462,
    474, 528, 533
Ionophore, 309
Iris, 309
Irradiation, 309, 437, 493
Irrotational, 309, 546
Irrotational fluid flow, 309
IS element, 309
Island model of migration, 310
Islet cell, 171
Islets of Langerhans, 310
ISO Certification, 310
Isoallele, 310
Isochromosome, 310
Isodiametric, 310
Isoelectric focusing gel, 310
    point, 310, 530
Isoform, 168, 310
Isogametes, 555
Isogamy, 310
Isogenic stock, 311
Isolating mechanism, 311
    medium, 311, 399
Isoleucine, 124, 155, 311
Isomer, 311
Isomerase, 311, 517
Isomorphicity, 311
Isomorphous, 311
Isopropyl-3-D-thiogalactopyranoside, 309
Isothermal, 311
Isotonic, 311, 384
Isotope, 13, 136, 246, 311, 316, 373, 437,
    519
Isotropic, 311, 341
Isozyme, 312
Iuxtaglomerular apparatus, 312

J

Jaundice, 145, 312
Jejunum, 312
Jiffy pots, 312
Joining segment, 312
Joule, 312, 547
Juvenile
    hormone, 312
    in vitro embryo technology, 312, 313
    reproductive stages, 524
Juvenility, 313

K

K anamycin-resistance gene, 313
Kappa chain, 313, 316
Karyogamy, 243, 313, 448
Karyokinesis, 313
Karyotype, 261, 303, 313, 402
Karyotypic factors, 436
Karyotyping, 313
Keratin, 314
Kidney, 27, 32, 65, 68–70, 73, 75, 76, 84,
    86, 118, 184, 239, 279, 307, 312, 314,
    367, 368, 391, 412, 445, 446, 535, 550
    stone, 239, 307, 314, 446
    transplantation, 314
Killer T cells, 314
Kilobase (kb), 192, 229, 313, 314, 547
    pairs, 143, 313, 314, 552
Kilobyte (kB), 314
Kilocalorie (kcal), 314
Kilodalton (kDa), 206, 314, 387
Kilohertz (kHz), 314
Kilojoule (kJ), 314
Kinase, 314
Kinesiology, 28, 30
Kinetic energy, 144, 172, 455, 475, 509
Kinetics, 28, 230, 314
Kinetikos, 314
Kinetin, 199, 314
Kinetochore, 315
Kinetoplastid protozoa, 530
Kinetosome, 315
Kinin, 315
Klenow fragment, 315
Klinefelter syndrome, 315
Knockout, 258, 315
    organisms, 258
Knoop, 161
Kupffer cells, 451, 501

L

L. biennium, 145
L. permeabilis, 396
L. petiolus, 396
L. pistillum, 404
L. plumula, 409
L. pollen, 410
L. primus, 419
L. quantum, 434
L. receptaculum, 440
L. species, 485
L. stamen, 490
L. turgidus, 531
L. vasculum, 539, 541
Label, 285, 315
Labeling, 20, 156, 230, 252, 316, 369, 434,
    435, 439, 496
Lac repressor, 316
Laceration injury, 431
Lacerations, 316
Lactase, 316
    deficiency, 316
Lactation, 424
Lactic acid, 56, 413
    bacteria, 56
Lactoferrin, 316
Lactose, 290, 316
    tolerance test, 316
Lag phase, 316, 318
Lagging strand, 316
Lambda
    chain, 313, 316
    phage, 69, 304, 317
Lamella, 317, 345
Lamina, 105, 317, 452, 493
Laminar (non-turbulent), 317
    air-flow cabinet, 317
Laminarin, 317
Lampbrush chromosomes, 317
Lanagement unit, 317
Landrace, 317
Landscape genetics, 317
Laparoscope, 317
Laparoscopy, 317, 318
Large intestine, 132, 134, 136, 142, 159,
    169, 181, 184, 195, 206, 209, 213, 228,
    283, 306, 318, 326, 443, 474, 479, 532,
    545
Larginal overdominance, 318
Larynx, 171, 231, 318, 398
Laser, 13, 73, 85, 104, 152, 247, 318, 343,
    381, 400, 480, 527
Laser
    ablation therapy, 318
    trimming, 318
Latch probability, 318
Latent
    agent, 318

bud, 318
phase, 316, 318
Lateral
bud, 114, 140, 318, 370, 476, 499, 504
meristem, 318
Laternal effects, 318
Law of,
independent assortment, 336
segregation, 336
Lawn, 319
Laxatives, 136
Layering, 319
Lead compound, 216, 319
Leader
peptide, 319, 474
sequence, 319
Leading strand, 319
Leaf
blade, 319
bud cutting, 319
margin, 320
roll, 320
scar, 320
Leakage, 226, 252, 320, 344, 385, 496
Leaky mutant, 320
Lectin, 320, 455
Left
atrium, 248, 347, 431, 541
ventricle, 131, 347, 541
ventricular assist device, 73
Legume, 320
Lek, 320
Leptonema, 320
Leptotene, 320, 555
stage, 555
Leucocyte, 143, 305
Leukaemia, 320
Leukocyte, 321, 369
Lewis
acid, 321
base, 321
Life cycle, 23, 128, 133, 145, 253, 321,
327, 328, 489, 541
Lifetime, 160, 321
Ligand, 272, 321, 350, 383, 429, 440
Ligase, 77, 319, 321, 442
chain reaction, 319, 321
Ligation, 67, 125, 158, 179, 207, 321, 393,
506

Light
emitting diode (LED), 320, 435
microscopy observation, 389
Lightaddressable potentiometric sensor
(LAPS), 223
Lignification, 321
Lignin, 321, 322, 352
Lignocellulose, 322
Likelihood statistics, 322
Limb deformities, 538
Limbic system, 322
Limit of detection, 322
Lineage, 126, 180, 205, 212, 266, 322, 402,
441, 485
sorting, 322
Linear phase, 202, 322
Linearity, 322
Linearized vector, 321, 322
Linkage
disequilibrium, 253, 322, 323
equilibrium, 253, 322
map, 90, 127, 203, 263, 323, 442
mapping, 323
Linker, 323
Lipases, 323
Lipid molecules, 145, 400
Lipids, 269, 323, 340, 392, 403
Lipid-soluble molecule, 167
Lipoamide, 156
Lipofection, 323
Lipophilic, 125, 323
Lipopolysaccharide, 323, 326
Liposomes, 218, 228, 316, 323
Liquefaction, 323, 460
Liquid
crystal display, 343
media, 324
medium, 161, 170, 324, 470, 472, 493,
503
membrane, 324
nitrogen, 324
Lithography, 6, 303, 360
Litmus, 324
paper, 324
Live
recombinant vaccine, 324
vaccine, 324
Liver, 18, 73, 75, 84, 96, 119, 135, 145,
171, 174, 179, 185, 209, 219, 232, 253,

279, 280, 288, 291, 299, 312, 324, 399,
408, 416, 437, 438, 451, 466, 484, 501,
521, 533
  cancer, 437
  damage, 119, 145
  function tests, 324
Living modified organism, 324
Lobectomy, 324
Local
  adaptation, 324
  scale, 325
Locus, 116, 121, 123, 133, 136, 138, 139,
175, 207, 214, 216, 221, 231, 245, 254,
257–259, 261–263, 266, 278, 282, 283,
286, 297, 325, 352, 355, 414, 416, 433,
448, 449, 454, 471, 490, 551
Lod score, 325
Log
  odds ratio, 325
  phase, 325
Logarithmic phase, 275, 325
Long
  day plant, 325
  interspersed nuclear element, 322, 325
  template, 325
  terminal repeat, 325, 326, 451
Loop bioreactors, 326
Low voltage current, 326
Lower
  esophageal sphincter, 326
  GI series, 326
Low-intensity pulsed ultrasound, 431
Luciferase, 315, 326
Luciferase/luciferins, 315
Lumen, 228, 326, 409
Luminescence, 174, 247, 326, 400, 462
Lung, 27, 56, 61, 66–68, 70, 73, 77, 79, 83,
97, 102, 106, 124, 161, 324, 326, 387,
409
  capacity, 326
  volume, 326
Luteinizing hormone, 326
Lux, 248, 326
Luxury consumption, 327
Lyase, 327
Lymph, 164, 228, 327, 451, 469
  node, 327, 451, 469
Lymphatic
  system, 327

  vessels, 228, 327
Lymphocyte, 118, 200, 289, 327, 506
Lymphokine, 327
Lymphoma, 92, 327, 508
Lympho-reticular sites, 327
Lyophilization, 327
Lyophilize, 250, 327
Lysine (Lys) , 124, 166, 309, 427, 529
Lysis, 207, 268, 327, 328, 408
Lysogen, 327
Lysogenic, 327, 328, 521, 543
  bacteria, 327
  cycle, 328
  infection, 521, 543
Lysogeny, 317, 327
Lysosome, 327
Lytic, 317, 327, 328, 509
  cycle, 328

M

M13 strand, 328
Macerate, 328
Mach number, 328
Macromolecular
  drugs, 228
  structures, 349
Macromolecule, 81, 130, 328, 330, 374,
413, 426
Macronutrient, 328
Macro-organic molecule, 309
Macrophage system, 450
Macrophages, 278, 329, 451
Macropropagation, 329
Macroscopic films, 360
Mad cow disease, 159, 329, 421
MADS box, 329
Magenta, 329
Magnetic
  field, 275, 295, 329, 330, 342, 373, 389,
395
  flux, 320, 329, 330, 548
  lenses, 223
  properties, 373
  resonance imaging, 21, 27, 81, 252, 330
  stirrer, 472
  susceptibility (χ), 330
Magnetization (M), 330
Magnetostrictive material, 330

Major histocompatibility
  antigen, 330
  complex, 270, 330, 339
Malabsorption syndrome, 330
Malarial infection, 406
Malignant tumors, 197, 330
Malnutrition, 330
Malt extract, 330, 534
Malting, 330
Mammalian, 58, 76, 79, 92, 168, 224, 236,
  270, 272, 330, 340, 373, 385, 396, 406,
  486, 489
  cells, 270, 340, 373
  embryos, 224
  species, 396
Mammary
  glands, 331
  tumors, 331
Management of,
  farm animal genetic resources, 331
  unit, 354
Mannitol, 331, 384
Mannose, 253
Map, 64, 70, 85, 88, 103, 172, 182, 185,
  190, 263, 331, 402
Map
  distance, 182, 331
  unit, 172, 331
Mapping, 175, 179, 203, 258, 263, 331,
  354, 452, 460, 519, 520
Marfan syndrome, 331
Mariculture, 332
Marine
  bioresources, 332
  biotechnology, 332
  organisms, 332
Marker, 84, 125, 128, 129, 134, 257, 262,
  263, 315, 323, 332, 350, 367, 416, 465,
  472, 475, 490, 511, 540, 547
  gene, 129, 332, 367, 465
  peptide, 332
Markov chain Monte Carlo, 333
Marsupial, 332
Marsupium, 332
Martensite, 332
Mask, 332
Mass
  selection, 332, 371
  spectroscopy, 332

transfer, 27, 332
Massive chronic intervillositis, 333
Masterbatches, 490
Materials science, 4, 18, 39, 113, 361, 363
Maternal
  decidua, 200, 542
  effect, 333
  endometrial changes, 202
  inflammation, 542
  inheritance, 333
  ovary, 504
Mathematical modeling, 38, 39
Matric potential, 333
Matrigel, 106, 333
Maturation, 149, 165, 231, 296, 312, 333,
  398, 486–488, 506, 518
Maximal voluntary ventilation, 333
Maximum likelihood estimate, 237, 333
Mean, 160, 173, 206, 212, 234, 246, 334,
  409, 466, 539, 546
Measles, 461
Mechanical
  effects, 334
  valves, 334
Mechano-signal transduction, 383
Mechanosynthesis, 5, 6
Mechatronics, 42, 334
Meconium exposure, 334
Media, 56, 64, 119, 135, 139, 140, 150,
  157, 224, 237, 300, 324, 328, 334, 341,
  344, 382, 399, 415, 443, 470, 480, 533,
  544
Median, 334, 501
Medical
  devices, 3, 19, 21–23, 26, 27, 78, 148,
    151, 358, 364
  diagnostics, 152, 299, 343
  imaging, 38, 97, 148, 152, 191, 228, 246
  information systems, 152
  technology, 24, 37, 93, 431
Medium, 61, 137, 139, 143, 163, 169, 174,
  181, 188, 190, 191, 193, 202, 208, 237,
  239, 255, 274, 275, 290, 291, 301, 302,
  311, 312, 316, 324, 334, 341, 346, 376,
  380, 389, 396, 415, 443, 445, 447, 451,
  458, 468, 476, 480, 486, 493, 498, 522,
  531, 533, 541, 554
  formulation, 334
Mega yeast artificial chromosome, 334

Megabase, 331, 335
  cloning, 331, 335
Mega
  byte, 335
  dalton, 335
  gametophyte, 335
  hertz, 335
  karyocytes, 278, 408
  sporangium, 373, 488
Meiosis, 126, 132, 135, 179, 195, 199, 207,
  210, 212, 258, 263, 285, 310, 320, 335,
  339, 371, 387, 410, 424, 429, 442, 443,
  448, 449, 465, 489, 500, 511, 534, 555
Meiotic
  analysis, 335
  drive, 335
  product, 213, 335
  prophase, 136, 175, 335, 504, 510, 555
Melanin, 120, 335
Melanocytes, 335
Melanoma, 89, 335, 508
Melting temperature, 335
Membrane, 6, 17, 65, 67, 70, 72, 73, 77, 78,
  83, 124, 176, 208, 228, 231, 234, 240,
  255, 269, 270, 294, 310, 317, 327, 328,
  335, 340, 345, 360, 362, 370, 372, 375,
  381–384, 392, 395, 396, 406, 407, 409,
  422, 424, 426, 440, 464, 480, 481, 484,
  507, 538, 548
  bioreactor, 335, 507
  oxygenator, 70, 78, 83, 335
Memory cell, 336
Mendelian
  inheritance, 336
  population, 336
  segregation, 257, 264, 336
Mendelism, 336
Meninges, 134, 336
Meningitis, 336
Menopause, 233
Menstruation, 537
Mental
  deficiency, 127, 250, 372, 404, 418, 459
  deterioration, 288, 508
Mericlinal, 336
Mericloning, 337, 344
Meristele, 337, 492
Meristem, 301, 318, 337, 344, 382, 420,
  424, 473, 504, 513, 544

culture, 337, 344, 513
  tip, 337, 382, 473, 513
Meristematic, 132, 162, 164, 205, 301, 304,
  337, 374
Meristemoid, 337
Meristic character, 337
Merozygote, 227, 236, 337
Mesemchymal stem cell, 337
Mesh bioreactor, 244, 337
Mesoderm, 405, 409, 420
Mesophile, 338
Mesophyll, 176, 338
Metabolic
  acidosis, 338, 549
  cell, 338
  pathways, 197, 444
  process, 338, 423
  rate, 338
Meta
  bolism, 63, 73, 126, 171, 199, 229, 233,
  236, 316, 338, 339, 390, 392, 394, 404,
  440, 515, 543
  bolite, 126, 228, 286, 299, 301, 338
  bolome, 338
  bonomics, 338
  centric, 310, 338, 457
  centric chromosome, 310, 338, 457
  genomics, 338
Metal
  affinity chromatography, 119, 339
  nanoshells, 339
  oxide semiconductor, 354
  oxide semiconductor field effect transistor, 354
Metallo
  enzyme, 168, 339
  thionein, 339
Meta
  phase, 126, 335, 339, 425, 433, 554
  population, 339
  population scale, 339
  stasis, 339, 466, 469
  static cancer, 437
  zoan genomes, 237
Methionine, 301, 339, 491
Methylation, 231, 339, 449, 450
  activities, 449, 450
Methylcytosine, 201
Methylmalonic acidemia, 339

Micelle, 339, 340
Michaelis constant, 315, 340
Micro optical electro mechanical systems,
    348
Microalgal culture, 340
Micro-array, 340
Microbe, 61, 340
Microbial fermentation, 459
Microbiology, 2, 18, 42, 340
Microbody, 340
Microbubbles, 340
Micro-carriers, 340
Microchemistry, 340, 346
Microchromosomes, 340
Microcirculation, 39, 51
Micro-CT, 340, 341
Microdroplet array, 341
Microelectro mechanical systems (MEMS),
    152, 336
Microelectronic devices, 342
Microelectronics, 341, 365, 435
Micro-element, 334, 341
Microemboli, 162
Microemulsions, 341
Micro-encapsulation, 341
Microengineering, 341
Micro-environment, 341
Microfabrication, 341, 358, 361, 364
    processes, 341
Microfibrillar structure, 463
Microfibrils, 342
Microfluidics, 38, 342, 346
Microgametophyte, 342
Micrograft, 342, 473
Microinjection, 342
Micro-isolating system, 342
Micromachined
    inkjet printheads, 342
    nozzles, 342
    sensors, 341
Micromachinery, 6, 162
Micromachining, 6, 342, 346
Micromanipulation, 307, 342
Micromaterials, 342
Micrometre scales, 341
Micro-moles, 418
Micro-MRI, 342
Micron, 102, 343, 377
    scale, 377

Micronutrient, 343
Microorganism, 129, 130, 149, 161, 173,
    237, 256, 343, 372, 476
Microparticles, 309, 343, 344
Micro-PAT, 343
Micro-PET, 343
Microphone, 343
Microplast, 344
Micropore filter, 376
Microprocessor, 88, 99, 344
Micropropagation, 329, 344, 419, 490
Micropropulsion, 342
Micropyle, 344
Microsatellite, 344, 452, 484, 493
    loci, 452
Microscope blood vessels, 165
Microscopic
    dielectric objects, 381
    level, 171, 435
    mass, 169
Microscopy techniques, 360
Microshock, 344
Micro-SPECT, 344
Microspheres, 344
Microsporangium, 410, 488
Microspore, 344, 388, 410
Microstrip and microgap radiation detec-
    tors, 342
Microstructures, 344
Microsystem, 345
MicroTAS, 345
Microtransponder, 345
Microtuber, 345
Microtubules, 172, 345, 530
Micro-ultrasound, 345
Microwave diathermy, 132
Micturition, 345
Middle cerebral artery, 133
Middle lamella, 345
Mid-parent value, 345
Migration, 190, 257, 276, 345, 387, 439,
    480, 481, 492, 493
Military applications, 8, 90, 95
Miller indices, 345
Millions of Instructions per second, 347
Mimotope, 390
Mineralization, 345
Miniature devices deliver compounds, 171
Miniaturization, 4, 346, 363, 435

Minimally
    invasive procedure, 228, 346
    invasive surgery, 87
Minimum
    effective cell density, 346
    inoculum size, 346
    viable population, 346
Mini-prep, 347
Minisatellite, 201, 347
Minitubers, 347
Mismatch repair, 347
Missense mutation, 347
Mist propagation, 347
Mitochondria, 115, 199, 239, 347, 382, 422, 424
Mitochondrial DNA, 347
Mitochondrion, 347
Mitogen, 347
Mitosis, 60, 61, 124, 126, 169, 172, 184, 199, 212, 232, 310, 339, 347, 371, 449, 487
Mitral valve, 137, 347
Mixed bud, 347
Mixoploid, 348
Mobile phase, 177
Mobility, 7, 348, 444, 512
Mobilization, 163, 348
Mobilizing functions, 348, 528
Modal class, 348
Mode, 87, 348, 428
Model, 3, 36, 39, 42, 43, 46, 47, 68, 70, 78, 113, 133, 185, 190, 201, 218, 237, 245, 278, 282, 288, 295, 310, 314, 322, 343, 348, 366, 396, 434, 445, 493, 543, 549
MODELER, 351
Modern biotechnology, 155, 348
Modification, 65, 131, 224, 257, 263, 348, 417, 441
Modulation, 90, 348, 435
Modulus of elasticity (E), 348
Mohs scale, 276
Molar concentration, 232, 311, 383
Molarity, 349
Mold cavity, 349
Molding, 349
Mole, 241, 277, 349, 403, 553
Molecular
    assembler, 5, 6
    biology, 4, 13, 67, 218, 223, 265, 287, 364, 411, 427, 494, 535
    conformational geometries, 350
    dipoles, 538
    dynamics, 13, 188, 475
    electronics, 5, 101, 435, 501
    engineering, 1, 17, 55, 70, 81, 107
    genetics, 442
    imaging, 53, 342
    level, 2, 3, 87, 295, 349
    marker, 211, 213, 306, 350, 469, 475, 490
    mass, 314, 386
    mechanics, 188, 248, 350, 351
    property, 434
    scale, 3, 5, 155
    self-assembly, 5, 501
    weight, 166, 323, 328, 338, 349, 363, 394, 408, 412, 413, 415, 474, 476, 480, 530
Molecule, 351
Molecule manipulation technologies, 153
Monitoring, 37, 18, 147, 151, 352, 358, 520
Monoclonal antibody, 114, 280, 328
Monocotyledenous plants, 184
Monoculture, 352
Monocytes, 352, 379, 451
Monoecious, 352
Monogastric animal, 352
Monogenic, 352, 355
Monohybrid
    cross, 352
    heterozygous, 352
Monokine, 352
Monolayer, 5, 6, 352
Monolignols, 352
Monomer, 192, 205, 352, 413, 454, 533
Monomorphic, 352, 498
Mononuclear phagocyte system, 450, 451
Monophosphates, 375, 411
Monophyletic, 352, 353, 402, 441
Monophyly, 353, 485
Monosaccharide, 205, 353, 379, 414, 454
Monosomic, 127, 353
Monosomy, 194, 353, 523
Monotypic, 353
Mono-unsaturates, 353
Monozygotic twin, 293, 353
Monte Carlo technique, 353

Moratorium, 78, 353
Morbidity, 353, 537
Morphogen, 353
Morphogenesis, 353
Morphogenic response, 353
Morphological
    characters, 539
    features, 6, 361
Morphology, 197, 353, 519
Mosaic, 62, 66, 164, 270, 354, 551
Mosaicism, 354, 539
Mother plant, 216, 237, 354
Motif, 329, 354, 554
Motility, 354
Motorola Nanofabrication Research
    Facility, 38
Mouse embryonic stem cells, 287
Movable genetic element, 354
Mucosa and submucosa, 409
Mucus, 198, 354, 495
Mucus accumulation, 198
Multicellular organisms, 132, 386, 468,
    519, 533
Multi-copy, 354
Multidimensional scaling, 333
Multi-disciplinary teams, 46
Multifactorial, 180, 354
Multigene family, 354
Multigenic, 352, 355
Multilocus genotypes, 136
Multimer, 355
Multiple alleles, 355
Multiple
    arbitrary amplicon profiling, 328, 355
    cloning site, 333, 355, 413
    drop array, 333, 341, 355
    factor sex determination, 354
    myeloma, 357, 508
    ovulation, 225, 348, 349, 355
    ovulation and embryo transfer, 348, 355
    sclerosis, 88, 91, 536
Multiplex, 355
Multipotent, 278, 355, 492
    stem cells, 355, 492
Multivalent vaccine, 355
Muscle cell type, 171
Muscular
    dystrophy, 355
    flexions, 488

movements, 173
    region, 495
Musculoskeletal system, 29, 148, 383
Mutable gene, 356
Mutagen, 174, 356, 488
Mutagenisis, 356
Mutant, 13, 120, 139, 186, 249, 258, 262,
    307, 320, 356, 378, 379, 396, 398, 448,
    452, 483, 501, 520, 539
    chromosomes, 186
Mutation, 63, 67, 125, 138, 141, 160, 167,
    174, 188, 196, 201, 203, 217, 249, 250,
    260, 263, 276, 279, 285, 288, 292, 300,
    320, 347, 356, 369, 371, 373, 375, 386,
    387, 409, 410, 428, 439, 452, 483, 488,
    493, 501, 524, 528
Mutation pressure, 356
Mutualistic, 356
Mycoplasma elimination, 513
Mycoprotein, 356
Mycorrhizae, 357
Mycorrhizal, 457, 503
Mycotoxin, 357
Myelin sheath, 357
Myelinated
    axons, 548
    nerve fibers, 357
Myeloid progenitor cell, 379
Myeloma, 357
Myo inositol, 357
Myocardium, 277, 438
Myoepithelial cells, 357
Myoglobine-like units, 279
Myometrium, 405
Myotonic dystrophy, 357
Myxoma, 357
Myxomatosis, 357

N

Naked bud, 358
NAND (NOT-AND), 358
Nano
    abacus, 6
    architectonics, 4
    array, 358
    barcode, 358
    bioengineering/bioelectronics lab, 39
    bioprocessors, 358

biotechnology center, 4, 7, 16, 27, 39, 155, 358
bots, 3, 14
cages, 7
capsules, 362
chemistry, 363, 358
chip, 359
chondria, 6
circle, 359
composites, 359
crystal, 7, 98, 104, 359
crystalline, 81
detectors, 359
devices, 7, 16, 17, 359
electrical components, 39
electromechanical systems, 6
electronics, 5, 365
electrospray, 359
elements, 361
engineering, 4, 358, 359
fabrication, 38, 101, 359, 360
fibers, 360
filtration, 360
fluidics, 360
foam, 7, 360
harvesting, 360
imaging, 360
imprint lithography, 6
imprinting, 360
knot, 6, 7
labels, 361
liters, 359
lithography, 5, 94
manufacturing, 11, 361
materials, 3, 5, 6, 8, 10, 102, 361, 363, 365
    research, 361
mechanics, 4, 5, 17
mesh, 7
metals, 361
meter scale, 2, 4, 359, 362, 364
metersize molecule, 359
motor, 6, 362
newton forces, 362
newtons, 362
patterning technique, 210
pharmaceuticals, 3
photonics, 4, 9, 16, 362
physics, 362, 363

pillar, 7
pin film, 7
plumbing, 362
pore, 6, 362
positioning, 362
precipitation, 362
prism, 362
ring, 7
robotics, 5, 6
rod, 7
scale dimensions, 6, 361
scale, 2, 4, 6, 8, 9, 12, 16, 98, 99, 102, 103, 105, 351, 358, 361–364, 367, 464, 501
spheres, 17, 363
structure, 7, 365
sulphur, 7
toxicological studies, 3
toxicology, 3, 5, 7, 365
tubes, 8, 12, 87, 95, 96, 98, 99, 102, 394, 365
wire, 99, 104, 365
Narrow sense heritability, 365, 280
National
    Institutions of Health, 365
    Nanotechnology Initiative, 10, 11, 94, 95, 366
    Oceanic Atmospheric Administration, 370
    Science Foundation, 34, 94, 101, 358, 366
Native
    organisms, 366
    protein, 366
    species, 366
Natural
    catastrophes, 366
    Health Products, 276
    killer cells, 379
    selection, 60, 117, 213, 251, 261, 324, 366, 439, 445, 467, 490
Nausea, 130, 229, 277, 298, 366
Navier-Stokes equations, 234, 366
Nearly exact test, 366
Necrosis, 199, 367, 399, 544
Necrotic cell, 542
Necrotizing enterocolitis, 367
Negative
    autogenous regulation, 367

control system, 367
selection, 367
self-regulation, 367
Neighborhood, 367
Nematode, 367
Neo-formation, 367
Neologisms, 146
Neomycin, 367, 373
phosphotransferase, 367, 373
Neoplasm, 367, 438
Neoteny, 367
Nephrectomy, 367
Nephritis, 368
Nephrolithotomy, 368
Nephrology, 32, 368
Nephron, 159, 184, 268, 368
Nephrotic syndrome, 368
Nernst equation, 368
Nerve, 115, 118, 119, 160, 192, 204, 208,
254, 255, 345, 357, 368, 369, 378, 451,
487, 521, 545
cells, 357, 368, 369, 378
tract, 487
Nervous
system, 19, 57, 96, 105, 136, 368, 369,
409
tissue, 69, 271, 368, 472, 548
Nested clade analysis, 366
Neural
plate fails, 487
stem cell, 368
systems, 20, 54, 151
technology, 19
Neuro
blastoma, 368
endocrine tumor, 437
fibromatosis, 368
pathic bladder, 368
science, 39
transmission and cell proliferation, 483
transmission, 369, 483
transmitter, 229, 372, 369
Neurogenic bladder, 368
Neuroglia cells, 144
Neuroimaging procedure, 252
Neurologic condition, 220
Neurological
disability, 39
disorder, 161

Neuromuscular blocking agent, 368
Neurons, 206, 229, 267, 368, 369
transmit information, 369
Neutral
allele, 284, 369
mutation, 369
organic substances, 382
position, 380
theory, 369
Neutrophil, 369
Newtonian fluid, 366, 369, 371
Nick, 369
Nicked circle, 369
Nif gene cluster, 369
Nissen fundoplication, 370
Nitrate, 370
Nitrification, 370
Nitro cellulose, 172, 222, 370, 372, 484
filter, 372, 484
Nitrogen
assimilation, 370
fixation, 62, 370, 453
Nitrogenous, 175, 205, 370
Nod box, 370
Nodal culture, 370
Node, 137, 193, 370, 469, 478
Nodular, 370
Nodulation, 370, 371
Nodule, 371
Non-additive genetic variation, 371
Non-autonomous, 371
Non-coding strand, 371
Noncovalent bonding, 371
Noncrystalline material's, 544
Non-disjunction, 371
Nonhomologous chromosomes, 433
Non-Mendelian mode, 412
Non-metallic materials, 4
Nonsense mutation, 371, 495
Non-sex chromosome, 138
Nonstoichiometric microcrystalline magne-
tite cores, 501
Non-target organism, 371
Non-template strand, 372
Non-virulent agent, 372
Noonan syndrome, 372
Norepinephrine, 118, 372
Northern
analysis, 372

blot, 158, 372, 548
 hybridization, 372
Novel
 food, 372
 trait, 373
N-type, 373
Nucellar embryo, 373
Nucellus, 373, 385
Nuclear
 chain reaction, 373
 DNA, 373
 gene, 199, 373, 486
 magnetic resonance, 373
 reactor, 373
 transfer, 118, 197, 224, 373, 425, 448,
  512
Nuclease, 373, 374, 460
Nucleation, 374
Nucleic acid, 17, 64–66, 69, 112, 143, 145,
 153, 170, 182, 184, 186, 204, 207, 213,
 225, 226, 273, 286, 296, 304, 308, 313,
 314, 319, 321, 328, 346, 374, 375, 387,
 411, 413, 422, 427, 432, 447, 454, 464,
 469, 477, 480, 484, 506, 507, 512, 520,
 522, 531, 543, 544
 probe, 184, 296, 374
Nuclein, 60, 61, 374
Nucleoid, 248
Nucleolar
 organizer region, 374
 zone, 374
Nucleolus, 207, 374
Nucleoplasm, 374
Nucleoprotein, 375
Nucleoside, 117, 198, 204, 205, 253, 273,
 297, 375, 394, 514, 515, 536
 analogue, 375
 cytidine, 198
 guanosine, 273
 thymidine, 515
Nucleosome, 375
Nucleotide, 74, 112, 116, 117, 121, 134–
 137, 156–158, 165, 179, 186, 187, 190,
 198, 204, 206, 209–211, 213–215, 241,
 250, 261, 262, 272, 273, 286, 288, 308,
 309, 340, 348, 351, 354, 369, 371, 375,
 380, 383, 409, 412, 414, 421, 439–441,
 449, 450, 454, 456, 469, 478, 479, 506,
 510, 522, 524, 525, 529, 533, 536, 545

complementary, 134
 sequence, 112, 121, 137, 157, 179, 187,
  190, 210, 211, 214, 215, 262, 288, 308,
  340, 354, 369, 371, 383, 412, 421, 441,
  449, 450, 469, 479, 510, 522, 545
Nucleus, 114, 123, 124, 127, 141, 169, 172,
 197, 199, 227, 229, 234, 239, 254, 273,
 281, 288, 343, 347, 373–375, 382, 415,
 422, 424, 437, 448, 449, 482, 505, 511,
 518
Null allele, 375
Nullisomy, 376
Nurse culture, 376
Nusselt number, 376
Nutraceutical, 251, 376
Nutrient
 cycle, 376
 deficiency, 376
 film technique, 376
 gradient, 102, 376
 medium, 143, 212, 237, 249, 376, 383,
  384, 458, 480
Nutritional
 deficiency, 539
 effect, 251
 elements, 139

O

Obstruction, 127, 132, 226, 308, 376, 446
Occipital lobes, 376
Occult blood, 377
Occupational Safety and Health
 Administration, 383
Ochrobactrum anthropi, 269
Octoploid, 377, 414
Oestrogen, 377
Oestrous cycle, 377
Office of the Gene Technology Regulator,
 377
Offset, 377, 553
Offshoot, 377
Offspring, 132, 160, 237, 243, 245, 256,
 257, 264, 265, 289, 298, 301, 317, 318,
 336, 352, 355, 377, 409, 424, 436, 442,
 448, 466, 486, 493, 501, 523
Ohm (Ω), 378
Ohmmeter, 378
Okazaki fragment, 378, 421

Olfactory cell, 143
Oligodendrocyte, 378
Oligohydramnios, 378
Oligomer, 378
Oligonucleotide, 116, 121, 133, 134, 201,
    323, 377–379, 394, 420, 421, 439, 498
    ligation assay, 378
    probe, 121, 378
Oligopotent, 278
    progenitor cells, 379
Oligosaccharide, 268, 269, 379
Oligospermia, 486
Oncogene, 76, 216, 379, 428, 441, 522, 543
Oncogenesis, 216, 379
Oncogenic, 379
Oncology, 379, 438
Onco-mouse, 379
Oncovirus, 379
Ontogeny, 379
Oocyte, 296, 307, 317, 379, 410, 463
Oogenesis, 379
Oogonium, 379, 410
Oosphere, 380, 385
Oospore, 380
Opal stop codon is related to stop codon,
    380
Open
    continuous culture, 380, 530
    pollination, 380
    reading frame, 259, 380, 381, 429
Opening pressure, 380
Operational definition, 380
Operator, 192, 299, 380, 444, 447
Operon, 71, 167, 192, 380, 447, 448
Opiate activity, 144
Opine, 380, 381
Opsonin, 381
Optical
    diagnostic technology, 53
    imaging
        instrumentation, 39
        laboratory, 39
    mapping, 381
    trapping, 381
    tweezers, 381
Optoelectronic devices, 381
Optoelectronics, 381, 435
Ordinate, 381

Organ, 61, 62, 65, 69, 74, 104, 113, 119,
    122, 138, 142, 150, 156, 163, 167, 169,
    182, 227, 228, 235, 245, 250, 266, 270,
    280, 283, 297, 299, 302, 304, 306, 318,
    324, 326, 335, 353–355, 379, 381–383,
    385, 390, 395, 404, 419–421, 424, 446,
    447, 482, 487, 488, 497, 511, 515, 519,
    521, 526, 527, 535–537, 539
    culture, 381
    system, 382
Organellar gene, 382
Organelle, 115, 138,154, 172, 176, 245,
    265, 286, 340, 347, 374, 382, 448
Organic, 41, 42, 126, 132, 139, 143, 147,
    150, 153, 156, 165, 167, 174, 182, 183,
    203, 233, 242, 272, 286, 302, 309, 314,
    323, 330, 338, 345, 360, 362, 363, 374,
    376, 382, 387, 400, 401, 406, 426, 434,
    435, 453, 456, 471, 480, 481, 508, 515,
    534, 544, 552
    chemistry, 382, 387
    complex, 382, 534, 552
    compounds, 153, 203, 330, 345, 382,
        401, 481
    co-solvent, 382
    evolution, 382
    molecule, 126, 132, 167, 174, 183, 314,
        508
Organized
    growth, 382
    tissue, 382
Organizer, 370, 374, 382
Organogenesis, 96, 97, 301, 367, 382, 407,
    420, 444
Organoid, 383
Organoleptic, 383
Origin of replication, 383
Orphan
    gene, 383
    receptor, 383
Ortet, 181, 383, 439
Orthologous, 383
Orthopaedic
    bioengineers, 383
    surgery, 148
    technology, 19
Osbourne Reynolds, 452
Oscillator, 383
Oscilloscope, 547

Osmolarity, 383
Osmosis, 384, 383, 392
Osmotic potential, 291, 292, 311, 383, 384, 419
Osmoticum, 384
Osseo-integration, 69
Osteoarthritis, 384
Osteoblast, 171
Osteogenesis imperfecta, 384
Ostomy, 384, 417
Otoliths, 384
Outbreeding, 236, 265, 384
    depression, 265, 384
Outflow, 181, 380, 384, 491, 536
Outlet valve, 131
Outlier loci, 385
Output traits, 385
Outsourcing, 196, 385
Ovary, 167, 192, 248, 266, 326, 379, 385, 404, 424, 429, 498
Overdominance, 385
Overflow incontinence, 385
Overlapping
    generations, 385
    reading frames, 385
Ovulation, 192, 349, 385
Ovule, 224, 254, 304, 344, 373, 385, 465, 488
Ovum, 165, 273, 296, 379, 381, 385, 410, 486
    pickup, 296, 381, 385
Oxidation, 130, 139, 167, 282, 399, 385, 403
    reactions, 130
Oxidative phosphorylation, 347, 385
Oxidizing agent, 292, 385
Oxygen
    atom, 174, 205, 290
    molecules, 279
Oxygenation, 71, 77, 386, 431
Oxygenator, 64, 386, 540
    system, 540
Oxygen-electrode-based sensor, 386
Oxyhemoglobin, 279, 386

P

P element, 200, 386
P53 gene, 386, 387

Pacemaker, 21, 66, 68, 69, 71–73, 75, 77, 80, 102, 387
Pachynema, 387
Pachytene stage, 210, 555
Package, 210, 226, 333, 387, 521
Packaging cell line, 387
Packed cell volume, 387, 393
Paclitaxel, 387, 426, 508
    total synthesis, 387
Pad electrodes, 387
Paired
    end mapping, 387
    interspersed nuclear elements, 403
Pairing (synapsis), 388
Pair-rule gene, 388
Paleontology, 388
Palindrome, 388
Palindromic sequence, 388
Palisade parenchyma, 388
Palpitation, 388
Pancreas, 18, 76, 171, 209, 215, 303, 310, 388, 546
Pancreatic juice, 166, 388
Panicle, 388
    culture, 388
Panmictic, 221, 297, 367, 388, 546
    population, 221, 297, 388
Panmixis, 388
Papilla, 388
Papillary layer, 388, 389
Par gene, 389
Paracentric inversion, 389
Paraffin, 117, 389
    bath, 389
Parafilm, 389
Parahormone, 389
Parallel evolution, 389
Paralogous, 389
Paramagnetism, 389
Parameter, 191, 254, 272, 322, 333, 390, 449, 469, 491
Paramyxovirus family, 461
Paraphyletic, 390
Parasexual
    cycle, 390
    hybridization, 390
Parasite, 267, 286, 300, 390, 402, 546
Parasitism, 390
Parasporal crystal, 390

Parathyroid
    gland, 390
    hormone, 390
Paratope, 390
Parenchyma, 163, 176, 338, 390, 399, 539
    tissue, 338
Parenchymal cells, 390
Parenchymatous, 280, 390
Parentage analysis, 391
Parental
    generation, 386
    phenotypic value, 524
    taxa, 261, 265, 426
Parenteral, 79, 391, 518
    nutrition, 391, 518
Parsimonious hypothesis, 377
Parsimony, 391
Parthenocarpy, 391
Parthenogenesis, 127, 273, 391
Partial digest, 186, 298, 391, 472
Partial nephrectomy, 391
Particle radiation, 391
Parts per million, 391, 418
Parturition, 267, 391, 406
Passage, 137, 157, 159, 207, 208, 244, 245,
    255, 300, 376, 391, 432, 464, 521
    number, 391
    time, 391
Passive
    immunity, 295, 392
    transport, 240, 392
Pat gene, 392
Patent, 67, 68, 80, 82–86, 92, 173, 392
Paternal, 156, 210, 392, 443, 476, 551
Paternity exclusion, 393
Pathogen, 92, 140, 212, 214, 286, 291, 294,
    301, 318, 324, 392, 418, 427, 436, 537,
    543
    attack, 392, 418
    free, 392
Pathogenic micro-organisms, 266, 469
Pathogenicity, 392, 515
Pathotoxin, 392
Pathovar, 392
PCoA stands, 393
Pectin, 393, 412
Pectinase, 393
Pediatric gastroenterologist, 393
Pedigree, 257, 325, 394

Pelvic bones, 156
Penetrance, 394
Penicillin beta-lactamase, 393
Penicillins, 144
Penis, 18, 142, 181, 397, 419, 467, 527
Peptic, 394
Peptic ulcer, 394
Peptidase, 319, 394, 426
Peptide, 124, 132, 144, 161, 166, 176, 182,
    201, 227, 238, 303, 319, 343, 350, 355,
    380, 394, 395, 409, 414, 426, 474, 483,
    524
    bond, 166, 227, 394, 395, 426, 524
    chain, 355, 394
    expression library, 394
    hormone, 161, 303, 483
    nanotubes are nanotubes, 394
    nucleic acid, 394, 409
    synthesis, 343
    vaccine, 395
Peptidyl
    transferase, 395
    tRNA binding site, 395
Percutaneous, 84, 395
Perfect fluid, 395
Perforation, 395
Perfusion, 61, 62, 64, 65, 173, 395
Pericentric inversion, 395
Periclinal, 130, 395
    chimera, 395
Pericycle, 192, 227, 395
Peripatric, 122
Peripheral membrane, 428
Periplasm, 395
Periplasmic space, 395, 464
Peristalsis, 354, 395
Peritonitis, 395
Permeability, 330, 384, 392, 395, 445
Permeable, 74, 208, 335, 337, 396, 468,
    480
Permeable membrane, 335, 337, 480
Permitivity, 396
Peroxidase, 156, 315, 316, 342
Persistence, 396, 415, 469, 501
Persistent, 196, 396, 434
Persistent right umbilical vein, 396
Personalized medicine, 396
Pessary, 396
Pesticide, 114, 116, 161, 371, 396, 485

application, 371
Petal, 160, 192, 396
Petiole, 436, 396
Petite mutant, 396
PH electrode, 398
Phage, 68, 141, 193, 317, 347, 397, 478, 509, 521, 522, 540, 543
Phagemids, 397
Phagocytic cells, 450
Phagocytosis, 226, 381, 403
Pharmaceutical, 19, 20, 116, 118, 148, 151, 196, 235, 252, 397, 398, 464
    agent, 397
    industry, 196
    properties, 252
Pharmacodynamics, 397
Pharmacogenomics, 398, 397
Pharmacokinetic properties, 146
Pharmacokinetics, 398
Pharmacology, 3
Pharming, 153, 351, 398
Pharynx, 398
Phase
    change, 398
    shift, 398
    state, 193, 398
    transformation, 374, 398
PH-electrode-based sensor, 398
Phenetics, 398
Phenocopy, 398
Phenogram, 398
Phenolic
    compounds, 399
    oxidation, 161, 398, 399
    products, 157
Phenomics, 399
Phenotype, 64, 136, 142, 160, 164, 183, 186, 190, 216, 251, 252, 257, 262, 298, 303, 310, 318, 320, 394, 398, 399, 412, 416, 441, 452, 475, 481, 489, 490, 507, 520, 522, 523, 530, 539
Phenotypic, 161, 180, 183, 188, 190, 193, 211, 261, 280, 345, 365, 398, 433, 451, 501, 514, 524
    characters, 180, 188
    effect, 183, 451, 501
    trait, 190, 211, 261, 398
    variance, 365
    variation, 161, 280, 365, 433

Phenylalanine, 73, 124, 399, 404
Phenylketonuria, 399, 404
Phenylmercury acetate, 131
Pheromone, 399
Philopatry, 399
Phloem, 164, 185, 192, 395, 399, 461, 464, 474, 539
Phonon, 400
Phosphatase, 316, 400, 463
    group, 112, 117, 136, 137, 205, 375, 400, 427, 506, 533
Phosphatidylinositol signal pathway, 270
Phospho-diester bond, 158, 214, 321, 400, 456, 490
Phosphodiester linkages, 205
Phospholipase A2, 400
Phospholipid, 302, 323, 400
Phosphoprotein p53, 386
Phosphorescence, 247, 400
Phosphorolysis, 400
Phosphorylation, 117, 400
Photo
    acoustic imaging, 400
    bioreactor, 400
    bleaching, 435
    chemotherapy, 401
    diode, 401
    dynamic therapy, 401, 402
PHOTOFRIN, 401
Photographic emulsion, 138
Photolithography, 6, 360
Photon, 220, 254, 401, 417, 477
Photoperiod, 298, 401
Photoperiodism, 401
Photophosphorylation, 401
Photoreactivation, 401
Photoresist layer, 502
Photosensitizer, 401, 402
Photosynthate, 401
Photosynthesis, 166, 389, 401, 403, 408, 418, 455
Photosynthetic, 150, 389, 401, 417
    activity, 401
    conversion, 150
    efficiency, 401
    organisms, 389
    photon flux density, 401, 417
Photosynthetically active radiation, 389, 401

Phototherapy, 401
Photothermal therapy, 402
Phototransistor, 402
Phototropism, 402, 529
Phylogenetic, 160, 333, 388, 402, 441
    tree, 160, 333, 441
Phylogeny, 159, 180, 391, 402, 504
Phylogeography, 402
Physical
    map, 127, 258, 381, 402, 437, 452, 482
    mapping, 258, 437, 452, 482
Physiological process, 229, 267, 272, 407
Phyto, 402
Phytochemical, 402
Phytohormones, 402
Phytoparasitic, 367, 402
Phytopathogen, 402
Phytoremediation, 403
Phytosanitary, 403
Phytostat, 403
Phytosterol, 403
Picoengineering, 403
Picomole, 403
Piconewton, 403
Piezoelectric material, 403
Pigment, 120, 166, 324, 335, 402, 403, 429,
    549
Pinhole, 403
Pinocytosis, 226, 397, 403, 404
Pin-out, 404
Pipette, 404
Pistil, 385, 404
Pitting, 404
Pituitary
    dwarfism, 484
    gland, 161, 272, 288, 292, 483
Pixel, 404
PK value, 404
Placenta, 123, 176, 177, 202, 244, 264, 279,
    283, 289, 368, 404–406, 424, 511, 518,
    529, 542, 545
    accrete, 404
    increta, 405
    previa, 405
Placental
    blood, 114, 405, 542
    cord, 396, 405, 406
    infection, 542
    membranes, 406

    surface, 542
    tissue, 492
    vein, 405, 406
    villi, 284, 405, 504
Placentation, 370
Placentophagia, 406
Planar waveguides, 342
Plant
    cell culture, 406
    diseases, 5
    genetic resources (PGR) , 258, 317, 406
    genetics, 406
    growth regulator, 114, 139, 191, 199,
    233, 267, 376, 406, 407, 410
    hormone, 334, 407
    molecular farming (PMF), 407
    pathogenic species, 120
    pathology, 301
    Variety Act (PVA), 532
    variety protection (PVP), 241, 304, 432,
    407
    Variety Protection Act (PVPA), 407
    variety rights, 392, 432
Plantibody, 407
Plantlet, 225, 407
Plaque, 296, 396, 397, 407
Plasma, 157, 158, 180, 181, 199, 227, 239,
    244, 253, 270, 278, 288, 291, 338, 357,
    407, 408, 440, 445, 446, 468, 474, 483
    cell myeloma, 357
    cells, 158, 181, 278, 357, 407, 483
    clearance, 407
    membrane, 253, 270, 440, 468
    proteins, 158, 288
Plasmalemma, 407
Plasmid, 69, 77, 112, 125, 141, 145, 161,
    167, 181, 184, 189, 192, 193, 197, 211,
    240, 257, 278, 282, 322, 347, 348, 354,
    365, 369, 378, 388, 389, 393, 397, 404,
    407, 412, 430, 445, 449, 453, 471, 473,
    497, 500, 515, 522, 528, 530, 540, 542,
    551, 552
    containing a galactosidase gene, 430
    segregation, 389
    vector, 141, 193, 528, 552
Plasmodesma, 408
Plasmodium, 62
Plasmolysis, 291, 328, 408
Plasmonic-magnetic silica nanotubes, 502

Plastic deformation, 219, 408, 478
Plasticity, 399, 408
Plasticizer, 408
Plastid, 120, 166, 199, 408
Plastoquinone, 408
Plate, 62, 68, 165, 208, 317, 345, 407, 408, 425, 437, 447
Platelets, 158, 278, 379, 408
Platform technology, 408
Plating efficiency, 408
Pleiotropy, 408
Plethysmography, 408
Pleura, 409
Plicae circulares, 409
Ploidy, 409
Plumule, 409
Pluripotent, 225, 242, 299, 409, 492, 509, 511
    cells, 409, 509
Pneumococcus bacteria, 67
Pneumonectomy, 409
Point
    defect, 409
    mutation, 499, 524
    mutation, 409
Poisson distribution, 409
Polar
    mutation, 410
    nuclei, 228, 410
    transport, 410
Polarity, 410
Polarization (P), 410
Pole cells, 410
Pollen, 57, 129, 195, 196, 199, 217, 235, 244, 254, 261, 266, 298, 342, 344, 410, 411, 467, 470, 488, 491, 494, 498, 511, 550
    culture, 410
    grain, 254, 266, 298, 342, 344, 410
Pollination, 5, 56, 380, 411
Poly
    (A) polymerase, 411
    adenylation, 259
Polyacrylamide, 112, 115, 223, 293, 387, 406
    gel electrophoresis (PAGE), 112, 223, 387, 411
    gels, 411
Polyadenylation, 411

Polyanhydrides, 411
Polycistronic, 411, 525
Polyclonal antibodies, 328, 411
Polycloning, 411
Polycrystalline
    material, 271
    silicon, 414
Polycystic kidney disease (PKD), 412
Poly-deoxyribo nucleotides, 200
Polyembryony, 412
Polyester, 412
Polyether compound, 166, 394
Polyethylene, 166, 218, 394, 412
    glycol, 166, 218, 394
    terephthalate (PET), 412
Polyfunctional stabilizers, 490
Polygalacturonase (PG), 84, 397, 412
Polygene, 412
Polygenic, 355, 412, 524
Polyhydramnios, 412
Polyhydroxybutyrate (PHB), 398, 412
Polylactic-*co*-glycolic acid (PLGA), 412
Polylinker, 355, 412, 413
Polymer, 42, 150, 153, 162, 195, 204, 208, 237, 244, 267, 286, 290, 308, 352, 408, 409, 411–413, 415, 420, 453, 481, 490, 513
    chemistry, 413
    matrix, 208, 308
    membrane, 208, 413
Polymerase, 80, 82, 92, 121, 180, 211, 214, 315, 325, 328, 355, 369, 393, 413, 420, 421, 425, 429, 457, 473, 484, 489
    chain reaction (PCR) , 80, 121, 180, 211, 214, 325, 328, 355, 393, 413, 473, 484
    chain reaction technique, 211, 328, 355
Polymeric nano delivery systems, 50
Polymerization, 352, 413, 421
Polymery, 413
Polymethine group, 197
Polymorphic, 323, 414, 422, 439, 469, 498
Polymorphism, 142, 180, 196, 263, 414, 439, 449, 450, 452, 472, 476, 479, 489, 490, 524
Polynucleotide, 112, 128, 274, 286, 289, 414, 421, 477
Polyoxyethylene (POE), 166, 394

Polypeptide, 140, 142, 179, 204, 224, 252, 285, 286, 301, 371, 380, 395, 411, 414, 426, 453, 455, 495, 497, 510, 524, 525
   chain, 140, 204, 224, 285, 301, 371, 395, 453, 455, 495, 525
Polyphyletic, 414
Polyploid, 122, 138, 414, 549
Polyploidy, 178, 414
Polysaccharide, 119, 121, 126, 153, 165, 175, 176, 206, 256, 317, 323, 414, 420
   capsule, 414
Polysilicon, 414
Polysome, 415
Polyspermy, 415
Polytene
   chromosome, 415
   tissues, 177
Polyunsaturated fat, 415
Polyunsaturates, 415
Polyvalent vaccine, 415
Polyvinyl pyrrolidone (PVP), 415
Population
   density, 415
   genetics, 245, 317, 403, 415
   viability analysis (PVA), 415
Porcine endogenous retrovirus (PERV), 396, 415
Porcinexenografts, 150
Portal
   hypertension, 416
   vein, 416
Position effect, 416
Positional
   assembly, 416
   candidate gene, 416
Positive
   control system, 416
   selectable marker, 416
   selection, 416
Positron emission tomography (PET), 21, 246, 417
Potassium channel blocking agent, 417
Potentiometric device, 417
Pouch, 132, 213, 332, 334, 417, 530
Power, 8, 99, 212, 345, 347, 409, 417, 437, 526, 547
PR protein, 392, 418
Prader-Willi syndrome, 418
Prandtl Number, 418

Precautionary principle, 418
Precision, 71, 77, 81, 318, 322, 334, 364, 418, 445
Preclinical
   imaging, 418
   rodent model, 52
Precocious germination, 418
Predator, 418
Predisposition, 418
Preeclampsia, 418
Pre-filter, 418
Preimplantation, 157, 419
   stage, 157
Premature
   ejaculation (PE), 419
   germination, 418
   termination, 137, 371
Prenatal, 124, 177, 183, 419
   diagnostic technique, 183
   medical care, 419
Pressure
   coefficient, 419
   potential, 333, 419, 531
Presymptomatic diagnosis, 419
Pre-transplant, 419
Preventive immunization, 537
Priapism, 419
Pribnow box, 419, 508
Primary, 16, 39, 120, 192, 207, 228, 233, 236, 241, 244, 255, 258, 266, 283, 294, 308, 379, 391, 411, 419, 420, 422, 428, 436–438, 463, 487, 507, 510, 520, 539, 542, 549
   antibody, 419, 463
   cell, 294, 391, 420
   culture, 420
   germ layers, 255, 266, 420
   growth, 420
   immune response, 420
   meristem, 420, 422, 428
   structure, 420, 510
   tissue, 192, 420
   transcript, 236, 308, 411, 420, 520
Primer, 121, 133, 230, 325, 375, 413, 420, 421, 439, 473, 477, 546, 547
   DNA polymerase, 420
   walking, 421, 546, 547
Primordial germ cells, 421
Primordium, 320, 421

Primosome, 421
Principal component analysis, 393
Printed circuit board (PCB), 421
Prion (PrP) , 161, 421, 427, 544
Private allele, 421
Probability, 22, 41, 116, 144–146, 242, 297,
    318, 325, 333, 366, 393, 403, 409, 417,
    421–423, 442, 448, 466, 470, 493, 499,
    531
  density function, 393
  of identity (PI), 403, 422
Proband, 422
Probe, 11, 12, 195, 210, 220, 225, 236, 247,
    319, 347, 355, 422, 484, 485, 548
Procambium, 422, 539
Process intensification, 422
Processed
  food, 422
  pseudo-gene, 322, 422
Prodrug, 423
Product rule, 423
Production
  environment, 116, 423
  trait, 423
Productivity, 85, 189, 243, 262, 327, 423,
    503, 529
Pro-embryo, 423
Progenitor cell, 184, 379, 423, 521
Progeny, 195, 209, 233, 240, 282, 355, 377,
    378, 423, 424, 466, 474, 490, 506, 510,
    521, 524
  testing, 424
Progestational agent, 424
Progesterone, 202, 424
Prognosis, 396, 424
Programed cell death (PCD), 424
Prokaryote, 83, 422, 424
Prokaryotic
  cell, 347, 530
  equivalents, 173
  genes, 419
  organism, 271
  signal proteinase, 319
Prolactin, 424
Prolapse, 424
Pro-meristem, 424
Promoter, 163, 164, 167, 175, 190, 205,
    299, 380, 393, 422, 424, 425, 444, 497,
    498

sequence, 425
Pro-nuclear micro-injection, 425
Pronucleus, 425
Proofreading, 425
Propagation, 140, 217, 337, 425, 444, 448,
    465, 466, 529, 540
Propagule, 425
  pressure, 425
Propeptidase, 319
Prophase, 207, 210, 306, 320, 339, 387,
    388, 425, 433, 504
Proportion of admixture, 425
Prostatalgia, 426
Prostate, 89, 144, 209, 426, 508, 527
  cancer, 89, 426, 508
  gland, 209, 426, 527
  specific antigen (PSA), 426
Prostatectomy, 426
Prostatism, 426
Prostatitis, 426
Prosthetics, 20, 33, 39, 444
  devices, 148, 152
Protamine, 426
Protease, 85, 426
Protein
  crystallization, 427
  denaturation, 162
  engineering, 4, 140, 427
  kinase, 427
  metabolic step, 427
  sequencing, 427
  synthesis, 135, 170, 173, 176, 198, 272,
    427, 454, 459, 495, 511, 529
Proteinuria, 427
Proteolysis, 185, 427
Proteolytic
  cleavage, 532
  enzyme, 529
Proteome, 427
Proteomics, 171, 265, 295, 399
Proteus syndrome, 428
Protoclone, 428
Protocorm, 428
Protoderm, 428
Protogyny, 428
Protomeristem, 428
Proton pump inhibitors, 428
Protooncogene, 428
Protoplasm, 62, 169, 356, 428, 531

Protoplasmic, 375, 408
   membranebound region, 375
Protoplast, 136, 407, 408, 412, 428, 429,
   487
Prototroph, 429
Pro-toxin, 390, 429
Protozoa, 57, 156, 245, 343, 408, 429
Protozoan, 179, 340, 429
Protruding ends, 429
Provenance, 429
Provirus, 415, 429
Pseudo
   affinity chromatography, 119, 429
   autosomal region, 429
Pseudocarp, 241, 429
Pseudogene, 429
Pseudomonas, 269, 429, 499
Pseudo-overdominance, 136
P-site, 395, 430
Psychrophile, 430
Public
   health, 277, 430, 503
   Health Service Act, 248, 430
   participation, 8
   policy, 430
Pulmonary, 65, 150, 430, 431, 505, 541
   artery, 430, 431, 541
   autograft valves, 430
   autografts, 150
   circulation, 431
   hypertension, 431
   minute volume, 431
   trunk, 505
   valve, 430, 431
   vein, 431
   ventilation, 431
Pulsed
   ultrasound, 431
   field gel electrophoresis (PFGE) , 82,
   397, 431
Pump, 65, 66, 74, 79, 85, 92, 140, 291, 428,
   431, 538, 541
Punctuated equilibrium, 431
Pure
   culture, 61, 431
   line, 289, 431
Purging, 245, 432
Purine, 117, 143, 204, 205, 273, 375, 432,
   524, 528

nucleoside guanine, 204
Pyloric
   sphincter, 432
   stenosis, 432
Pyloroplasty, 432
Pylorus, 432
Pyrethrins, 432
Pyrimidine, 143, 198, 200, 205, 375, 432,
   506, 515, 524, 528, 532, 535
   biosynthesis, 198
   deoxynucleoside, 205
   derivative, 200, 532
Pyroelectricity, 432
Pyrogen, 432
Pyrophosphate, 137, 432
Pyruvate dehydrogenase complex, 156

Q

Q factor, 432
Q-beta replicase, 433
QMS certification, 310
Quadrivalent, 433
Quadruplex, 433
Quantization, 434
Quantum, 4, 6, 7, 83, 88, 95, 113, 188, 198,
   223, 254, 318, 349, 350, 358, 365, 373,
   400, 401, 417, 418, 434–436, 468
   chemical calculations, 113, 188, 434
   computer, 434
   computing, 7, 434
   dots, 83, 88, 95, 434, 435
   heterostructure, 7
   mechanical
      calculations, 113
      methods, 113, 468
   mechanics, 4
   meter, 418
   nanophysics, 435
   physics, 435
   point contact, 6
   speciation, 435, 436
   supercomputer, 434
Quaranta, 436
Quarantina, 436
Quarantine, 403, 436
Quaternary structure, 436, 510
Quercus suber, 146
Quiescent, 436, 492

R

R genes, 140, 436

Rabbit calicivirus disease (RCD), 163

Rabies, 61, 436

Race, 140, 227, 236, 256, 362, 436

Raceme, 436

Rachilla, 436

Rachis, 436

Radiation
  hybrid cell panel (RH), 437
  therapy, 437, 438, 466

Radicle, 184, 437

Radio frequency, 400

Radioactive, 145, 149, 224, 226, 246, 316,
  417, 422, 437, 438, 484, 519

Raft culture, 438

Ramet, 439

Random
  amplified polymorphic DNA (RAPD),
  439
  genetic drift, 439

Range, 2, 23, 115, 142, 144, 147, 154, 189,
  192, 218, 220, 223, 239, 250, 260, 268,
  318, 324, 338, 339, 341, 344, 347, 359,
  361–366, 385, 389, 406, 412, 435, 438,
  439, 461, 467, 469, 513, 518, 524

Rapid prototype, 439

Rareearth compound, 304

RasMol, 351

Rate-limiting enzyme, 439

Rational drug design, 350, 439

Reactance, 432, 439

Reactant, 440

Reactive oxygen species (ROS), 440

Read-through, 440

Reca, 440

Recalcitrant, 440

Receptacle, 429, 440

Receptor, 13, 83, 118, 120, 157, 167, 216,
  252, 270, 279, 280, 350, 368, 381, 383,
  403, 440, 441
  binding screening, 440
  mapping, 440

Recessive, 121, 138, 142, 186, 193, 196,
  216, 217, 220, 336, 432, 441, 448, 451,
  471, 510
  allele, 193, 217, 432, 441
  gene, 196, 441

oncogene, 441

Reciprocal
  crosses, 441
  monophyly, 441

Reciprocating shaker, 441

Recognition
  sequence, 323, 441, 449, 450
  site, 348, 388, 441, 450, 455

Recombinant, 5, 17, 77–79, 84, 85, 89, 117,
  137, 142, 170, 175, 252, 260, 297, 394,
  411, 427, 436, 441, 442, 450, 452, 478,
  488, 494, 522, 538

Recombinant DNA, 442
  human (rh), 442, 452
  protein, 117, 137, 170, 427, 442
  RNA, 442
  toxin, 442
  vaccine, 84, 85, 442

Recombinase, 442

Recombination, 257, 259, 263, 282, 283,
  285, 304, 323, 325, 331, 417, 440, 442,
  443, 478, 484, 506, 511, 520, 521
  fraction, 323, 325, 331, 442, 511
  frequency, 331, 442

Recombinational hot spot, 443

Reconstructed cell, 443

Recrystallization, 443

Rectal manometry, 443

Rectifier, 249, 443, 451

Rectum, 128, 159, 184, 209, 228, 254, 318,
  326, 443, 474, 495, 532, 537

Red blood cells (RBC), 126, 127, 145, 171,
  278, 379, 406, 407, 474, 518, 536

Re-differentiation, 443

Reducing agent, 443

Reduction, 91, 136, 159, 167, 168, 203,
  218, 219, 297, 346, 384, 436, 443, 454
  division, 443

Reference daily intake, 443

Reflection, 139, 233, 443

Reflux, 443
  esophagitis, 443

Refraction, 101, 443
  index, 443

Refractory, 443

Refugium, 444

Regeneration, 16, 96, 260, 344, 444, 518

Regenerative medicine, 17, 102, 107, 444,
  515

Regulation, 5, 22, 42, 71, 76, 99, 114, 192, 242, 248, 259, 260, 279, 370, 377, 444, 532
Regulator, 89, 377, 444
Regulatory
  gene, 444
  issues, 22, 98, 103
  sequence, 444
Regurgitation, 444
Rehabilitation, 39, 148, 152
  engineering, 148, 152, 444
Reintroduction, 444
Rejuvenation, 444
Relative
  fitness, 444, 466
  humidity, 191, 445, 495
  magnetic permeability, 445
  circle (nicked circle), 445
  circle plasmid, 445
  plasmid, 445
  factors, 445
Reliability, 445
Remediation, 445
Renal
  abnormalities, 538
  angiography, 445
  arteriography, 445
  blood flow (RBF), 307, 445
  pelvis, 184
  plasma flow, 445
  ultrasound, 446
Renaturation, 446, 477, 479
Renature, 446
Renewable biomass, 2
Rennin, 312, 446
Repeat unit, 344, 347, 411, 446, 454, 493
Repeatability, 446
Repetitive DNA, 446
Replacement, 18, 21, 27, 62, 69–72, 75, 76, 96, 103, 105, 106, 143, 204, 235, 266, 307, 368, 383, 446, 482
  therapy, 446
Replica plating, 446
Replicase, 447
Replication, 6, 71, 119, 125, 133, 138, 181, 305, 316, 319, 378, 379, 383, 407, 414, 415, 417, 421, 425, 439, 447, 467, 468, 473, 479, 484, 509, 517, 540, 543
  fork, 447, 517

Replicative form (RF), 452, 447
Replicon, 393, 447
Replisome, 447
Repolarize, 447
Reporter gene, 260, 447
Repressible
  enzyme, 447
  gene, 447
Repression, 447, 551
Reproduction, 56, 62, 63, 89, 116, 122, 134, 135, 141, 181, 212, 225, 227, 233, 243, 245, 247, 300, 328, 366, 382, 385, 399, 424, 428, 448, 466, 490, 544
Reproductive
  cloning, 448
  cycles, 233
  materials, 448
  organs, 207, 280, 385, 428, 534
  stage, 401
Repulsion, 193, 398, 448, 554
Rescue effect, 448
Research labs, 43, 44
Residual
  stress, 448, 513
  volume, 448
Residues, 237, 285, 339, 387, 394, 411, 413, 417, 420, 427, 448, 474, 554
Resistance, 69, 95, 129, 131, 133, 140, 144, 154, 195, 207, 276, 300, 301, 305, 313, 350, 367, 373, 378, 380, 392, 393, 418, 432, 436, 448, 449, 490, 503, 511, 513, 517, 543, 544
Resistance factor, 449
Resistivity, 500, 449
Resolution, 66, 101, 105, 241, 307, 322, 340–343, 360, 449
Resonant frequency, 449
Respiration, 114, 119, 126, 149, 292, 449, 515
Respiratory
  infections, 198
  system, 431
Response time, 449
Rest period, 449
Restitution nucleus, 449
Restriction
  endonuclease, 116, 123, 179, 180, 186, 197, 209, 238, 250, 291, 303, 323, 388, 413, 441, 449, 450, 478

enzyme, 76, 77, 125, 260, 348, 381,
  391, 393, 412, 441, 450, 452, 478
  exonuclease, 450
  fragment, 82, 321, 449, 450, 452, 460,
  484
  map, 450
  nuclease, 450
Reticulocyte, 450
Retina, 376, 451, 457, 546
Retinitis pigmentosa, 451
Retinoblastoma, 451, 530
Retro-element, 451
Retroposon, 451
Retroviral infection, 473, 476
Retrovirus, 120, 284, 325, 415, 428, 429,
  451, 452
Reversal transfer, 451
Reverse
  bias, 451
  genetics, 451, 515
  mutation rate, 452
  transcriptase, 76, 169, 185, 423, 429,
  451, 452, 473, 476
  transcription, 451, 452
Reversion, 444, 452, 528
Reynolds number, 452, 497
Rh disease, 452
Rhizobacterium, 453
Rhizobia, 453
Rhizobium, 370, 453, 505
Rhizosphere, 453
Ri plasmid, 453, 457
Ribbon diagrams, 453
Ribonuclease (RNAse), 453
Ribonucleic acid (RNA) , 259, 374, 419,
  453, 510
Ribonucleotide triplets, 529
Ribose, 64, 65, 112, 117, 182, 198, 204,
  205, 375, 454, 456, 514, 533, 536
Ribosomal
  binding site, 454
  DNA, 454
  RNA (rRNA), 454
Ribosome, 69, 135, 173, 236, 374, 395,
  455, 491, 495, 522, 524, 529
  binding site, 236
  precursors, 374
Ribozyme, 259
Ribulose, 455, 459

  biphosphate (RuBP), 455, 459
Richardson number, 251, 455
Rights, 241, 304, 309, 392, 406, 407, 455
Rinderpest, 85, 455
Ripening process, 397
Risk, 11, 22, 23, 87, 89, 118, 144, 154, 160,
  261, 418, 455, 456
  analysis, 455, 456
  assessment, 23, 456
  communication, 455, 456
  management, 22, 456
Ritual purification, 456
R-loops, 456
RNA
  chain termination, 510
  editing, 273, 457
  polymerase, 131, 374, 421, 425, 433,
  447, 457, 474, 508, 510, 520, 521
  sequence, 84, 128, 131, 422, 433
Robertsonian
  fission, 457
  translocation, 457
Robotic radiosurgery, 197
Robust, 38
Rochelle salt, 242
Rockwell (HR), 276
Rod, 141, 188, 451, 457, 542, 543
Roentgen, 62, 69, 437, 457
Rol genes, 457
Roller pump, 66, 79, 457
Root
  cap, 457
  culture, 457
  cutting, 457
  hairs, 458
  mean square, 456
  nodule, 458
  stock, 462
  tuber, 458
  zone, 458
Rossby number, 458
Rotary
  evaporator, 458
  shaker, 458
Rotating drum artificial kidney, 67
Rotational fluid flow, 458
Rotavirus, 458
Rotting process, 397
Rouleaux formation, 459

Rubinstein-Taybi syndrome, 459
Ruminant, 459
Runner, 459
Russian dolls, 363
Rust, 459

S

Saccaromyces cerevisiae, 473, 552
Saccharamyces pombe, 552
Saccharification, 460
Sacrificial anode, 460
Sacrificial layer, 460
Safety, 21, 22, 46, 74, 94, 102, 104, 105,
    154, 180, 377, 460, 514, 516, 519
Saline resistance, 460
Saliva, 125, 230, 460
Salivary glands, 177, 209
Salmonella, 460
Salt tolerance, 460
Salting out, 460
Sanger sequence, 461
Saprophyte, 461
Satellite
    DNA, 461
    RNA (viroids), 461
Saturated fat, 461
Scaffold, 18, 186, 357, 462
Scale up, 462
Scanning
    electron microscope (SEM), 462, 467
Scanning
    probe microscope, 6
    tunneling microscope, 5, 6, 12, 81, 93
Scarification, 462
Schizomycetes, 141
Schrödinger equation, 113, 434
Scintillator, 462
Scion, 271, 462, 534
    stock interaction, 462
Sclerenchyma, 462
Sclerosing agent, 462
Scrapie, 427, 462
Screen, 13, 64, 98, 281, 295, 404, 418, 446,
    462
Screening, 20, 82, 97, 264, 283, 295, 296,
    405, 430, 462, 466, 483, 543
Scrotum, 463, 530
Sebaceous glands, 463

Secondary
    antibody, 463
    cell wall, 463
    growth, 463, 464
    immune response, 336, 463
    messenger, 463
    metabolism, 463
    metabolite, 463
    oocyte, 379, 463
    phloem, 464, 540
    plant products, 464
    structure, 371, 464, 510, 529
    thickening, 464
    vascular tissue, 464
    vasodilatation, 464
    xylem, 464, 540, 551
Secretion, 122, 289, 351, 390, 428, 464,
    555
Seed, 59, 65, 95, 114, 115, 120, 134, 143,
    203, 212, 224, 228, 241, 244, 247, 264,
    266, 271, 289, 304, 344, 385, 393, 395,
    407, 410, 437, 449, 462, 465, 486, 488,
    503, 510, 529, 541
Seed storage proteins, 465
Segment polarity gene, 465
Segmental duplication, 465, 549
Segregant, 465
Segregation, 179, 465
Selectable, 193, 367, 416, 465, 514
    marker, 193, 367, 416, 465, 514
Selection, 60, 116, 129, 134, 142, 148, 175,
    183, 201, 216, 251, 262, 264, 276, 283,
    284, 332, 367, 369, 385, 389, 429, 432,
    461, 462, 466, 471, 475, 490, 528, 540,
    543
    coefficient, 466
    culture, 466
    differential, 466
    pressure, 183, 389, 466
    response, 466
    unit, 466
Selective
    agent, 291, 466, 476
    breeding, 5, 262, 466
    internal radiation therapy (SIRT), 466
    sweep, 467
Selectivity, 188, 467
Self
    fertilization, 467

incompatibility, 467
pollination, 467
reconfigurable, 6
renewal, 467, 482
replicating elements, 467
replication, 467
sterility, 467
Semen, 56, 68, 134, 235, 238, 467
sexing, 467
Semi
conductor, 467
conservative replication, 468
continuous culture, 468
empirical methods, 468
permeable membrane, 468
solid, 468
sterility, 468
Seminal
fluid, 468
vesicles, 468
Seminiferous tubules, 266
Semipermeable membrane, 169, 335, 360
Semi-solid medium, 468
Senescence, 130, 149, 468
Sense RNA, 468
Sensitivity, 4, 195, 343, 344, 469, 544
testing, 469
Sentinel lymph node, 469
Sentinel node procedure, 469
Sepsis, 131, 469
Septum, 469
Sequence, 469
characterized amplified region (SCAR)
, 462, 469
divergence, 469
hypothesis, 470
tagged site (STS), 470
tandem repeat, 495, 507
Sequencing, 6, 79, 82, 83, 91, 165, 175,
179, 201, 213, 215, 259, 339, 387, 421,
452, 469, 470, 473, 516, 546, 548
DNA molecules, 470
Sequential Bonferroni correction, 470
Serial
divisions, 470
float culture, 470
Serological reaction, 119
Serology, 470
Serum, 84, 276, 392, 470

albumin, 470
Sewage treatment, 65, 470, 471
Sex
chromosome, 122, 138, 278, 281, 285,
471, 552
determination, 139, 471
factor, 471
influenced dominance, 471
limited, 471
linkage, 471
linked locus, 471
mosaic, 273, 471
Sexduction, 471
Sexed embryos, 471
Sexual
reproduction, 125, 160, 246, 289, 310,
313, 425, 448, 471, 482, 555
selection, 320, 471
Shadow effect, 472
Shake culture, 472, 491
Shaker, 472
Shawnn cells, 472
Shear, 251, 369, 371, 472, 478, 494, 544
Shine Dalgarno sequence, 472
Shoot, 132, 140, 162, 184, 337, 344, 347,
382, 407, 409, 420, 470, 473, 494, 499,
510, 531, 534, 544, 554
Shoot
differentiation, 473
tip, 337, 382, 473, 544
Short
day plant, 473
interspersed nuclear element (SINE),
473
tandem repeats, 475, 489
template, 473
Shuttle vector, 145, 473
SI units, 473
Sib mating, 462, 474
Sickle cell anemia, 474
Siderophore, 474
Sieve
cell, 185, 474
element, 474
plate, 474
tube, 399, 474
Sievert (Sv), 474, 457
Sigma factor, 474
Sigmoid

colon, 474
    pattern, 143
Sigmoidoscopy, 474
Signal
    sequence, 189, 474
    transduction, 270, 472, 474
Signals, 27, 42, 96, 197, 258, 260, 270,
    302, 305, 356, 380, 398, 475, 492, 493,
    510, 516, 521
Signal-to-noise ratio, 475
Silencer, 475
Silencing, 295, 475, 479
Silent mutations, 475
Silicon sensors, 223
Simple sequence repeat, 312, 489
Simplicity, 475
Simulated annealing, 475
Single
    cell line, 476
    cell protein (SCP), 476
    copy, 216, 409, 441, 476, 497
    crystal, 476, 548
    domain antibody, 200
    electron devices, 359
    node culture, 476
    nucleotide polymorphism (SNP), 275,
    378, 476
    primer amplification reaction (SPAR),
    477, 484
    strand conformational polymorphism
    (SSCP), 477
Sinoatrial node, 137, 387
Sintering, 478
Sister chromatid exchange (SCE), 462, 478
Site
    directed mutagenesis, 478
    specificmutagenesis, 478
Six base cutter, 478
Skeletal muscle, 368
Skin, 18, 27, 65, 76, 100, 120, 145, 166,
    171, 197, 205, 217–219, 227, 230, 264,
    277, 293, 301, 308, 312, 335, 346, 356,
    368, 395, 428, 463, 478, 482, 498, 514,
    521
Skull fracture, 114
Slip, 181, 478
    casting, 478
Slipped strand mispairing, 479
Small

interfering RNA (siRNA), 479
intestine, 145, 185, 195, 209, 219, 232,
    253, 293, 301, 306, 312, 316, 326, 394,
    409, 432, 479, 495, 497, 535
nuclear ribonucleoprotein (snRNP), 479,
    488
nuclear RNA (snRNA), 479
Smart sensor, 479
Smooth muscle, 479
Social hierarchy, 479
Sodium, 157, 223, 292, 460, 463, 480
    channel blocking agent, 480
    dodecyl sulphate (SDS) , 223, 463, 480
        polyacrylamide gel electrophoresis
        (SDS-PAGE), 223, 463, 480
    hypochlorite, 157, 292
Soil
    amelioration, 480
    fertility, 56
Soilless culture, 480
Solanine and tomatine, 268
Solar
    energy, 150
    radiation, 389, 519
Solid
    free-form fabrication, 104
    hydrocarbon, 389
    medium, 340
Solubility, 196, 204, 311, 480
Solvent, 209, 210, 252, 283, 340, 349, 359,
    362, 383, 384, 480, 481
    diffusion, 481
    drag, 481
    evaporation, 481
    extraction, 481
Somaclonal variation, 481
Somatic, 75, 134, 135, 142, 162, 169, 170,
    192, 210, 266, 299, 335, 373, 423, 437,
    448, 481–483, 492, 505, 506, 512, 518,
    523, 551
    cell, 135, 142, 169, 266, 299, 437, 448,
    482, 483, 492, 505, 512, 518, 551
    hybridization, 483
    hypermutation, 483
    mutation, 162, 483
    reduction, 483
    tissue, 373, 482
Somatomedin, 484
Somatostatin growth, 483

Somatropin, 484
Sonication, 484, 532
Sonogram, 533
Sonography, 532
SOS response, 484
Sound barrier, 484
Southern
    blot, 158, 372, 484, 548
    blotting, 372, 484
    hybridization, 347, 484
Spacer sequence, 484
Spallation, 484
Span, 484
Sparger, 485
Spasm, 485
Spatial autocorrelation statistics, 485
Speciation, 122, 431, 485
Species, 485
    concepts, 485
    scale, 485
Specific
    combining ability (SCA), 461, 462, 485
    heat, 485
    modulus, 485
    strength, 485
Specificity, 165, 286, 395, 438, 485, 539
Spectral karyotyping, 313
Spectroscopic
    method, 366
    performance, 198
Spectroscopy, 13, 220, 366, 485, 501
Speed of sound, 485, 486, 501, 526
Spent medium, 181, 486
Sperm, 57, 60, 63, 136, 157, 178, 196, 228,
    235, 241, 248, 253, 266, 289, 296, 307,
    409, 410, 415, 421, 426, 448, 467, 471,
    482, 486, 487
Sperm
    cells, 296, 426
    sexing, 467
Spermatid, 487
Spermatocyte, 487
Spermatogenesis, 487
Spermatogenic stem cells, 534
Spermatogonium, 487
Spermatozoa, 115, 231, 486, 511
Spermatozoon, 165, 486, 487
Spheroplast, 487
Sphincter, 443, 487

Sphygmomanometer, 487
Spike, 436, 487
Spikelet, 436, 487
Spina bifida, 487
Spinal
    column, 541
    cord, 57, 172–174, 271, 329, 336, 487,
    488, 505, 538, 548
Spindle, 126, 172, 184, 232, 254, 339, 488,
    509
Spine, 346, 488, 541
Spinning, 154, 172, 488
Spirochaete, 141, 488
Spirogram, 488
Spirometer, 488
Spirulina spp, 121
Spleen, 327, 416, 451, 488, 501
Spliceosome, 488
Splicing, 76, 258, 273, 461, 465, 488, 523
    junction, 488
Split gene, 488
Spongiform encephalopathy, 161, 329, 421,
    462
Spontaneous mutation, 287, 488, 489
Sporangia, 488, 489
Sporangium, 488
Spore, 188, 329, 335, 380, 488, 489, 555
Sporocyte, 489
Sporophyll, 489
Sporophyte, 489
Sport, 29, 162, 489
Spririllum, 489
Squamous cells, 142
Squeeze-film damping, 489
Stability, 198, 386, 431, 435, 490, 506, 552,
    554
Stabilizer, 490
Stabilizing selection, 490
Stable polymorphism, 490
Stacked genes, 490
Stages of culture (I-IV), 490
Staggered cuts, 490
Stamen, 129, 280, 490, 491
Standard
    atmosphere, 491
    deviation, 491
    error, 491
    hydrogen electrode (SHE), 491

Starch, 125, 126, 206, 256, 323, 330, 337, 413, 414, 491
Starter culture, 491
Stationary
    culture, 491
    phase, 177, 202, 351, 463, 491
Statistic, 175, 322, 366, 385, 416, 491
Steady state, 380, 491
Steatorrhea, 492
Stele, 492
Stem, 17, 93, 96, 103, 105, 132, 143, 159, 160, 162, 164, 169, 192, 193, 195, 211, 225, 234, 242, 247, 271, 272, 275, 278, 285, 287, 292, 304, 306, 319, 320, 322, 326, 337, 355, 368, 370, 377, 394, 405, 408, 409, 423, 424, 459, 463, 467, 473, 476, 481, 482, 492–495, 499, 504, 509, 511, 512, 516, 518, 521, 531, 533, 534
    cell, 17, 93, 105, 143, 159, 164, 169, 211, 225, 242, 247, 278, 285, 287, 326, 337, 355, 368, 405, 408, 409, 423, 424, 467, 481, 482, 492, 493, 509, 511, 512, 516, 518, 521, 533, 534
    tissue, 162, 370, 473, 476
Stenosis, 493
Step response, 493
Stepping stone model of migration, 493
Stepwise mutation model (SMM) , 452, 493
Steric hindrance, 493
Sterile, 135, 142, 159, 302, 317, 493
Sterility, 199, 212, 288, 493, 551
Sterilization, 38, 66, 140, 244
Sterilize, 58, 137, 503, 493
    utensils, 137
Sternotomy, 493
Sternum, 493
Steroid hormones, 118, 202, 233, 471
Steroidogenesis, 199
Steward bottle, 493
Sticky end, 211, 303, 450, 460, 494
Stiction, 494
Stigma, 167, 404, 411, 494
Stimulated emission of radiation, 318
Stirred-tank fermenter, 494
Stochastic, 245, 262, 494
Stock
    plant, 490, 494
    solution, 376, 494
Stoichiometric quantities, 494

Stoichiometry, 494
Stolon, 459, 494
Stoma, 273, 495
Stomach, 129, 171, 179, 209, 219, 232, 254, 255, 274, 277, 290, 326, 352, 354, 370, 394, 409, 428, 432, 443, 446, 459, 479, 488, 497, 499, 519, 535, 545
Stomatal
    complex, 495
    index, 495
Stool, 159, 226, 228, 242, 306, 318, 377, 384, 417, 495
Stop codon, 173, 183, 371, 495, 510, 525
Strain, 91, 137, 139, 141, 181, 221, 269, 282, 283, 348, 369, 392, 431, 436, 443, 445, 494–496, 499, 518, 553
Stratification, 496
Stratum
    basale, 496
    corneum, 496
Streamlines, 309, 317, 458, 496, 531
Streptavidin, 156
Streptomyces aureofaciens sp., 137, 392
Streptomycin, 67
Stress, 59, 113, 114, 155, 194, 222, 277, 292, 326, 348, 383, 410, 448, 466, 495–497, 500, 503, 509, 517, 518, 544, 547, 553
    concentration, 496
    corrosion, 496
    incontinence, 496
    protein, 277, 496
    ulcer, 497
Strict one-step, 493
Stricture, 497
Stringency, 121, 497
Stringent plasmid, 497
Stroke, 158, 536
Stroma, 176, 497
Stromal cells, 159, 497
Strouhal number, 497
Structural gene, 192, 380, 475, 476, 488, 497
Structure
    analysis, 351, 497
    functionalism, 498
Style, 167, 404, 494, 498
Sub
    cellular structures, 171, 200

clone, 498
cloning, 494, 498
culture, 274, 391, 420, 498
    interval, 498
    number, 498
Subculturing, 498
Subgenomic promoter, 498
Sublingual immunotherapy, 498
Subpopulations, 339, 499
Subspecies, 238, 499
Substantial derivative, 499
Substitution, 475, 499, 528
Sub-strain, 499
Substrate, 103, 154, 160, 166, 168, 198,
    216, 231, 271, 299, 313–315, 327, 334,
    371, 373, 374, 385, 460, 489, 494, 499,
    506
Subtelomeric region, 546
Subunit vaccine, 499
Sucker, 499
Suckering, 499
Sucrose density gradient centrifugation, 499
Sum rule, 499
Super resolution microscopy, 5
Superbug, 499
Supercoil, 500
Supercoiled plasmid, 500
Superconductivity, 500
Supercontig (scaffold), 500
Supergene, 500
Superior vena cava, 500
Supernatant, 500
Supernumerary chromosome, 74, 500
Supersonic flow, 501
Supportive breeding, 501
Suppressor, 386, 387, 501, 530
Suppressor mutation, 501
Supramolecular
    assembly, 5, 6
    chemistry, 4, 364, 501
    electronics, 501
Surface
    markers, 502
    micromaching technology, 502
    plasmon, 502
    plasmon resonance, 502
    tension, 206, 502, 548
Surfactant, 83, 339, 341, 502, 548
Surrogate, 503

Surveillance, 8, 15, 20, 277, 503
Susceptible, 85, 196, 222, 251, 291, 321,
    329, 472, 503
Suspension culture, 143, 170, 190, 202,
    233, 302, 376, 382, 476, 498, 530, 503
Sustainable
    development, 503
    intensification of animal production
    systems, 503
SWOT analysis, 503
Symbiont, 503
Symbiosis, 356, 503
Symbiotic association, 503
Sympatric speciation, 504
Symplast, 504
Sympodial, 504
Synapomorphy, 504
Synapsis, 320, 429, 504, 511, 555
Synaptonemal complex, 461
Synchronous culture, 504
Syncitiotrophoblast, 200, 504
Syncitiotrophoblast cells, 200
Syncope, 504
Syncytiotrophoblast cells, 177
Syncytium, 504
Syndrome, 89, 112, 120, 219, 279, 288,
    421, 461, 504, 538, 552
Synergid, 505
Synergism, 505
Syngamy, 505
Synteny, 505
Syringohydromyelia, 505
Systematic error, 505
Systemic, 65, 72, 89, 131, 146, 418, 500,
    505, 521, 541
    artery, 131
    circulation, 65, 146, 500
    circulation, 505
Systole, 505
Systolic pressure, 334, 505

T

T cell
    antigens, 506
    receptor, 185, 506, 508
T4 DNA ligase, 321, 506
Tailing, 186, 200, 411, 472, 506
Tanacetum cinerariifolium, 432

Tandem repeat, 470, 507, 528, 539, 545
Tank bioreactor, 507
Tap
    root, 244, 507
    polymerase, 507
Target, 20, 115, 117, 121, 132, 134, 162,
    172, 175, 186, 195, 214, 216, 218, 222,
    226, 237, 318, 321, 332, 358, 359, 363,
    378, 391, 413, 419, 435, 438, 441, 455,
    463, 469, 484, 485, 507, 524, 527, 528,
    553
    site duplication, 507
Targeted
    drug delivery, 20, 507
    image guided therapy, 52
Targeting vector, 508
TATA box, 508
Tautomeric shift, 508
Tautomerism, 508
Taxol and Onxol, 387
Taxonomic group, 353
Taxoprexin, 508
Tay-Sachs disease, 508
Technetium, 438
Telemeter, 508
Telomerase, 89, 508
Telomere, 457, 509
Telophase, 126, 306, 374, 509
Temperate phage, 509
Temperature, 124, 142, 149, 150, 164, 173,
    188, 191, 196, 197, 219, 243, 245, 250,
    267, 276, 277, 290, 293, 295, 298, 299,
    311, 318, 335, 338, 349, 389, 437, 440,
    445, 448, 465, 478, 485, 486, 496, 497,
    509, 512, 513
Template, 96, 130, 165, 169, 182, 185, 214,
    217, 236, 273, 379, 413, 419, 447, 452,
    457, 468, 477, 479, 509, 520, 521
    strand, 130, 182, 379, 509
Tensile strength, 102, 106, 485, 509
Teratogens, 509
Teratoma, 509
Term finalization, 509
Terminal
    bud, 114, 459, 473, 504, 510
    transferase, 506, 510
Terminalization, 510
Termination
    codon, 173, 380, 445, 495, 510

signal, 131, 440, 510
Terminator, 14, 510
    codon, 510
    gene, 510
    region, 510
Tertiary structure, 510
Tesla, 329, 510
Test cross, 510
Test tube, 212, 296, 511, 541
Testis, 230, 248, 266, 326, 487, 511, 552
Testosterone, 315, 511
Test-tube fertilization, 511
Tetracycline, 393, 511
Tetrad, 213, 511
Tetrahydrothiophene, 155
Tetraodan nigroviridis, 236
Tetraploid cells, 511
Tetrasomic, 511
Tetratype, 511
Thalassaemia, 512
Thallus, 356, 512
Theranostics (Rx/Dx), 513
Therapeutic
    agent, 93, 218, 249, 295, 397, 512
    serum, 150
Thermal, 4, 223, 255, 300, 342, 403, 418,
    465, 511–513
    conductivity (κ), 512
    diffusivity, 418
    effects, 465
    expansion coefficient, linear, 512
    fatigue, 512
    gel gradient electrophoresis (TGGE)
    ,223, 512
    shock, 513
    stress, 512, 513
Thermistor, 513
Thermocouple, 513
Thermodynamic quantity, 250
Thermodynamics, 150
Thermoelastic expansion, 400
Thermolabile, 513
Thermophile, 513
Thermophilic bacterium, 507
Thermoplastic polymer, 513
Thermoregulation, 291
Thermosensitivity, 513
Thermoset polymer, 513
Thermostability, 513

Thermostable, 513
Thermotherapy, 277, 291, 513
  techniques, 291
Thick skin, 163, 496, 514
Thin skin, 496, 514
Thinning, 514
Threshold character, 514
Thrombophilias, 514
Thrombosis, 514
Thrombus, 514
Thymidine, 205, 514, 515, 517, 529
  5'-triphosphate, 529
Thymidine
  kinase (tk), 514, 517
  triphosphate, 514
Thymine, 68, 143, 186, 200, 205, 370, 375,
  432, 470, 506, 508, 514, 515, 531, 532,
  535
Thyroid gland, 188, 390, 515
Thyroid hormones, 515
Ti plasmid, 508, 515, 530
Tidal volume, 237, 303, 431, 515
Time
  constant, 435, 515
  since the most recent common ancestor,
  517
Timeline, 1, 6, 34, 55, 74, 107, 109
Tissue
  culture, 118, 135, 137, 139, 143, 163,
  174, 196, 202, 224, 233–235, 239, 253,
  274, 289, 293, 301, 302, 317, 329, 334,
  337, 345, 382, 393, 415, 419, 444, 457,
  464, 480, 490, 515, 526, 533, 541, 544,
  553
  engineering, 4, 5, 17–19, 28, 34, 38, 39,
  43, 55, 80, 86, 93, 95, 107, 515, 516
Titanium oxides, 8
Titre, 516
Tobacco
  mosaic virus, 66, 67, 70
  ringspot virus, 461
Tolerance, 503, 517
Tomographic images (virtual slices), 187
Tomography studies, 39
Tonoplast, 517
Tools, 4, 16, 17, 20, 38, 40, 46, 94, 106,
  114, 151, 259, 282, 351, 358, 361, 363,
  408, 505, 516
Top–down nanotechnology, 517

Topo-isomerase, 215, 517
Torr, 517
Total
  lung capacity, 517
  parenteral nutrition, 518
Totipotent, 492, 518
Totipotent
  cell, 518
  nucleus, 518
  stem cells, 492, 518
Toughness, 518
Toxic, 8, 78, 91, 119, 129, 131, 141, 154,
  156, 174, 252, 264, 268, 272, 280, 334,
  357, 396, 401, 442, 449, 455, 486, 518,
  519, 529
  gases, 8
  metabolic products, 486
  substance, 264, 357, 449
  Substances Control Act (TSCA), 518
Toxicity, 3, 8, 196, 260, 295, 319, 365, 515
Toxicogenomics, 519
Toxin, 141, 150, 237, 295, 429, 455, 519
Tracer, 417, 438, 519
Trachea, 101, 104, 161, 171, 318, 515, 519,
  520
Tracheid, 519
Tracheoesophageal fistula, 519, 538
Trait, 117, 139, 162–164, 185, 187, 188,
  193, 196, 221, 237, 238, 251, 265, 266,
  332, 337, 345, 352, 355, 357, 368, 412,
  416, 433, 441, 452, 465, 471, 504, 520,
  524
Trans
  acting, 520
  capsidation, 520
  configuration, 193, 448, 520
  heterozygote, 520
  test, 186, 520
Transcript, 131, 165, 169, 308, 354, 372,
  468, 475, 520
Transcription, 119, 130, 137, 163, 165, 167,
  168, 170, 182, 190, 205, 217, 218, 229,
  237, 259, 271, 299, 317, 329, 339, 380,
  387, 393, 410, 419, 420, 425, 440, 447,
  451, 453, 456, 474, 479, 508, 510, 517,
  521, 535
  factor, 299, 329, 387, 521
  unit, 237, 410, 521
  vector, 521

Transcriptional
   anti-terminator, 521
   roadblock, 521
Transcutaneous electrical nerve stimulation
   (TENS), 521
Transdermal, 521
   drug delivery, 521
   patch, 521
Transdifferentiation, 211, 521
Transducer, 446, 521
Transducing phage, 521
Transduction (t), 521
   mode, 522
Transfection, 116, 294, 387, 522, 523
Transfer, 70, 118, 120, 121, 130, 132, 134,
   149, 154, 158, 189, 193, 214, 222, 224,
   240, 250, 257, 265, 271, 277, 279, 280,
   283, 287, 293, 313, 314, 324, 348, 365,
   376, 408, 411, 419, 425, 451, 454, 456,
   457, 471, 482, 490, 498, 508, 521–524,
   528, 529, 540, 543, 550
   RNA (tRNA), 522
Transferase, 522
Transferrins, 522
Transformant, 522
Transformation, 20, 64, 67, 86, 120, 147,
   257, 260, 264, 289, 311, 332, 337, 373,
   376, 379, 408, 412, 425, 506, 522
Transformer, 523
Transforming oncogene, 523
Transgene, 237, 249, 259, 425, 490, 506,
   522–524
Transgenesis, 129, 193, 260, 264, 367, 425
Transgenic, 81, 86, 90, 214, 249, 291, 348,
   398, 510, 511, 516, 523
   animal, 81, 249, 523
   organism, 523
   plant, 291, 511, 523
   technology, 348
Transgressive segregation, 524
Transient, 220, 223, 400, 524
   expression, 524
   response, 220, 524
Transistor, 68, 71, 103, 104, 175, 243, 402,
   511, 524
Transition, 88, 168, 169, 211, 254, 524, 525
   stage, 524
   state intermediate, 168

Translation, 131, 259, 296, 301, 313, 319,
   369, 380, 395, 440, 445, 455, 472, 479,
   491, 510, 522, 524, 546
Translational
   medicine, 525
   stop, 525
Translocation refers, 525
Transmembrane carrier, 167
Transmission, 199, 243, 257, 301, 369, 509,
   526, 550
   electron microscope (TEM), 509, 526
Transplant, 21, 70, 73–75, 80, 84–86, 101,
   102, 105, 122, 161, 162, 526, 533, 550
Transplantation, 32, 69, 83, 89, 101, 122,
   138, 161, 181, 276, 278, 282, 415, 526,
   550
   biology, 526
Transposable genetic element, 526, 527
Transposase, 526
Transposition, 451, 526, 546
Transposon, 187, 258, 303, 386, 451, 526,
   527, 539
   activity, 303, 539
   tagging, 258, 527
Transrectal ultrasound of the prostate, 527
Transurethral
   hyperthermia, 527
   incision of the prostate (TUIP), 527
   laser incision of the prostate (TULIP),
   527
   resection of the prostate (TURP), 527
   surgery, 527
Transvaginal ultrasound, 405
Transverse colon, 527
Transversion, 528
Trauma, 497
Tribology, 528
Tribrid protein, 528
Trichome, 528
Tricuspid, 137, 528
   aortic valve, 528
Tri-hybrid, 528
Trinucleotide, 489, 510, 528
   repeats, 528
Trioventricular valve, 137
Tripartite mating, 528
Triplet, 183, 203, 380, 440, 510, 528, 529,
   534
   code, 203, 528

Triploblastic, 420
Triploid, 414, 529, 549
Trisomy, 118, 217, 529
Triticale, 529
Trophoblast, 176, 177, 405, 504, 529
Tropism, 529
True-to-type, 529
Trypanosoma brucei, 546
Trypsin, 64, 529
Trypsin inhibitor, 529
Tuber, 345, 347, 530
Tubulin, 345, 530
Tumble tube, 530
Tumor, 84, 93, 120, 164, 166, 199, 228,
    289, 367, 368, 379, 381, 386, 387, 391,
    437, 438, 453, 469, 507, 509, 515, 522,
    527, 530, 544
  cells, 368, 379, 381, 438
  protein p53, 386
  suppressor gene, 386, 530
  suppressor p53, 386
  tissue, 228, 453
  virus, 522, 530, 544
Tunica, 192, 530
  vaginalis, 530
Turbidostat, 530
Turbulent, 366, 452, 531
Turgid, 531
Turion, 531
Turner syndrome, 353, 523
Turn-on-voltage, 531
Twin, 69, 70, 531
Type
  I statistical error, 470, 531
  II statistical error, 531
Tyrosinase, 120

U

U.S. Department of Agriculture, 242, 532,
    536
Ubiquitin, 532
Ulcer, 278, 497, 532
Ulcerative colitis, 532
Ultramicroscopic spaces, 293
Ultrasensitivity, 532
Ultrasonic
  bath, 532
  emission, 400

  foil, 532
  frequencies, 532
  probe, 532
  transducers, 400
Ultrasonication, 532
Ultrasound, 21, 27, 97, 152, 196, 340, 343,
    405, 406, 431, 532, 533
  imaging, 343, 533
Ultrathin, 339, 362
Umbelliferones, 316
Umbilical
  ligament, 405
  vein, 406
Understock, 534
Undifferentiated cells, 202, 206, 225, 390,
    534
Unequal crossing over, 534
Unicellular (organisms), 141, 189, 410,
    533, 534, 553
Uniform flow, 534
Uniparental inheritance, 534
Unipotent stem cells, 278, 534
Unisexual, 534
Unit cell, 534
United States Department of Agriculture
    (USDA), 407
Univalent, 534
Universal
  constructor, 534
  donor cells, 534
Universality, 534
Unorganized growth, 535
Unspecialized, 118, 202, 208, 474, 535
Untranslated regions on mRNA (UTRs),
    525
Upper GI series, 535
Upstream, 8, 380, 424, 447, 508, 525, 535
  processing, 535
Uracil, 73, 143, 156, 186, 200, 370, 375,
    432, 456, 515, 531, 533, 535, 536
Urea breath test, 535
Ureido (tetrahydroimidizalone), 155
Ureter, 535, 536
Ureterocele, 535
Ureteroscope, 536
Urethritis, 536
Urge incontinence, 536
Uridine, 453, 533, 536, 537
  5'-monophosphate, 536

Uridylic acid, 533, 536
Urinalysis, 536
Urinary
    bladder, 536
    system, 239, 538
    tract, 92, 198, 290, 536, 541
    tract infection, 92, 541
Urination, 345, 426, 536
Urine flow test, 536
Urogenital, 536
    orifices, 274
    sinus, 405,
    pluripotent, 409
Urology, 536
Usenet, 537
Uterine
    endometrium, 202
    muscle, 405
    serosa, 405
    stromal cells, 202
    wall, 405, 504, 537
Uterus, 134, 157, 173, 202, 213, 221, 225,
    396, 419, 424, 448, 492, 511, 537
Utilization of,
    farm animal genetic resources, 537

V

Vaccination, 58, 419, 537
Vaccine, 61, 63, 85, 92, 215, 302, 324, 355,
    415, 442, 537
Vaccinia, 538
Vaconstrictors, 128
VACTERL, 538
Vacuole, 170, 172, 461, 517, 531, 538
Vacuum, 65, 208, 250, 300, 443, 445, 538
Vagina, 164, 173, 396
Valence
    band, 538
    electrons, 468, 538
Valeric acid, 155
Valine, 124, 155, 301, 491
Value-added traits, 538
Van der Waals forces, 538
Vapor pressure, 538
Vaporization, 113
Variable domain, 140, 476
Variable
    domain, 538
    expressivity, 539

number tandem repeat (VNTR), 539,
    545
    surface glycoprotein, 546
Variance, 365, 409, 489, 539
    effective number, 539
Variant, 121, 134, 205, 257, 262, 285, 310,
    312, 329, 421, 466, 482, 539, 545, 546
Variation, 117, 128, 180, 191, 192, 211,
    226, 229, 231, 262, 265, 280, 300, 335,
    337, 348, 365, 385, 387, 399, 412, 491,
    524, 529, 539, 546
Variegated, 539
Variety, 17, 43, 89, 99, 147, 215, 220, 264,
    272, 284, 294, 298, 340, 342, 343, 348,
    359, 361, 392, 393, 403, 406, 407, 421,
    426, 429, 432, 474, 481, 510, 515, 539,
    544
Vascular, 39, 79, 185, 219, 225, 227, 241,
    305, 317, 318, 337, 399, 405, 422, 463,
    464, 492, 539, 540
    bundle, 241, 305, 539
    cambium, 422, 463, 464, 540
    connection, 219, 225
    cylinder tissue, 337
    endothelial growth factor, 405
    headache, 540
    plant, 185, 399, 464, 492, 540
    system, 540
    tissue, 227, 422, 463, 464, 540
Vascularization, 516
Vasoconstrictions, 540
Vasodilatation, 184, 191, 540
Vasodilator, 540
Vector, 145, 181, 184, 193, 211, 238, 256,
    303, 304, 321, 322, 465, 494, 496, 498,
    528, 540, 543, 548, 552
Vegetative propagation, 181, 319, 499, 540
Vehicle, 97, 124, 181, 540
Vein, 71, 202, 300, 307, 391, 406, 416, 462,
    500, 540, 541
Velocity circulation, 458
Velocity density gradient centrifugation,
    540
Velogenetics, 540
Venous reservoir, 540
Ventilation, 540
Ventricle, 74, 137, 430, 431, 459, 528, 541
Ventricular
    assist device, 74

fibrillation, 66, 541
Venturi, 541
Venule, 541
Vermiculite, 541
Vernalization, 267, 541
Vertebral
  defects, 538
  disc, 431
Vertebrates, 124, 205, 232, 274, 310, 314,
  327, 330, 420, 487, 498
Vesico ureteral reflux (VUR), 541
Vesicular internalization, 397
Vessel, 120, 126, 133, 153, 162, 199, 219,
  228, 335, 337, 346, 405, 494, 498, 507,
  539–541
Veterinary
  medicine, 3
  profession, 3
Viability test, 541
Viable, 188, 199, 233, 245, 272, 541, 443,
  449, 464, 493, 550
Vibration, 541
Vibrio, 141, 541
Villitis, chronic, 542
Villous stroma, 542
Villus, 542
Vir genes, 542
Viral
  attack, 193
  infection, 396, 407, 455, 539
  pathogen, 543
  symptoms, 544
Virion, 543, 544
Virtual screening, 543
Virulent, 137, 324, 327, 499
Virulent phage, 543
Viruliferous, 543
Virus particles, 516
Viscoelastic property, 544
Viscoelasticity, 544
Viscosity, 234, 369, 371, 395, 544
Viscous fluid, 458, 494
Visible light, 222, 300, 381, 485, 533, 544
Visual disturbance, 277
Vital capacity, 544
Vitamin B complex, 544
Vitamin B12, 339
Vitamin C, 130
Vitrification, 544

Vitrified, 544
Viviparous animals, 267
Vivipary, 247, 544
Volatilization, 545
Volt, 71, 279, 299, 329, 378, 448, 452, 526,
  545, 548
Voltage, 66, 165, 222, 237, 251, 269, 295,
  310, 343, 383, 401–403, 417, 448, 523,
  531, 545, 553
Voltage sensitive permeability, 545
Voltmeter, 310, 417
Volume contraction, 545
Volvulus, 545
Vomiting, 130, 277, 545
Von Hippel-Lindau syndrome, 545
Vortex principle, 172
Vorticity, 546

W

W chromosomes, 471
Wafer, 175, 332, 502, 546
Wahlund principle, 546
Walking, 57, 102, 546
Wall pressure, 547
Wash-out, 380, 547
Water
  environments, 132
  intoxication, 547
  potential, 333, 384, 419, 531, 547
  soaked, 544, 547
  stress, 320, 547, 549
Watt (W), 547
Wave drag, 547
Wave
  form, 283
  length, 547
Wax, 197, 389, 412, 547
Weber, 329, 510, 548
Weed, 406, 548
Weediness, 548
Western
  blot, 158, 230, 548
  blotting analysis, 548
Wet
  nanotechnology, 4
  weight, 251, 548
Wetting agent, 502, 548
Whisker, 548

Whitaker Foundation, 28, 31, 93, 94, 97, 99, 107
White
    blood cells, 129, 140, 320, 321, 327, 329, 407, 411, 488, 518, 536
    matter, 548
    cell analysis, 51
    genome shotgun sequence, 549
Wild type, 249, 356, 378, 466, 520, 539, 548
Wilt, 549
Wobble hypothesis, 549
Wobble method, 217
World Wide Web (WWW), 549
Wright–Fisher model, 549

X

X chromosome, 142, 187, 241, 355, 471, 486, 550–552
Xanthophyll, 549
Xenia, 550
Xenobiotic, 550
Xenogeneic, 122, 282, 516, 550
    cells, 516
    organs, 550
    tissues, 550
Xenograft valves, 550
Xenopus, 532
Xenosis, 550
Xenotransplantation, 415, 550
Xerophyte, 550, 551
X-inactivation, 551
X-linked, 144, 219, 250, 279, 336, 451, 551
    disease, 551
    recessive, 219, 279, 551
    trait, 219, 250
X-ray, 17, 21, 64–67, 127, 138, 142, 152, 176, 187, 198, 223, 228, 247, 307, 309, 326, 381, 437, 445, 533, 535, 551

    crystallography, 66, 67, 551
    film, 138
    technique, 127
XYY syndrome, 551

Y

Y chromosome, 63, 187, 241, 429, 486, 550–552
Yeast, 56, 59, 62, 63, 80, 85, 91, 112, 133, 154, 160, 162, 164, 256, 287, 306, 334, 382, 396, 412, 551, 552
    artificial chromosome (YAC), 85, 287, 334, 551
    extract, 382
    strength, 553
Yielding, 242, 553
Yoctomole, 553

Z

Zener diode, 553
Zeptomole, 553
Zero offset, 553
Zeta potential, 553, 554
Zig-zag pattern, 554
Zinc finger, 554
Zone refining, 554
Zoo blot, 554
Zoonosis, 555
Zoospore, 555
Zygonema, 387
Zygospore, 555
Zygote, 118, 125, 157, 175, 221, 225, 243, 253, 261, 337, 380, 443, 486, 505, 518, 531, 555
Zymogen, 555

Printed and bound by CPI Group (UK) Ltd, Croydon, CR0 4YY

23/10/2024

01777702-0019